国际材料前沿丛书
International Materials Frontier Series

Lorraine F. Francis

金属、陶瓷和聚合物的加工方法

Materials Processing:
A Unified Approach to Processing of Metals,
Ceramics and Polymers

影印版

· 长 沙 ·

图字:18 - 2017 - 164 号

Materials Processing: A Unified Approach to Processing of Metals, Ceramics and Polymers
Lorraine F. Francis
ISBN: 9780123851321
Copyright © 2016 by Elsevier Inc. All rights reserved.
Authorized English language reprint edition published by the Proprietor.
Copyright © 2017 by Elsevier (Singapore) Pte Ltd. All rights reserved.
Elsevier (Singapore) Pte Ltd.
3 Killiney Road
#08 - 01 Winsland House I
Singapore 239519
Tel: (65) 6349 - 0200
Fax: (65) 6733 - 1817
First Published <2017>
<2017>年初版

Printed in China by Central South University Press under special arrangement with Elsevier (Singapore) Pte Ltd. This edition is authorized for sale in China only, excluding Hong Kong SAR, Macao SAR and Taiwan. Unauthorized export of this edition is a violation of the Copyright Act. Violation of this Law is subject to Civil and Criminal Penalties.

本书英文影印版由 Elsevier (Singapore) Pte Ltd. 授权中南大学出版社在中国大陆境内独家发行。本版仅限在中国境内(不包括香港、澳门以及台湾)出版及标价销售。未经许可之出口,视为违反著作权法,将受民事及刑事法律之制裁。

本书封底贴有 Elsevier 防伪标签,无标签者不得销售。

内容简介

本书主要介绍金属、陶瓷和聚合物三种材料加工的基本原理，着重探讨加工的基本概念及其在典型工艺中的应用，主要内容包括材料加工概述、原材料、熔融加工固态法、粉末法、分散体与溶液法、气相法等。

本书采用的统一分类法是基于新形状形成时的物质状态。这种分类法有助于探究和应用先前已有的知识，帮助建立加工工艺和物质结构之间的联系。

本书可供材料、冶金等领域的科研人员、工程技术人员使用，同时也可作为高等院校材料、冶金等相关专业学生的参考书。

作者简介

Lorraine F. Francis 博士，美国明尼苏达大学教授，在明尼苏达大学任教25余年。主要研究方向是涂层、陶瓷和复合材料的材料科学与加工，研究重点是材料的加工及其微观结构的控制。1985年获得美国阿尔弗雷德大学纽约州立陶瓷学院的学士学位，1987年获得美国伊利诺伊大学厄巴纳-香槟分校陶瓷专业的硕士学位，1990年获得美国伊利诺伊大学厄巴纳-香槟分校陶瓷专业的博士学位。

序

材料加工是材料科学与工程领域的4个关键构成要素之一。材料的成形方式对其结构(即晶体结构、相、微观组织)相当重要,从而对其性能与服役表现相当重要。例如,冷变形工艺(如轧制)提高位错密度,从而提高金属的屈服强度。反之亦然,材料的结构与性能决定其用某种方法加工的难易程度。例如,典型的聚合物熔体黏度过高,不适宜用重力驱动流动的成形操作(如熔体铸造),但非常适宜用压力驱动流动的成形操作(如挤压和注射成型)。加工-结构-性能的相互关系广泛存在于各种工程材料中。加工对决定最终产品的价格起重要作用,对材料选择和设计至关重要。因此,研究材料的加工自然而然地是为了弄懂结构-性能关系,是材料选择与设计的重要组成部分。

本书介绍材料加工的基本原理,广泛涉及材料科学与工程的基本原理及具体工艺细节。其目的当然并非涵盖所有的细节,而是探讨基本概念及说明其在典型工艺中的应用。这些典型工艺包括传统工艺(如金属的砂型铸造)和新型的加成工艺(如熔融沉积工艺,即3D打印)。本书涉及用于3类主要工程材料(即金属、陶瓷和聚合物)的加工基本原理。本书采用的统一分类法是基于新形状形成时的物质状态。例如,关于"熔体法"的章节探讨金属流动与凝固的基本方面及其在金属熔体铸造和聚合物注射成型中的应用。这种分类法有助于探究和应用先前已有的知识。

本书可供材料科学与工程及相关领域的本科生使用。完成了材料科学与工程专业的入门课程,以及微积分、物理学和化学课程的学生,便具备了学习本书所需的背景知识。例如,只要具备了上述必备知识,本书可直接用作大学一年级或二年级学生的课程教学,或者用作高年级学生的"旗舰"课程。研究生和实习工程师也可通过本书扩展其知识基础,增进其对基本概念的理解。

本书正文分为7章。第1章介绍材料加工涉及的领域,概述金属、陶瓷和聚合物的加工。第2章探讨用于加工的原材料的制备、成形和表征。其余5章分别介绍不同的加工工艺,根据最终形状的材料性质可分为：熔体法、固体法、粉末法、分散体或溶液法以及气相法。重要的加工后处理操作(例如烧结)已融入各节中。每章包括材料科学与工程基本原理和工艺。工艺部分包括工艺描述、建模分析方法和实例。每章(除第 1 章外)以参考文献、习题和问题结束。

目 录

序 ⋯⋯⋯⋯⋯⋯⋯⋯⋯⋯⋯⋯ xiii

致谢 ⋯⋯⋯⋯⋯⋯⋯⋯⋯⋯⋯ xv

1 材料加工概述 ⋯⋯⋯⋯⋯⋯ 1
  1.1 材料加工：定义与适用范围 ⋯⋯⋯⋯⋯⋯⋯⋯⋯⋯⋯⋯ 1
  1.2 材料加工的三种方法 ⋯⋯ 4
  1.3 材料加工的步骤 ⋯⋯⋯⋯ 7
  1.4 金属的加工 ⋯⋯⋯⋯⋯⋯ 10
  1.5 陶瓷的加工 ⋯⋯⋯⋯⋯⋯ 13
  1.6 聚合物的加工 ⋯⋯⋯⋯⋯ 16
  1.7 总结 ⋯⋯⋯⋯⋯⋯⋯⋯⋯ 19
  延伸阅读 ⋯⋯⋯⋯⋯⋯⋯⋯⋯ 19
  参考文献 ⋯⋯⋯⋯⋯⋯⋯⋯⋯ 20

2 原材料 ⋯⋯⋯⋯⋯⋯⋯⋯⋯ 21
  2.1 什么是原材料？ ⋯⋯⋯⋯ 21
  2.2 金属 ⋯⋯⋯⋯⋯⋯⋯⋯⋯ 22
    2.2.1 引言 ⋯⋯⋯⋯⋯⋯ 22
    2.2.2 块状金属原材料 ⋯ 25
    2.2.3 粉末金属原材料 ⋯ 35
  2.3 陶瓷 ⋯⋯⋯⋯⋯⋯⋯⋯⋯ 50
    2.3.1 引言 ⋯⋯⋯⋯⋯⋯ 50
    2.3.2 陶瓷粉末原材料 ⋯ 51
    2.3.3 玻璃原材料 ⋯⋯⋯ 60
  2.4 聚合物 ⋯⋯⋯⋯⋯⋯⋯⋯ 68

    2.4.1 引言 ⋯⋯⋯⋯⋯⋯ 68
    2.4.2 热塑性聚合物原材料 ⋯⋯⋯⋯⋯⋯⋯⋯⋯⋯⋯ 74
    2.4.3 热固性聚合物原材料 ⋯⋯⋯⋯⋯⋯⋯⋯⋯⋯⋯ 89
  2.5 总结 ⋯⋯⋯⋯⋯⋯⋯⋯⋯ 95
  延伸阅读 ⋯⋯⋯⋯⋯⋯⋯⋯⋯ 97
  参考文献 ⋯⋯⋯⋯⋯⋯⋯⋯⋯ 98
  习题与问题 ⋯⋯⋯⋯⋯⋯⋯⋯ 98
  习题 ⋯⋯⋯⋯⋯⋯⋯⋯⋯⋯⋯ 98
  问题 ⋯⋯⋯⋯⋯⋯⋯⋯⋯⋯⋯ 100

3 熔融加工 ⋯⋯⋯⋯⋯⋯⋯⋯ 105
  3.1 引言 ⋯⋯⋯⋯⋯⋯⋯⋯⋯ 105
  3.2 基本原理 ⋯⋯⋯⋯⋯⋯⋯ 107
    3.2.1 熔体结构与表面张力 ⋯⋯⋯⋯⋯⋯⋯⋯⋯⋯⋯ 107
    3.2.2 熔体流变学 ⋯⋯⋯ 115
    3.2.3 流变原理 ⋯⋯⋯⋯ 129
    3.2.4 热传导原理 ⋯⋯⋯ 139
    3.2.5 凝固 ⋯⋯⋯⋯⋯⋯ 144
  3.3 成型铸造 ⋯⋯⋯⋯⋯⋯⋯ 153
    3.3.1 概述 ⋯⋯⋯⋯⋯⋯ 153
    3.3.2 金属熔体的制备 ⋯ 155
    3.3.3 砂模铸造 ⋯⋯⋯⋯ 156
    3.3.4 金属模铸造 ⋯⋯⋯ 171
    3.3.5 压力铸造 ⋯⋯⋯⋯ 175
    3.3.6 铸件的加工后处理 ⋯ 178

3.4 平板铸造 ················ 183
  3.4.1 概述 ················ 183
  3.4.2 玻璃熔体制备 ········ 183
  3.4.3 浮法玻璃工艺 ········ 184
  3.4.4 融合下拉工艺 ········ 189
  3.4.5 玻璃板的加工后处理
         ······················· 190
3.5 挤压 ···················· 192
  3.5.1 概述 ················ 192
  3.5.2 单螺杆挤出机中的
       熔化与流动 ········ 192
  3.5.3 模流 ················ 201
  3.5.4 单螺杆挤出机操作图
         ······················· 201
  3.5.5 双螺杆挤压 ·········· 205
  3.5.6 模具出口的影响 ····· 207
  3.5.7 挤压制品与凝固 ····· 211
3.6 注射成型 ················ 213
  3.6.1 概述 ················ 213
  3.6.2 注射成型机械与周期
         ······················· 214
  3.6.3 模内流动性 ·········· 217
  3.6.4 包装与凝固 ·········· 221
  3.6.5 反应注射成型 ········ 223
3.7 吹塑成型 ················ 226
  3.7.1 概述 ················ 226
  3.7.2 玻璃的吹塑成型 ····· 227
  3.7.3 高分子的吹塑成型
         ······················· 229
3.8 熔体增材制造工艺 ······· 232
  3.8.1 概述 ················ 232
  3.8.2 融合沉积成型 ········ 233
  3.8.3 熔体的喷墨打印 ····· 237
3.9 总结 ···················· 238

延伸阅读 ···················· 240
参考文献 ···················· 242
习题与问题 ·················· 244
习题 ························ 244
问题 ························ 245

4 固态法 ················· 251
4.1 引言 ···················· 251
4.2 基本原理 ················ 252
  4.2.1 单向拉伸下的变形与
       塑性流动 ············ 252
  4.2.2 温度与应变速率对
       变形的影响 ········ 267
  4.2.3 三向应力下的变形与
       屈服 ················ 272
  4.2.4 摩擦 ················ 282
  4.2.5 效率与温升 ·········· 284
4.3 固态加工 ················ 285
  4.3.1 概述 ················ 285
  4.3.2 拔丝 ················ 287
  4.3.3 挤压 ················ 299
  4.3.4 锻造 ················ 308
  4.3.5 轧制 ················ 318
  4.3.6 弯曲 ················ 325
  4.3.7 形变热处理 ·········· 328
  4.3.8 超塑性成形 ·········· 331
4.4 总结 ···················· 333

延伸阅读 ···················· 335
参考文献 ···················· 336
问题与习题 ·················· 337
习题 ························ 337
问题 ························ 338

附录：球面压力容器中的应力
         ······················· 341

5 粉末法 343

5.1 引言 343
5.2 基本原理 346
5.2.1 粉末特性与流动性 346
5.2.2 烧结与微观组织演变 353
5.2.3 致密化过程中的尺寸变化 367
5.3 压制成形 370
5.3.1 概述 370
5.3.2 粉末制备 372
5.3.3 单向压制 376
5.3.4 等静压制 389
5.3.5 半成品的后期成型过程 391
5.3.6 热压与热等静压 392
5.4 回转成形 396
5.4.1 概述 396
5.4.2 粉末制备 397
5.4.3 回转成形工艺步骤 398
5.5 粉末增材制造 400
5.5.1 概述 400
5.5.2 选择性激光烧结(熔炼) 401
5.5.3 喷墨打印(3D打印) 406
5.6 总结 407
延伸阅读 409
参考文献 410
习题与问题 411
习题 411
问题 412

6 分散体与溶液法 415

6.1 引言 415
6.2 基本原理 418
6.2.1 胶态分散体 418
6.2.2 聚合物溶液 442
6.2.3 分散体与溶液的流变学 448
6.2.4 分散剂和溶液中挥发性液体的特性 455
6.2.5 干燥 455
6.2.6 液态单体的固化 462
6.3 成型铸造法 464
6.3.1 概述 464
6.3.2 毛细管作用 466
6.3.3 铸造层厚度预测 467
6.3.4 注浆成型法 472
6.3.5 加工后处理 473
6.4 涂层与流延成型 473
6.4.1 概述 473
6.4.2 涂层方法 475
6.4.3 聚合物涂层 486
6.4.4 陶瓷的流延成型 489
6.5 挤压与注射成型 491
6.5.1 概述 491
6.5.2 浓缩分散体的挤压 491
6.5.3 粉末注射成型 494
6.6 液态单体的增材制造法 495
6.6.1 概述 495
6.6.2 光固法 495
6.6.3 液态单体喷墨打印 500
6.7 总结 504
延伸阅读 505
参考文献 507

习题与问题 …………………… 508
习题 …………………………… 508
问题 …………………………… 509

7 气相法 …………………… 513

7.1 引言 ………………………… 513
7.2 基本原理 …………………… 515
7.2.1 气体动力学原理及其与气相过程的关系 …………… 515
7.2.2 薄膜微观组织 ……… 532
7.2.3 单晶薄膜的外延生长 …………………………… 539
7.3 蒸发 ………………………… 540
7.3.1 概述 ………………… 540
7.3.2 蒸发的热力学 ……… 542
7.3.3 合金与化合物的蒸发 …………………………… 546
7.3.4 传输现象与薄膜的均匀性 ………………… 547
7.4 喷涂法 ……………………… 555
7.4.1 概述 ………………… 555
7.4.2 等离子体物理学 …… 557
7.4.3 磁控溅射 …………… 559
7.4.4 射频溅射 …………… 561
7.4.5 反应溅射 …………… 561
7.4.6 溅射速率优化 ……… 563
7.5 化学气相沉积 ……………… 566
7.5.1 概述 ………………… 566
7.5.2 形成反应的热力学 … 568
7.5.3 反应类型 …………… 570
7.5.4 化学气相沉积动力学 …………………………… 571
7.5.5 沉积速率与均匀性 … 573
7.6 沉积薄膜的后处理 ………… 578
7.6.1 退火 ………………… 578
7.6.2 图案结构 …………… 578
7.7 总结 ………………………… 580
延伸阅读 …………………………… 582
参考文献 …………………………… 583
习题与问题 ………………………… 583
习题 ………………………………… 583
问题 ………………………………… 585

附录 A ……………………………… 589

索引 ………………………………… 591

Materials Processing
A Unified Approach to Processing of Metals,
Ceramics and Polymers

Materials Processing
A Unified Approach to Processing of Metals, Ceramics and Polymers

Lorraine F. Francis
University of Minnesota

With contributions from

Bethanie J. H. Stadler
University of Minnesota

Christine C. Roberts
Sandia National Labs

AMSTERDAM • BOSTON • HEIDELBERG • LONDON
NEW YORK • OXFORD • PARIS • SAN DIEGO
SAN FRANCISCO • SINGAPORE • SYDNEY • TOKYO
Academic Press is an imprint of Elsevier

Academic Press is an imprint of Elsevier
125, London Wall, EC2Y 5AS.
525 B Street, Suite 1800, San Diego, CA 92101-4495, USA
50 Hampshire Street, 5th Floor, Cambridge, MA 02139, USA
The Boulevard, Langford Lane, Kidlington, Oxford OX5 1GB, UK

Copyright © 2016 Elsevier Inc. All rights reserved.

Chapter 6: Copyright © 2016 Elsevier Inc. All rights reserved. The contributions made by Ms. Christine Cardinal Roberts is under Sandia Corporation, Contract No. DE-AC04-94AL85000

No part of this publication may be reproduced or transmitted in any form or by any means, electronic or mechanical, including photocopying, recording, or any information storage and retrieval system, without permission in writing from the publisher. Details on how to seek permission, further information about the Publisher's permissions policies and our arrangements with organizations such as the Copyright Clearance Center and the Copyright Licensing Agency, can be found at our website: www.elsevier.com/permissions.

This book and the individual contributions contained in it are protected under copyright by the Publisher (other than as may be noted herein).

Notices

Knowledge and best practice in this field are constantly changing. As new research and experience broaden our understanding, changes in research methods, professional practices, or medical treatment may become necessary.

Practitioners and researchers must always rely on their own experience and knowledge in evaluating and using any information, methods, compounds, or experiments described herein. In using such information or methods they should be mindful of their own safety and the safety of others, including parties for whom they have a professional responsibility.

To the fullest extent of the law, neither the Publisher nor the authors, contributors, or editors, assume any liability for any injury and/or damage to persons or property as a matter of products liability, negligence or otherwise, or from any use or operation of any methods, products, instructions, or ideas contained in the material herein.

ISBN: 978-0-12-385132-1

British Library Cataloguing-in-Publication Data
A catalogue record for this book is available from the British Library.

Library of Congress Cataloging-in-Publication Data
A catalog record for this book is available from the Library of Congress.

For Information on all Academic Press publications
visit our website at http://store.elsevier.com/

Dedicated to Mark and Carolyn

Contents

Preface — xiii
Acknowledgements — xv

1. Introduction to Materials Processing — 1

- 1.1 Materials Processing: Definition and Scope — 1
- 1.2 Three Approaches to Materials Processing — 4
- 1.3 Materials Processing Steps — 7
- 1.4 Processing of Metals — 10
- 1.5 Processing of Ceramics — 13
- 1.6 Processing of Polymers — 16
- 1.7 Summary — 19
- Bibliography and Recommended Reading — 19
- Cited References — 20

2. Starting Materials — 21

- 2.1 What is a Starting Material? — 21
- 2.2 Metals — 22
  - 2.2.1 Introduction — 22
  - 2.2.2 Bulk Metal Starting Materials — 25
  - 2.2.3 Metal Powder Starting Materials — 35
- 2.3 Ceramics — 50
  - 2.3.1 Introduction — 50
  - 2.3.2 Ceramic Powder Starting Materials — 51
  - 2.3.3 Glass Starting Materials — 60
- 2.4 Polymers — 68
  - 2.4.1 Introduction — 68
  - 2.4.2 Thermoplastic Polymer Starting Materials — 74
  - 2.4.3 Thermoset Polymer Starting Materials — 89
- 2.5 Summary — 95
- Bibliography and Recommended Reading — 97
- Cited References — 98
- Questions and Problems — 98
  - Questions — 98
  - Problems — 100

viii   Contents

3. Melt Processes — 105
- 3.1 Introduction — 105
- 3.2 Fundamentals — 107
  - 3.2.1 Melt Structure and Surface Tension — 107
  - 3.2.2 Melt Rheology — 115
  - 3.2.3 Flow Fundamentals — 129
  - 3.2.4 Heat Transfer Fundamentals — 139
  - 3.2.5 Solidification — 144
- 3.3 Shape Casting — 153
  - 3.3.1 Process Overview — 153
  - 3.3.2 Metal Melt Preparation — 155
  - 3.3.3 Sand Casting — 156
  - 3.3.4 Permanent Mold Casting — 171
  - 3.3.5 Die Casting — 175
  - 3.3.6 Post-Processing of Cast Metal Parts — 178
- 3.4 Casting of Flat Sheets — 183
  - 3.4.1 Process Overview — 183
  - 3.4.2 Glass Melt Preparation — 183
  - 3.4.3 Float Glass Process — 184
  - 3.4.4 Fusion Downdraw Process — 189
  - 3.4.5 Post-Processing Operations for Glass Sheets — 190
- 3.5 Extrusion — 192
  - 3.5.1 Process Overview — 192
  - 3.5.2 Melting and Flow in a Single Screw Extruder — 192
  - 3.5.3 Die Flow — 201
  - 3.5.4 Single Screw Extruder Operating Diagram — 201
  - 3.5.5 Twin Screw Extrusion — 205
  - 3.5.6 Die Exit Effects — 207
  - 3.5.7 Extruded Products and Solidification — 211
- 3.6 Injection Molding — 213
  - 3.6.1 Process Overview — 213
  - 3.6.2 The Injection Molding Machine and Cycle — 214
  - 3.6.3 Mold Flow — 217
  - 3.6.4 Packing and Solidification — 221
  - 3.6.5 Reaction Injection Molding — 223
- 3.7 Blow Molding — 226
  - 3.7.1 Process Overview — 226
  - 3.7.2 Blow Molding of Glass — 227
  - 3.7.3 Blow Molding of Polymers — 229
- 3.8 Melt-Based Additive Processes — 232
  - 3.8.1 Process Overview — 232
  - 3.8.2 Fused Deposition Modeling (FDM) — 233
  - 3.8.3 Inkjet Printing of Melts — 237
- 3.9 Summary — 238
- Bibliography and Recommended Reading — 240
- Cited References — 242
- Questions and Problems — 244

			Questions	244
			Problems	245

4. Solid Processes 251

4.1 Introduction 251
4.2 Fundamentals 252
- 4.2.1 Deformation and Plastic Flow under Uniaxial Tension 252
- 4.2.2 Effects of Temperature and Strain Rate on Deformation 267
- 4.2.3 Deformation and Yielding under Triaxial Stresses 272
- 4.2.4 Friction 282
- 4.2.5 Efficiency and Temperature Rise 284

4.3 Solid Processes 285
- 4.3.1 Process Overview 285
- 4.3.2 Wire Drawing 287
- 4.3.3 Extrusion 299
- 4.3.4 Forging 308
- 4.3.5 Rolling 318
- 4.3.6 Bending 325
- 4.3.7 Thermoforming 328
- 4.3.8 Superplastic Forming 331

4.4 Summary 333
Bibliography and Recommended Reading 335
Cited References 336
Questions and Problems 337
Questions 337
Problems 338
Appendix: Stress in a Spherical Pressure Vessel 341

5. Powder Processes 343

5.1 Introduction 343
5.2 Fundamentals 346
- 5.2.1 Powder Characteristics and Flow 346
- 5.2.2 Sintering and Microstructure Development 353
- 5.2.3 Dimensional Changes during Densification 367

5.3 Pressing 370
- 5.3.1 Process Overview 370
- 5.3.2 Powder Preparation 372
- 5.3.3 Uniaxial Pressing 376
- 5.3.4 Isostatic Pressing 389
- 5.3.5 Post-Forming Processes for Green Parts 391
- 5.3.6 Hot Pressing and Hot Isostatic Pressing 392

5.4 Rotational Molding 396
- 5.4.1 Process Overview 396
- 5.4.2 Powder Preparation 397
- 5.4.3 Rotational Molding Process Steps 398

	5.5	Powder-Based Additive Processes	400
		5.5.1 Process Overview	400
		5.5.2 Selective Laser Sintering (Melting)	401
		5.5.3 Inkjet Binder Printing ("3D Printing")	406
	5.6	Summary	407
		Bibliography and Recommended Reading	409
		Cited References	410
		Questions and Problems	411
		Questions	411
		Problems	412

6. Dispersion and Solution Processes — 415
Lorraine F. Francis and Christine C. Roberts

6.1	Introduction		415
6.2	Fundamentals		418
	6.2.1	Colloidal Dispersions	418
	6.2.2	Polymer Solutions	442
	6.2.3	Rheology of Dispersions and Solutions	448
	6.2.4	Characteristics of Volatile Liquids for Dispersions and Solutions	455
	6.2.5	Drying	455
	6.2.6	Curing of Liquid Monomers	462
6.3	Shape Casting		464
	6.3.1	Process Overview	464
	6.3.2	Capillary Action	466
	6.3.3	Predicting Cast Layer Thickness	467
	6.3.4	Slip Casting Process Considerations	472
	6.3.5	Post-Processing Operations	473
6.4	Coating and Tape Casting		473
	6.4.1	Process Overview	473
	6.4.2	Coating Methods	475
	6.4.3	Polymer Coatings	486
	6.4.4	Tape Casting of Ceramics	489
6.5	Extrusion and Injection Molding		491
	6.5.1	Process Overview	491
	6.5.2	Extrusion of Concentrated Dispersions	491
	6.5.3	Powder Injection Molding	494
6.6	Liquid Monomer-Based Additive Processes		495
	6.6.1	Process Overview	495
	6.6.2	Stereolithography (SLA)	495
	6.6.3	Inkjet Printing with Liquid Monomers	500
6.7	Summary		504
	Bibliography and Recommended Reading		505
	Cited References		507
	Questions and Problems		508
	Questions		508
	Problems		509

7. Vapor Processes — 513
Bethanie Joyce Hills Stadler

- 7.1 Introduction — 513
- 7.2 Fundamentals — 515
  - 7.2.1 Kinetic Theory of Gases and Its Relationship to Vapor Processes — 515
  - 7.2.2 Thin Film Microstructures — 532
  - 7.2.3 Epitaxial Growth of Single Crystal Films — 539
- 7.3 Evaporation — 540
  - 7.3.1 Process Overview — 540
  - 7.3.2 Thermodynamics of Evaporation — 542
  - 7.3.3 Evaporation of Alloys and Compounds — 546
  - 7.3.4 Transport Phenomenon and Film Uniformity — 547
- 7.4 Sputtering — 555
  - 7.4.1 Process Overview — 555
  - 7.4.2 Plasma Physics — 557
  - 7.4.3 Magnetron Sputtering — 559
  - 7.4.4 Radio Frequency (RF) Sputtering — 561
  - 7.4.5 Reactive Sputtering — 561
  - 7.4.6 Optimizing Sputtered Rates — 563
- 7.5 Chemical Vapor Deposition — 566
  - 7.5.1 Process Overview — 566
  - 7.5.2 Thermodynamics of Formation Reactions — 568
  - 7.5.3 Types of Reactions — 570
  - 7.5.4 Kinetics of CVD — 571
  - 7.5.5 Deposition Rate and Uniformity — 573
- 7.6 Post-Processing of Films after Deposition — 578
  - 7.6.1 Annealing — 578
  - 7.6.2 Patterning — 578
- 7.7 Summary — 580
- Bibliography and Recommended Reading — 582
- Cited References — 583
- Questions and Problems — 583
  - Questions — 583
  - Problems — 585

Appendix A — 589
Index — 591

Preface

Materials processing is recognized as one of the four key components of the field of Materials Science and Engineering (MSE). How a material is made into its final form has great importance to a material's structure (i.e., crystal structure, phases, microstructure) and therefore to its properties and performance. For example, cold deformation processes, such as rolling, increase the dislocation density and hence the yield strength of metals. The reverse is also true: a material's structure and properties determine its ability (or inability) to be processed easily by a given method. For example, the viscosity of a typical polymer melt is too high for forming operations involving gravity-driven flow, such as melt casting, but is well suited for processes involving pressure-driven flow, such as extrusion and injection molding. Processing-structure-property interrelationships abound in all types of engineering materials. Processing also plays a significant part in determining the cost of the final item, and is central to materials selection and design. Hence, the study of materials processing builds naturally from a base understanding of structure-property relationships and is an essential component of materials selection and design.

This book introduces the fundamentals of materials processing. The area is broad both in the scientific and engineering principles and in the details involved in the practical processes. The intent here is not to cover all the details, but to explore fundamental concepts and show their application in example processes. The examples range from traditional processes, such as sand casting of metals, to newer additive processes, such as fused deposition modeling (i.e., 3D printing). The book covers processing fundamentals that apply to the three main classes of engineering materials: metals, ceramics, and polymers. The unified approach used here considers processes in categories according to their state of matter as the new shape is formed. For example, the chapter on "melt processes" explores the fundamentals aspects of melt flow and solidification and applies them to processes such as metal melt casting and polymer injection molding. This approach lends itself to exploration and application of prior knowledge.

The book is intended for undergraduates in MSE and related fields. Students who have completed an introductory materials science and engineering course, as well as calculus, physics, and chemistry courses have the background needed for this book. For example, the book could be used in a course offered in the junior or even the sophomore year directly after these prerequisites are completed, or in a course for seniors as a capstone. Graduate students and practicing engineers may also find this book useful to broaden their knowledge base and add to their understanding of fundamental concepts.

There are seven chapters in this text. The first chapter introduces the field of materials processing and provides an overview of the processing of metals, ceramics, and polymers. The second chapter deals with the preparation, formulation, and characterization of the starting materials for processing. The remaining chapters are devoted to different processing routes. These routes are grouped by the nature of the material as the final form is created: melt, solid, powder, dispersion or solution, and vapor. Important post-processing operations, such as sintering, are integrated into these chapters. Each chapter includes sections dealing with scientific and engineering fundamentals, followed by sections on the processes, including descriptions, analytical approaches to process modeling and worked examples. Each chapter ends with a bibliography, review questions, and problems.

Acknowledgements

This book is the culmination of over 12 years of effort on and off. I would like to first thank God for providing strength and inspiration. There are many people to thank and acknowledge. I would like to thank my colleagues in the Department of Chemical Engineering and Materials Science at the University of Minnesota for encouraging the development of a course in materials processing (MatS 4301) and for supporting this book. I am especially grateful to Frank Bates, who was Department Head during the time when this book was initiated and most of it written, for his support and encouragement, and to Dan Frisbie, the current Department Head, for his support during the final push to finish. Thanks to all the students and teaching assistants in MatS 4301 for their questions and suggestions over the years. Their input has shaped and improved the text immensely; I am grateful for the opportunity to teach such wonderful students! Thanks to Chris Macosko, who taught MatS 4301 with me during its first offering, for his encouragement and valuable input on polymer processing. I would also like to acknowledge the late L. E. "Skip" Scriven, who taught me to think broadly about process fundamentals through our years of collaborating on coating processes. I am grateful to Beth Stadler for writing the chapter on vapor processes and Christine Roberts for her contributions to the chapter on dispersions and solution processes. Their expertise and contributions strengthened the book considerably and it was wonderful to work with them. Thanks also to Penn State University and Gary Messing for hosting a semester stay during the sabbatical that launched the project and for discussions about processing. I would like to acknowledge the L. E. Scriven Chair and the Taylor Professor Fund for support of my educational activities. Thanks to Tiffany Smith, Carolyn Francis, Tho Kieu, David Fischer, Connie Dong, Phil Jensen and Jacquelyn Hoseth for assistance with figures, references, and proof reading. Thanks also to Eray Aydil, Frank Bates, Marcio Carvalho, Xiang Cheng, Yuyang Du, Vivian Ferry, Bill Gerberich, Cindie Giummarra, Russ Holmes, Satish Kumar, Efie Kokkoli, Robert Lade, Chris Macosko, Ankit Mahajan, Sue Mantell, Michael Manno, Ashok Mennon, Alon McCormick, Luke Rodgers, Jeff Schott, Wieslaw Suszynski, Yan Wu, and Jenny Zhu for commenting on various sections and chapters of the book and providing information. Thanks to Steve Merken, Jeff Freeland, and Christina Gifford of Elsevier Publishing as well as Kiruthika Govindaraju and the Elsevier book production team for their patience and assistance. Lastly, very special thanks to family, especially my husband, Mark, and daughter, Carolyn, for their love and support.

Lorraine F. Francis
June 2015

Chapter 1

Introduction to Materials Processing

1.1 MATERIALS PROCESSING: DEFINITION AND SCOPE

Materials processing is used to create all the manmade items that we know from everyday life. Products ranging from simple items, such as plastic wrap and paper clips, to complex multipart designs, such as automotive engines and electronic devices, are made by the same general sequence of events. Ultimately, the materials in these items originate from the earth and its resources. As shown in Figure 1.1, raw materials from the earth are converted, by mechanical, chemical, and thermal processes, to more refined *starting materials*. These starting materials are then processed into useful products. For example, mineral ores are mined from the earth; metals are then extracted from the ores and processed further to make metal alloys in standard forms, such as slabs, which can be converted into products or components. Likewise, ceramics originate from ores, which are mined and then refined into chemicals and ceramic powders, the starting materials for the creation of ceramic objects. Polymers also follow a similar sequence, often originating from oil. Oil is refined to a monomer (e.g., ethylene), processed into a polymer starting material (e.g., polyethylene pellets), and eventually converted into a polymer part. The initial steps in this complex series of events (i.e., mining and refining of raw materials) are of interest to materials scientists and engineers, but are the primary domain of mining engineers and chemical engineers. Materials engineers are more concerned with the latter steps of synthesizing and formulating the starting materials, especially processing them into useful products. The operations that fit into the box labeled *Materials Processing* are the focus of this book. That is, the emphasis is on the conversion of starting materials to new forms or shapes that are used in products.

Throughout the complex sequence of events shown in Figure 1.1, engineers search for ways to minimize environmental impact, and develop more *sustainable* practices and process routes. Recycling as products are manufactured and after the end of their useful lives is one effort. Minimizing waste and emissions of pollutants is another important effort. Lastly,

2 Materials Processing

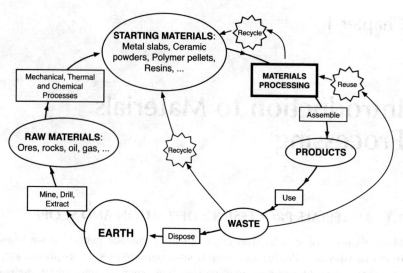

FIGURE 1.1 The materials cycle. *Materials Processing* takes place in the conversion of starting materials to products. *Adapted and updated from National Research Council (1974).*

engineers seek to develop processes that minimize energy consumption and the emission of greenhouse gases. The move toward sustainability is also leading scientists and engineers to develop new materials that are derived from renewable resources, such as polymers synthesized from biomass feedstocks. One mechanism for tracking progress on these fronts and others is *life cycle assessment (LCA)*. LCA provides a framework for following a product or family of products from "cradle to grave" and determining how to lower environmental impact. See Figure 1.2.

Materials processing is the series of steps that converts a starting material into a useful form with controlled structural features and properties. To successfully complete a materials processing sequence, engineers need to understand a number of fundamental topics, including heat and mass transport, flow, deformation, and phase change. Historically, the study of materials processing has been subdivided according to the type of engineering material. Ceramic processing, metal fabrication and forming, and polymer processing are covered in separate texts. However, common scientific and engineering principles unify the processing of all three classes of engineering materials. These unifying principles are explored throughout this book.

Materials processing is a key element in the field of materials science and engineering (MSE). The foundation of MSE is a set of interrelationships that govern the behavior of all engineering materials, including metals, ceramics, polymers, electronic materials, and composites. See Figure 1.3. One important relationship is between *structure* and *properties*. For example, the structure of a metal determines its mechanical strength, and the structure of a

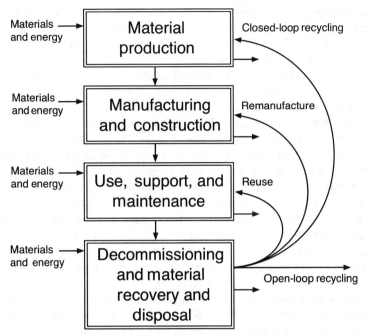

FIGURE 1.2 Schematic of the typical steps in a life cycle assessment (LCA). LCA is used to track products from production to the end of their useful life. *From Cooper and Vigon (2001).*

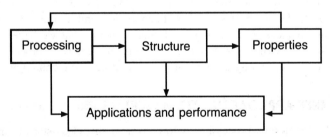

FIGURE 1.3 Interrelationships between processing, structure, and properties of engineering materials.

semiconductor determines its electrical conductivity. The single term *structure* covers a broad range of length scales, including atomic structure, interatomic bonding, crystal structure, nanostructure, microstructure, and macrostructure (i.e., size and shape). But how is structure itself determined? There are two important factors. The first is the chemical composition of the material, which controls the atomic structure and the interatomic bonding and affects other levels of structure as well. The second is processing. Materials with the same chemical composition can be processed in different ways to create materials with vastly different microstructures and even crystal structures. Interestingly, there is a connection between a material's properties, such as melting point or glass transition temperature, and its ability to be processed by different methods. But materials processing is much more than just a scientific pursuit!

Materials processing is at the heart of product design. The design of the size, shape, and features of a product hinges on the selection and control of the processing method. Therefore, processing influences the applications that are possible for a given material. Economic factors invade each process step and impact the selection of the processing route. For example, if a company needs to produce 1 million widgets a year, the processing route chosen is different from that chosen for the expectation of 1000 widgets a year and different still from the route selected to make one widget. In addition, the combination of processing method and material is not necessarily the one that results in the best properties, but rather the one that provides acceptable properties and the highest profit. Similarly, the *performance* of a material, or its ability to retain properties and function over time, can also be affected by the processing route chosen. So, the engineer weighs a myriad of factors in choosing a material and a processing route.

Materials processing should be recognized as one part of the field of *manufacturing*, commonly defined as the making of goods and articles. In addition to materials processing, manufacturing also includes surface treatment and finishing, the assembly and joining of multiple parts, automation, quality control, and the coordination of multiple steps to create a finished product in an economically viable manner. Manufacturing is a significant segment of the United States and world economies, and a vital component in continuing advances in technology and quality of life that we have come to expect. New developments and improvements in materials processing are essential to the continuing advance of manufacturing and technology.

1.2 THREE APPROACHES TO MATERIALS PROCESSING

The goal of materials processing is to convert a shapeless starting material into a useful object with complexity and function. There are three basic approaches to achieving this goal. Examples of these approaches are shown in Figure 1.4.

The first approach is to form the object. In a forming operation, a starting material is converted into a useful shape by a sequence of events. The starting material *flows*, a new *shape is defined*, and finally the *shape is retained*. In forming, external forces cause the flow, or motion of the starting material, as well as the creation of the new shape. A "tool," "mold," or some sort of fixture is involved in the shaping, and there is an identifiable mechanism for shape retention. For example, in powder pressing, a powder is poured into a die—the flow step. Next, pressure is applied via a set of punches to compact the powder and define the shape. Lastly, the powder compact is ejected and

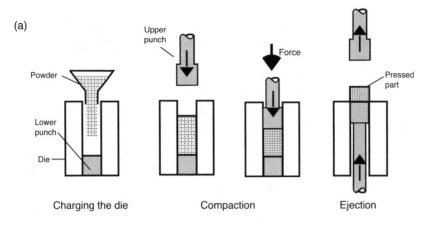
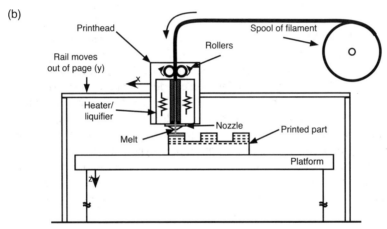
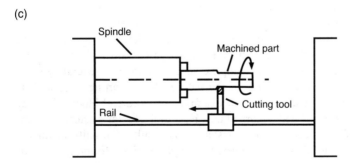

FIGURE 1.4 Examples of the three approaches to materials processing: (a) Forming, as illustrated by powder pressing; (b) Additive processing, as illustrated by fused deposition modeling (FDM) [adapted from Stratasys, www.stratasys.com and Gibson et al. (2010)]; and (c) Subtractive processing, as illustrated by turning. Diagrams are schematics only and not to scale.

the shape is retained due to bonds that are created during compaction. There is tremendous variety in forming operations, including casting of melts, molding of melts, and forging of solid materials. In forming operations, the object, after forming, may need some additional refinement through post-processing operations before it is complete.

The second approach of materials processing uses addition. In the additive processing mode, a 3D object is constructed by sequentially building 2D layers. There are several additive techniques. One of the easiest to envision is fused deposition modeling (FDM). In FDM, a spool of thermoplastic polymer filament is fed into a heated nozzle, where it is converted into a viscous melt and extruded out of the nozzle and onto the growing part, where it cools and solidifies. The nozzle moves in a two-dimensional (2D) pattern dictated by a computer file containing the details of the part shape, resulting in a layer of the part. The platform moves incrementally so that the part is built in layers from the ground up. Because printing is a term used to describe an additive process that makes a 2D layer, additive processes like FDM are also known as three-dimensional (3D) printing processes. In additive methods, we can also identify the sequence of flow, shape definition, and shape retention, similar to forming operations. There are also post-processing steps for some additive processes. However, unlike forming, additive processes are flexible, capable of creating unique parts with much greater ease. Therefore, additive approaches are used for rapid prototyping, a step in product or part design in which the size and shape of a proposed part are tested with a prototype before a final material and production strategy is selected. In recent years, the use of additive processes has been on the rise for customized parts as well as for larger production runs. One limitation of additive processes, however, is that the materials selection is not always extensive.

The last approach is subtraction. The starting point for subtractive processes is a block of solid engineering material. From this solid, material is removed, subtracted, to leave behind the shape of interest. In materials processing, subtractive processes are commonly known as machining. Machining processes frequently involve the action of hard tools that are precisely controlled to produce a complex shape. There are several types of machining processes, depending on the nature of the starting solid and the requirements of the final object. For example, a cylindrical block of a starting material can be rotated or turned as a hard tool acts on its surface to create features. The object, product, or shape that results is usually in final or near final form. In this text, however, the emphasis is on forming and additive processes.

The discussion above concerned 3D objects, but we could just as readily consider how thin films or coatings, which are 2D in nature, are formed from melts, powders, or vapors, how they might be changed into more useful forms by subtractive processes, such as etching, and how they might be built, layer-by-layer, by additive processes, such as inkjet printing. These are also topics of interest for materials processing.

1.3 MATERIALS PROCESSING STEPS

Materials processing is most easily viewed as a series or sequence of steps (Figure 1.5). Starting materials begin the series. Unlike *raw* materials, starting materials have the desired level of purity as well as properties and characteristics that are required for the forming process and final end use. However, before forming, starting materials may need to be prepared further and formulated with additives in order to enhance either their ability to process the material or the final properties of the product. For example, a plasticizer may be mixed with a polymer to create a starting material that is then extruded into a shape, such as a tube. In this case, the plasticizer enhances the process (improves flow in the extruder) and the product (improves flexibility). Or, a binder may be added to a ceramic powder starting material to improve the strength of a pressed powder part. Characterization of starting materials, before and after any formulation or preparation step, is vital to the development of the forming operation, and the final structure and properties of the product. Lastly, the cost of starting materials must be considered in view of the entire processing cost and the cost of the finished product. Chapter 2 covers starting materials.

At the heart of materials processing is the processing operation itself. During this stage, the size and shape of the product are defined. During processing, the material flows as a liquid into a mold or through a die, or deforms as a solid in such a way that a shape is achieved. In addition, the new shape or form must be retained. For example, a liquid metal fills a mold and is then converted to a solid shape by crystallization on cooling. In thin film processes, a vapor formed by evaporation or bombardment by ions is directed onto a surface where it condenses into a solid film. The shape is the solid film. The last stage, post-processing, may follow to refine the product. For example, a pressed powder part is heated to densify the part by a process known as sintering.

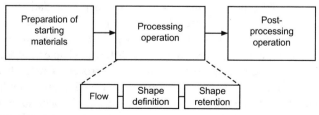

FIGURE 1.5 Materials processing steps or stages. The inset shows the sequence of events that occurs during forming and additive processes.

In this book, the fundamentals of processing operations are addressed according to the nature or state of the material as the shape is created:

*Melt Processes (*Chapter 3*):* Formation of a 3D shape, a constant cross-section shape, such as a tube or rod or sheet, from a melt. The melt is poured or forced under pressure into a mold, through a die, or onto a surface, followed by solidification on cooling or reaction. Both forming operations and additive operations are used to make products from melts. In a forming operation, the melt can be further deformed or shaped as it cools, but before it solidifies.

*Solid Processes (*Chapter 4*):* Formation of a 3D shape or a constant cross-section shape by plastically deforming a solid under the application of a mechanical stress transmitted by a tool, such as a die or roll. Solid deformation may be carried out at room temperature or elevated temperature. Alternatively, solids can be converted to shapes by subtractive, machining processes.

*Powder Processes (*Chapter 5*):* Formation of a 3D shape by filling a die or mold with powder and applying pressure, uniaxially or isostatically, to compact the powder with or without concurrent heating. Parts formed by powder processes typically require post-processing (e.g., sintering) for densification. Additionally, several additive processes also involve powders.

*Dispersion and Solution Processes (*Chapter 6*):* Formation of a 3D shape, sheet, or coating from a dispersion of particles in a simple liquid (or polymer melt), or a solution of polymer in a solvent. Solidification typically occurs as liquid is removed (e.g., by drying). This chapter also includes processes that convert liquid monomer resins to polymer coatings or 3D objects.

*Vapor Processes (*Chapter 7*):* Formation of a thin film of engineering material by transformation of vapor phase to a solid. Processes include evaporation and sputtering, which are physical vapor deposition processes, and chemical vapor deposition. Subsequent subtractive processes can be used to define 2D regions in the films and with repetition build 3D structures.

Table 1.1 shows how common processing methods for metals, ceramics, and polymers fit into the categories listed above. Like any classification system, it is not perfect. Some processes are a bit more difficult to place than others. For example, in rotational molding, powder flow and shaping give way to melt flow during the process. Reactive polymer processes like reactive injection molding and UV curing of polymer coatings also have no obvious place and so are incorporated in two different chapters. Additionally, Table 1.1 gives specific examples of important processes in each category and across materials systems. It is far from an exhaustive list. A multivolume encyclopedia would be needed to cover the entirety of materials processing. This text is designed to teach fundamental concepts and principles using a select group of processes as examples.

TABLE 1.1 Classification of Common Processes in Terms of the Five Categories

Category	Metal	Ceramic	Polymer
Melt	Sand casting Permanent mold casting Die casting	Float glass process Fusion draw process Blow molding	Extrusion Injection molding Reactive injection molding Blow molding Fused deposition modeling
Solid	Extrusion Forging Rolling Wire drawing	Rare	Thermoforming
Powder	Uniaxial and isostatic pressing Hot pressing Hot isostatic pressing Selective laser melting	Uniaxial and isostatic pressing Hot pressing Hot isostatic pressing Selective laser sintering 3D printing with binder	Rotational molding Selective laser sintering
Dispersion and Solution	Powder injection molding Coating	Slip casting Tape casting Extrusion Powder injection molding	Coating Stereolithography
Vapor	Sputtering Evaporation Chemical vapor deposition	Sputtering Evaporation Chemical vapor deposition	Rare

Some processing operations require additional steps to further refine structure and shape. Some common post-processing operations are listed in Table 1.2. Machining, to create additional features, and surface finishing are two common operations. In metal casting, for example, the extraneous solidified metal must be removed from the cast piece, the surface must be polished, and sometimes holes and other features added. Another common post-processing operation is firing, or sintering. Some parts made from powdered ceramic or metal are not dense after forming, and a high temperature sintering treatment is needed for densification and improvement in properties. Heating is also used to refine crystal structure and microstructure of metal alloys produced by a variety of methods. In general, the need for post-processing

TABLE 1.2 Common Post-Processing Operations

Post-forming operation	Purpose	Materials
Sintering	Densification of a object prepared from powder	Ceramics
		Metals
Heat-treatment	Modification of microstructure and mechanical properties	Metals
Binder removal	Thermal or chemical removal of polymer used in a forming operation	Ceramics
		Metals
Machining	Definition of macrostructural features by material removal; removal of extra material from part	Metals
		Ceramics[a]
		Polymers[b]
Thermal annealing	Removal of thermal stresses, modification of crystallinity	Metals
		Ceramics (glasses)
		Polymers
Surface finishing	Smoothing of surface, decreasing of surface roughness	Metals
		Ceramics[b]
		Polymers[b]

[a] Most typically in the green state (before binder removal and sintering).
[b] Rare.

operations depends on the original process, material type, and application. In this text, post-processing operations are introduced at appropriate points within the chapters outlined above.

1.4 PROCESSING OF METALS

Metals are perhaps the most important engineering material in terms of production quantity, use, and economic impact. They have the exceptional mechanical properties of high strength and fracture toughness, which makes them useful in structural applications. In addition, their high electrical and thermal conductivities lead to applications in electronics and communications. Pure metals are most often used in applications that require exceptionally high electronic conductivity. For most other applications, a metal alloy, with a composition tailored to provide specific properties, is used. Metal alloys are grouped into ferrous alloys (e.g., steel, cast iron) and nonferrous alloys (e.g., aluminum alloys, copper alloys).

In terms of their processing, the two most important general traits of metals are their melting points and their ability to undergo large amounts of uniform, plastic deformation. While the melting points of metallic elements range from below room temperature to in excess of 2000°C, most metals can be melted at moderate temperatures (i.e., below 1600°C). Entering into the molten state has two advantages: (1) alloys with uniform chemical composition can be prepared easily, and (2) the melt can be solidified into a shape by pouring and cooling in a mold. The second important characteristic of most metals is their ability to deform plastically. A metal changes shape under a mechanical load and then retains its new shape when the load is removed. Solid deformation processes, such as extrusion, forging and rolling, rely on this characteristic. Some metals are able to undergo plastic deformation easily, but others are more brittle and must be processed by alternative methods.

The processing of metals begins with the extraction of the metallic elements from ores mined from the ground. The elemental metals are then formulated into alloys in the melt and cast into standard shapes (e.g., ingots). Cast pieces may then be further processed by solid deformation operations, such as hot rolling and hot extrusion, into smaller standard shapes (e.g., slabs, bars, and sheets). Bulk metal processing begins with these standard shapes or the original cast pieces as starting materials. Melt-based forming operations begin with remelting the alloy and adjusting its composition. The melt is then poured into a sand or metal mold, or it may be forced under pressure into a metal mold. Solidification occurs on cooling, and complex 3D shapes are possible. Alternatively, bulk metal starting materials can be mechanically deformed and shaped by solid deformation operations, including forging and extrusion. These processes occur at elevated temperatures (hot deformation or hot working) or at lower temperatures (cold deformation or cold working). During hot deformation processes, the metal recrystallizes as it is deformed and much more deformation can take place. Often, deformation processes are carried out in a series until the final desired shape is attained. Post-processing operations for metals include heat-treatment, machining, and grinding or surface finishing. The final properties and appearance of the metal are greatly affected by these treatments.

Powder metallurgy is a term used to describe the processing of metals from metal powder starting materials. Metal powders can be prepared by atomization of a melt or by chemical processes such as reduction of metal oxide powders. The metal powders must be carefully sized and prepared for the forming operation. The most common forming operation is powder pressing, in which metal powders are compacted under pressure into a shape. In this process, the metal powder is formulated to flow easily into the die or mold, and to pack and compact to high density. Metal powders can also be combined with thermoplastic polymers, heated (to melt the polymer), and then forced under pressure through a die or into a mold. The polymer is removed thermally or chemically in a post-forming operation. Sintering, densification of a porous powder compact during heating, is also an essential post-forming operation. A main advantage of powder metallurgy routes, as compared with bulk metal forming operations, is that much less machining is typically necessary. Powdered metals can also be converted to complex shapes using an additive process known as selective laser sintering or selective laser melting, as illustrated in Figure 1.6.

Thin films of metal are formed from the vapor phase by processes such as sputtering and evaporation. The starting materials for these processes are solid metal targets that may be made by casting, solid deformation or powder metallurgy, or they may be powders. Vapor phase processes remove atoms from the starting material, known as the source or target material, and send them through a vacuum to be collected as a thin film on a substrate, such as a silicon wafer. In evaporation, the atoms are removed thermally, usually using by heating with a current-carrying filament. In sputtering, the atoms are removed by gas ions that bombard the target, which is negatively charged. See Figure 1.7. By having several source materials in one vacuum chamber, layers of different materials can be built up to form micrometer and nanometer sized devices. The high surface area to volume ratio of the end shapes means that surface energies between the different layers and the substrate play an important role in processing. Because of this, and the

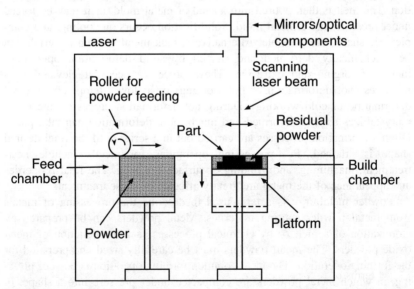

FIGURE 1.6 Selective laser sintering process. The platform of the feed chamber is raised by an increment, while the platform for the build chamber is lowered by an increment. A counter-rotating roller pushes powder from the feed chamber into the build chamber. Then a laser scans over the layer of fresh powder in the build chamber, causing the powder to locally sinter (densify). This layer-by-layer process repeats until the part is finished. *Adapted from Gibson et al. (2010).*

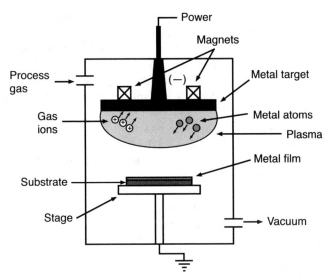

FIGURE 1.7 Sputtering process. Process gas fed into a vacuum chamber is ionized in the electric field created by the bias on the metal target. A plasma containing the ions is created near the target; magnets tune the plasma position. The ions bombard the oppositely charged target, knocking out metal atoms, which deposit as a thin film on the grounded substrate.

nonequilibrium conditions of the vacuum chamber, the grain sizes tend to be small, similar in size to the thickness of the films, and there tends to be a higher concentration of defects, which can affect the film properties. However, the atomistic nature of the processes can also allow layers of single crystals with no grain boundaries or defects to be grown atom by atom if the conditions are controlled properly. In this case, called "epitaxy," the growing film crystalline lattice matches to the substrate lattice.

1.5 PROCESSING OF CERAMICS

Ceramics are important engineering materials. They have the advantages of high hardness, chemical inertness, thermal stability, high compressive strength, and useful electrical, magnetic, and optical properties. Ceramics, however, are weak in tension and brittle (i.e., they undergo little or no plastic deformation before failure). Ceramics may be crystalline or noncrystalline (glass).

In terms of processing, the two most significant characteristics of crystalline ceramics are their high melting points and their brittle nature. The high melting points (e.g., in the range of 1000–3000°C) make melt processing of ceramics difficult and costly. Their inability to plastically deform extensively prohibits the shaping of dense polycrystalline ceramics by solid deformation processes. Ceramic glasses, on the other hand, can be formulated into melts that can be formed into shapes or sheets easily. The viscosity of a glass melt increases as temperature decreases; hence, a melt can be cast or deformed at elevated temperature and the new shape can be locked in by decreasing the temperature (i.e., increasing the viscosity).

The starting materials for the fabrication of polycrystalline ceramics are ceramic powders. The powders themselves are prepared by a variety of methods, ranging from mining and purifying ores to chemical synthesis operations, such as precipitation. The powders are sized, prepared, and formulated in different ways, depending on the forming operation. As in powder metallurgy, pressing is common way to form powder into a shape; in this case, the powder is prepared into granules containing ceramic particles, small amounts of liquid (typically water), and polymer binder. During compaction, the granules deform and the particles pack more efficiently. In contrast, dispersion casting (e.g., slip casting, see Figure 1.8) requires the powder to be dispersed in a liquid, along with additives, which stabilize the dispersion and improve the strength of the as-formed piece. The dispersion is poured into a porous mold and liquid is removed by capillary action, or it is coated onto a substrate or carrier and the liquid removed by drying, creating a thin layer. Interestingly, ceramic powders, notably clays, can be made into a plastic state by the addition of the correct amount of water (enough to cover the particle surfaces but not so much as to make a fluid dispersion). These clay bodies are plastic—they deform under a shear stress and retain their deformed shape when the stress is removed. Plasticity can be imparted to non-clay ceramic by adding clay or appropriate amounts of a polymer binder and a liquid to the ceramic powder. Alternatively, ceramic particles can be

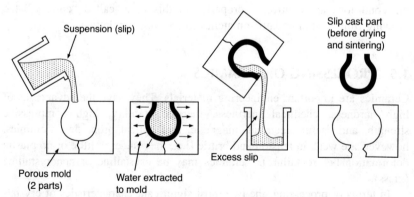

FIGURE 1.8 Slip casting process. A suspension of ceramic particles in water (also known as a "slip") is poured into a porous gypsum mold. The water from the suspension is pulled into the mold by capillary action, leaving behind a layer of consolidated particles on the mold surface. This layer thickens with time and when it reaches the desired dimension, the excess slip is poured out and the part is removed. The slip cast part must be dried and fired (sintered). *Adapted from Richerson (1992).*

compounded with thermoplastic polymer and then formed by melt-based processes, again similar to powder metallurgy. For all forming operations, the as-formed ceramic is porous and a post-forming firing or sintering treatment is need for densification. Machining or polishing of ceramics after they are fired is not as common as it is in metals; the higher hardness and brittleness of ceramics makes these processes costly and time consuming.

Ceramic glasses are fabricated from melts. The starting material is a glass batch created from ceramic powders and minerals, such as feldspars (naturally occurring alkali aluminosilicates). The glass batch is melted and its composition adjusted in the melt to provide the processing characteristics and properties required in the final glass object. Glass melts are cast to form shapes or sheets. See Figure 1.9. Flowing molten glass onto a bath of molten tin, for example, is the first step in making window glass. The molten glass sheet cools and its viscosity rises until it is solid; the float process produces glass with uniform thickness and a smooth surface that requires no further polishing or grinding. Glass bottles are made by pressing molten glass into a mold and deforming it by air pressure in another mold to create the hollow shape. Lastly, the post-forming operations for glass pieces are critical. Oftentimes, a glass shape is cooled rapidly and hence thermal stresses develop. Annealing at moderate temperatures removes the thermal stress without destroying the shape or amorphous nature of the glass.

For ceramic thin films, atoms are removed from a source or target material, similar to metal film processing. Due to the high melting points of ceramics, the starting material for evaporation has to be heated by a high energy source, such as an electron beam, rather than by simple thermal filaments. For sputtering, higher sputter powers are required and, most importantly, the target cannot hold the negative charge that causes the gas ions to bombard it. The solution is to use radio frequencies to bias the target so that

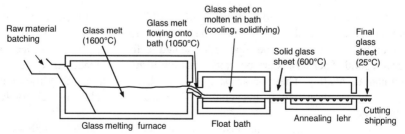

FIGURE 1.9 Float glass process. Powdered raw materials are fed into a glass melting furnace and melted. The melt, which is homogeneous at the end of the furnace, flows onto a bath of molten tin, creating a uniform thickness layer with smooth surfaces. The glass melt cools and solidifies as it progresses across the tin bath. On exit it is solid enough to be pulled by rollers into an annealing lehr, where it is slowly cooled, relieving thermal stresses. The final glass is cut into sheets at the end of the production line.

it has a net negative charge. In both evaporation and sputtering, metal source materials can be used if a reactive gas, such as oxygen, is fed into the vacuum chamber so that the substrate gathers both metal and oxygen atoms together. In addition to these physical vapor deposition processes, ceramic films can be grown by chemical vapor deposition (CVD). In CVD, the source or starting materials are vapor phase (volatile) chemicals that react to form a solid product and usually other volatile products. The solid product becomes the thin film as it builds up on the substrate. The reaction is usually thermally driven so that the substrate is held at the reaction temperature, and the rest of the chamber is either heated or cooled to a temperature outside the reaction zone. During any of these physical or chemical processes, the substrate can be heated, but usually post-anneals are required to form the desired crystalline phase.

1.6 PROCESSING OF POLYMERS

Of all the materials classes, polymers have experienced the greatest growth in applications in the last several decades. Polymers exhibit a range of mechanical behaviors, from elastomeric, to plastic, to brittle, depending on their structures and the temperature. Polymers have low densities compared with metals and ceramics, and they have high strength-to-weight ratios, which make them outstanding candidates for many applications. Polymers are generally insulating thermally and electrically. These properties have been part of the reason that production and use of polymers has grown so much in recent years, but the ability to process polymers into complex shapes is of at least equal importance to this trend. A single, complex polymer part may be designed to replace a multi-part metal assembly, resulting in a huge cost savings.

In terms of processing, the low melting points and the ability to tailor the cross-linking reactions are two of the more important characteristics of polymers. Thermoplastic polymers are polymer solids that can be heated into a molten state and then solidified on cooling. While in the molten state, they are easily shaped. Thermoplastics may be amorphous (glassy) or semi-crystalline, and they may be repeatedly heated to form a melt and cooled to form a solid. Thermosets are polymers that are formed by reactions, typically initiated on heating. They too are shaped in the liquid state, but solidification occurs by reaction and is irreversible. Thermosets are almost exclusively amorphous.

Most polymers are thermoplastics. Starting materials for these polymers are pellets, granules, and particles, which have been synthesized to their desired structures and molecular weights, and often modified with additives, such as colorants and flame retardants. Most commonly, a molten polymer is forced under pressure into a mold (e.g., injection molding) or through a die (e.g., extrusion). Solidification of the new shape takes place on cooling. These melt-based methods are considered the workhorses of polymer processing. Furthermore, an important segment of additive manufacturing makes use of polymer melts extruded locally to create objects, as shown in Figure 1.4b. Another process involving thermoplastic polymers is coating. Here, a polymer solution is coated onto a surface and then solidified by drying into a continuous polymer coating in a "roll to roll" process. See Figure 1.10. A similar route involves applying a dispersion of polymer particles in a liquid onto a substrate and forming a coating by drying-induced coalescence of the particles into a dense coating. Sequential forming operations may also be used. For example in thermoforming, a sheet of polymer prepared by extrusion is reheated and deformed against a mold to make a thin walled shape.

Thermoset polymer starting materials are monomers, oligomers, or prepolymers, and curing agents. These are formulated into liquids or low melting point solid resins, which are then formed by similar (but not identical) melt-based methods as thermoplastics. The solidification of the melt in the mold takes place by reaction and typically, the molded part is kept at elevated temperature to complete solidification, and then cooled. Figure 1.11 shows the reaction injection molding process; two liquid reactants are mixed and then injected into a mold to create a shape. Since the viscosity of the resin depends on both the temperature and the extent of reaction, controlling the temperature and the rate of reaction is vital. The mold is opened after the reaction is complete, and the part is removed. Reactive monomers and polymers are also coated and made into objects by selective curing in an additive manufacturing process known as stereolithography.

Post-forming operations are minimal with polymers. Some forming operations create a series of polymer parts in a single mold. Each part must

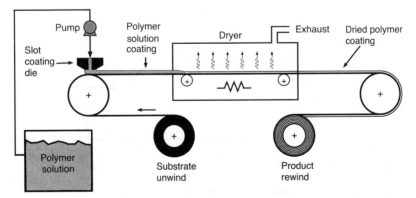

FIGURE 1.10 Polymer coating process. Polymer solution is pumped at a controlled rate into a slot die, which is stationed at a precise position beneath a back-up roll. Flexible substrate, known as web, is guided at a controlled rate beneath the die such that a thin polymer solution coating is applied onto the substrate. The coated substrate is carried through a dryer, where solvent is removed, and the final coated product is then collected in a roll at the end of the production line.

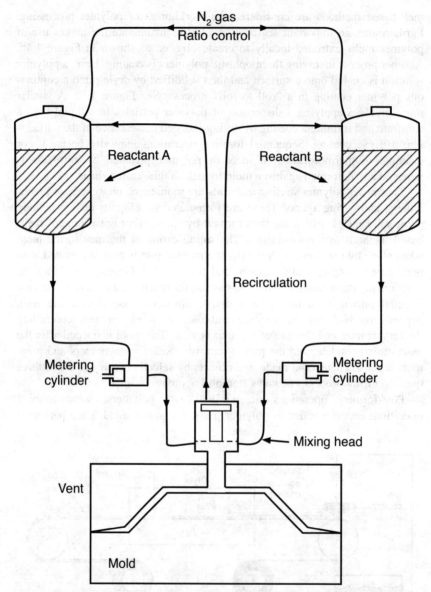

FIGURE 1.11 Reactive injection molding process. In this cyclic process, two reactant liquids are metered into a mixing head by high pressure metering cylinders. On meeting and mixing, the reaction begins and the mixture continues to flow into the mold. Reaction (curing) is completed in the mold, forming a solid part with dimension set by the mold. The part is removed and the cycle repeats. *Adapted from Johnson (1988) and Macosko (1989).*

be separated and sometimes undergo a surface polishing at the separation point. Polymers may be heated or stretched to induce crystallinity. Machining is usually not necessary as details can be created during forming.

1.7 SUMMARY

Materials processing involves a complex series of chemical, thermal, and physical processes that prepare a starting material, create a shape, retain that shape, and refine the structure and shape. The goal of materials processing is to develop the structural features (e.g., crystal structure, microstructure, size, and shape) needed for the product to perform well in its intended application. Materials processing is central to the field of materials science and engineering, and is a vital step in manufacturing.

The conversion of the starting material to the final product occurs in three steps: preparation of the starting material, processing operation, and post-processing operation(s). The processing operations can be divided into five categories based on the state of matter most important to the process: melt, solid, powder, dispersion or solution, and vapor. Metals, ceramics, and polymers are formed by operations in each of the categories so that common scientific and engineering principles can be understood and applied to various types of materials.

BIBLIOGRAPHY AND RECOMMENDED READING

Beddoes, J., Biddy, M.J., 1999. Principles of Metal Manufacturing Processes. John Wiley & Sons, New York, NY.

Creese, R.C., 1999. Introduction to Manufacturing Processes and Materials. Marcel Dekker, New York, NY.

DeGarmo, E.P., Black, J.T., Kohser, R.A., 2003. Materials and Processes in Manufacturing, ninth ed. Wiley, New York, NY.

Groover, M.P., 1996. Fundamentals of Modern Manufacturing: Materials, Processes and Systems. Prentice-Hall, Upper Saddle River, NJ.

Hosford, W.F., Caddell, R.M., 1983. Metal Forming: Mechanics and Metallurgy. Prentice-Hall, Inc., Englewood Cliffs, NJ.

Kalpakjian, S., 1997. Manufacturing Processes for Engineering Materials, third ed. Addison Wesley, Menlo Park, CA.

Kingery, W.D. (Ed.), 1958. Ceramic Fabrication Processes. John Wiley & Sons, New York, NY.

Morton-Jones, D.H., 1989. Polymer Processing. Chapman and Hall, New York, NY.

Ohring, M., 1992. Materials Science of Thin Films. Academic Press, Boston, MA.

Osswald, T.A., 1998. Polymer Processing Fundamentals. Hanser/Gardner Publishing, Cincinnati, OH.

Rahman, M.N., 1995. Ceramic Processing and Sintering. Marcel Dekker, New York, NY.

Reed, J.S., 1995. Principles of Ceramic Processing, second ed. Wiley Interscience, New York, NY.

Ring, T.A., 1996. Fundamentals of Ceramic Powder Processing and Synthesis. Academic Press, New York, NY.
Schey, J.A., 2000. Introduction to Manufacturing Processing, third ed. McGraw Hill, New York, NY.
Schneider, S.J. (Ed.), 1991. ASM Engineered Materials Handbook: Ceramics and Glasses (Vol. 4). ASM International, Metals Park, OH.
Smith, D., 1995. Thin-Film Deposition: Principles and Practice. McGraw Hill, New York, NY.
Strong, A.B., 2000. Plastics Materials and Processing, second ed. Prentice-Hall, Upper Saddle River, NJ.
Tadmor, Z., Gogos, C.G., 1979. Principles of Polymer Processing. Wiley Interscience, New York, NY.

CITED REFERENCES

Cooper, J.C., Vigon, B., 2001. Life Cycle Engineering Guidelines. (EPA Publication No. 600/R-01/101). U.S. Environmental Protection Agency, Cincinnati, OH.
Gibson, I., Rosen, D.W., Stucker, B., 2010. Additive Manufacturing Technologies: Rapid Prototyping to Direct Digital Manufacturing. Springer, New York, NY.
Johnson, C.F., 1988. In: Dostal, C.A. (Ed.), ASM Engineered Materials Handbook, Vol. 2: Engineering Plastics. ASM International, Materials Park, OH, pp. 344–351.
Macosko, C., 1989. RIM: Fundamentals of Reaction Injection Molding. Hanser Publishing, New York, NY.
National Research Council, 1974. Materials and Man's Needs: Materials Science and Engineering. The National Academies Press, Washington, DC.
Richerson, D., 1992. Modern Ceramic Engineering. Marcel Dekker, New York, NY.

Chapter 2

Starting Materials

2.1 WHAT IS A STARTING MATERIAL?

A starting material is ready to be formed into a shape or used in an additive manufacturing operation. Raw materials taken directly from natural sources must undergo several stages of mechanical and chemical processes to prepare them for processing and for the demands of the final product. The science and engineering of the amazing transformation from a natural, raw material to a useful starting material encompasses the fields of extractive metallurgy, organic and inorganic chemistry, polymer synthesis, and ceramic synthesis to name a few. While comprehensive coverage of these fields is outside the scope of this book, an overview and examples of the preparation of starting materials are provided in this chapter in order to set the stage for the materials processing methods that follow in subsequent chapters.

Starting materials are readily available. Some companies specialize in the production of them. These companies, often called "suppliers," sell starting materials to other companies—the manufacturers—who carry out the processes to make the final desired shapes and forms. The connections between suppliers and manufacturers vary with the type of material and sometimes the industry involved. In some cases, the supplier is internal; for example, one division in a company produces starting materials for another division. In other instances, raw material is converted into a starting material and then into the final product at one location or facility. Many ceramic glass companies and some steel mills are set up in this comprehensive manner.

Starting materials are frequently formulated with multiple components or with additives designed to improve the subsequent processing steps or to enhance the properties of the final product. The need for additives and special formulations varies from material to material. Thermoplastic polymers, for example, often contain additives designed to enhance the flow characteristics of the polymer melt, while metal powders are prepared with lubricants to prevent friction problems during forming. Additives for processing are sometime transient; for example, organic binders, used to enhance strength of powder compacts before they are sintered, are removed during a post-processing heating step. Additives and compositional adjustments to starting materials are introduced in this chapter. More details, particularly on additives that improve processing, are given in subsequent chapters.

The characteristics of starting materials impact the processing steps as well as the properties and utility of the final shape. The connections between the starting material and the process are often deliberate. One can purchase a "casting alloy" that is designed specially for melt casting processes, a thermoplastic polymer specifically designed for the additive manufacturing process of fused deposition modeling, or a ceramic powder designed to be pressed and sintered to high density. The characteristics of the starting material also have an indelible effect on the properties of the final product. A small amount of impurity in a starting material can poison the electrical properties of a ceramic insulator, for example. Thus, characterization of the starting material is imperative. Suppliers provide information about their materials in the form of data sheets (also called product sheets). Examination of data sheets reveals the important characteristics and properties of the starting materials. For example, data sheets for ceramic powders contain information on the size and purity of the powders. Data sheets for thermoplastic polymers often include information on the temperatures needed for different processes, such as extrusion.

This chapter is divided into sections according to the materials class: metals, ceramics, and polymers. After a brief overview, the main types of starting materials for each materials class are described along with some examples of their synthetic origin, data sheets, additives, and characterization methods.

2.2 METALS

2.2.1 Introduction

Over two-thirds of the elements on the periodic table are classified as metals, based on their electronic structure, bonding, and properties. Pure elemental metals, like copper, find applications that exploit their high electrical conductivity, thermal conductivity, and ductility. More commonly, metallic elements are combined with each other, and sometimes with nonmetallic elements (e.g., Si, C) to make metal alloys. The compositions and microstructures of alloys are tailored to achieve high mechanical strength and fracture resistance, and useful functional properties such as magnetism. Scientists and engineers have discovered important relationships between alloy composition (or elemental metal purity) and properties. Compositional control, therefore, is a goal of any process used to create a metal starting material. In addition to composition, the microstructure and properties of alloys are altered by the processing steps used to create the final shape and by subsequent heat treatment.

Structurally, metals and alloys are predominately crystalline. Single crystal metals and alloys are prepared using special techniques; however, most alloys are used in the polycrystalline state. That is, their microstructures consist of many crystalline grains. Further, the compositional complexity and strategies to develop mechanical properties frequently result in microstructures that contain several phases. The development of these complex microstructures depends on the alloy composition and processing conditions. Noncrystalline metallic glasses are a special category of alloys prepared by adjusting alloy composition and processing conditions. Metallic glasses have unusual mechanical and magnetic properties, but they are produced in much lower volumes than crystalline metals.

In terms of composition, metal alloys are classified in two broad categories: ferrous (or iron-based) and nonferrous. This classification is not only based on chemical composition, but also on the distinctive properties and uses of the alloys. Ferrous alloys include steels and cast irons. Carbon is the most important alloying element in these materials. Steel compositions have carbon contents below around 2 wt%, while cast irons have higher carbon contents. Other alloying elements, such as nickel, molybdenum, and chromium, are added to influence steel properties. For example, stainless steels have a high concentration of chromium, which imparts corrosion resistance. Steel and other ferrous metals are the workhorses of many industries, including construction. Common nonferrous alloys include aluminum, magnesium, and copper-based alloys. The use of nonferrous alloys has been climbing steadily in recent decades. One driving force is the lower density of these materials relative to ferrous alloys, leading the potential for reduced weight and fuel savings in applications such as automotive components. The properties of nonferrous alloys also depend on the alloying elements that are used in their composition. For all metals, impurities, which often originate from raw materials, also impact properties.

Designation codes for alloys are largely based on composition. Ferrous alloys follow systems developed jointly by the American Institute of Steel and Iron (AISI) and the Society for Automotive Engineers (SAE), as well as the Unified Numbering System (UNS), which is used for both ferrous and nonferrous alloys. Table 2.1 shows some examples of designation codes for steels. Likewise, nonferrous alloys have designations set up by various organizations as well as the UNS system. Examples are given in Table 2.2. For some alloys, heat-treatment is specified in the coding system. Compositionally based designation codes are one basis for classification of alloys. There are also codes developed based on the mechanical properties of the alloy as well as their final structural forms (e.g., strips, bars). These rigorous designation systems facilitate the selection and use of metals in applications, ensuring predictable properties and performance. In some cases, metals are also known by trade names. For example, Nitinol is a NiTI alloy with shape memory properties.

TABLE 2.1 Examples of AISI-SAE and UNS Designation Codes for Steels

AISI-SAE code	UNS code	Steel composition (given in wt% with remainder Fe)
Plain Low Carbon Steels[a]		
10XX[b]	G10XX0	Mn 1.00 (max)
15XX	G15XX0	Mn 1.00–1.65
Nickel Steels[a]		
23XX	G23XX0	Ni 3.50
25XX	G25XX0	Ni 5.00
Nickel-Chromium Steels[a]		
31XX	G31XX0	Ni 1.25, Cr 0.65 and 0.80
32XX	G32XX0	Ni 1.75, Cr 1.07
33XX	G33XX0	Ni 3.50, Cr 1.50 and 1.57
Molybdenum Steels[a]		
40XX	G40XX0	Mo 0.20 and 0.25
44XX	G44XX0	Mo 0.40 and 0.52
Nickel-Chromium-Molybdenum Steels[a]		
43XX	G43XX0	Ni 1.82, Cr 0.50 and 0.80, Mo 0.25
47XX	G47XX0	Ni 1.05, Cr 0.45, Mo 0.20 and 0.35
81XX	G81XX0	Ni 0.30, Cr 0.40, Mo 0.25
Stainless Steels		
403	S40300	C 0.15, Mn 1.00 (max), Si 0.5 (max), Cr 11.5–13.0, P 0.04 (max), S 0.03 (max)
440B	S44003	C 0.75–0.95, Mn 1.00, Si 1.0 (max), Cr 16.0–18.0, P 0.04 (max), S 0.03 (max)

[a]All have maximum limits for P and S of approximately 0.04 and 0.05, respectively. Alloy steels also have limits on Si contents (typically 0.15–0.35)
[b]XX represents the amount of carbon (wt% × 100). For example, 1010 steel has a carbon content of 0.10%.

In terms of starting materials for materials processing operations, a different sort of classification comes about naturally. Metal starting materials are most often either bulk pieces of metal or metal powders. Bulk metal starting materials come in a variety of forms, including ingots that can weigh up to several tons. Other regular bulk shapes include slabs that are ~1–3" thick and 24–40" wide and a variety of lengths. Likewise, blooms and billets are regular shapes with square cross-sections. These bulk shapes are further

TABLE 2.2 Examples of Industry and UNS Designation Codes for Nonferrous Alloys

Industry code	UNS code	Alloy composition (given in wt% with remainder main alloy element)
Aluminum Alloys[a]		
1050 (wrought)	A91050	Si 0.25, Fe 0.40, Cu 0.05, Mn 0.03, Mg 0.03, Zn 0.05, V 0.05, Ti 0.03
1350 (wrought)	A91350	Si 0.10, Fe 0.40, Cu 0.05, Mn 0.01, Cr 0.01, Zn 0.05, Ga 0.03, B 0.05, V + Ti 0.02
6061 (wrought)	A96061	Si 0.6, Mn 0.28, Mg 1.0, Cr 0.20
100.1 (cast)	A01001	Si 0.15, Fe 0.60-0.80, Cu 0.10, Zn 0.05
305.2 (cast)	A03052	Si 4.5-5.5, Fe 0.14-0.25, Cu 1.0-1.5, Mn 0.05, Zn 0.05, Ti 0.05
Magnesium Alloys[b]		
AZ80A (wrought)	M11800	Al 8.5, Mn 0.12, Zn 0.15
EZ33A (cast)	M12330	Zn 2.6, rare earths 3.2, Zr 0.7
Copper Alloys		
electronic grade	C10100	0.01 impurities (99.99% Cu)
cartridge brass	C26000	Zn 30
free cutting brass	C36000	3.0 Pb, 35.5 Zn

[a] *Aluminum Association of America codes.*
[b] *American Society for Testing Materials (ASTM) codes.*

formed by solid deformation processes into plates, sheets, beams, and bars. These products of deformation processes are termed "wrought." Bulk metals, either cast or wrought, are the starting materials for melt processes, such as shape casting, as well as solid processes, such as forging. Metal powders are the starting materials for powder metallurgy operations, such as uniaxial pressing, and additive manufacturing processes, such as selective laser melting. These two general categories of starting materials are explored more in the next two sections.

2.2.2 Bulk Metal Starting Materials

Almost all metal starting materials originate from ores mined from the earth. An ore contains a metal compound, such as a metal oxide or a metal sulfide. As natural materials, ores are not pure. The metal compound must be separated from other constituents in the ore, reduced to the metal, and purified. The science and engineering of this important conversion from ore to metal is a field unto itself: extractive metallurgy. There are a variety of extractive metallurgy methods, based on the diversity of metals. All of these methods begin with some initial ore processing steps, such as crushing and separation. The subsequent extractive metallurgy processes generally fall into one of three categories:

Pyrometallurgy: Pyrometallurgical processes use heat. The metal-containing compounds in the ore (e.g., metal oxides, sulfides, carbonates) are reduced by reaction with a source of carbon at high temperature. The reduction reaction requires strict control of temperature and atmosphere composition. Iron is extracted from iron ores by pyrometallurgy processes.

Hydrometallurgy: Hydrometallurgy involves extracting metal from aqueous solutions containing metal ions. The ore is first placed in an aqueous acid solution to leach the metal ions to the aqueous phase. The solution is then purified. Next, the metal is deposited from the purified aqueous solution by one of several processes, including metal ion replacement reactions and reduction by bubbling hydrogen (or another reducing gas) though the aqueous solution. Hydrometallurgy processes are carried out at relatively low temperatures, from ambient to the boiling point of the aqueous solution. Hydrometallurgy processes are used in the production of copper.

Electrometallurgy: Electrometallurgical processes use electrical energy to convert metallic ions into metals. Similar to hydrometallurgy, the ions are first formed by leaching them from ore, either in aqueous solution or in a molten salt. Then an electrolysis process is used to form the metal. Aluminum is made by this method.

Scrap metal is also an important source of metal for the processing of bulk metal starting materials. Scrap is added at different stages of metal production, such as during the initial ore to metal conversion. In some cases, the bulk metal starting material is made almost entirely from scrap. The use of scrap has economic and environmental advantages. Scrap originates from within a company (e.g., excess metal solidified with a casting) or it is purchased from outside sources, in which case control of scrap composition poses some challenges.

Example 1–Steel. An overview of the steel making process is shown in Figure 2.1. The first step is the conversion of a concentrated iron ore to an impure form of iron known as pig iron. The iron ore is reduced in a blast furnace by reaction at high temperature with carbon. The raw materials are specially prepared for the blast furnace. Iron ore is mined and prepared into "sinter" or pellets for the steel making process. For example, taconite, a common low grade iron ore, is roughly 28 wt% iron in the form of Fe_2O_3, Fe_3O_4, and $FeCO_3$. Using crushing, magnetic separation and other processes,

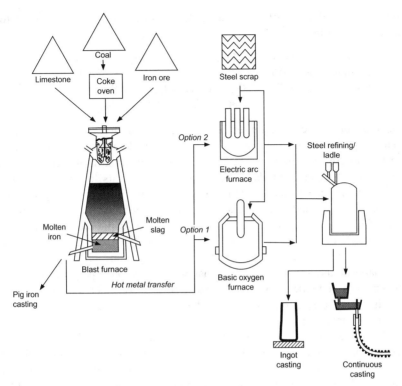

FIGURE 2.1 Flow diagram for the production of steel from iron ore and scrap. *Adapted from steel making flowlines from www.steel.org.*

marble-sized pellets with about 50–60% iron are made. Likewise the carbon used in the reaction is prepared before it is charged into the blast furnace; coal is converted to coke, a more concentrated form of carbon, by a thermal treatment. The third ingredient for iron and steel making is limestone, $CaCO_3$. This ingredient is called a flux. In the furnace, the limestone promotes the formation of impurity-rich oxide melt called "slag." The ore pellets, coke, and limestone are added to the top of the blast furnace. Hot air (~1000°C) is supplied from below via channels called tuyeres. The hot air and the carbon react to generate the heat needed for the reduction reaction and melt formation. The formation of iron and steel is therefore a pyrometallurgy process.

The chemistry in the blast furnace is complex due to gradients in temperature and composition throughout the furnace cavity. The primary reactions are (i) the reaction of carbon from the coke with oxygen in the hot air to produce heat and carbon monoxide:

$$2C_{(s)} + O_{2(g)} = 2CO_{(g)} + \text{Heat} \tag{2.1}$$

and (ii) the reduction of the iron oxides to iron liquid via reaction with carbon monoxide:

$$\begin{aligned}
&\text{Beginning at } 450°C: 3Fe_2O_{3(s)} + CO_{(g)} = CO_{2(g)} + 2Fe_3O_{4(s)} \\
&\text{Beginning at } 600°C: Fe_3O_{4(s)} + CO_{(g)} = CO_{2(g)} + 3FeO_{(s)} \\
&\text{Beginning at } 700°C: FeO_{(s)} + CO_{(g)} = CO_{2(g)} + Fe_{(l)} \\
&\quad or\ FeO_{(s)} + C_{(s)} = CO_{(g)} + Fe_{(l)}
\end{aligned} \quad (2.2)$$

The reduction reaction requires temperatures of at least 1600°C to complete. Impurities from the ore (Al, Si, Mn, O) and the coke (S) react with the flux to create a molten, metal oxide slag. The liquid iron collects at the base of the furnace with slag floating on top due to its lower density. The blast furnace is operated continuously. Periodically, it is charged from the top, and slag and iron melt are tapped from different locations near the base. The iron exiting the blast furnace is known as pig iron, which has a composition of 3–5 wt% C, 1–3% Si, 0.15–0.25% Mn, 0.2–2% P, and 0.05–0.1% S in addition to iron. The term "pig iron" reputedly refers to the shape of the small cast pieces that were commonly prepared from the stream of iron tapped from the furnace. In modern steel making, the pig iron is not solidified, but instead transferred in the molten state to the steel making furnace.

There are two main types of steel making furnaces: the basic oxygen furnace and the electric arc furnace. Both are shown schematically in Figure 2.2. In the basic oxygen furnace (BOF), a lance delivers pure oxygen gas at high velocity into a mixture of molten pig iron, scrap steel (~30%) and flux ($CaCO_3$). The oxygen reacts with carbon, and the products (CO and CO_2) are removed as gases from the top of the furnace. Waste gas is removed through large hoods. Other impurities (Al, Si, P, Mn) also react with oxygen to form oxides that are then trapped in a slag layer on top of

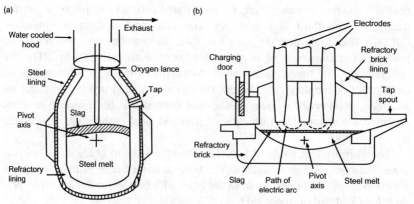

FIGURE 2.2 Schematic diagrams of steel making furnaces: (a) basic oxygen furnace and (b) electric arc furnace. *Adapted from Neely (1994), original source Bethlehem Steel.*

the molten iron. Reactions between impurities and oxygen are highly exothermic and so scrap is added as a coolant; the heat generated by the reaction is used to raise the temperature of the scrap and melt it. Scrap prevents overheating that might damage the refractory lining of the furnace, and introduces alloying elements that are designed for the target steel composition. Again, slag is removed and melt (at about 1650–1700°C) tapped into a ladle; the furnace is typically tilted for this purpose. After the furnace is tapped, a new charge is added and the process repeated, without cool down of the furnace. The BOF is very efficient, converting pig iron to steel in a short time (e.g., ~40 min for 250 ton charge). A challenge in BOF steel making is control of the quantity of dissolved oxygen in the melt. When the carbon content of the melt drops, the oxygen that is added begins to dissolve more readily into the melt. Ultimately, this oxygen needs to be removed in a subsequent production step, as described below.

The electric arc furnace (EAF) contains carbon (graphite) electrodes that are used to heat a charge of pig iron and scrap. The large carbon electrodes (up to 70 cm in diameter and 2.8 m in length) are grouped three to a column and attached to the furnace lid. The lid is removable, so that the charge can be added and then the lid and electrodes put in place as a second step. After charging, the lid is slowly lowered until the electrodes are just a few centimeters above the charge, and a large current is applied. A current passing from the electrodes through the charge is responsible for heating. The tip of the electrode reaches about 3000°C! Once a melt is attained, oxygen is introduced by injection and oxide additives to oxidize impurities, as described above. Calcium carbonate flux is added near the end of the melting phase. After melting and oxidation, the electrodes are raised and the furnace tilted for tapping of the molten steel. The cycle time from charging to tapping is typically less than one hour, and the process is repeated without furnace cool down.

An EAF can accommodate a larger fraction of scrap than a BOF, even to the extreme of operating on scrap alone. Because of the high scrap content, a wider range of steel compositions is accessible and the production cost drops, because scrap is cheaper than pig iron. In addition, use of scrap is good for the environment, removing metal from landfills. Another advantage of the EAF is that it can be decoupled from the blast furnace operation and incorporated into a so-called mini-mill, which operates exclusively on scrap.

The molten metal from the steel making furnace is transferred to a ladle. The ladle may accommodate large quantities of metal, and is the site for the final adjustments to the composition. Alloying elements are introduced, and deoxidants such as Al and Si are added to react with the dissolved oxygen and thus create a slag. The steel composition is adjusted in the ladle, leading to the term "ladle metallurgy" to describe this stage of the process. Steel making companies use precise recipes and procedures for making different grades of steel. Control of the composition, including alloying elements and impurities, is essential to the quality of the steel produced. When scrap is used, the scrap composition is chosen according to the desired target composition of the batch. Samples are taken at several stages of production and analyzed to ensure the composition is within specifications.

The molten steel is then cast into large ingots or continuously cast into smaller shapes. Ladles are often equipped with bottom pour options to facilitate separation from the slag. In ingot casting, the melt is poured into large metal molds with open tops. Ingot molds are frequently filled by pouring into a central conduit that sends the melt through ceramic channels to the bottoms of ingot molds. The melt then rises up into the molds. This bottom up filling operation prevents splashing and associated defects in the casting. The melt cools in the ingot mold and solidifies from the mold surface inward. Ingots range in size from several hundred to several thousand pounds. Due to their large size, heat removal and solidification are slow, resulting in segregation and lack of chemical homogeneity. To reverse this process, the ingot may be transferred to a soaking pit where it is held at high enough temperatures for solid state diffusion to remove compositional gradients. The large ingot must then be reduced to smaller sized pieces by a hot deformation process and cutting. As might be imagined, the size reduction of ingots requires large forces and massive equipment. The smaller metal pieces that are the products of this step serve as starting materials for other forming processes. Continuous casting circumvents many of the problems of ingot casting.

A continuous casting line is shown in Figures 2.3 and 2.4. In continuous casting, the ladle fills a holding tank called a tundish, which meters the melt

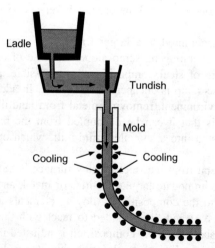

FIGURE 2.3 Schematic diagram of continuous casting process. *Adapted from steel making flowlines on www.steel.org.*

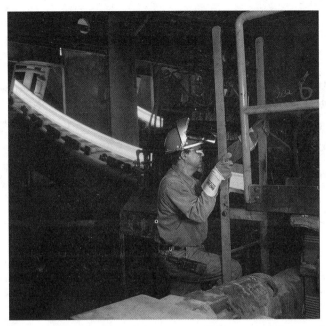

FIGURE 2.4 Continuous casting of alloy steel. *Photograph courtesy of Carpenter Technology Corporation, www.cartech.com.*

through a chilled cylindrical mold. As the melt contacts the cold mold, it solidifies locally as it continues to travel downward. An outer shell of solid steel encases the still liquid core as the material moves out of the mold and is then subjected to multiple water sprays. Solidification continues and the steel, still hot enough to be easily deformed, is straightened before it is cut into smaller pieces with an oxyacetylene torch. Continuous casting is capable of producing smaller shapes without the need for hot deformation processes. In addition, the smaller size of the material and faster solidification lead to fewer problems with chemical inhomogeneity. These advantages make continuous casting more common than ingot casting.

An integrated steel mill also produces wrought starting materials from the cast pieces. Wrought metal starting materials are created by hot deformation processes, such as rolling and forging. These products are also known as semi-finished pieces. The mechanics of deformation processes is discussed in Chapter 4.

Example 2−Aluminum. Aluminum metal is made by electrolytic reduction of aluminum oxide (Al_2O_3, alumina). The first step in the process is converting bauxite ore (an aluminum-rich ore) to alumina by a series of chemical and thermal steps known as the Bayer Process. Since the Bayer process is a good example of the synthesis of a ceramic powder (Al_2O_3)

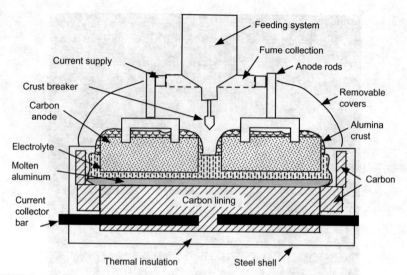

FIGURE 2.5 Schematic diagram of a Hall-Heroult cell used to produce aluminum. The aluminum melt is removed periodically by a siphon (not shown). *Adapted from schematic on www.alcoa.com.*

from an ore, it is described in detail later (Section 2.3.2). Alumina from the Bayer process is about 99.8% pure and is ideal for the preparation of aluminum.

The process to electrolytically reduce alumina to aluminum was developed in the late 1800s by C. M. Hall in the United States and P. Heroult in France. The electrolytic reduction takes place in the Hall-Heroult cell, which is shown in Figure 2.5. The aluminum formation process is usually called smelting, which is a general term for the preparation of a liquid metal by chemical reaction. Aluminum smelting is a continuous process carried out in a series of electrolytic cells (called pots). A modern plant may have several "pot lines" operating. Carbon rods serve as the anode of the electrolytic cell and the carbon lining on the cell is the cathode. A key step in the smelting process is creating a melt for electrolytic reduction. Alumina has a very high melting point (2010°C) and it is therefore impractical to form a pure melt. Instead, alumina is dissolved in a mixture of cryolite (Na_3AlF_6), aluminum fluoride (AlF_3), and a small amount of fluorspar (CaF_2) at about 940°C. When current is passed between the anode and the cathode, the conductive molten electrolyte is heated resistively and electrochemical reactions occur at the cathode and anode:

$$\text{cathode:} \quad 2Al^{3+} + 6e^- = 2Al_{(\ell)}$$
$$\text{anode:} \quad 3O^{2-} = \frac{3}{2}O_{2(g)} + 6e^- \tag{2.3}$$

Aluminum smelting requires a large amount of energy. About 13 kW-hours per kilogram of Al produced are required. Liquid aluminum forming at the cathode is denser than the electrolyte mixture and hence collects at the bottom of the cell, where it is periodically removed to a ladle. At the anode, oxygen forms and immediately reacts with the carbon electrode to form carbon dioxide and carbon monoxide gas, which are removed by a fume collection system. The process is continuous with alumina added periodically. The formation of gases at the anode leads to the generation of bubbles, which help keep the composition of the electrolyte uniform. Anodes are consumed in the process and must be replaced periodically.

Molten aluminum is cast into shapes by ingot casting or, more commonly, continuous casting. Like steel processing, the Al cast pieces undergo hot, solid deformation processing to form smaller shapes such as slabs. From the Hall-Heroult cell, the metal is about 99.5—99.7% Al with Fe and Si as the main impurities. To enhance the purity further, electrolytic refining is used. To formulate Al alloys, a melt is prepared from Al ingot (from the Hall-Heroult process) and alloying elements are added. The product alloy melt is then cast into standard shapes and used as a starting material for melt casting or deformation processes.

Recycling is very important to the aluminum industry. Scrap aluminum is remelted and reused. The remelting process uses only about 5% of the energy required to form aluminum in the Hall-Heroult cell. Beverage cans are an important source of scrap. The Aluminum Association of America reported that 67% of cans were recycled in the United States in 2012, making Al cans the most recycled beverage container. For recycling, cans are shredded, delaquered by a thermal process, and then heated to a temperature high enough to separate the two main alloy types used in the can (e.g., 5182 for the lid and 3001 for the can body). After separation, melts of the two alloys are cast and rolled into sheets. Aluminum mini-mills are used for this type of processing, which is not electrochemical in nature. It is important to note that recycling produces alloys rather than pure Al metal.

Bulk Metal Data Sheets and Characterization. A data sheet or specification sheet tells the consumer about the important characteristics and properties of a material. Data sheets can be as short as one page or as long as five or more pages. A condensed data sheet for a stainless steel alloy is shown in Figure 2.6 as an example. Several categories of information are displayed on a data sheet for a bulk metal.

Composition — The chemical composition of the alloy is given along with the appropriate compositional codes or designations.

Properties — Physical, thermal, and mechanical properties of the metal are provided. These basic properties are necessary for choosing a material for a particular application. These properties are also essential for designing a forming operation. For example, the yield stress and work hardening behavior of an alloy determine the magnitude of the forces needed for

444 STAINLESS STEEL DATA SHEET

AK Steel

PRODUCT DESCRIPTION

AK Steel Type 444 is a low-carbon, low nitrogen, dual-stabilized ferritic stainless steel that provides pitting and crevice corrosion resistance superior to most ferritic stainless steels and some austenitic stainless steels. AK Steel 444 is dual stabilized with titanium and niobium (columbium). AK Steel 444 is equivalent to Type 304 in many corrosion environments and has lower coefficient of expansion and better thermal conductivity. It has superior chloride stress corrosion cracking resistance to Types 304 and 316.

SPECIFICATIONS

AK Steel Type 444 Stainless Steel sheet and strip is covered by the following specifications:
ASTM A268
ASTM A240

AVAILABLE FORMS

AK Steel produces Type 444 Stainless Steel in coils and cut lengths in thicknesses 0.010 – 0.090 in. (0.25 – 2.29 mm) and widths up to and including 48 in. (1219 mm).

Values shown in this bulletin were established in U.S. customary units. The metric equivalents may be approximate.

COMPOSITION		(wt %)
Carbon	(C)	0.025 max.
Manganese	(Mn)	1.00 max.
Phosphorus	(P)	0.04 max.
Sulfur	(S)	0.03 max.
Silicon	(Si)	1.00 max.
Nitrogen	(N)	0.035 max.
Chromium	(Cr)	17.5 – 19.5
Nickel	(Ni)	1.00 max.
Molybdenum	(Mo)	1.75 – 2.5
Titanium + Niobium	(Ti + Nb)	0.20 + 4(C+N) min., 0.80 max.

PHYSICAL PROPERTIES

Density, lbs./in.^{3} (g/cm^3)	0.28 (7.75)
Electrical Resistivity, µΩ-in. (µΩ•cm) 70 °F (21 °C)	22.50 (57)
Specific Heat, BTU/lbs./°F (kJ/kg/K) 32 – 212 °F (0 – 100 °C)	0.102 (0.427)
Thermal Conductivity, BTU/hr./ft.^{2}/ft./°F (W/m/K) 212 °F (100 °C)	15.5 (26.8)
Coefficient of Thermal Expansion, in./in./°F (µm/m/K) 32 – 212 °F (0 – 100 °C)	6.1 x 10^{-6} (11.0)

PROPERTIES

TABLE 1 – TYPICAL MECHANICAL PROPERTIES

UTS ksi. (MPa)	0.2% YS ksi. (MPa)	Elongation % in 2" (50.8 mm)	Rockwell Hardness
75 (515)	52 (360)	30	B85

TABLE 2 – PROPERTIES ACCEPTABLE FOR MATERIAL SPECIFICATION

UTS ksi. (MPa)	0.2% YS ksi. (MPa)	Elongation % in 2" (50.8 mm)	Rockwell Hardness
60 (414) min.	40 (276) min.	27 min.	B90 max.

CORROSION RESISTANCE

AK Steel Type 444 provides excellent corrosion resistance in moderately severe environments of many types that include chlorides and organic acids found in foods and beverages. It also provides satisfactory resistanc to dilute sulfuric acid solutions and to highly concentrated sodium hydroxide.

FORMABILITY

AK Steel Type 444 provides excellent formability and can be readily deep drawn and spin formed. Stretch forming, however, is limited in application for this material.

WELDABILITY

The ferritic class of stainless steels is generally considered to be weldable by the common fusion and resistance techniques. Special consideration is required to avoid brittle weld fractures during fabrication by minimizing discontinuities, maintaining low weld heat input, and occasionally warmin the part somewhat before forming. This particular alloy is generally considered to have poorer weldability than the most common alloy of the stainless class, Type 409. A major difference is the higher chromium plus molybdenum and columbium additions for this alloy which requires even greater care to avoid brittle weld fractures during forming in cold weather When a weld filler is needed, AWS E/ER 316L is most often specified. Type 444, or 18 Cr-2 Mo, is well known in reference literature and more information can be obtained in this way.

1. ANSI/AWS A5.9, A5.22, and A5.4 (stainless welding electrode specifications).
2. "Welding of Stainless Steels and Other Joining Methods," SSINA, (www.ssina.com).

AK Steel is a world leader in the production of flat-rolled carbon, stainless and electrical steel products, primarily for automotive, infrastructure and manufacturing, construction and electrical power generation and distribution markets. Headquartered in West Chester, Ohio (Greater Cincinnati), the company employs approximately 8,000 men and women at eight steel plants, two coke plants and two tube manufacturing plants across six states: Indiana, Kentucky, Michigan, Ohio, Pennsylvania and West Virginia. Additional information about AK Steel is available at www.aksteel.com.

The information and data in this document are accurate to the best of our knowledge and belief, but are intended for general information only. Applications suggested for the materials are described only to help readers make their own evaluations and decisions, and are neither guarantees nor to be construed as express or implied warranties of suitability for these or other applications.

Data referring to material properties are the result of tests performed on specimens obtained from specific locations of the products in accordance with prescribed sampling procedures; any warranty thereof is limited to the values obtained at such locations and by such procedures. There is no warranty with respect to values of the materials at other locations.

AK Steel
AK Steel Corporation
9227 Centre Pointe Drive
West Chester, OH 45069
800.331.5050
www.aksteel.com

AK Steel and the AK Steel logo are registered trademarks of AK Steel Corporation. 02.19.15

444 STAINLESS STEEL

FIGURE 2.6 A condensed version of a three-page data sheet for a stainless steel alloy. *Courtesy of AK Steel, www.aksteel.com.*

deformation. For some alloys, other properties, such as electrical and magnetic properties, and corrosion resistance, are also cited. The characterization tools and methods needed to determine these properties are typically performed using an ASTM standard so that properties can be compared from alloy to alloy and company to company.

Processing Information — Most data sheets give information that is useful for the further shape casting, deformation processing, machining, and joining of the material. Qualitative or semi-quantitative information on "workability" and "machinability" is sometimes given. This information along with the fundamental mechanical properties of the alloy is needed to design a deformation process. Alloys designed for casting from the melt include other information, such as fluidity at a particular temperature and information on shrinkage during solidification. The techniques for determining processing related characteristics are described in later chapters. Heat treatment conditions for the final alloy part are also provided in some data sheets.

Other — The supplier may also give information on the available forms for the alloy (e.g., sheets, rods, coils), and sometimes, the dimensions.

Suppliers of bulk metal starting materials characterize their materials extensively, as evidenced by the data sheets. They provide detailed information on composition and properties to their customers. Many customers use a starting material with no further characterization, while others perform chemical analysis to check the composition and carry out mechanical tests to confirm the supplier's data. Some bulk alloys are prepared under tightly controlled conditions at their customer request; no deviation from the process route occurs without the customer's approval. Significant characterization data are supplied with each order so that the customer can verify that the alloy meets their standards.

2.2.3 Metal Powder Starting Materials

Metal powders have several uses in materials processing. First, they are the starting materials in powder metallurgy forming methods and powder-based additive manufacturing methods. In addition, metal powders are used to adjust alloy compositions in the melt. They are also used as conductive fillers in composites and coatings. The focus of this section is on the powders used in powder metallurgy and additive manufacturing.

Metal powders are prepared by a variety of processing routes. Before discussing these routes, an overview of particle and powder characteristics is given. Then, the methods used to make powders are described, followed by additives, characterization methods, and a data sheet.

Powder Characteristics. Individual metal particles have shapes and sizes, features that are important to the processing of the particles into dense metal products. A particle is either polycrystalline (i.e., composed of many crystal grains) or less commonly, a single crystal. Within the microstructure of a polycrystalline particle, there may be pores and multiple crystalline phases. Figure 2.7 shows some of the myriad of particle shapes that are found in metal powders. Defining a characteristic particle size is straightforward for a spherical particle. Other shapes are a challenge. Assigning a "size" for these

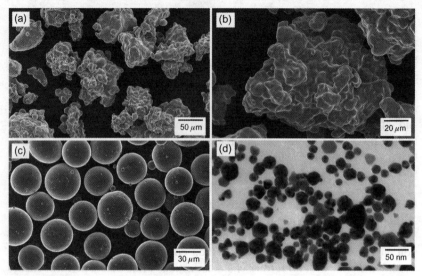

FIGURE 2.7 Electron micrographs of metal powders: (a) Scanning electron micrograph (SEM) of water atomized iron powder, low magnification. *Courtesy of T. Murphy, GKN Hoeganaes, www.GKN.com/hoeganaes.* (b) SEM of water atomized iron powder, high magnification. *Courtesy of T. Murphy, GKN Hoeganaes, www.GKN.com/hoeganaes.* (c) SEM of plasma atomized Ti alloy powder. *Courtesy of Advanced Powders and Coatings, www.advancedpowders.com,* and (d) Transmission electron microscopy of silver nanoparticles made by reducing silver nitrate in solution. *Courtesy of B.Y. Ahn and J.A. Lewis, Harvard University.*

shapes requires a convention. For example, the size of an irregular particle may be equated to the diameter or a sphere that has the same volume as the irregular particle, or the diameter of a sphere that has the same projected area as the particle, and so on. These equivalent spherical diameters are typically related to the method of particle size measurement.

A powder is a collection of many, many individual particles. For example, a 1 kg quantity of steel powder might contain about 10^9 particles. Only in rare instances are all of the particles nearly the same. Therefore, particle size distribution is a key characteristic of a powder. Powder flow properties, packing, and compaction behavior, as well as sintering characteristics, depend on the particle size distribution. There are several types of characterization methods available. See Table 2.3. Depending on the characterization technique, particle size data is usually measured as the number, volume, or weight of particles in particular size intervals.

Figure 2.8 is a schematic example of a particle size distribution based on number. The number of particles in a given size interval, n_i, is plotted as a histogram and also in a cumulative plot that shows the number of particles finer than a certain size. Both the histogram and the cumulative plot have merits. The histogram shows the breadth of the distribution most clearly,

TABLE 2.3 Particle Size Measurement Methods[a]

Method	Principle and description	Approximate size range (μm)[b]	Basis for distribution
Sieving	Powder is passed through a stack of sieves with decreasing screen opening size (down the stack). Amount of powder retained on each sieve is weighed to give the weight of particles in the size interval defined by the sieve openings. Sample may be wet or dry.	37–4000	Weight
Microscopy	General particle shape and size are directly observed. Assigning a dimension to an irregularly shaped particle is a challenge and no one method is accepted over another. Image analysis can be used to gather data more efficiently.	0.8–150 (light microscopy); 0.001–5 (electron microscopy)	Number
Sedimentation (by X-ray adsorption)	Sedimentation rate for particles in a liquid under gravity depends on the particle size. X-ray adsorption measurements are used to determine particle concentration at a particular location in the suspension as a function of time. Size range is extended using centrifugation.	2–100; 0.01–5 (with centrifuge)	Weight
Laser Diffraction	Diffraction of laser light from particles suspended in liquid depends on particle size. Spatial distribution of diffracted intensity measured. Scattering is related to particle size using Fraunhofer diffraction or Mie theory, depending on instrument and conditions. Some instruments combine the two to cover a broad size range.	1–1000 (Fraunhofer); 0.1–40 (Mie)	Volume

(Continued)

TABLE 2.3 (Continued)

Method	Principle and description	Approximate size range (μm)[b]	Basis for distribution
Photon Correlation Spectroscopy	A suspension of particles in a liquid is illuminated with coherent light and light scattered in a prescribed direction is measured. Brownian motion of small particles modulates the scattered intensity, according to particle size. Technique is also known as quasi-elastic light scattering.	0.003–3	Number

[a] Adapted from (Svarovsky, 1990).
[b] Size range depends also on particular instrument used.

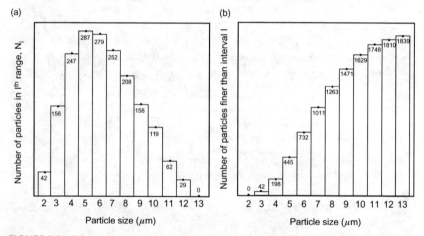

FIGURE 2.8 Schematic example of particle size data in which the number of particles in 1 μm intervals is counted. (a) Histogram with number of particles in each interval noted. The total number of particles is 1839. (b) Cumulative number finer than plot, showing that all 1839 particles are finer than 13 μm and none are finer than 2 μm.

while the cumulative plot can be used to find other important benchmarks of the distribution as illustrated below.

For statistical analysis, data such as those shown in Figure 2.8 are converted to a number fraction, x_i, of particles in the various size intervals.

$$x_i = \frac{n_i}{\sum_i n_i} \tag{2.3}$$

Each size interval is characterized by a middle size, d_i. The number-average diameter, $\bar{d}_n$, is calculated from these data:

$$\bar{d}_n = \sum_i x_i d_i \tag{2.4}$$

One way to characterize the breadth of the distribution is to find the standard deviation, which would be a number-average standard deviation, σ_n, in this context.

$$\sigma_n = \left[\sum_i x_i (d_i - \bar{d}_n)^2 \right]^{1/2} \tag{2.5}$$

This approach is used for data that follows a Gaussian distribution. Another common distribution is known as log-normal; the log of the particle size is given on the x-axis of distribution plots. For data gathered on a weight or volume basis, a similar approach is used.

Several other characteristics are also found from a particle size distribution. Most commonly, a median size or d_{50}, indicating that 50% of the particles are greater in size and 50% are smaller. Likewise, d_{10} and d_{90} are frequently specified; 10% of the particles are smaller than d_{10} and 90% are smaller than d_{90}. With sufficient data (i.e., at least ~8–10 intervals), size frequency functions can be fit to the data to facilitate statistical analysis.

Examples 2.1 and 2.2 illustrate particle size analysis for a powder analyzed using sieves, a common method for characterizing powder metallurgy starting materials. Sieves contain a mesh or screen comprised of interwoven metal wires with square openings. The mesh number is the number of openings per inch; hence, higher mesh numbers correspond to smaller openings. The smallest practical opening size for a metal sieve is 45 μm (mesh number 325). To find a particle size distribution, powder is loaded into the top of a stack of sieves that is arranged from top to bottom in increasing mesh number (decreasing mesh opening size) with the last sieve sitting on a pan. See Figure 2.9. The stack is vibrated and then the material caught on each mesh is weighed.

There are several other powder characteristics. The specific surface area (surface area per gram of powder) is a good predictor of the reactivity and sintering characteristics due to the role of surfaces in these processes. Based on geometry, finer particles have higher specific surface area than coarser particles. The apparent density is the weight of loose powder divided by its volume. The tap density is the density of powder after it is exposed to a mechanical tapping or vibration, which acts to rearrange the particles. Both the apparent density and tap density depend on the particle size distribution, particle shape, and the frictional characteristics of the powder. The chemical composition and crystalline phase content are characteristics determined for

40 Materials Processing

EXAMPLE 2.1 Plot the particle size distribution from the data below as a histogram and as a cumulative weight % finer than plot. Estimate d_{10}, d_{50} and d_{90}.

Mesh number	Opening size (μm)	Weight (%)
60	250	0
100	150	3
150	105	10
200	74	20
270	53	26
325	45	31
Pan	0	10
Total		100

To find the histogram, we need to define the size interval and middle size for each sieve. A realistic histogram is constructed with bars that are the appropriate width and typically the midpoint of the interval is located on the plot. Sometimes a frequency function can be fit to the midpoints.

Mesh number	Opening size (μm)	Weight (%)	Size interval (μm)	Middle size (μm)	Cumulative weight % finer than
60	250	0	>250		100
100	150	3	150–250	200	97
150	105	10	105–150	127.5	87
200	74	20	74–105	89.5	67
270	53	26	53–74	63.5	41
325	45	31	45–53	49	10
Pan	0	10	<45	22.5	0
Total		100			

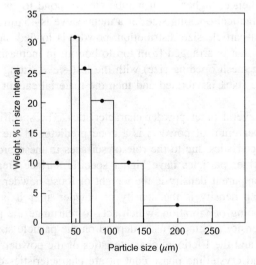

FIGURE E2.1A

For the cumulative plot, the appropriate reference points on the x-axis are opening sizes. For example, all the powder passed through the sieve with the 250 μm opening so 100% of the particles are finer than that size, and 3% was captured on the next sieve with the 150 μm opening so 97% are finer than 150 μm. The benchmarks d_{10}, d_{50}, and d_{90} are found on the plot.

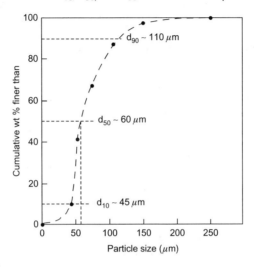

FIGURE E2.1B

---

the powder as a whole, and are extremely important to the processing and the final properties of the material formed from the powder.

Powder Fabrication. Metal powders are prepared by a variety of routes, the choice of which depends on the properties of the particular metal or alloy and the desired physical characteristics of the powder, including size and size distribution, shape, porosity, and surface area. Below is a summary of metal powder production methods, followed by two examples that illustrate the steps in producing a powder.

Mechanical Methods — Particulate metals can be made by mechanical size reduction of small pieces of metal, such as chips and shavings collected from machining operations. These metal pieces are size reduced by communition methods, such as ball milling (see Section 2.3.2). Since these methods rely on impact and fracture, they are only effective with brittle metal. Ductile metals are deformed rather than fractured by impact, and they are subject to cold welding (i.e., the formation of strong particle-particle contacts as particles are pressed together). Powders produced by mechanical methods are irregular in shape, can be contaminated by the grinding media, and are typically work hardened and therefore less deformable during further process steps. While mechanical methods are not as commonly employed compared

EXAMPLE 2.2 Convert the data from Example 2.1 to plot the distribution as a histogram based on number % and a cumulative number % finer than plot. Assume the particle density is 7.85 g/cm^3 and that the particles are spherical. Comment on the differences between displaying data in weight % and number %

To make this conversion, we need to make several approximations. To convert a weight of particles to a number of particles, we first assume that all particles in the size interval are spherical and have the same diameter, the middle size diameter, d_i, and then we can calculate the weight of a single spherical particle in that size interval. For example, the −60/+100 fraction has a middle size of 200 μm; therefore, the weight of a single particle is:

$$w_p = \rho_p \left[\frac{4}{3}\pi\left(\frac{d_i}{2}\right)^3\right] = 7.85 \text{ g/cm}^3 \left[\frac{4}{3}\pi\left(\frac{0.02 \text{ cm}}{2}\right)^3\right] = 3.29 \times 10^{-5} \text{ g}$$

Therefore, the number of particles in this size fraction is the total weight (3 g, assuming 100 g to start) divided by the weight of a particle:

$$N_p = \frac{\text{Total Weight in Size Interval}}{w_p} = \frac{3 \text{ g}}{3.29 \times 10^{-5} \text{ g/particle}} = 9.12 \times 10^4 \text{ particles}$$

To find the number % or number fraction, all the particles in the different fractions are summed, as shown in the data table below. Then the histogram and cumulative number % finer than plots are constructed similarly to Example 2.1.

Mesh number	Opening size (μm)	Weight %	Size interval (μm)	Middle size (μm)	Number of particles	Number %
60	250	0	>250			
100	150	3	150–250	200	9.12×10^4	0.03
150	105	10	105–150	127.5	1.17×10^6	0.38
200	74	20	74–105	89.5	6.79×10^6	2.19
270	53	26	53–74	63.5	2.47×10^7	7.96
325	45	31	45–53	49	6.41×10^7	20.65
Pan	0	10	<45	22.5	2.14×10^8	68.80
Total		100			3.10×10^8	

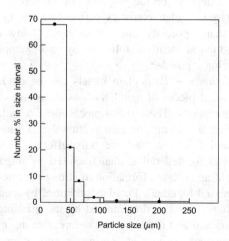

FIGURE E2.2A

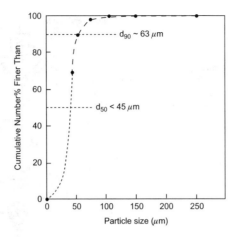

FIGURE E2.2B

This example clearly shows the differences between a weight basis and a number basis. The weight-based median is more than twice the number-based median, showing that big particles have more importance on a weight (or volume) basis than do small particles. Another way to think about it is that for a given weight of powder, the number of particles in that weight increases as the particle size decreases. Therefore, the number % histogram is shifted to a smaller size.

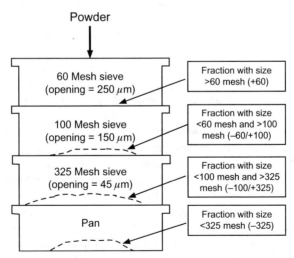

FIGURE 2.9 Example of a sieve analysis of a metal powder to find a particle size distribution. The notation for the fractions is also given. For example, the fraction of powder that passes through mesh number 60 and is caught on mesh number 100 is known as $-60/+100$.

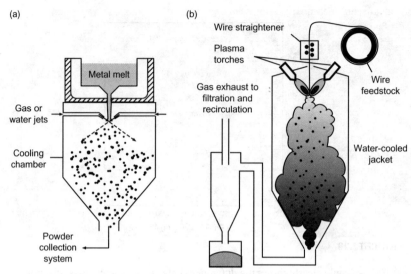

FIGURE 2.10 (a) Schematic diagram of melt-atomization by gas or water jets. *Adaped from German (1984).* (b) Schematic diagram for plasma atomization. *Courtesy of Advanced Powders and Coatings, www.advancedpowders.com.*

to others in this list, the use of reduced temperature (cryogenic) and controlled atmosphere grinding are improving the quality of mechanically prepared powders.

Melt Atomization − In melt atomization processes, a stream of liquid metal is broken up (atomized) into droplets and then the droplets solidify into particles as they cool. The stream of liquid is typically created as molten metal flows under gravity from a reservoir through an orifice. Atomization is accomplished by a number of methods, including impingement of gas or water jets. See Figure 2.10a. Powders prepared by melt atomization with a water jet are cooled very rapidly and have irregular shapes, while those atomized with gas jets are cooled more slowly and therefore have time to adopt a more spherical shape. The speed of cooling influences microstructure and properties of the powder; faster cooling leads to metal powders with finer microstructures and higher yield strengths. In general, particle size decreases as the liquid or gas pressure increases and the water (or gas) to metal flow ratio increases (over a limited ranges). One problem with melt atomization is the interaction between the particle surfaces and the gas or water; this interaction can produce entrapped gas and oxide films. Gas atomization with an inert gas leads to fewer of these problems. Another approach is plasma atomization. See Figure 2.10b. This process uses a wire feedstock and therefore avoids a separate melting step, which is especially advantageous for the production of high purity and high melting point metals. The plasma torches converge on the wire, melting and atomizing the metal in an argon gas environment. By controlling the wire feed rate and conditions of the plasma and gas, spherical powders of controlled size are prepared.

Another variation on melt atomization, the rotating electrode method, involves arcing a current from a stationary tungsten cathode to a rotating bar of the metal of interest (anode). The current melts the consumable anode and the rotation sends the melt produced radially away from the spinning anode, forming droplets that solidify into powder. The particle size of metal powder prepared by this method is inversely proportional to the product of the rotational speed and the square root of the electrode diameter.

Oxide Reduction — Oxide reduction methods begin with high purity, size-controlled metal oxide (ceramic) powders. The ceramic powder is exposed to a combination of heat and a reducing atmosphere to form a metal powder. Thermodynamic and kinetic considerations govern the choice of conditions for the oxide reduction. First, the reduction reaction must be energetically favorable, and second, the rate of the reaction must be high enough for an efficient process. The reaction rate increases with temperature, but the temperature must not be too high or the metal powder will begin to sinter, which is not desired for most applications. Oxide reduction is particularly attractive to the production of metals with high melting points, such as tungsten and molybdenum. Metal powders produced by this technique can be porous or spongy due to the density difference between metal oxides and metals. The particle size of metals made by this method is related to the oxide particle size, but also to the nature of the chemical reaction. An example of tungsten powder production is described below.

Chemical and Electrochemical Processes — Metal powders may be formed by precipitation due to chemical or electrochemical reduction of aqueous salt solutions. Copper powder is commonly prepared by electrolysis. A brittle, porous, and very pure copper deposit forms on the cathode of an electrolytic cell containing a copper salt solution. The powder is scraped from the cathode periodically. Chemical methods can produce powders of a wide variety of sizes and shapes, included nanosized particles. These powders typically have a high degree of purity.

Many powder metallurgy processes are based on the use of elemental metal powders that are blended to produce the desired alloy composition after forming and sintering. The rationale for this choice is that elemental metals have lower yield stresses and are more easily compacted into high-density pieces, which therefore shrink less on sintering. If alloy powders are desired, then alloy melts can be prepared and powders produced by melt atomization. Alternatively, elemental metals can be alloyed in a ball mill. The fracture and mechanical action, which heats the powders, leads to conditions suitable for alloy formation.

Powders are sometimes thermally annealed after they are produced. Powders prepared by mechanical processes work harden during fabrication and melt-atomized powders may have very fine grains and hence high yield strengths. To lower the yield strength, thermal annealing is used to recrystallize the work hardened powders and grow the grains of melt-atomized powders. The conditions of thermal annealing depend on the alloy composition, and are chosen so as to avoid any sintering.

Example 1 — Low Alloy Steel Powder. Steel powders are used in powder metallurgy forming operations. Steel is a complex material that becomes even more complicated in the context of powder metallurgy. Low alloy steel refers to steel with less than 5 wt% of alloying elements. The powder composition is chosen to deliver required properties in the final metal, and the powder characteristics are tuned such that the powder is formable by powder metallurgy methods, such as pressing. Hence, the dilemma: high hardness and strength are typically required of the final product, but such properties in the powder frustrate the compaction process. The majority constituent in steel, iron, is easily prepared by melt atomization or by a process known as the sponge iron process, reduction of iron oxide at high temperatures, leading to porous particles. Melt atomization is typically the preferred starting point for production of steel powders. The issue then is how to incorporate the alloying elements. Depending on the composition of the steel, the forming operation, and the specifications of the final product, one of three routes is chosen.

Prealloyed powders are made from a melt of the desired steel composition. In this case, the composition of the powder is uniform throughout, which is an advantage for developing the desired uniformity in the final product, but a disadvantage in that alloys have higher yield strengths than elemental metals. Some prealloyed powders are formulated to contain only a few of the alloying elements and the others are mixed in as powders with the prealloyed powder to achieve the desired composition on sintering. Nickel, molybdenum, and chromium do not have severe effects on the yield strength and can be included in the prealloyed powder composition, while carbon increases the yield strength significantly and is mixed in later as graphite.

Admixed powders are comprised of a physical mixture of elemental powders. The desired alloy composition forms via diffusion during sintering. This route allows for good compressibility due to the low yield strengths of the powder components, but attaining a uniform chemical composition may be difficult due to the inherent challenges in mixing and the need for extensive diffusion during sintering.

Bonded powders are composed of elemental constituents, but the alloying elements are bonded as fine particles onto the outside of larger iron particles. This option results in better chemical homogeneity as it eliminates segregation during mixing. The bonding is accomplished by heating the fine elemental powders with iron powder in a reducing atmosphere (to prevent oxidation); diffusion and partial sintering of fine particles to coarse particles takes place to create the composite particles.

Example 2 – Tungsten Powder. Tungsten is a refractory metal, which means that it has a high melting point (3422°C). Casting a melt of tungsten is not feasible; therefore, powders are used to form tungsten parts. The classic example of a tungsten part is the filament in an incandescent light bulb. The bulb filament is in fact the first major example of powder metallurgy, having been developed early in the 20th century. Tungsten is also used as an alloying element in steels; in this application it is used in powder form.

Tungsten metal powder is formed by reduction of tungsten oxide (WO_3); the tungsten oxide is first formed from tungsten-rich ores, such as wolframite [$(Mn,Fe)WO_4$]. The ore is dissolved in a basic water solution, purified, and crystallized as ammonium paratungstate, which is then calcined into tungsten oxide. The tungsten oxide is reduced at high temperatures (600–1100°C) under flowing hydrogen. Reduction is carried out in a push type furnace in which the oxide powder is loaded into shallow metal boats that are pushed through heated tubes, or in a rotary furnace in which the tungsten oxide powder is charged and moved through a heated rotating tube. The overall chemical reaction is:

$$WO_{3(s)} + 3H_{2(g)} = W_{(s)} + 3H_2O_{(g)} \qquad (2.6)$$

The reaction above occurs gradually with partially reduced tungsten oxide intermediates (WO_{3-x}) and hydroxide intermediates before forming the final metal powder. The morphology and particle size of the tungsten powder itself is governed by the conditions of the reduction, particularly the temperature and water vapor. Hydroxide intermediates have some volatility, which leads to evaporation and condensation processes that change particle size and shape.

Blending, Mixing, and Additives. Blending and mixing are key processing steps in the preparation of metal powders. Blending refers to the combination of powders of the same composition, but different sizes, so as to tailor the particle size distribution or change other properties such as moisture content or color. Mixing, on the other hand, is the term used for the combination of a powder with another powder of a different composition or with an additive, such as a lubricant. Mixing is sometimes called premixing. Blending and mixing are accomplished by mechanical means, typically by tumbling in a rotating vessel or by agitation in by a rotating paddle or screw in a stationary vessel. A few examples are shown in Figure 2.11.

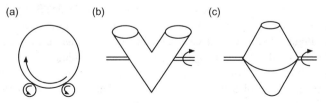

FIGURE 2.11 Schematic diagrams mixing equipment: (a) roller mill, (b) V-mixer, and (c) cone mixer.

The goal of a blending or mixing operation is to achieve a uniform distribution and avoid segregation. On a microstructural level, the goal is to increase the number of contacts between dissimilar materials. Mixing is accomplished by some combination of tumbling induced movements of individual particles (diffusion), and groups of particles (convection and shear), and affected by a host of variables, including the characteristics of the particles and additives and process conditions (e.g., tumbling rate, mixing time). In general, mixing is most easily accomplished in systems that contain powders of similar densities and sizes. In systems with disparate densities but similar size, the low-density particles migrate to the top of the mass and high-density particles to the bottom; likewise with disparate size and similar density, smaller particles segregate the top and larger particles to the bottom.

In powder metallurgy, additives are needed to enhance the forming process, improve uniformity of a mix, and adjust composition and properties. Solid lubricants are added in small quantity (0.5–1.5 wt% relative to the powder) to improve powder flow, uniformity of compaction, and ease of ejection. These materials are added directly during the mixing stage. Binders are additives designed to bind smaller additives onto larger metal particles. They are liquids (e.g., kerosene, glycerin) or solutions of polymers or lower molecular weight organics in a solvent, and are added after some dry mixing of metal powders and other additives. The liquid forms a thin film on the particle surfaces, which has the tendency to prevent segregation of finer particles by adhering them to larger ones. When solvents are used, they are often evaporated from the mix, recovered by condensation and removed.

Metal Powder Data Sheets and Characterization. Data sheets for metal powders reveal the characteristics most important to those using the powders in powder metallurgy forming operations or other applications. Figure 2.12 shows an example. Data sheets vary widely in the type and amount of information provided. Information provided on a data sheet falls into four categories:

Chemical Characteristics — Composition is determined by wet chemical or spectroscopic means and reported as an elemental analysis or as % purity, in the case of elemental powders. In some cases, X-ray diffraction is used for crystalline phase content.

Particle Size Data — Of all the physical attributes of the powder, the average particle size and particle size distribution are arguably the most important to powder processing operations. Several characterization methods are available, as noted in Table 2.3. Data sheets usually provide detailed information on the particle size and distribution. For example, data from sieve analysis is shown in Figure 2.12. In other case, benchmarks, such as an average particle size and standard deviation or d_{10}, d_{50}, and d_{90}, are provided.

Other Physical Characteristics — Some manufacturers include the specific surface area of a powder, the particle density, and comments about particle shape.

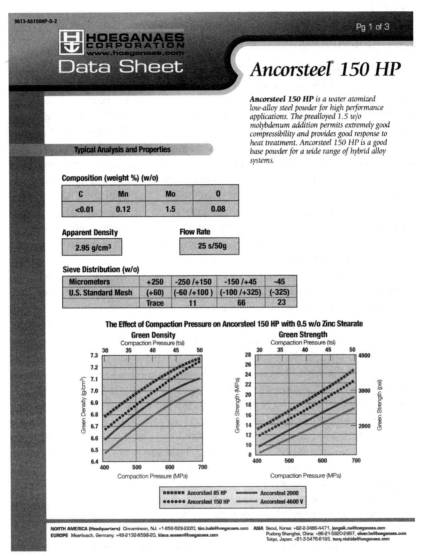

FIGURE 2.12 First page of a three-page data sheet for a steel powder. *Courtesy of GKN Hoeganaes, www.GKN.com/hoeganaes.*

Processing Information — For powders used in powder metallurgy, there may be compaction data included and densities achieved for particular conditions. Recommended sintering conditions can be included along with data for properties of parts made using recommended sintering conditions.

2.3 CERAMICS

2.3.1 Introduction

Ceramics are inorganic, nonmetallic materials. Many ceramics are simply compounds of a metallic element (or semiconducting element) with a nonmetallic element. Some well-known ceramics are oxide compounds, such as alumina (Al_2O_3), silica (SiO_2), and zirconia (ZrO_2). Others are known as nonoxide ceramics, as they are compounds based on other nonmetallics, like carbon, nitrogen, and boron. Silicon nitride (Si_3N_4), silicon carbide (SiC), and titanium diboride (TiB_2) are examples of nonoxide ceramics. These oxide and nonoxide ceramic examples are prepared and used as polycrystalline materials. Single crystal ceramics, such as sapphire (single crystal Al_2O_3), have applications as well. Noncrystalline ceramics (glasses) have a wide variety of compositions, mainly based on metal oxides. Window glass, glass containers, and glass fiber optic filaments are each designed with particular compositions in order to develop required properties. And, some important ceramics have microstructures containing a mixture of glassy and crystalline phases; these include glass-ceramics, which are used as cookware and stovetops, and many traditional ceramics, such as porcelain.

Ceramics are categorized by their role as traditional or advanced (or functional). Traditional ceramics include those primarily based on clay and other natural materials. Whitewares, porcelains, structural clay products, and refractory bricks are considered traditional ceramics. Commodity glasses (e.g., window glass, containers) are also in the traditional category. By contrast, advanced ceramics have specialized functions (e.g., electronic, mechanical, chemical), and demand stricter control of composition, microstructure, and properties to be successful. Some applications of advanced ceramics are highly visible, such as the glass on cell phone touch screens and insulators on high tension lines, while others are invisible to most people (e.g., capacitors in electronic components).

There are different conventions for naming and identifying compositions of ceramics. Polycrystalline ceramics are known by their common names and chemical formulas, as listed in Table 2.4. These ceramics are mostly stoichiometric compounds, though some have ranges of solid solubility. Many traditional ceramics are defined by a general name rather than a name that relates to their specific chemical compositions. Examples of names for traditional ceramics are earthenware, whiteware, stoneware, and porcelain. Glasses compositions are not based on the stoichiometries of crystal structures, but rather on a combination of glass network forming oxides and modifiers. For example, window glass is a combination of SiO_2 as a glass network former and Na_2O and CaO as modifiers. This common glass is known as soda-lime-silica glass or simply soda-lime glass. Other glasses are also known by a name, such as "E glass." Because of the broad compositional variation, Corning Corporation developed a coding system that is

TABLE 2.4 Examples of Polycrystalline Ceramic Materials

Ceramic name	Chemical formula
Alumina	Al_2O_3
Aluminum Nitride	AlN
Barium Titanate	$BaTiO_3$
Beryllia	BeO
Boron Carbide	B_4C
Boron Nitride	BN
Cerium Oxide	CeO
Chromium Oxide	Cr_2O_3
Cordierite	$Mg_2Al_4Si_5O_{18}$
Hafnia	HfO_2
Lead Zirconate Titanate (PZT)	$PbZr_{1-x}Ti_xO_3$
Magnesia	MgO
Mullite	$Al_6Si_2O_{13}$
Quartz	SiO_2
Silicon Carbide	SiC
Silicon Nitride	Si_3N_4
Spinel	$MgAl_2O_4$
Strontium Titanate	$SrTiO_3$
Titania	TiO_2
Tungsten Carbide	WC
Yttrium Barium Copper Oxide	$YBa_2Cu_3O_{7-x}$
Zirconia	ZrO_2

employed in some industries. Other glass companies also have codes for their products. Table 2.5 provides glass compositions, codes, and uses.

In terms of starting materials, ceramics are either processed from powder materials or from molten glass, which is formulated from a batch of powdered raw materials. These two classes of starting materials are described in the sections below.

2.3.2 Ceramic Powder Starting Materials

The starting materials for polycrystalline ceramics are ceramic powders. Advanced ceramics have strict requirements for powder starting materials.

TABLE 2.5 Examples of Glasses

Glass name or code	Use	Composition (wt%)
Soda-Lime		
Generic	Multipurpose	73 SiO_2, 2 Al_2O_3, 15 Na_2O, 10 CaO
Corning 0080	Lamp bulbs	73 SiO_2, 1 Al_2O_3, 17 Na_2O, 0.5 K_2O, 4 MgO, 6 CaO
Typical Container	Green tinted	73 SiO_2, 0.5 Al_2O_3, 16 Na_2O, 9.3 CaO, 0.5 Fe_2O_3, 0.2 Cr_2O_3, 0.5 SO_2
Typical Flat Glass	Clear float	73.1 SiO_2, 0.1 Al_2O_3, 13.7 Na_2O, 0.1 K_2O, 3.8 MgO, 8.9 CaO, 0.5 Fe_2O_3, 0.5 TiO_2, 0.2 F
Borosilicate		
Corning 0211	Microsheet	65 SiO_2, 2 Al_2O_3, 9 B_2O_3, 7 Na_2O, 7 K_2O, 7 ZnO, 3 TiO_2
Corning 7740	Multipurpose	81 SiO_2, 2 Al_2O_3, 13 B_2O_3, 4 Na_2O
Corning 9741	UV Transmitting	65 SiO_2, 2 Al_2O_3, 9 B_2O_3, 7 Na_2O, 7 K_2O, 7 ZnO, 3 TiO_2
Aluminosilicates		
Corning 7059	Electron. displays	49 SiO_2, 10 Al_2O_3, 15 B_2O_3, 25 BaO, 1 As_2O_3
E-type Fiberglass	Continuous fiber	54 SiO_2, 14 Al_2O_3, 10 B_2O_3, 4.5 MgO, 17.5 CaO
High Silica		
Typical fused quartz	Fused quartz	>99.9 SiO_2
Corning 7913	Vycor brand	96.5 SiO_2, 0.5 Al_2O_3, 3 B_2O_3
Corning 7971	Ultralow expansion	93 SiO_2, 7 TiO_2

Like metal powders, these materials are prepared with controlled chemical composition and particle size. There are many different methods for the production of powders. The choice of production method depends on the required particle characteristics. A key goal in the production of ceramic powders for advanced ceramic applications is attaining a fine particle size, typically less than 1 μm in diameter. The fine size is required for two reasons. First, powder processing of ceramics requires a high temperature sintering treatment in order to make a dense material. The sintering process is greatly enhanced with fine particles. (Sintering is also required for most powder metallurgy operations; however, powdered metals can be densified to a greater degree during forming due their ability to plastically deform.) Second, the mechanical properties, as well as many other properties, of the final ceramic are enhanced with fine grain size. The dimensions of the grains in the final microstructure cannot be smaller than the size of the original particle; therefore, fine particles are needed for a fine-grained polycrystalline ceramic.

Traditional ceramics require a different class of powder starting materials. For some traditional ceramics, raw materials from the earth can be used after some mechanical and chemical processing steps to prepare them. Traditional ceramic compositions are based on batching together these raw materials in different proportions. Many are so-called triaxial bodies, made by combining clay, feldspar (an alkali aluminosilicate), and flint (silica). The final microstructures of triaxial bodies contain several crystalline phases as well as a significant noncrystalline or glassy component. Other traditional ceramics include metal oxide powders, such as alumina, in their formulation to further refine properties. Traditional ceramics are also densified in a sintering operation, but the high content of glass makes the densification less sensitive to the particle size.

Figure 2.13 shows example microstructures of ceramic powders. A range of sizes and shapes are possible in these powders. Most technical ceramics, however, are prepared at the micron or submicron level in order to facilitate densification during sintering and to improve properties of the final ceramics. Ceramic powders are characterized by the same quantities described in the previous section on metal powders. The median particle size, the particle

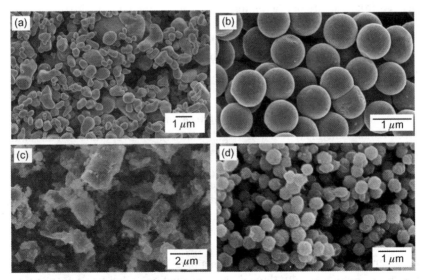

FIGURE 2.13 Scanning electron microscope images of ceramic powders: (a) Titania powder. *Courtesy of C. C. Roberts, University of Minnesota.* (b) Silica powder. *Courtesy of K. A. Price, University of Minnesota.* (c) Silicon carbide powder. *Courtesy of Superior Graphite, www.superiorgraphite.com.* and (d) Barium titanate powder. *Courtesy of TPL Inc., www.tplinc.com.*

size distribution, and the specific surface area are important characteristics of ceramic powders.

Ceramic Powder Fabrication. There are many different methods for synthesizing ceramic powders; most involve a combination of chemical, thermal, and mechanical steps. The choice of fabrication method is based on the required characteristics for the powder, which in turn, depend on the intended processing method and the desired properties for the final ceramic. For example, polycrystalline alumina used as an electronic substrate must be prepared with a high purity, submicron-sized alumina powder in order to achieve a dense microstructure and the required electrical and thermal properties. A refractory brick, on the other hand, has less stringent requirements and can be made with a less pure, coarser starting powder. Powder fabrication methods can be placed in four broad categories: mineral processes, chemical solution processes, solid state reaction processes, and gas phase processes.

Mineral Processes — Some ceramic powders are mined from the ground, but typically these minerals require extensive mechanical, chemical, and thermal processing before they are ready to be used as a starting material. Clay minerals, quartz, feldspars, limestone, and dolomite are examples of materials that are mined and then used to make traditional ceramics and glasses. These materials are crushed and undergo separation processes to remove impurities and make the particle size smaller and more uniform. Advanced ceramic powders are made from minerals after more extensive processing. For example, zirconia is prepared from zircon ore (nominally, $ZrSiO_4$), and alumina is derived from bauxite ore (a mixture of a several aluminum hydroxide minerals). Purification often involves dissolution and precipitation, separation of impurities, heating to define the crystalline phase, and size reduction.

Chemical Solution Processes — Chemical solution processes begin with inorganic chemicals that are dissolved in water or an organic solvent. Metal chlorides, nitrates, oxalates, and alkoxides are examples of inorganic chemicals involved in chemical processes. Metal oxides, hydroxides, or other metal intermediate compounds are produced from metal-containing solutions by precipitation, chemical reaction (e.g., gelation), or evaporation. The solid product is typically not in the final composition or crystal structure required for the powder, and must be heated to produce the oxide powder. This thermal process is known as calcination. For example, a hydroxide prepared by precipitation is calcined to convert it to the oxide. In other cases, the process is called thermal decomposition as it involves a more complex decomposition of a metal organic compound, such as acetate or oxalate. Multicomponent ceramic oxide compositions are prepared with chemical homogeneity by creating solutions containing multiple metal ions and then retaining the homogeneity through the precipitation, gelation, or evaporation process. There are two other important advantages of chemical solution processes over mineral-based processes: the particle size can be adjusted to finer size, and the purity of the final product is higher. In a related type of process, nonoxide ceramics are prepared by thermal decomposition of metal containing polymers. For example, SiC can be made by thermal decomposition of polycarbosilane polymers.

Solid State Reaction Processes — Multicomponent ceramic oxide compositions (e.g., $BaTiO_3$) can be prepared by solid state reaction of materials created by mineral-based or chemical solution processes. For example, mullite, an aluminosilicate compound, can be prepared by reacting, at high temperatures, alumina and silica powders. Nonoxide ceramics can be prepared by carbothermal reduction of oxides created by mineral processes. For example, silicon carbide is prepared by reaction of silica and carbon. To obtain a complete reaction, the reaction product is often milled to expose more reactant surfaces, and then the mixture is heated again to complete the reaction.

Gas Phase Processes — Gas phase processes involve gases or vapors as one or more of the reactants. A common vapor phase route involves reaction between a metal chloride vapor and the water vapor in a heated reactor. Gas–solid reactions involve passing a reactant gas over a solid reactant at high temperature.

Example 1 — Alumina (Al_2O_3). Alumina powder is produced in large quantities using a mineral process. The production begins with the mining of bauxite. Bauxite is an aluminum hydroxide-based ore that contains a variety of impurities. In 1888, an Austrian chemist named Karl Bayer designed a process to obtain purer alumina hydrate from this ore; the process, known as the Bayer Process, is still used today. The alumina hydrate is calcined to create aluminum oxide. Alumina has a variety of applications in advanced ceramics, but the vast majority (~90%) of the alumina produced by the Bayer process is used to produce aluminum metal. The demand for aluminum metal has made the production of alumina a large scale, economical process. The process is described in Figure 2.14.

In the first step of the Bayer process, bauxite ore is crushed, milled, and fed into a digester with a concentrated, aqueous sodium hydroxide solution. The conditions in the digester depend on the grade of bauxite; typical temperatures are 150–250°C and pressures are in the range of 150–400 psi. The result of the digestion is the formation of an aqueous solution of sodium aluminate ($NaAlO_2$); impurities in the ore, such as iron oxide, silica, and titania, are not soluble. These impurities, known collectively as "red mud", are separated from the sodium aluminate solution. The concentrated sodium aluminate solution is then cooled and alumina trihydrate ($Al_2O_3 \bullet 3H_2O$) seed crystals, which instigate precipitation of the same, are added. The conditions of the precipitation, including the seed size and quantity, and the temperature and agitation, influence the characteristics of the precipitate and are closely controlled. The alumina trihydrate precipitate is washed and separated from the liquid, which is recycled back into the process. The soda content of the alumina trihydrate from the Bayer process ranges from 0.05% to 0.5%, depending on the processing steps carried out.

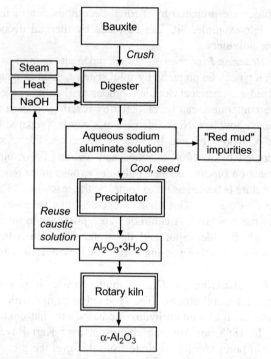

FIGURE 2.14 Schematic flow diagram of Bayer process to convert bauxite to aluminum oxide.

The alumina trihydrate, which is also known as aluminum hydroxide [Al(OH)$_3$], is then calcined to create aluminum oxide. Calcination eliminates water and creates an anhydrous aluminum oxide phase. In production, rotary kilns are typically used for calcination. A rotary kiln is a large, heated cylinder (up to 10 ft in diameter) that is at a slight angle off horizontal. The aluminum hydrate is fed into the kiln at the end at higher elevation and slowly travels down its length. During the journey, water is removed and the hydrate transforms to α-Al$_2$O$_3$, via several transitional alumina phases. Peak temperatures in the kiln are in the range of 1100–1200°C. The goal is close to 100% conversion to α-Al$_2$O$_3$. The conditions needed to achieve the transformation lead to the formation of agglomerates of fine crystals. A size reduction step is needed before the powders are suitable for processing. Different grades of alumina are available, based on impurity content (chiefly Na$_2$O) and particle size. Powder cost increases with increasing purity and decreasing particle size.

The Bayer process has limitations. If very high purity (>99.9% Al$_2$O$_3$) is required, then an alternative processing route must be used. For example, ultrahigh purity alumina is prepared by the thermal decomposition of high purity aluminum salts.

Example 2 – Silicon Nitride (Si_3N_4). Silicon nitride is an example of a ceramic that has no natural, mineral source and must be prepared synthetically. Silicon nitride has excellent thermal and chemical stability as well as high elastic modulus and hardness. Si_3N_4 has two polymorphs (α and β), and powders are typically some mixture of the two. Of the several routes used to the prepare silicon nitride powder, direct nitridation and the diimide route are two of the most used methods. Both of these routes make use of gas or liquid phase reactants to produce powder.

The direct nitridation process involves the reaction of silicon powder with nitrogen or a mixture of ammonia and nitrogen gas under conditions of high temperature (1100–1400°C). The reaction for silicon and nitrogen is given by:

$$3Si_{(s)} + 2N_{2(g)} = Si_3N_{4(s)} \tag{2.7}$$

The reaction is exothermic; conditions need to be closely controlled to prevent melting of the silicon powder, which occurs at 1412°C. The product of this reaction is predominately the α-Si_3N_4 phase unless the reaction temperature gets high enough for silicon melt to form; the presence of melt encourages the formation of the β polymorph. The powder production is carried out in a batch process or a continuous process using a rotary kiln. The product powder is agglomerated and must be milled to break down agglomerates and reduce particle size. The purity and size of the Si_3N_4 powder produced by this method depends on the characteristics of the Si powder. High purity powders are prepared from semiconductor-grade Si powder, while metallurgical grade Si powder produces less pure powders. Particle sizes range from several 100 nanometers to 10 microns, depending on processing and milling conditions.

The diimide process involves a reaction at a liquid/liquid or liquid/gas interface. Silicon tetrachloride dissolved in an organic solvent is mixed with liquid ammonia for the liquid/liquid version. At the liquid/liquid interface, a reaction forms a diimide intermediate [$Si(NH)_2$], which is washed and then heated to be converted to silicon nitride and crystallized on heating in a nitrogen or ammonia atmosphere. The reactions are as follows:

$$\begin{array}{ll} -40-0°C & SiCl_{4(\ell)} + 6NH_{3(\ell)} = Si(NH)_{2(s)} + 4NH_4Cl \\ 900-1200°C & 3Si(NH)_{2(s)} = Si_3N_{4(s)} + 2NH_{3(g)} \\ 1300-1500°C & Si_3N_4(amorphous) = Si_3N_4(cryst.) \end{array} \tag{2.8}$$

In addition to the liquid/liquid reaction step, a gas/liquid reaction can be used to create the diimide. Ammonium gas is bubbled through an organic $SiCl_4$ solution. The diimide processes allow the creation of high purity, fine (i.e., submicron) powder with mostly α phase.

Communition. Communition is the general term for particle size reduction by crushing, milling, or grinding. There are two main reasons for including communition in the processing of a powder starting material: (1) to

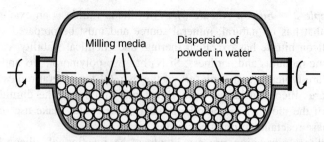

FIGURE 2.15 Schematic cross-section of a ball mill operating as a wet mill. Rotation causes tumbling of the hard milling media (balls) along with the powder, which is dispersed in a liquid such as water.

break up agglomerates, and (2) to reduce the primary particle size. Agglomerates are clusters of primary particles that are held together either weakly, for example by van der Waals forces, or strongly, for example by chemical bonds formed during calcination. Sometimes agglomerates with particles held together by strong bonds are called aggregates. In either case, breaking up these agglomerates is frequently a necessary step. Another reason for carrying out a milling process is to adjust the particle size distribution (commonly, reducing the size of the largest particles in a distribution). In addition, communition processes carried out with the powder dispersed in a liquid (so-called wet milling) facilitate mixing (if more than one type of powder is involved) and create suspensions or dispersions that are needed for some forming operations.

One of the most common methods of communition of ceramic particles is ball milling. See Figure 2.15. The powder, either dispersed in a liquid (wet) or dry, is placed in a vessel partially filled with hard media (balls). Wet milling is more common than dry milling, because the presence of the liquid inhibits the reagglomeration of particles after size reduction. The vessel and its contents are then set into motion (e.g., by rotating the vessel). When the mill is in motion, the particles are exposed to multiple stresses as they are impacted by media and other particles, and as they hit the vessel walls. The milling process is inefficient: only about 10% of the input energy into a ball mill is used for the breaking down of agglomerates and particles. Most of the energy is spent lifting the media and is dissipated as heat. Tumbling ball mills are commonly used for breaking down particles to the micron scale and creating uniform dispersions.

On a microscopic level, communition is largely a fracture process, which results in particle and agglomerate breakage into smaller fragments. The stress needed for fracture is imparted on a particle by impact as described above for ball mills, or in other ways. The level of stress needed for fracture depends on the mechanical properties and microstructure of the particles and agglomerates. For brittle materials, like ceramics, Griffith's theory is used to understand the fracture process. The stress needed for fracture, σ_f, is given by:

$$\sigma_f = \frac{K_{IC}}{Y\sqrt{c}} \tag{2.9}$$

where K_{IC} is the fracture toughness, c is the characteristic flaw size, and Y is a geometry-dependent constant. This equation shows that materials with high fracture toughness and small flaw sizes require more stress in order to be fractured and size reduced. Of interest is the effect of particle size on the fracture process. After a particle is fractured, the resulting fragments tend to require a higher stress for further size reduction, because the characteristic flaw size in the newly formed fragments is less than that of the parent particle. So, there is a limit to the effectiveness of a communition method depending on the level of stress that is imparted and the materials properties. In addition to size reduction by fracture, attrition or wear from contact with other particles and media can reduce particle size.

In addition to tumbling ball mills, there are several other methods of particle size reduction. Ball mills can also be operated with media in motion due to vibration of the vessel (vibratory ball mill) or stirring (agitation or attrition mills). These methods are more efficient than tumbling mills and result in faster particle size reduction. Attrition mills, in particular, are able to reduce particle size quickly and can produce submicron particles. Their efficient action is due to the rapid motion of small media, which gives more contacts per particle. Lastly, fluid energy mills employ particle-particle impact to break down agglomerates and primary particles. High velocity streams of dry particles impact each other, leading to fracture and attrition.

Many variables influence the size reduction performance of communition methods. The media size and shape, the size distribution and properties of the incoming feed, the ratio of feed to media, and the power or speed used to drive the mill are among the factors. In general, changes in variables that allow for more media collisions per unit time and greater probability of media impacting a particle lead to faster milling. For example, attrition and vibratory mills create more motion in the media as compared with tumbling mills, leading to more efficient milling.

Communition processes can change more than the particle size. Wearing of the media and vessel walls leads to contamination of the powder. Media materials are chosen based not only on their high hardness and density (both qualities make the media better at impact fracture), but also for their effect as potential contaminants. The possibility of contamination is a serious limitation of milling powders for applications requiring high purity ($>99.9\%$). Also, communition can cause some particles to change structure. For example, certain types of modified zirconia can undergo phase transformations during milling.

Additives. The types of additives used with ceramic powders depend on the forming operations that are planned. It is very common to have the powder dispersed in a liquid at some stage in its processing. Additives that help the liquid wet the particle surfaces and create conditions for a uniform, stable dispersion are used. These additives are known as surfactants and dispersants or deflocculants, respectively. These additives are covered in Chapter 6. Another important type of additive is a binder. Depending on the processing operation, polymer binders are needed in different concentrations to assist in the forming operation or to provide strength after forming, but before sintering. The binder quantities and types for each forming operation are provided in subsequent chapters. It is common for the manufacturer to add the surfactants, dispersants, and binders needed for processing in-house rather than buying powders already formulated with the additives. Lastly, additives to assist the sintering process can be included. These additives may enhance the rate of sintering or allow densification to occur at lower temperatures.

Ceramic Powder Data Sheets and Characterization. Ceramic powder data sheets resemble metal powder data sheets in their content. An example is given in Figure 2.16. A general description and applications are given along with chemical composition and information on the crystalline phase. Information on particle size is featured, including particle size distribution data characterized using a laser diffraction method (Microtrac). Specific surface area, determined by nitrogen gas adsorption (BET), is provided as well. Other information that can be found on some ceramic powder data sheets includes loss on ignition (LOI), which is the weight lost by the powder on heating, processing related information, such as pressed density and sintering temperature, and typical properties and applications of the powder.

2.3.3 Glass Starting Materials

For glass processing, the starting material is a glass batch. The glass batch is formulated from a variety of materials, some mined from the earth and used with only a few preparation steps, and some more refined (e.g., metal oxide powders). When heated, the powdery glass batch is converted to a homogeneous glass melt. Batch mixing, melting, and forming are typically carried out in sequence at a single facility. The schematic in Figure 2.17 shows typical raw materials used in a batch to create a soda-lime-silica glass, and the mixing process, which includes the addition of cullet (broken glass scraps). Other process steps, features, and terms are discussed below and in later chapters. The integrated approach is driven by economics. Forming a glass melt requires considerable energy, so the direct formation of the product from the melt is preferred.

The alternative to the integrated route is to create a melt, like the scheme above, followed by forming a glass powder, instead of a product. The glass powder is then shipped as a starting material to customers who carry out the forming processes. The costs of shipping and remelting are too great for this route to be used in the production of commodity glass products, such as flat

A Grain Zirconia

Zircoa A-grain (zirconium oxide, ZrO_2, or zirconia) is synthesized from zircon sand ($ZrO_2 \cdot SiO_2$) using a solid-state reaction process. The Zircoa A Grain process yields a consistently high purity zirconia composed of monoclinic phase particles. Mean particle size is 2.1 microns by Microtrac analysis. Surface area is 1 m²/g by BET analysis method. Application of A Grain in manufacturing of products is straightforward because of its low surface area and unique spheroid shaped particles.

Some Applications Include:

- Ceramic Color
- Opacifiers
- Refractories
- Catalysts
- Sensors
- High Temperature Insulation
- Electronic Ceramics
- Wear Resistant Products
- Zirconium Metal Production
- High Temperature Filler

Typical Chemical Analysis (Wt.%)

ZrO_2*	SiO_2	CaO	MgO	Fe_2O_3	Al_2O_3	TiO_2
99.6	≤0.3	0.2	<0.1	<0.1	<0.1	<0.1

Other grades can be formulated to satisfy special requirements.

Typical Lot Size .... 12,000 lb.

Standard Packaging .... 500 lb. recycled steel drums with plastic bag liners

* Includes naturally occuring HfO_2 2.5% maximum

Typical Particle Size Distribution

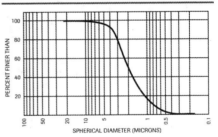

31501 Solon Rd., Solon, OH 44139
Tel: (440)248-0500 Fax: (440)248-8864
Email: sales@zircoa.com http://www.zircoa.com

NOTICE: Recommendations, property values, and application information we publish are based on various sources including measurements by us and others, and estimates of experience. We intend this to be a reliable guide, but we do not guarantee the applicability, completeness, or accuracy of the information. Users should make their own tests to determine the suitability of any product for their application.

Zircoa and Zyttrion are registered trademarks of Zircoa, Inc. The "Flame Graphic Symbol" and the "Stylized Zircoa Logo" incorporating the "Flame Graphic Symbol" are trademarks of Zircoa, Inc. All other trademarks remain the property of their registered owners.

12/2011 - PDF only for Website

FIGURE 2.16 Data sheet for zirconia powders. *Courtesy of Zircoa, www.zircoa.com.*

glass, fiberglass, and containers. However, some specialty glass items are made in this way. Glass powders, sometimes known as frit, are prepared and used for forming as well as a variety of other applications. For example, glass powders are combined with metal powders to make conductive inks for thick film circuits. They are also used as additives to enhance the sintering behavior of polycrystalline ceramics.

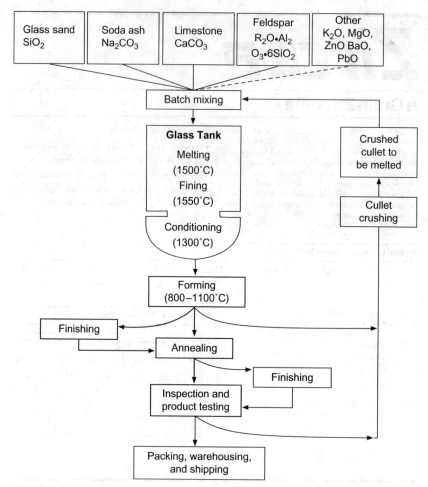

FIGURE 2.17 Glass plant layout showing the integration of batching, melting, and forming operations in sequence. Starting materials are for the example of a soda-lime-silica glass. *From US DOE/OIT (2002).*

Glass Batching. The materials in a glass batch come from a variety of sources and have different functions. Common glass batch ingredients are listed in Table 2.6. The ingredients either supply a metal oxide for the glass composition or they are additives that help produce a uniform melt. In the latter category, fining agents are used to remove gas bubbles from the melt, and fluxes and melt accelerators form a low temperature melt that is needed to dissolve refractory components, such as silica. In the second case, the flux helps in melting and acts as a source of a critical component in the glass composition. More on the glass melting process is presented in Chapter 3.

TABLE 2.6 Common Raw Materials for Glass Batches[a]

Material name	Purpose
Barite ($BaSO_4$)	Flux and fining agent (helps remove bubbles during melting), source of barium oxide
Borate materials	Source of B_2O_3
Sodium tetraborate ($Na_2O \cdot 2B_2O_3 \cdot 10H_2O$)	
Anhydrous borax ($Na_2O \cdot 2B_2O_3$)	
Boric acid ($B_2O_3 \cdot 3H_2O$)	
Caustic Soda (NaOH)	Used as an aqueous solution (50%) for batch wetting
Chromite ($FeO \cdot Cr_2O_3$)	Colorant for green bottles
Cobalt oxide (CoO)	Strong blue colorant
Cullet	Crushed or powdered glass may be internal from the plant or from other sources
Dolomite ($CaCO_3 \cdot MgCO_3$)	Source of calcium and magnesium oxides
Feldspars	Sources of alumina, which improves durability of silicate glasses
Albite ($Na_2O \cdot Al_2O_3 \cdot 6SiO_2$)	
Anorthite ($CaO \cdot Al_2O_3 \cdot 2SiO_2$)	
Microcline ($K_2O \cdot Al_2O_3 \cdot 6SiO_2$)	
Gypsum ($CaSO_4 \cdot 2H_2O$)	Flux and fining agent
Iron oxides	Colorants (e.g., amber glass)
FeO, Fe_2O_3, Fe_3O_4	
Lead oxides	Source of PbO for lead glasses and "crystal"
Litharge (PbO), Red lead (Pb_3O_4)	
Limestone ($CaCO_3$)	Source of CaO, an important ingredient for durability of soda-lime-silica glasses
Lithia materials	Source of Li_2O, flux, melting accelerator
Lepidolite ($LiF \cdot KF \cdot Al_2O_3 \cdot 3SiO_2$)	
Spodumene ($Li_2O \cdot Al_2O_3 \cdot 3SiO_2$)	
Manganese dioxide (MnO_2)	Colorant
Potash (K_2O) and potassium carbonate (K_2CO_3)	Source of potassium oxide

(Continued)

TABLE 2.6 (Continued)

Material name	Purpose
Salt cake (Na_2SO_4)	Melting and fining agent
Silica sources	Source of silica (glass former)
Sand/quartz	
Feldspatic sand (mix of feldspar and sand)	
Soda ash (Na_2CO_3)	Major flux used in all soda-lime glasses
Sodium nitrate ($NaNO_3$)	Oxidizing and fining agent

[a]Adapted from Bauer and Baily (1990).

The main ingredients in soda-lime-silica glasses are mined from the earth and then used after a few steps, such as washing and crushing. The major constituent, silica, is the glass former in important commodity glasses (e.g., flat glass, container glass) and most specialty glasses as well. Silica is found in sand and sandstone deposits in the form of quartz. After a few crushing and separations steps, the starting material is >99% pure SiO_2 and in an appropriate particle size range (see below). In silica sands, the impurity with the greatest consequence is iron oxide, which causes a brown color in the glass. Soda ash is the second important ingredient in the glass batch. It acts as a flux, lowering the melting point and viscosity of the glass. Salt cake may also be used as a sodium source and as a melt accelerator in combination with soda ash. Salt cake and soda ash form a low melting point eutectic melt that dissolves silica. Both soda ash and salt cake are found in natural deposits and are used after a few preparatory steps to reduce particle size and increase purity. Lastly, limestone provides the calcium oxide, another modifier, to the glass. Calcium oxide improves the chemical durability of sodium silicate glasses. Limestone (and other carbonate starting materials) decompose on heating, forming carbon dioxide gas. The gas bubbles help stir and homogenize the melt, but eventually, these bubbles must be removed before the glass forming operation.

The formulation of the batch (i.e., the relative amounts of the ingredients) is based on the desired final composition of the glass as well as the need to produce a homogenous glass melt. It might seem logical to pick a single starting material for each of the desired metal oxide components in the glass, but due to the availability and cost of raw materials, more complex formulations are routine. For example, aluminum oxide is part of some glass compositions to provide strength. Alumina is added to the glass batch in the form of feldspars instead of alumina from the Bayer process; feldspars are more easily incorporated into the glass melt and are less expensive. The use of feldspar also provides silica and alkali or alkaline earth oxides. Batch formulations also include considerations of melting behavior and oxidation states in the melt.

Batch calculations are typically based on the need to produce a certain weight of final glass. An example batch calculation for an E-type fiberglass is given in Example 2.3. In this example, cullet is not included for simplicity, but in most glasses, cullet is an important ingredient. As shown in Figure 2.17, cullet generated in a forming operation is fed back into the glass batch. This type of internal recycling was noted for metals as well. Cullet additions to the batch not only provide cost savings, they also improve the melting behavior of the batch.

The mixing of the glass batch presents challenges due to segregation that occurs during the weighing and mixing process. As with the mixing of metal powders, glass batch uniformity is encouraged by mixing starting materials with similar particle size and density. Glass batch ingredients are particulate with sizes in the range of around 100 μm on the fine end and 1 mm on the coarse end of the distribution. Cullet pieces are typically around a millimeter in size. This size range prevents excessive dusting and allows easy flow into the mixer from the individual storage hoppers for each ingredient. However, the wide range in sizes and densities of the ingredients leads to segregation on mixing and flow. One method of combating this segregation is to add enough of a wetting liquid, such as water or a 50% solution of caustic soda (NaOH), to form a thin liquid film on the particles. The thin film prevents the free flow of the particles, which is necessary for their segregation by size or density. Rotary mixers such as those shown in the section on metal powder starting materials and paddle mixers are used. A batch with good chemical homogeneity results from this wet mixing. The next stages in most glass plants are melting, conditioning, and forming.

Glass Powder or Frit Production. Glass powder or frit is prepared by quenching a uniform glass melt, and can also be considered a glass starting material. The glass batch is fed into a furnace, typically called a melter, and then converted into a homogeneous melt. The melting operation is discussed in Chapter 3. Quenching the glass melt from high temperature into water or between metal rollers results in thermal shock and fracture of the glass into fragments. The fragments can be broken down further using ball milling to create a frit of the desired particle size.

Glass Data Sheets and Characterization. Many of the starting materials used in the glass batch are mined from the earth and minimally refined, and hence data sheets for these materials are somewhat different than those of the controlled powders used in processing of polycrystalline ceramics. A data sheet for unground silica is given in Figure 2.18. The key pieces of information on this and data sheets for other glass batch ingredients are the chemical composition, melting temperature, and the particle size distribution.

While the individual data sheets are important to the formulating and processing of a glass batch into a melt, information about of the behavior of the

EXAMPLE 2.3 Formulate a glass batch to create 1000 g of E-type Fiberglass. Refer to Table 2.5 for the glass composition and Table 2.6 for raw materials. Also find the total weight loss on converting the raw materials to glass.

We can first specify the composition and determine the moles of each of the oxide constituents needed in the final 1000 g of glass.

Component	Weight %	Weight in final glass (g)	Molecular weight (g/mol)	Moles in final glass
SiO_2	54	540	60	9.0
Al_2O_3	14	140	102	1.4
B_2O_3	10	10	69.6	1.4
MgO	4.5	45	40.3	1.1
CaO	17.5	175	56.1	3.1
Total	100	1000		

Now the selection of the raw materials is made. While it might seem reasonable to choose one raw material for each of the oxides, it is typically more effective to seek out lower cost materials that contain more than one component. For example, anorthite, $CaO \cdot Al_2O_3 \cdot 2SiO_2$ (molecular weight = 278 g/mol) is a good choice for supplying the Al_2O_3. This raw material also provides some CaO and SiO_2. We need 1.4 moles of anorthite (389 g) to supply all 1.4 moles of Al_2O_3 since there is one mole of Al_2O_3 per mole of anorthite. In this amount, we also add 1.4 moles of CaO (a portion of the required 3.1 moles) and 2.8 moles of SiO_2 (a portion of the required 0 moles). The first line in the batch table below summarizes the anorthite addition.

Ingredient	Moles	MW (g/mol)	Wt. (g)	Moles SiO_2	Moles Al_2O_3	Moles B_2O_3	Moles MgO	Moles CaO	Other
Anorthite ($CaO \cdot Al_2O_3 \cdot 2SiO_2$)	1.4	278	389	2.8	1.4	–	–	1.4	–
Dolomite ($CaCO_3 \cdot MgCO_3$)	1.1	184	203	–	–	–	1.1	1.1	2.2 moles CO_2
Boric Acid ($B_2O_3 \cdot 3H_2O$)	1.4	124	173	–	–	1.4	–	–	4.2 moles H_2O
Limestone ($CaCO_3$)	0.6	100	60	–	–	–	–	0.6	0.6 moles CO_2
Silica sand (SiO_2)	6.2	60	372	6.2	–	–	–	–	–
Total			1197	9.0	1.4	1.4	1.1	3.1	123 g CO_2 76 g H_2O

Next, we can consider a source for MgO. The only one in Table 2.6 is dolomite, $CaCO_3 \cdot MgCO_3$ (molecular weight = 184 g/mol). All of the 1.1 moles of MgO originate from dolomite. Dolomite decomposes to MgO and CaO on heating, forming CO_2 gas; two moles of CO_2 gas form per mole of dolomite. Since one mole of dolomite supplies one mole of MgO, we need 1.1 moles of dolomite in the batch. Note that another portion (1.1 moles) of the CaO originates from dolomite. See the second line in the batch table.

The only raw material available for B_2O_3 is boric acid, $B_2O_3 \cdot 3H_2O$, which loses water on heating. Therefore, 1.4 moles of boric acid are need. Finally, the amount of CaO from anorthite and dolomite does not quite add up to the 3.1 moles needed in the final glass. Hence, some limestone, $CaCO_3$, is also added. And, glass sand is used to provide the remainder of the silica.

The total loss on ignition (LOI) or weight loss during melting of the batch is 123g CO_2 + 76 g H_2O = 199 g.

#1 DRY
UNGROUND SILICA

PLANT: BERKELEY SPRINGS, WEST VIRGINIA

TYPICAL VALUES
(% RETAINED ON SIEVE)

U.S.A. SIEVE ANALYSIS

USA STD SIEVE SIZE		TYPICAL VALUES		
		% RETAINED		% PASSING
MESH	MILLIMETERS	INDIVIDUAL	CUMULATIVE	CUMULATIVE
20	0.850	0.0	0.0	100.0
30	0.600	0.0	0.0	100.0
40	0.425	4.4	4.4	95.6
50	0.300	40.6	45.0	55.0
70	0.212	35.0	80.0	20.0
100	0.150	15.5	95.5	4.5
140	0.106	4.0	99.5	0.5
200	0.075	0.3	99.8	0.2
270	0.053	0.1	99.9	0.1
Pan		0.1	100.0	0.0

TYPICAL PHYSICAL PROPERTIES	
Grain Shape	Subangular
Hardness (Mohs)	7
Melting Point (Degrees F)	3100
Mineral	Quartz
pH	6.5
Specific Gravity	2.65

TYPICAL CHEMICAL ANALYSIS, %	
SiO_2 (Silicon Dioxide)	99.7
Fe_2O_3 (Iron Oxide)	0.024
Al_2O_3 (Aluminum Oxide)	0.07
TiO_2 (Titanium Dioxide)	0.01
CaO (Calcium Oxide)	0.01
MgO (Magnesium Oxide)	<0.01
Na_2O (Sodium Oxide)	<0.01
K_2O (Potassium Oxide)	0.01
LOI (Loss On Ignition)	0.2

December 1, 2009

U.S. Silica Company
8490 Progress Drive, Suite 300
Frederick, MD 21701
(301) 682-0600 (phone)
(800) 243-7500 (toll-free)
ussilica.com

DISCLAIMER: The information set forth in this Product Data Sheet represents typical properties of the product described; the information and the typical values are not specifications. U.S. Silica Company makes no representation or warranty concerning the Products, expressed or implied, by this Product Data Sheet.

WARNING: The product contains crystalline silica – quartz, which can cause silicosis (an occupational lung disease) and lung cancer. For detailed information on the potential health effect of crystalline silica - quartz, see the U.S. Silica Company Material Safety Data Sheet.

FIGURE 2.18 Data sheet for an unground silica. *Courtesy of U. S. Silica Company, www.ussilca.com.*

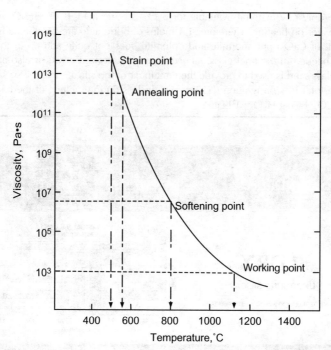

FIGURE 2.19 Effect of temperature on the viscosity of a silicate glass, showing the key points (see Table 2.7). See Chapter 3 for more information.

target glass composition is also critical to the forming of that glass into a product (e.g., glass sheet, container). The thermal behavior is most important. Specifically, the changes in viscosity with temperature depend on the glass composition and data are needed to develop a good forming operation for the glass. The effect of temperature on the viscosities of several glasses is shown in Figure 2.19. To compare glasses, several reference points are found, as described in Table 2.7. These reference points are listed in tables on glass compositions. The rheology and processing of glass melts is discussed in Chapter 3.

An example of a glass powder data sheet is given in Figure 2.20. Here, the key pieces of information are the composition (given in general and as Corning codes), the thermal behavior (softening point, etc.), and properties. The thermal behavior is needed to form the glass. The company also provides glass powders with controlled particle sizes.

2.4 POLYMERS

2.4.1 Introduction

A polymer is composed of an assembly of high molecular weight molecules known as macromolecules. A basic unit (mer) is repeated hundreds

TABLE 2.7 Definitions of Reference Point Temperatures for Glasses

Temperature reference point	Definition	Viscosity
Working Point	Temperature at which glass is able to be formed easily	10^3 Pa·s
Softening Point	Temperature at which the glass object deforms under its own weight; found with an ASTM standard method (C338) that involves monitoring elongation of a fiber under its own weight	Approx. $10^{6.6}$ Pa·s
Annealing Point	Temperature at which thermal stresses are relaxed in about 15 minutes	Approx. 10^{12} Pa·s
Strain Point	Temperature at which thermal stresses are relaxed in about 4 hours	Approx. $10^{13.5}$ Pa·s

Glass powders data sheet

Corning glass powders are offered in a variety of compositions that are low expansion, high temperature, electrically resistive and durable. These glasses are available in standard US mesh sizes 4 through 325 and can be customized to meet your needs.

Glass code	Composition	Softening point°C	Annealing point°C	Strain point°C	CTE 0–300°C	Density g/cc
7052	Borosilicate	712	484	440	47.0	2.27
7056	Alkali borosilicate	718	512	472	51.5	2.29
7070	Borosilicate	755	496	456	32.0	2.13
7740	Soda borosilicate pyrex®	821	560	510	32.5	2.23
9013	Alkali barium	656	462	423	88.5	2.64
1890	Soda zirconium silicate	868	662	574	74	2.62

Other glasses may be available on request

Minimum order requirement for stock items is 30 kg for glass powder and 10 kg for cullet

FIGURE 2.20 Data sheet for crushed glasses. *Courtesy of Corning Incorporated, www.corning.com.*

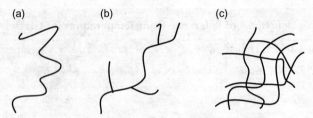

FIGURE 2.21 Molecular structures of polymers: (a) linear, (b) branched, and (c) crosslinked (network).

or thousands of times in one macromolecule or chain. One way a polymer is described is based on the chemical structure of its mer unit. In the simplest case, a polymer has a single mer unit (homopolymer); alternatively, a polymer may consist of two or more repeat units (copolymer). In a copolymer, the repeat units may be arranged randomly, in alternating fashion, or in "blocks." At another level, the individual macromolecules have structural features. Macromolecules may be linear, branched, or crosslinked, as shown in Figure 2.21. In the solid polymer, individual macromolecules interact with each other in various ways. Linear and branched molecules become entangled, and if their structure permits, develop regions of crystalline order (i.e., they are semi-crystalline). Crosslinked or network polymers have covalent bonds connecting the macromolecular chains; these bonds limit rearrangement and crystallinity. In a further level of structure, polymers of different chemical structures can be mixed to create a polymer blend.

Polymer molecules are built from relatively few elements (e.g., C, H, N, O, Cl, F, Si). Polymers are found in nature (e.g., rubber, cotton, wood), and some commercially important polymers (e.g., cellulose acetate) are derived from the chemical processing of these natural sources. Most polymers, however, are considered synthetic; they are created by purposeful chemical reactions of precursors that are mainly derived from natural gas or petrochemicals. The refining of natural gas, oil, and other petrochemicals into useful fuel and chemical products is a major industry, requiring the expertise of chemists and chemical engineers.

Polymerization reactions are classified as either chain growth polymerization (also known as addition polymerization) or step growth polymerization (also known as condensation polymerization). Briefly, chain growth polymerization involves three steps: initiation, propagation, and termination. For example, in free-radical chain growth polymerization, a free radical (species with an unpaired electron) is formed by decomposition of an initiator molecule (e.g., a peroxide), before attacking the $C = C$ bond in the monomer, leading to the formation of an active center on a $C-C$ bond (with unpaired electron). During propagation, this active center attacks other monomers and in so doing, attaches them to a growing chain. The growth of chains is

TABLE 2.8 Examples of Different Names for Polymers

Common name	Polystyrene	Polymethyl methacrylate
IUPAC[a]	Poly(-1-phenyl ethylene)	Poly[1-(methoxycarbonyl)-1-methyl ethylene]
Abbreviation[b]	PS	PMMA
Trade Names	Styrofoam (Dow), Dylene (Acro)	Perspex (ICI), Plexiglas (Atoglas), Lucite (DuPont)

[a] *International Union of Pure and Applied Chemistry.*
[b] *ASTM D 1600–83.*

limited by termination reactions; for example, active centers at the ends of two growing chains may annihilate each other. Addition polymerization can also take place by creating ionic groups on the growing chains (anionic and cationic polymerization). Control of the molecular weight distribution of a polymer created by addition polymerization requires attention to the kinetics of the reactions involved in the three steps. Step growth polymerization involves the reactions of two chemically different monomers (or oligomers). The two species are typically difunctional (i.e., they can react at both ends) so that chains grow as reaction proceeds. The reaction creates a by-product that is typically removed as polymerization proceeds. Molecular weight is controlled by managing reaction kinetics and adjusting the composition of the starting mix of monomers.

Polymers are known by many names as shown in the examples of Table 2.8. The chemical structure of a polymer has a name assigned by the International Union of Pure and Applied Chemistry (IUPAC). This chemical name is not widely used. The common names, which are based on the mer structures, abbreviations, and trade names, are more prevalent.

In terms of processing, polymers are best divided into two groups: thermoplastics and thermosets. On heating, thermoplastic polymers become soft and eventually flow as melts; they are formed into a shape at elevated temperature and then retain their shape on cooling. No chemical reactions (e.g., crosslinking) take place during forming; hence, thermoplastics can be reheated and reused. Thermosets, on the other hand, are formed from a prepolymer mix that is a liquid or a low melting point solid; crosslinking during or after forming results in shape retention. Note that elastomers (e.g., polymers that are able to undergo extensive elastic deformation) are considered as a subset of either thermoset or thermoplastic polymers. Most elastomers are composed of highly coiled chains that are lightly crosslinked.

Thermoplastic polymers are based on linear and branched molecules that are synthesized before the forming operation. Table 2.9 shows some common thermoplastic polymers and distinguishes them as amorphous or semi-

TABLE 2.9 Examples of Common Thermoplastic Polymers

Polymer name and abbreviation	Solid structure	Mer structure
Polystyrene (PS)	Amorphous	$-\text{C}(\text{H})(\text{H})-\text{C}(\text{H})(\text{C}_6\text{H}_5)-$
Polymethyl methacrylate (PMMA)	Amorphous	$-\text{C}(\text{H})(\text{H})-\text{C}(\text{CH}_3)(\text{COOCH}_3)-$
Polycarbonate (PC)	Amorphous	$-\text{O}-\text{C}_6\text{H}_4-\text{C}(\text{CH}_3)(\text{CH}_3)-\text{C}_6\text{H}_4-\text{O}-\text{C}(=\text{O})-$
Polyvinyl chloride (PVC)	Amorphous	$-\text{C}(\text{H})(\text{H})-\text{C}(\text{H})(\text{Cl})-$

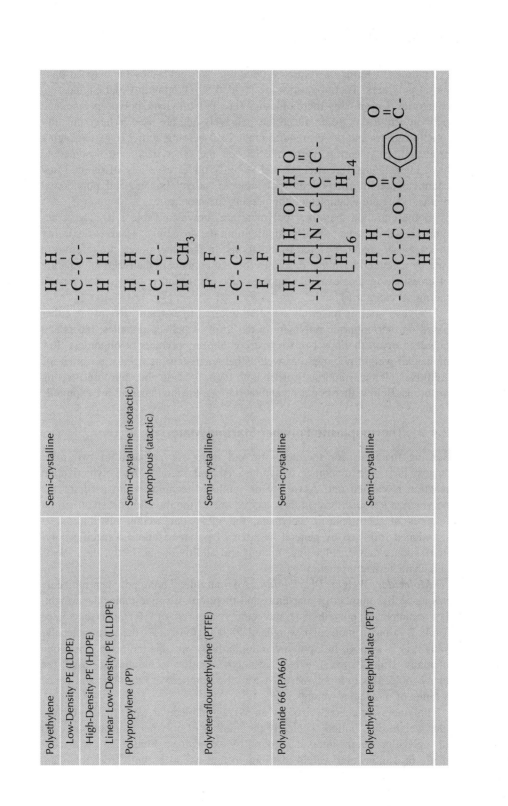

crystalline. In terms of volume production, thermoplastics far outpace thermoset polymers. Thermoplastics are divided into commodity and engineering categories. Commodity thermoplastics (i.e., polyethylene, polypropylene, and polyvinylchloride) are produced inexpensively and are used in large quantity in applications that do not require much load bearing ability (e.g., packaging products). Engineering thermoplastics, on the other hand, are designed to have high mechanical strength and stability under a variety of thermal and chemical conditions. Nylons (polyamides), polycarbonate, and polymethyl methacrylate are examples of engineering thermoplastics.

Some common thermoset polymers are shown in Table 2.10. The structures are shown schematically, highlighting the crosslinks. There is flexibility in thermoset chemistries and formulations. In fact, thermosets are referred to first by a general term that describes the components in the reaction or nature of crosslinking. For example, epoxies, phenolics, and polyesters are general thermoset categories.

Polymer starting materials are often referred to as resins. Resins are either based on thermoplastic polymers in the form of pellets, granules, flakes and powders, or on thermoset mixes that contain prepolymers or oligomers (low molecular weight polymers) as well as hardeners, curing agents, or initiators. Additives play an important role in both thermoplastic and thermoset starting materials. Below, thermoplastic and thermoset starting materials are explored.

2.4.2 Thermoplastic Polymer Starting Materials

Starting materials for thermoplastic polymers are prepared in two main stages. First, the monomer is synthesized and second, the polymerization reaction is carried out. A variety of synthetic techniques are used for the polymerization reactions. In this section, one of the most important structural features of thermoplastic polymers, the molecular weight distribution, is introduced followed by general methods of polymerization, a specific polymerization example, blending, mixing and additives, and lastly, data sheets and characterization methods.

Molecular Weight Distribution. An individual homopolymer molecule has a specific number of mer units and therefore, a single molecular weight. For example, a polyethylene molecule composed of 1000 mer units (see Table 2.9) has a molecular weight of 28,000 g/mol (i.e., the mer molecular weight is 28 g/mol). In a polymer sample, however, this molecule is just one of many. Unless special, expensive polymerization techniques are used, the many molecules in a polymer have varying numbers of mer units and hence a range of molecular weights.

Polymers are characterized by an average molecular weight and a molecular weight distribution. Typically the number of polymer molecules in each of molecular weight interval is found by experiment. Figure 2.22 is a schematic of a molecular weight distribution that would be built from this data.

TABLE 2.10 Examples of Common Thermoset Polymers

Polymer name and abbreviation	Crosslinked structure[a]
Phenol Formaldehyde (PF)	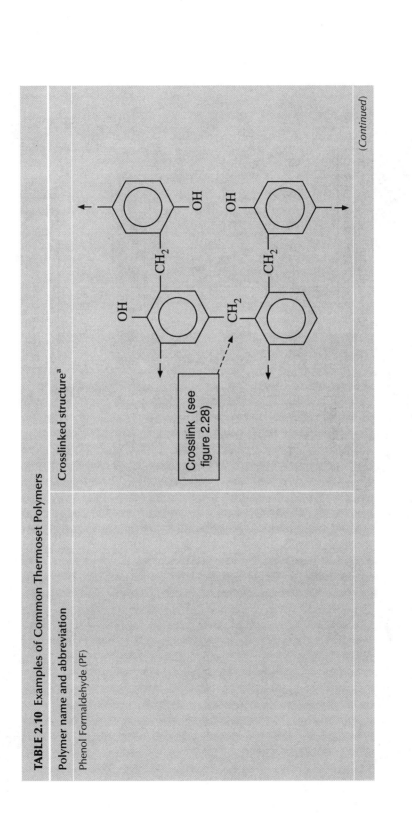

(Continued)

TABLE 2.10 (Continued)

Polymer name and abbreviation	Crosslinked structure[a]
Epoxy (EP)	$\boxed{\text{Epoxy group}}$ $\boxed{\text{Crosslink from reaction of epoxy groups with amine hardener}}$ $\underset{\displaystyle \text{CH}_2\!-\!\text{CH}_2\!-\!\text{CH}_2\!-\!\text{R}\!-\!\text{CH}_2\!-\!\overset{\text{OH}}{\overset{|}{\text{CH}}}\!-\!\text{CH}_2\!-\!\underset{\displaystyle \underset{\displaystyle \;\;\;\text{OH}}{\overset{\displaystyle |}{-\!-\!\text{CH}\!-\!\text{CH}_2}}}{\text{N}\!-\!\text{R}'\!-\!\text{N}}\!<\!\!\begin{array}{l}\text{CH}_2\!-\!\overset{\text{OH}}{\overset{|}{\text{CH}}}\!-\!\!\!\longrightarrow\\ \text{CH}_2\!-\!\overset{|}{\underset{\displaystyle \text{OH}}{\text{CH}}}\!-\!\!\!\longrightarrow\end{array}}{\overset{\displaystyle \text{O}}{\triangle}}$
Unsaturated Polyester (UP)	$\boxed{\text{Polyester}}$ $\boxed{\text{Crosslink from reaction with styrene}}$ $\boxed{\text{Free radical (originally formed at C=C in polyester)}}$ (structure showing polyester chains crosslinked through styrene with phenyl group)

[a] Arrows show extensions of the crosslinked network.

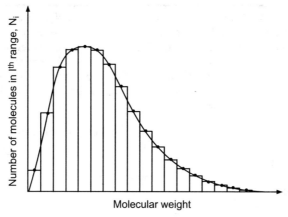

FIGURE 2.22 Schematic diagram of a molecular weight distribution showing molecular weight ranges with the number of molecules in each range i as a function of the molecular weight. The midpoint molecular weights are noted.

The data appear in molecular weight intervals with the number of molecules in any given interval, i, is N_i. The data are converted to a number fraction, x_i, of molecules in the various molecular weight intervals.

$$x_i = \frac{N_i}{\sum_i N_i} \tag{2.10}$$

The number average molecular weight, M_n, is calculated from these data:

$$M_n = \sum_i M_i x_i \tag{2.11}$$

where M_i is the molecular weight in the middle of interval i. A similar approach is used for data presented as a weight fraction, w_i, leading to a weight average molecular weight, M_w, of:

$$M_w = \sum_i M_i w_i \tag{2.12}$$

The degree of polymerization (the average number of mer units in a molecule) is also found in terms of the number-based distribution, n_n, or the weight-based distribution, n_w.

$$n_n = \frac{M_n}{M_o} \quad \text{and} \quad n_w = \frac{M_w}{M_o} \tag{2.13}$$

where M_o is the mer molecular weight. Example 2.4 shows the differences between molecular weight on a number basis and a weight basis.

Polymerization Methods. Polymerization methods are chemical engineering approaches to synthesizing polymers from monomers on a large scale.

EXAMPLE 2.4 Experimental data for a polyethylene polymer sample is given below as the number of polyethylene molecules in each molecular weight interval. From this data find the number average molecular weight, the number average degree of polymerization, the weight average molecular weight and the weight average degree of polymerization. Also plot histograms of the distributions based on number fraction and weight fraction in the molecular weight interval.

Molecular weight range (g/mol)	Number of molecules in range, N_i
0–5,000	1,000
5,000–10,000	2,000
10,000–15,000	3,500
15,000–20,000	4,700
20,000–25,000	8,900
25,000–30,000	10,500
30,000–35,000	12,000
35,000–40,000	7,000
40,000–45,000	2,000
45,000–50,000	1,000

The mer unit for polyethylene is C_2H_4. Therefore the mer molecular weight is:

$$M_o = 2(12.01 \text{ g/mol}) + 4(1.01 \text{ g/mol}) \cong 28.0 \text{ g/mol}$$

The number fraction in each interval is found using Eq. 2.10. To find the weight fraction, the number of molecules in each interval, N_i, is converted to a weight of molecules using the weight at the middle of the interval, M_i. For example, consider the first interval. There are 1000 molecules with molecular weights in the interval 0–5000 g/mol, which has a middle molecular weight of 2500 g/mol. One polymer molecule with a molecular weight of 2500 g/mol weighs $2500/N_{Av}$ g, where N_{Av} is Avogadro's number; therefore the approximate weight of polymer in this particular interval is $2,500,000/N_{Av}$. Or more generally, the weight in each interval is $M_i N_i / N_{Av}$. (Note that since N_{Av} is a constant, it is not necessary to carry this along in the computation of weight fraction.)

Molecular weight interval (g/mol)	Number of molecules in interval, N_i	Number fraction x_i	Molecular weight midpoint, M_i(g/mol)	$M_i x_i$ (g/mol)	$M_i N_i$(g/mol)	Weight fraction, w_i	$M_i w_i$ (g/mol)
0–5,000	1,000	0.019	2,500	48	2,500,000	0.002	4.41
5,000–10,000	2,000	0.038	7,500	285	15,000,000	0.011	79.3
10,000–15,000	3,500	0.067	12,500	838	43,750,000	0.031	386
15,000–20,000	4,700	0.089	17,500	1,558	82,250,000	0.058	1,015
20,000–25,000	8,900	0.169	22,500	3,803	200,250,000	0.141	3,179
25,000–30,000	10,500	0.200	27,500	5,500	288,750,000	0.204	5,602
30,000–35,000	12,000	0.228	32,500	7,410	390,000,000	0.275	8,942
35,000–40,000	7,000	0.133	37,500	4,988	262,500,000	0.185	6,944
40,000–45,000	2,000	0.038	42,500	1,615	85,000,000	0.060	2,549
45,000–50,000	1,000	0.019	47,500	903	47,500,000	0.034	1,592
Sum of Column	52,600	1.00		26,945	1.42×10^9	1.00	30,292

The average molecular weights and degrees of polymerization are found using the information in the table.

$$M_n = \sum_i M_i x_i = 26{,}945 \text{ g/mol} \quad \text{and} \quad M_w = \sum_i M_i w_i = 30{,}292 \text{ g/mol}$$

$$n_n = \frac{M_n}{M_o} = \frac{26{,}949}{28} = 962 \quad \text{and} \quad n_w = \frac{M_w}{M_o} = \frac{30{,}292}{28} = 1082$$

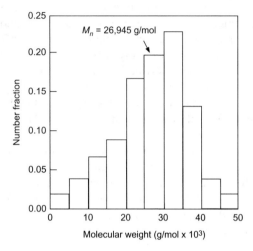

FIGURE E2.4A

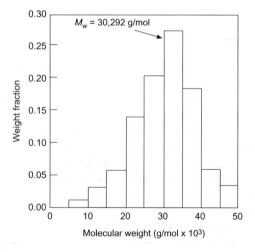

FIGURE E2.4B

This example shows that when the distribution is considered on a number basis, the average molecular weight is smaller than when it is considered on a weight basis. This contrast is similar to the particle size distributions in which the distribution histogram shifts to lower size when considered on a number basis and higher size on a weight basis.

Methods are developed and adopted for particular polymers based on the nature of the polymerization reaction, the required final form and molecular weight of the polymer, the amount of polymer required, and economic factors. Some of the more common polymerization methods are listed below.

Bulk Polymerization — In bulk polymerization methods, reactants (monomers, oligomers) are combined directly with initiators (if required) in a vessel at controlled temperature and pressure. Bulk polymerization methods are used to synthesize polymers by chain addition reactions, and less commonly, by step growth reactions, which usually require a means of removing reaction by-products. Free-radical polymerization reactions are exothermic and generate a lot of heat. Bulk polymerization has the advantage of producing high purity polymers at high yield. Disadvantages are the difficulty in controlling the temperature of reaction, which impacts the ability to control molecular weight.

Solution Polymerization — Solution polymerization makes use of a solvent to remove the heat generated by the polymerization reaction. All components needed for the reaction must be soluble in the solvent. The reactions are carried out in a vessel. While this approach is effective for removing the heat of reaction, it does have a lower yield than bulk polymerization and requires solvent removal.

Suspension Polymerization — Like solution polymerization, a liquid is used to remove the heat of reaction, but in suspension polymerization methods, the reactants are suspended as droplets in the liquid. The individual droplets contain all the reactants necessary to produce the polymer. The suspension is stirred and additives are used to help keep the dimensions of the reacting droplets small. The product is in the form of small polymer spheres or beads, which can be used as is or formed into larger pellets by extrusion. Suspension polymerization has similar advantages and disadvantages as solution polymerization, but has the additional disadvantage of low purity due to the use of additives to control droplet size.

Emulsion Polymerization — Emulsion polymerization methods use water as a liquid medium for a suspension of well-controlled monomer droplets. The initiator, in this case, is dissolved in the water phase (different than suspension polymerization). A surfactant is typically used to stabilize the monomer droplets and the resulting polymer after reaction, and to form micelles (small cells of surfactant). The reaction takes place when the monomer, which has a low solubility in water, diffuses to a micelle created by surfactant and reacts with a radical generated by the water soluble initiator. More and more monomers diffuse to the micelle and participate in the reaction. The result is the formation of solid polymer particles suspended in water. These particles are ordinarily submicron in size. This suspension is called latex, and is used in paints, adhesives, and other products. The suspension can be coagulated to make a solid product; this route is also used to create synthetic rubber.

TABLE 2.11 Comparison of Types of Polyethylene[a]

Type of PE	Structure	Density (g/cm^3)	Crystallinity (%)
Low-Density PE (LDPE)	Branched	0.910–0.925[a]	40–70[a]
High-Density PE (HDPE)	Linear (no branches)	0.935–0.960[a]	80–95[a]
Linear Low-Density PE (LLDPE)	Linear with short branches	0.915–0.925[b]	30–45[b]

[a]Data from Fried (1995).
[b]Data from Kissin (2000).

Example – Polyethylene. Polyethylene polymers, including low-density polyethylene (LDPE), high-density polyethylene (HDPE), and linear low-density polyethylene (LLDPE), lead all other polymers in terms of production and sales. LLDPE, HDPE, and LDPE are the three most PE common varieties. The differences between these polymers are outlined in Table 2.11. The linear structure of HDPE allows the polymer to form more crystalline regions, which leads to higher density. LLDPE also has a linear structure, but its structure contains small branches that restrict the crystallinity. LDPE is branched but the branches are long and irregular, which also limits crystallization.

Polyethylene is formed by polymerization of ethylene (C_2H_4), a gas at room temperature and pressure. To control the structure of PE, different methods of polymerization are used. Further, LLDPE is formed by a copolymerization of the ethylene and an α-olefin, such as 1-butene, with the olefin forming the short branches. Ethylene (C_2H_4) is an important monomer, serving as a basis for not only PE, but also as a starting point for the synthesis of other important monomers, such as vinyl chloride. As a result, much effort has gone into developing methods and facilities for producing high quality ethylene from a variety of sources, including molasses (a by-product of sugar production), crude oil, and natural gas.

Ethylene is most commonly extracted from crude petroleum oil. This oil is a heterogeneous mixture of hydrocarbon molecules of various lengths. During the initial refining of oil, a distillation process is used to separate fractions of different molecular weights (or different number of C atoms). Examples of the lower molecular weight products are ethane, C_2, and naphtha, C_{4-7}. Thermal processes are used to "crack" ethane to produce ethylene gas. Temperatures in excess of 700°C are needed. At these temperatures, the initial hydrocarbon feedstock is gaseous and fed through tubular reactors. The formation of solid carbon ("coking") is prevented by mixing steam in with the gas stream. Most facilities produce ethylene of a high enough quality so that no additional purification steps are needed. The impurities that may cause some problems are oxygen and water, which can interfere with free-radical polymerization, and hydrogen, which interferes with catalysts used in some polymerization routes.

PE is created by polymerization of ethylene. For example, Figure 2.23 shows schematic representation of a free-radical reaction sequence. During initiation, a free radical (R•) is formed and attacks the $C=C$ bond in ethylene to create a $C-C$ bond with the radical (•) attached. Propagation reactions add monomer units onto the growing chain. Radical transfer or chain transfer can take place during propagation with the radical moving to a carbon atom mid-chain. Such a change in radical position leads to the formation of a branch off the main chain. Lastly, termination occurs when radicals annihilate each other or through disproportionation, in which a free radical reverts back to a $C=C$ bond.

Several polymerization methods and reactor designs are used to create LDPE, LLDPE, and HDPE. To illustrate the diversity and structure control, two methods are discussed: (1) high pressure bulk polymerization processes used to make LDPE and (2) a gas phase, fluidized bed process used to make LLDPE and HDPE. Other industrially important processes, include slurry and solution polymerization of HDPE and LLDPE, and loop reactors to produce LDPE.

High pressure bulk polymerization processes are used to synthesize LDPE. Ethylene is pressurized and fed into an autoclave reactor (pressurized chamber) or through a pressurized tubular reactor (several hundred meters of thick-walled tubes roughly 60 mm in diameter). Reactors include ports for injection of an initiator (e.g., benzoyl peroxide, oxygen). The pressures in an autoclave-style reactor are 150–200 MPa, while those for a tubular reactor are 200–350 MPa. A schematic of the autoclave version of the process is shown in Figure 2.24. High pressure during the PE polymerization boosts the molecular weight of the product. The rate of propagation, addition of monomers to growing chains, increases with monomer concentration and hence

Initiation		$R\bullet + CH_2 = CH_2 \rightarrow RCH_2CH_2\bullet$
Propogation		$RCH_2CH_2\bullet + CH_2 = CH_2 \rightarrow RCH_2CH_2CH_2CH_2\bullet$
		$R(CH_2)_4\bullet + CH_2 = CH_2 \rightarrow R(CH_2)_6\bullet$
Radical transfer		$RCH_2CH_2CH_2CH_2\bullet \rightarrow R\overset{\bullet}{C}H CH_2CH_2CH_3$
Termination		$R(CH_2)_n\bullet + R'(CH_2)_m\bullet \rightarrow R(CH_2)_n(CH_2)_m R'$

FIGURE 2.23 Steps in polymerization of ethylene into polyethylene.

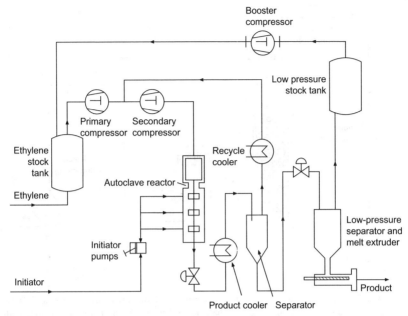

FIGURE 2.24 Schematic diagram of a high pressure autoclave process. E. S. Wilks, Ed.: *Industrial Polymers Handbook, Volume 2. p. 674. 2001.* Copyright Wiley-VCH Verlag GmbH & Co. KGaA. Reproduced with permission.

pressure. The reaction to form polyethylene is very exothermic (93.6 kJ/mol), and so, the geometry of the reactor is designed to have a large amount of cooling area per volume. Reactors run at temperatures in the range of 150–300°C. Overheating is particularly hazardous as ethylene decomposes at elevated temperature to gases prone to explosion (methane or hydrogen) and carbon. After exiting the reactor, the product is cooled. The polymer is separated from the unreacted monomer and extruded into pellets. The monomer is recycled. The conversion of monomer to polymer by this process is about 20% for an autoclave process and up to 35% for a tubular reactor process. The important variables in this process are the temperature and pressure of the reaction, the type of initiator, the location of initiator injection points into the reactor, and the use of chain transfer agents. While these variables allow control of polymer branching and molecular weight distribution, the polymer created in high pressure processes is limited to the LDPE variety.

Gas phase fluidized bed processes with engineered catalysts are used to synthesize both HDPE and LLDPE. A schematic of the process is in Figure 2.25. For HPDE, ethylene gas is fed into the bottom of the reactor vessel, which is packed with catalyst and polymer particles resting on a distribution plate. There are several possible catalyst systems. For example, Ziggler-Natta catalysts are based on a combination of a transition metal salt

84 Materials Processing

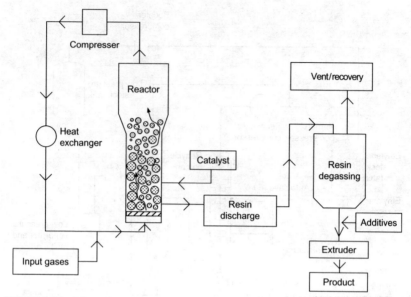

FIGURE 2.25 Schematic diagram of a fluidized bed gas phase polymerization process for synthesis of HDPE and LLDPE. In the reactor, particles in bed are shown (not to scale) along with arrow for example gas pathways. *Adapted from Kissin (2000), Whitely et al. (2001), Burdette (2008), and www.univation.com.*

(e.g., $TiCl_3$) and a group III alkyl (e.g., $Al(C_2H_5)_3$). The interaction between these two components creates a surface reactive site that launches initiation also influences the subsequent polymerization. As the gas input makes its way through the catalyst bed, polymer is formed. The details of the reaction sequence are complex, but a reactive site on the catalyst interacts with the monomer for only a fraction of a second and therefore is capable of forming of thousands of polymer molecules in a short time. Polymer essentially forms in and around the catalyst. The reactor operates at ~2 MPa and a temperature of 80–100°C. The upward flow of gas keeps the particles and catalyst mobile. Only about 2% of the monomer reacts as it transits through the bed, but the gas exiting the top of the bed, cycles back through a compressor and heat exchanger, and is reused. but the recycling polymer product or resin is removed at the base of the reactor and directed into a degassing chamber and lastly a pelletizer. Since catalyst is removed with the product, new catalyst is added to the reactor continuously. For LLDPE, the same reactor can be used but the catalyst is changed and an α-olefin co-monomer, such as 1-butene, is added to the input stream. Advances in catalyst systems, including the use of metallocenes, have had an impact in the synthesis of LLDPE.

Blends, Additives, and Mixing. Thermoplastic polymers are synthesized to have controlled molecular weight distributions and chemical structures (including copolymers), but there is often still a need for further modification of the polymer to tailor processing behavior or final properties. One approach is to create a polymer blend, a mixtures of two or more polymers with distinct chemistries. Blending is carried out in the molten state using equipment that provides a high shearing action. Forming a uniform polymer blend is often a challenge because most polymer melts are immiscible, and so necessarily consist of two phases. Special compounds, typically copolymers, are often added to enhance compatibility of the interface between the two phases. An example of a polymer blend is high impact polystyrene (HIPS), which is a blend of polystyrene and an elastomer (rubber). HIPS is prepared by mixing polystyrene and an elastomer such as polybutadiene, but a more effective way to blend these materials is to dissolve the elastomer directly into the styrene monomer before it is polymerized. The resulting HIPS blend has a better distribution of rubbery particles and a stronger interface between PS and rubbery regions as compared with HIPS prepared from a mechanical blending process.

Another common route to modifying a base polymer is though additives. Additives are non-polymeric materials that are used in a relatively small quantity. Table 2.13 provides a list of important additives for thermoplastics and their functions. The relative amounts of additives are tuned to achieve processing and property goals. Polymers vary in their need for additives. For example, plasticizers are used to make PVC flexible (a requirement for tubing, for example) and chemical additives, such as methyl tin mercaptide, reduce the rate of thermal degradation. Other polymers, such as polyethylene, do not require many additives.

Additives can be introduced and blends created at various stages in the processing of polymer products: at the end of the polymerization reaction, in a separate step, or during the polymer forming operation (e.g., in the extruder). Of interest to polymer starting materials are the modifications that occur as a separate operation, resulting in tailored thermoplastic pellets or granules ready to be used in forming operations, such as extrusion or injection molding.

Mixing of the additives with the base resin can be carried out by several methods. Granules of the thermoplastic polymer can be mixed with additives in a dry state using equipment similar to the rotating mixers used to blend metal powders. The use of intense shearing action on polymer melt/additive mixtures is another alternative. Special equipment is used to provide intense shearing action. More on mixing and compounding in extruders is presented in a Chapter 3.

Thermoplastic Polymer Data Sheets and Characterization. Thermoplastic polymer data sheets describe the general chemistry of the polymer, the processing conditions, and the final properties of the polymer. Figure 2.26 shows an example of an acrylonitrile butadiene styrene (ABS) designed for fused deposition modeling, a type of additive manufacturing.

TABLE 2.13 Additives Used in Thermoplastic Polymers

Additive type	Purpose(s)	Examples
Fillers	Increase stiffness of polymer, reduce cost, add color or special properties	Calcium carbonate particles, Glass fibers, Carbon black particles
Plasticizers	Lower the melt viscosity at a given temperature, lowers glass transition temperature and increases flexibility	High boiling point, low molecular weight liquids (e.g., dioctyl phthalate, tricresyl phosphate
Impact Modifiers	Increase the mechanical toughness	Acrylics, rubbery particles
Heat and UV Stabilizers	Prevent degradation from heat, oxidation, ultraviolet (UV) exposure	Organic compounds that scavenge free radicals created by heat or UV
Flame Retardants	Reduce flammability	Alumina trihydrate, organobromine compounds
Lubricants	Reduce friction with forming equipment, allows melting at lower temperatures	Oils, waxes
Release Agents	Prevent sticking to metal molds	Organofunctional silicones
Blowing Agents	Make foam during processing	Solids that decompose to form gases
Colorants	Add color	Soluble dyes, pigments

Note that key properties that are important to the application of the polymer are given (e.g., electrical, mechanical properties) along with the thermal behavior. However, important features of the polymer structure and composition are not included: the molecular weight distribution, the glass transition temperature, and the amount and type of other additives. If these features are important to the customer, they may be requested. Other data sheets include quite specific information on the processing of the polymer, including suggested melt temperature, and extrusion or injection molding conditions.

Characterization of thermoplastic polymers falls into three general categories: structure, thermal behavior and melt rheology, and final polymer properties. The most important feature of the structure category is molecular weight and molecular weight distribution of the base polymer. This characteristic is determined using one of several techniques outlined in Table 2.14. Other structural features of interest include the crystallinity, which may be determined by X-ray diffraction or differential scanning calorimetry, and

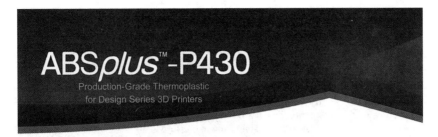

ABSplus is a true production-grade thermoplastic that is durable enough to perform virtually the same as production parts. When combined with Design Series 3D Printers, ABSplus is ideal for building 3D models and prototypes in an office environment.

Mechanical Properties	Test Method	English XZ Axis	Metric XZ Axis
Tensile Strength, Ultimate (Type 1, 0.125", 0.2"/min)	ASTM D638	4,700 psi	33 MPa
Tensile Strength, Yield (Type 1, 0.125", 0.2"/min)	ASTM D638	4,500 psi	8 MPa
Tensile Modulus (Type 1, 0.125", 0.2"/min)	ASTM D638	320,000 psi	2,200 MPa
Tensile Elongation at Break (Type 1, 0.125", 0.2"/min)	ASTM D638	6%	6%
Tensile Elongation at Yield (Type 1, 0.125", 0.2"/min)	ASTM D638	2%	2%
IZOD Impact, notched (Method A, 23°C)	ASTM D256	2.0 ft-lb/in	106 J/m

Mechanical Properties	Test Method	English XZ Axis	English ZX Axis	Metric XZ Axis	Metric ZX Axis
Flexural Strength (Method 1, 0.05"/min)	ASTM D790	8,450 psi	5,050 psi	58 MPa	35 MPa
Flexural Modulus (Method 1, 0.05"/min)	ASTM D790	300,000 psi	240,000 psi	2,100 MPa	1,650 MPa
Flexural Strain at Break (Method 1, 0.05"/min)	ASTM D790	4%	4%	2%	2%

Thermal Properties[2]	Test Method	English	Metric
Heat Deflection (HDT) @ 66 psi	ASTM D648	204°F	96°C
Heat Deflection (HDT) @ 264 psi	ASTM D648	180°F	82°C
Glass Transition Temperature (Tg)	DSC (SSYS)	226°F	108°C
Melt Point	---------	Not Applicable[3]	Not Applicable[3]
Coefficient of Thermal Expansion	ASTM E831	4.90E-05 in/in/°F	8.82E-05 mm/mm/°C

Electrical Properties[4]	Test Method	Value Range
Volume Resistivity	ASTM D257	2.6E15 - 5.0E16 ohm-cm
Dielectric Constant	ASTM D150-98	2.3 - 2.85
Dissipation Factor	ASTM D150-98	0.0046 - 0.0053
Dielectric Strength	ASTM D149-09, Method A, XZ Orientation	130 V/mil
Dielectric Strength	ASTM D149-09, Method A, ZX Orientation	290 V/mil

FIGURE 2.26 Data sheet for an acrylonitrile butadiene styrene thermoplastic polymer. The first page of a two-page data sheet is shown. *Courtesy of Stratasys, www.stratasys.com.*

microstructure features, which are often of interest in blends and polymers modified by fillers.

The thermal behavior, including the rheology of the polymer melt, is essential to the processing and use of thermoplastics. As temperature increases, the mechanical behavior of the polymer changes dramatically. Like ceramic glasses, there are critical temperatures that are encountered as

TABLE 2.14 Methods to Determine Polymer Molecular Weight

Method	Principle and description	Molecular weight determined
Osmometry	The osmotic pressure of a polymer solution depends on its number average molecular weight. Osmotic pressure is measured using a cell with a membrane separating pure solvent from the polymer solution. The method is effective only for higher molecular weights (>20,000) due to membrane limitations. Vapor phase osmometry makes use the relationship between the vapor pressure of a polymer solution and its molecular weight; this technique is capable if measuring molecular weights down to 10,000.	Number average molecular weight
Light Scattering	The light scattering behavior of polymer solutions depends on the size of the polymer molecules in the solution, which is a function of molecular weight. Scattering measurements are taken at several angles to fine the average molecular weight. Radius of gyration is also found.	Weight average molecular weight
Intrinsic Viscosity Measurements	The intrinsic viscosity of a polymer solution is found be taking viscosity measurements of polymer solutions with varying concentration. A standard relationship relates the intrinsic viscosity to the viscosity average molecular weight. If the solvent is a good solvent for the polymer, then the viscosity average molecular weight is approximately equal to the weight average molecular weight.	Viscosity average molecular weight
Gel-Permeation Chromatography	The concentrations of polymer molecules in different size fractions are found by passing a dilute polymer solution through a column containing porous gel beads. Larger molecules pass through the gel quickly as they cannot diffuse into the fine pores, while smaller molecules require more time to pass through the column. The result is a molecular weight distribution. The technique must be calibrated.	Molecular weight distribution based on number or weight

TABLE 2.15 Important Temperatures for Thermoplastic Polymers

Temperature	Definition	Characterization method
Glass Transition Temperature	Temperature above which cooperative motion of long chains occurs	Differential scanning calorimetry; dynamic mechanical analysis
Heat Distortion Temperature	Temperature above which material deforms significantly under a mechanical load	Deflection measurements under controlled temperature conditions
Melting Temperature	For semi-crystalline polymers, the melting temperature corresponds to a phase transition in the crystalline regions, leading to a drop in viscosity. The viscosity of amorphous polymers decreases gradually as temperature increases.	Differential scanning calorimetry, rheological measurements
Degradation Temperature	Temperature above which the polymer undergoes chemical reactions leading to degradation or decomposition, which is manifested in emission of volatile fragment	Thermogravimetric analysis

the polymer is heated. Table 2.15 shows these temperatures and how they are characterized. For processing of the thermoplastic polymer, there is a window of temperatures between the melting temperature and the decomposition temperature. In this "processing window," the polymer melt is easily processed. The rheological or flow behavior of a thermoplastic polymer melt determines the conditions needed for the forming operation. Hence, characterization of melt rheology is essential. This important topic is discussed in Chapter 3.

Lastly, the properties of the polymer appear on the data sheet. Characterization of mechanical, thermal, and electrical properties follows standard methods, some of which are elaborated in ASTM standards.

2.4.3 Thermoset Polymer Starting Materials

Starting materials for thermoset polymers are reactants that form a crosslinked polymer at the appropriate time. These reactants may be monomers, oligomers, or polymers. Thermoset starting materials or resins are either premixed and ready for curing (crosslinking reactions are typically triggered by heating) or they are multi-part kits that are mixed shortly before use. Thermoset starting materials may be liquid or solid. In this section, some general features of thermoset polymers are introduced followed by a specific example of thermoset polymer starting material and curing reaction, data sheets, and characterization methods.

Thermoset starting materials are designed to provide control of chemical reactions. These reactions take place before, during, and after the forming operation. This feature is in contrast to the processing of thermoplastic polymers, in which the chemical reactions that form the polymer are carried out before processing and further chemical reaction during and after processing is usually not desired. Some chemical reaction and growth of molecular weight can be designed into the starting material. Based on their extent of initial chemical reaction, thermoset starting materials are categorized by "stage." Stage A resins are those in which there has been no crosslinking and stage B resins are partly polymerized. Stage C resins are fully crosslinked and therefore, refer to the final polymer. Shelf life is an important property as it relates to staging. If there is little or no capacity for crosslinking reactions at storage conditions (e.g., stage A resin), then the starting material has a long shelf life.

A key feature of a thermoset polymer is the crosslink density. The chemical structure and composition of the starting materials, as well as the curing conditions, determine the crosslink density, which in turn governs many of the physical properties of the resulting polymer. For example, epoxy resin formulations are chosen to provide a highly crosslinked, glassy polymer or a less crosslinked, flexible polymer. The formation of a crosslinked structure on curing is typically a consequence of the functionality of one or more of the reactants. The functionality refers to the number of sites at which reaction can occur per molecule of reactant. Thermoplastic polymers are based on linear molecules that are formed from bifunctional molecules. A linear chain grows by reactions that take place at one end or the other. For thermosets, trifunctional and tetrafunctional reactants are common. The multiple sites for reaction lead to interconnections between chains or crosslinks.

Crosslinked, thermoset polymers are formed by both chain growth and step growth modes of polymerization. In chain growth polymerization, monomers are added onto one end or the other of a growing chain. Such a mechanism necessarily results in a linear polymer, which would be a thermoplastic; however, if $C = C$ bonds remain in the chain even after attack at the ends, then they are sites for further reaction and crosslinking. However, most thermosets are created by step growth polymerization, as illustrated in the following example.

Example – Phenolics. Phenolics have a rich history and current importance. Phenol-formaldehyde polymer was first produced in 1907 by L. Baekeland, who named the product Bakelite. In fact, this polymer is one of the first synthetic polymers. As a thermoset family, phenolics are those polymers produced by a polycondensation between phenol and an aldehyde, the most common of which is formaldehyde. Phenolic resins are still among the top selling thermosets in the United States. The popularity of this thermoset is based on the low cost of the reactants and the polymer properties. Phenolics are used in a variety of applications, including adhesives, paints and coatings, photoresists, binding agents for coated abrasives and laminates, and as matrices for composites.

There are three important components in phenolic resin starting materials: phenol [C_6H_5OH], formaldehyde [CH_2O] and hexamethylenetetramine [HEXA, $(CH_2)_6N_4$]. Phenol is crystalline solid at room temperature, but its melting point is low enough (41°C) that it is typically used as a liquid. Phenol is prepared from the distillation of coal tar or more commonly, by synthetic routes beginning with benzene [C_6H_6] or chlorobenzene [C_6H_5Cl]. In the Dow process, for example, chlorobenzene is reacted with sodium hydroxide at 300°C and high pressure (28 MPa) to produce phenol and sodium chloride. Phenol is a trifunctional molecule with respect to the reaction with formaldehyde, as shown in Figure 2.27. Formaldehyde is produced by the oxidation of methanol in the vapor state at 300–650°C in the presence of a catalyst. The oxidation reaction produces both formaldehyde and water; formalin, the aqueous solution of formaldehyde after some purification, is commonly used in preparation of phenolic resins. Lastly, HEXA is created by the reaction of formaldehyde solution and ammonia gas.

Phenolic resin starting materials are created by the controlled reaction of phenol and formaldehyde. The first step in the reaction, as shown in Figure 2.28a, is the formation of a hydroxymethyl-substituted molecule. This molecule then undergoes a condensation reaction with a phenol (Figure 2.28b). The extent of these reactions before the forming and final curing varies. There are two general categories of resins based on how these reactions are carried out: resoles and novolacs.

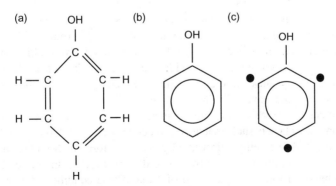

FIGURE 2.27 The structure of phenol (C_6H_5OH): (a) chemical structure, (b) shorthand version of structure, and (c) shorthand version with reactive sites shown.

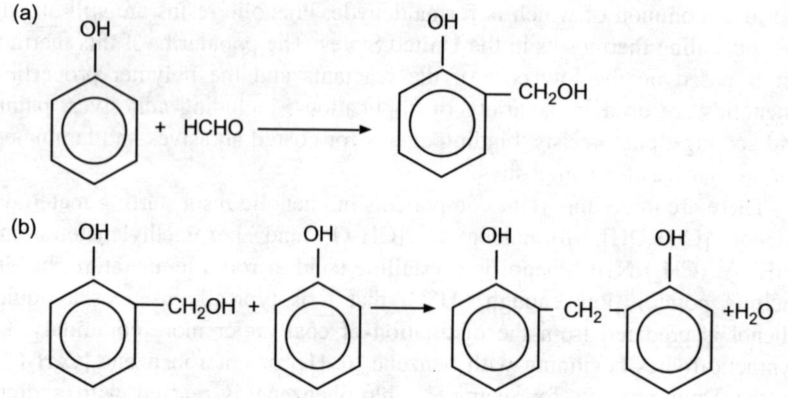

FIGURE 2.28 (a) The reaction between phenol and formaldehyde to form a hydroxymethyl-substituted intermediate and (b) condensation reaction between phenol and hydroxymethyl-substituted intermediate. *Adapted from Hesse and Kalle-Albert (2001).*

Resoles are liquid or solid resins prepared by reacting phenol and formaldehyde to build up oligomers (B stage) that are then capable of reacting, without further additives, to form a crosslinked polymer during and after the forming operation. The resin is formed by reacting formalin with phenol in a reaction vessel at ~100°C in the presence of an alkaline catalyst. See Figure 2.29. The phenol to formaldehyde ratio is chosen to be 1:1 to 1:1.5. Under basic conditions, the reaction to form the hydroxymethyl-substituted molecule (Figure 2.28a) is fast, but the condensation reaction (Figure 2.28b) is slow. The reaction vessel is equipped with a condenser so that the reaction by-product, water, can be removed. Cooling stops the reaction and the reaction mixture, a water solution containing the oligomers, is used as a starting material, or the water is removed by vacuum to create a neat liquid or solid, depending on the molecular weight of the oligomer. Because the condensation reactions are slow and the phenol to formaldehyde ratio is 1:1 or greater, the oligomeric product contains hydroxymethyl-substituted groups. The resole starting material, therefore, is capable of continued crosslinking without any additional crosslinking agent added. Resoles are known as "one-step" phenolics, and because of their chemical reactivity, they are produced in small quantity and have a limited shelf life.

Novolacs are created in a similar manner as resoles except the reaction conditions are chosen such that low molecular weight polymers containing no reactive hydroxymethyl-substituted groups are formed. Novolacs are prepared by reacting phenol and formalin in the presence of an acidic catalyst, with a phenol to formaldehyde ratio of 1:0.8. These conditions favor a slow reaction to form the hydroxymethyl-substituted phenol (Figure 2.28a), but a fast reaction to form the condensed product (Figure 2.28b). The reaction

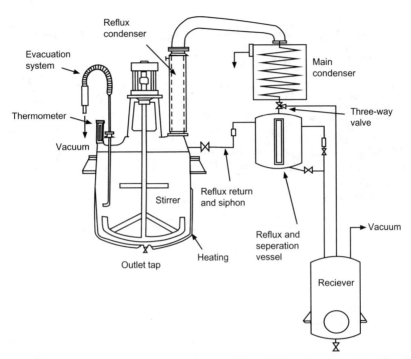

FIGURE 2.29 Phenolic resin production vessel. *E. S. Wilks, Ed.: Industrial Polymers Handbook, Volume 2. p. 1137. 2001. Copyright Wiley-VCH Verlag GmbH & Co. KGaA. Reproduced with permission.*

is carried out at ~160°C with the water by-product removed continually. The product in this case is a low molecular weight thermoplastic polymer (sometimes called a "prepolymer") that does not contain the hydroxymethyl-substituted groups. Hence, novolacs cannot form crosslinked polymers without the addition of a curing agent. To make a novolac resin, the thermoplastic solid polymer is crushed and mixed with HEXA curing agent in an amount of 10 to 15 parts per 100 parts by weight of resin. Because the HEXA curing agent is inactive under typical storage conditions, novolacs have an infinite shelf life.

Resole and novolac resins are converted into crosslinked polymers during and after the forming operation. In resoles, the curing reaction is essentially a continuation of the condensation shown in Figure 2.28b. The fact that excess formaldehyde is used results in more than one crosslinking point on the phenol molecule. The reaction is accelerated by heat and results in the production of water. Novolac resins contain HEXA, which creates crosslinks at about 150°C. While the mechanism of crosslinking is not well understood it appears that HEXA decomposes to formaldehyde and ammonia at high temperature, creating the potential for forming crosslinks.

TABLE 2.16 Formulation of a General Purpose Novolac Molding Resin[a]

Component	Purpose	Quantity (parts per hundred by weight)
Novolac resin	Prepolymer starting material	100
HEXA	Crosslinking agent initiated by heating	15
Wood Flour	Filler, enhances mechanical properties, reduces shrinkage.	120
Calcium oxide or magnesium oxide	Basic additive that accelerates cure	2
Magnesium stearate	Lubricant	2
Dyes or pigments	Colorants	4

[a]Adapted from Ibeh (1998).

Additives are commonly employed in phenolics. Table 2.16 shows the formulation of a typical novolac molding resin. In particular, inert fillers are used to improve mechanical properties and lower the overall shrinkage on curing. Wood flour, which is a purified form of saw dust, serves as an inert filler in many phenolic molding compounds. An accelerator, in this example, calcium oxide, is used to speed the HEXA-based curing reaction. Phenolic resins typically have good adhesive properties, which must be combated with lubricants in molding operations.

Mixing additives into phenolics and other thermoset starting materials is an easier process than mixing the same types of additives into thermoplastics. Thermoset resins are based on low molecular weight materials that are either low viscosity liquids at room temperature or they enter into a similar liquid state at relatively low temperatures.

Thermoset Polymer Data Sheets and Characterization. Data sheets for thermoset polymer starting materials contain information on recommended reaction and processing conditions and expected final properties of the polymers. Figure 2.30 shows an example of a phenolic resin data sheet. Information on two grades of material developed for two forming processes is provided, with properties being the central feature. In this example, particular molding conditions are not given but these details are often included. Notice that additives and the details of the chemistry are not included.

The characterization of thermoset starting materials is focused on the evolution of properties with curing. Data on the rate of curing under different molding conditions is needed in order to create a desirable final product. Such characterization may be specified in part by the supplier, or the molder carries out test runs.

DUREZ
00118
Black Phenolic

Durez 118 Black Phenolic is a two-stage, general purpose molding material. It exhibits improved impact strength and resistance to flexural fatigue for demanding automotive, electrical and appliance applications. Shrinkage and mechanical strengths are closely controlled to meet part reliability requirements.

Revision Date Dec 18, 2014

Plasticities available for compression, transfer, and injection molding.

Form of Material Granular   Feeding & Preforming Good   Storage Life One Year

Typical Properties		Compression		Injection Grade	
		International Units	English Units	International Units	English Units
Physical	Specific Gravity (D792)	1.40	1.40	1.40	1.40
	Apparent Density (D1895)	0.58 g/cc	0.58 g/cc	0.58 g/cc	0.58 g/cc
	Molding Shrinkage* (D6289)	0.007 m/m	0.007 in/in	0.0110 m/m	0.0110 in/in
	Water Absorption (D570)	0.50 %	0.50 %	0.50 %	0.50 %
Mechanical	Tensile Strength (D638)	55 Mpa	8,000 psi	48 Mpa	7,000 psi
	Flexural Strength (D790)	76 Mpa	11,000 psi	69 Mpa	10,000 psi
	Compressive Strength (D695)	207 Mpa	30,000 psi	207 Mpa	30,000 psi
	Tensile Modulus (D638)	9.6 Gpa	1.4×10^6 psi	8.3 Gpa	1.2×10^6 psi
	Izod Impact (D256)	18.1 J/m	0.34 ft lb/in	16.5 J/m	0.31 ft lb/in
Thermal	Deflection Temperature (D648)	171 °C	340 °F	149 °C	300 °F
	UL Flammability (UL-94) @	1.5 mm	HB	1.5 mm	HB
	For complete UL Listing for this material refer to the UL web Site www.ul.com	3.0 mm	V - 1	3.0 mm	V - 1
		6.0 mm	V - 0	6.0 mm	V - 0
	UL Temperature Index (Elect) @	3.0 mm	150 °C		150 °C
Electrical	Dielectric Strength (D149)				
	Short Time	15.7 MV/m	400 V/mil	8.8 MV/m	225 V/mil
	Step by Step	13.8 MV/m	350 V/mil	6.9 MV/m	175 V/mil
	Dissipation Factor (D150)1 MHZ	.05	.05	.06	.06
	Dielectric Constant (D150)1 MHZ	4.7	4.7	5.5	5.5
	Volume Resistivity(ohms)(D257)	10.0×10^{10} m	10.0×10^{12} cm	0.1×10^{10} m	0.1×10^{12} cm

Properties determined with test specimens molded at 340-350°F *Typical transfer-molded shrinkage is 0.009 in/in or m/m

Other Properties
ASTM D-5948 (former Mil-M-14G), Type CFG certification requires batch testing

RoHS2 (2011/65/EU) Compliant
REACh (EC 1907/2006) SVHC / Annex XIV Compliant
HR4173 Sec.1502 Conflict Materials Compliant
Halogen Free

IMPORTANT! The information presented herein, while not guaranteed, was prepared by technical personnel and is true and accurate to the best of our knowledge. No warranty or guaranty, expressed or implied is made regarding performance stability or otherwise. This information is not intended to be all inclusive as the manner and conditions of use, handling, storage, and other factors may involve other or additional safety or performance considerations. While our technical personnel will be happy to respond to questions regarding safe handling and use procedures, safe handling and use remains the responsibility of the customer. No suggestions for use are intended as and nothing herein shall be construed as a recommendation to infringe any existing patents or violate any Federal, State or local laws.

Sumitomo Bakelite North America, Inc.
Corporate Office and Technical Market Development Center
46820 Magellan Drive, Suite C, Novi, MI 48377
Tel. 1-248-313-7000

Durez Canada Company Ltd.
100 Dunlop Street, P.O. Box 100
Fort Erie, Ontario L2A 5M6
Tel. 1-905-346-8700

www.sbhpp.com

FIGURE 2.30 Data sheet for a phenolic molding resin. *Courtesy of Sumitomo Bakelite North America, Inc., www.sbhpp.com.*

2.5 SUMMARY

Starting materials are ready to be used in a forming operation and designed to provide the required properties in the final product. The synthesis of starting materials from raw materials found in nature (minerals, ores, oil, natural gas, etc.) encompasses a wide variety of techniques and methods. The common goal of all these methods is to create starting materials with the structure, composition, and properties required both by the shape forming operation and the final product. To this end, a starting material may be formulated with additives to modify its processing behavior or the final properties of the piece made from it. Characterization of starting materials is therefore essential to the forming operation and the final product.

Metal starting materials are either bulk pieces of alloys or powders. Bulk metals are starting materials for melt casting and solid deformation processes. Metal powders are used in powder metallurgy forming processes, such as pressing, and additive manufacturing. Both types of metal starting materials originate from ores that contain metal compounds. Extractive metallurgical processes convert the metal compounds to elemental metals or alloys. For both ferrous and ferrous alloys, the control of composition is essential and made rigorous by numbering or coding systems. In the final stage of most extractive metallurgy processes, the metal is molten. Bulk metal starting materials are then either cast into standard shapes, such as billets and slabs, or cast and then deformed into wrought semi-finished shapes, such as rods. Metal powders are created from bulk metals by melting the metal and atomizing it into droplets that are solidified. Alternatively, metal powders are synthesized by chemical route, such as reduction of metal oxide powders.

Ceramic starting materials originate from minerals and ores mined from the earth, and sometimes from specialty chemicals. Ceramics are classified as oxides and nonoxides, based on chemistry and polycrystalline or glass, based on structure. For polycrystalline ceramics, ceramic powder starting materials are synthesized by mechanical and chemical treatment of ores into higher purity powders or by chemical processing routes involving liquid or gas phase reactions. The powder composition is important and typically follows the chemical formula with information on the impurity content, rather than a numbering system. For ceramic glasses, a glass batch is the starting material. Some of the ingredients in a glass batch are minerals that have undergone some mechanical and chemical preparation steps, and more well-controlled metal oxides. The glass batch is mixed and heated to create molten glass that is then typically formed directly into shapes. Alternatively, glass powders or frit can be made from the melt and then used as starting materials for glass forming operations. Ceramic glasses have a wide range of compositions that is not restricted to the stoichiometry of a crystalline phase; numbering systems or names are given to glass compositions.

Polymer starting materials are mainly derived from oil and natural gas raw materials. Monomers, oligomers, and chemicals synthesized from these raw materials are then the starting point for polymerization reactions. Thermoplastic starting materials with controlled molecular weight distribution and macromolecular structure are synthesized by addition polymerization or step growth polymerization. Polymerization methods, such as bulk polymerization and solution polymerization, are chosen to create the controlled structure. The thermoplastic resin is sold as pellets, granules, or flakes and is typically referred to by a general name, such as polycarbonate, or a trade name such as Lexan. For some processes, thermoplastic filaments and sheets are the starting materials. Thermoset starting materials are reactants that are converted into a final crosslinked polymer structure during or after the forming process. Oligomers or prepolymers are often created by step growth polymerization and then combined with other starting materials in appropriate ratios to create the desired crosslinked structure by the end of the forming operation. Thermosets are known by their general family names, such as epoxies or phenolics.

BIBLIOGRAPHY AND RECOMMENDED READING

Metals

Beddoes, J., Bibby, M.J., 1999. Principles of Metal Manufacturing. John Wiley & Sons, New York, NY.

German, R.M., 1984. Powder Metallurgy Science. Metal Powder Industries Federation, Princeton, NJ.

Gilchrist, J., 1989. Extractive Metallurgy. Perganon Press, Inc, Elmsford, NY.

Davis, J.R. (Ed.), 1998. ASM Metals Handbook: Desk Edition. second ed. ASM International, Materials Park, OH.

Kaye, B.H., 1997. Powder Mixing. Chapman and Hall, New York, NY.

Lee, P.W. (Ed.), 1998. ASM Metals Handbook, Vol. 7: Powder Metallurgy Technologies and Applications. ASM International, Materials Park, OH.

Mangonon, P.L., 1999. The Principles of Materials Selection for Engineering Design. Prentice Hall, Upper Saddle River, NJ.

Neely, J.E., 1994. Practical Metallurgy and Materials of Industry. Prentice Hall, Englewood Cliffs, NJ.

Rhodes, M.J. (Ed.), 1990. Principles of Powder Technology. John Wiley & Sons, New York, NY.

Van der Voort, G.F. (Ed.), 2004. ASM Metals Handbook, Vol. 9: Metallography and Microstructures. ASM International, Materials Park, OH.

Ceramics

Boyd, D.C., Danielson, P.S., Thompson, D.A., 1994. Glass. In: fourth ed. Kroschwitz, J.I. (Ed.), Kirk-Othmer Encyclopedia of Chemical Technology, Vol. 12. Wiley, New York, NY, pp. 557–627.

Reed, J.S., 1995. Principles of Ceramic Processing, second ed. Wiley Interscience, New York, NY.

Ring, T.A., 1996. Fundamentals of Ceramic Powder Processing and Synthesis. Academic Press, New York, NY.

Hart, L.D. (Ed.), 1990. Alumina Chemicals. American Ceramic Society, Westerville, OH.

Schneider, S.J., 1991. ASM Engineered Materials Handbook: Ceramics and Glasses, Vol. 4. ASM International, Materials Park, OH.

Polymers

Brydson, J., 1999. Plastics Materials, seventh ed. Butterworth-Heinneman, Boston, MA.

Fried, J.R., 1995. Polymer Science and Technology. Prentice-Hall, Englewood Cliffs, NJ.

Goodman, S.H. (Ed.), 1998. Handbook of Thermoset Plastics. second ed. Noyes Publications, Westwood, NJ.

Grulke, E.A., 1994. Polymer Process Engineering. Prentice-Hall, Englewood Cliffs, NJ.

Rogers, M.E. (Ed.), 1990. Encyclopedia of Polymer Science and Technology. Wiley, New York, NY.

Simpson, D.M., Vaughn, G.A., 2001. Ethylene polymers, LLDPE. In: Encyclopedia of Polymer Science and Technology, John Wiley & Sons.

Strong, A.B., 2000. Plastics Materials and Processing, second ed. Prentice-Hall, Upper Saddle River, NJ.

Ulrich, H., 1982. Introduction to Industrial Polymers. Macmillan Publishing Co., Inc., New York, NY.

Wilks, E.S. (Ed.), 2001. Industrial Polymers Handbook—Products, Processes, Applications. Wiley & Sons, New York, NY.

CITED REFERENCES

Burdette, I., 2008. New innovations drive gas phase PE technology. Hydrocarbon Eng. 13, 67–76.

Bauer, W.C., Baily, J.E., 1990. Raw materials/batching. In: Schneider, S.J. (Ed.), ASM Engineered Materials Handbook, Vol. 4: Ceramics and Glasses. ASM International, Materials Park, OH, pp. 1115–1170.

Fried, J.R., 1995. Polymer Science and Technology. Prentice-Hall, Englewood Cliffs, NJ.

German, R.M., 1984. Powder Metallurgy Science. Metal Powder Industries Federation, Princeton, NJ.

Hesse, W., Kalle-Albert, W., 2001. Phenolic resins. In: Wilks, E.S. (Ed.), Industrial Polymers Handbook—Products, Processes, Applications. Wiley & Sons, New York, NY, pp. 1129–1151.

Ibeh, C.C., 1998. Phenol-formaldehyde resins. In: Goodman, S.H. (Ed.), Handbook of Thermoset Plastics, second ed. Noyes Publications, Westwood, NJ, pp. 23–71.

Kissin, Y., 2000. Polyethylene, linear low density. In: Kroschwitz, J.I. (Ed.), Kirk-Othmer Encyclopedia of Chemical Technology. Wiley, New York, NY, pp. 23–71.

Neely, J.E., 1994. Practical Metallurgy and Materials of Industry. Prentice Hall, Englewood Cliffs, NJ.

Svarovsky, L., 1990. Characterization of powders. In: Rhodes, M.J. (Ed.), Principles of Powder Technology. John Wiles & Sons, New York, NY, pp. 35–69.

US DOE/OIT, 2002. Energy and environmental profile of the U. S. glass industry, US DOE/OIT, April 2002. < http://www.energy.gov/sites/prod/files/2013/11/f4/glass2002profile.pdf >.

Whitely, K.S., Heggs, T.G., Kochm, H., Mawer, R., Immel, W., 2001. Polyolefins. In: Wilks, E.S. (Ed.), Industrial Polymers Handbook—Products, Processes, Applications. Wiley & Sons, New York, NY, pp. 641–787.

QUESTIONS AND PROBLEMS

Questions

1. What distinguishes a starting material from a raw material? Give examples of each from metals, ceramics, and polymers.

2. Why is characterization important for starting materials?
3. What is the UNS code? Why are coding systems important for metal alloys?
4. Find the compositions of the following alloys: (a) SAE-AISI 1055, (b) SAE-AISI 4620, (c) Aluminum Association No. 2024, (d) UNS Z13001, (e) UNS G12144, and (f) UNS N02200.
5. What are the two common forms for metal starting materials?
6. What do the three types of extractive metallurgy processes have in common? What are some differences?
7. Describe the purpose and operation of the blast furnace, basic oxygen furnace and electric arc furnace in steel making. List the inputs and outputs from each furnace.
8. Discuss the special roles of carbon and oxygen in steel making.
9. In steel, what is the difference between an impurity and an alloying element? List several impurities found in steel and several alloying elements.
10. What are the advantages of continuous casting as compared with ingot casting? What are the disadvantages?
11. What category of extractive metallurgy process is used to produce aluminum?
12. Describe two common methods for producing metal powders.
13. Give several types of information that might be found on a data sheet for a metal powder. For each describe why the information is important.
14. A data sheet shows a weight-based median diameter of 100 μm, but examination of the powder in an electron microscope shows a multitude of smaller particles (less than 50 μm). How can this be?
15. Give several classifications or ways of categorizing ceramic materials in general and ceramic starting materials.
16. What are the four different methods for producing ceramic powder?
17. Why are the particle sizes for ceramic powders much less than those typically found for metal powders?
18. List several characteristics of a glass that are important to the subsequent processing.
19. Powders of all types may need to be size reduced before they are processed. (a) Wet milling is typically preferred over dry milling for ceramics. Why? (b) What factors influence the final size and size distribution from a milling operation? (c) Why might it be difficult to size reduce some polymers and metals?
20. List the main differences between thermoplastic and thermoset polymers, including structure and properties.
21. Find the IUPAC name and trade names for the following polymers: (a) polypropylene (PP), (b) polytetrafluoroethylene (PTFE), and (c) polyethyletherketone (PEEK).

22. Sketch the effect of temperature on the viscosity of thermoplastic and thermoset starting materials (i.e., as they are used in making shapes).
23. List three examples of polymers produced by addition polymerization and three examples of polymers produced by condensation polymerization. Do not use examples from the text.
24. Describe the reactors and polymerization methods used to produce LDPE and HDPE. How is the structure controlled?
25. What is the difference between an A stage, B stage, and C stage thermoset materials? What are the advantages and disadvantages of working with A and B stage thermosets as starting materials?
26. What important information is typically not found on polymer data sheets?
27. Discuss the similarities between the following: (a) characteristics important to ceramic glass starting materials and thermoplastic polymer starting materials, (b) plant layouts for steel making and glass making, and (c) methods used for analyzing (not characterizing) particle size distribution and molecular weight distribution.
28. Data sheets vary considerably in their content for the different starting materials. (a) For bulk metals, ceramic powders, and thermoplastic polymers compare the amount and type of information found on the composition, structural characteristics, and properties of the starting material. Give reasons for the differences that you note. (b) Also compare the amount and type information given on the processing of the starting material into the final product. Give reasons for the differences that you note.
29. Additives are used in many starting materials. For each of the following starting materials, list two types of additives, discuss their purpose, and give a specific example: (a) thermoplastic polymers and (b) ceramic powders.
30. Give examples of recycling in the production of starting materials.

Problems

1. The synthesis of aluminum from aluminum oxide is energy intensive, and recycling (remelting) is less expensive. Calculate the energy required to heat 1 kg of Al to its melting point and compare that to the energy required to value quoted in the text for Al smelting. The constant pressure heat capacity and the heat of fusion of Al are:

$$C_p = 20.67 + 0.01238T \; (J \; mol^{-1}K^{-1}) \quad \Delta H_f = 10,700 \; (J \; mol^{-1})$$

2. Find information on the synthesis of nickel alloys from nickel ores. (a) What ores are used as the raw materials? (b) Describe one process for converting the ore into Ni alloy. Include the type of extractive

metallurgy process and a general description of the method, including any important chemical reactions. (c) List your source of information. (d) Cite a common use for nickel alloys.

3. Find information on the process for creating magnesium alloys from brine sources (e.g., seawater). (a) What are the raw materials? (b) Describe one process for converting the raw materials into Mg alloy. Include the type of extractive metallurgy process and a general description of the method, including any important chemical reactions. (c) List your source of information. (d) Cite a use for magnesium alloys.

4. Find data sheets for the following: (a) silver powder, (b) bulk copper alloy, (c) polypropylene pellets, (d) epoxy resin, and (e) strontium titanate powder. For each, list the important characteristics that are provided and indicate if the characteristics are important to the final product made from the starting material or to the processing or both.

5. Data for two commercial alumina powders are shown below. (a) Indicate the most probable method used to determine the particle size. For powder B, there is very little information; suggest a method for determining the particle size of this powder. (b) The data sheet for these alumina powders also explains that the average size of individual crystallites is 3–5 μm. Reconcile this fact with the data provided. (c) Powder bulk densities are provided. Compare these to theoretical (e.g., % theoretical density).

Chemical analysis, wt%	Powder A	Powder B
Al_2O_3 by difference	99.6	99.6
Na_2O	0.18–0.37	0.18–0.37
SiO_2	<0.011	<0.029
Fe_2O_3	<0.015	<0.025
CaO	<0.04	<0.04
Particle Size (cumulative)		
% >150 μm	5	NA
% greater than 75 μm	52–80	NA
% greater than 45 μm	85–95	NA
% less than 45 μm	NA	95–96
Physical Analysis		
Loose Bulk Density (lbs/ft^3)	48	57
Packed Bulk Density (lbs/ft^3)	81	95

6. Find information on the synthesis of silicon carbide from silicon powder and carbon using the Acheson Process. (a) Describe the process by creating a flow diagram. (b) Write down the key chemical reaction(s). (c) What are the impurities that are commonly found in this type of silicon carbide? (d) Which of the four types of ceramic synthesis processes is the Acheson process? (e) Cite a common use for silicon carbide. (f) List your source of information for this question.

7. Particle size data are shown in the table below. (a) Plot a histogram showing number fraction as a function of size. (b) Plot the distribution as

cumulative number % finer than. (c) Find the number-average particle size, number-average standard deviation and estimate d_{10}, d_{50}, and d_{50}.

Particle size interval (μm)	Number counted
0.1–1.0	5
1.0–2.0	50
2.0–5.0	175
5.0–10	80
10–20	10
20–40	2

8. Particle size distribution data for a zircon powder is given in the table below. The data shows the weight of powder, which is captured on sieves of different sizes. (a) Plot a histogram showing weight fraction as a function of size. (b) Plot the distribution as a cumulative weight % finer than as a function of size. (b) Estimate d_{10}, d_{50}, and d_{50}.

Sieve size (μm)	Weight on sieve (g)
1000	0
500	0.46
355	0.94
250	3.10
180	5.30
125	7.34
90	8.08
63	5.00
44	3.50
0	1.00

9. Convert the data from Problem 8 into number data by assuming a spherical particle size and a density of 4.7 g/cm^3. (a) Plot a histogram showing number fraction as a function of size and create a graph showing cumulative number % finer than as a function of size. (b) Compare the d_{50} values from number-based data and weight-based data and comment.

10. Formulate a glass batch using standard raw materials (see Table 2.6) in order to produce 1 ton of Corning 7740 glass. What is the loss on ignition (LOI)?

11. A glass has a composition of 54% SiO_2, 2% Al_2O_3, 7% Na_2O, 8% K_2O, 3.5% CaO, 2.5% MgO, and 23% PbO (by weight). To prepare the glass you must use the raw materials on hand which are shown below. Calculate the weights of the raw materials necessary to prepare a 1000 kg batch.

Material name	Formula
Feldspar	$K_2O \cdot 2Al_2O_3 \cdot 6SiO_2$
Fused Boric Acid	B_2O_3
Red lead	Pb_3O_4
Limestone	$CaCO_3$
Glass Sand	SiO_2

Glassmaker's Potash	$K_2CO_3 \cdot 1.5 H_2O$
Barium Carbonate	$BaCO_3$
Soda Ash	Na_2CO_3
Dolomite	$CaCO_3 \cdot MgCO_3$

12. One method to form TiO_2 powder is a vapor phase reaction between $TiCl_4$ and water vapor, according to the following reaction:

$$TiCl_{4(g)} + 2H_2O_{(g)} \rightarrow TiO_{2(s)} + 4HCl_{(g)}$$

(a) Describe a process to make 1 kg of titania powder. Specify the amounts of raw materials needed, the procedures to make the powder, and the amount of waste generated during the process. (b) Comment on the expected level of purity of the powder and the main types of impurity. (c) Find a data sheet for titania powder. Comment on whether you think the titania powder you found was made by this type of vapor phase process. State your reasons.

13. A polyethylene polymer sample is analyzed and the data is shown in the table below. (a) Plot a histogram of number fraction as function of molecular weight and find the number average molecular weight. (b) Plot a histogram of weight fraction as function of molecular weight and find the weight average molecular weight. (c) Comment on the differences between the two distributions. (d) Find the number average degree of polymerization and the weight average degree of polymerization.

Molecular weight range	Number of molecules
0–10,000	1,000
10,000–20,000	3,000
20,000–30,000	7,000
30,000–40,000	15,000
40,000–50,000	5,000
50,000–60,000	2,000
60,000–70,000	1,000

14. Find information on the synthesis of polyisoprene starting from isoprene monomer. (a) What are the raw materials and the chemical reactions for the synthesis? (b) Describe the process. Which of the four types of polymerization methods is used? (c) How is molecular weight controlled? (d) Cite a common use for polyisoprene. (e) List your source(s) of information for this question.

15. Find information on the synthesis of liquid epoxy resin from epichlorohydrin and bisphenol A. (a) Write down the chemical reaction that forms the resin. (b) How are the raw materials synthesized? (c) What is an "epoxy equivalent weight"? How is it controlled and why is it important? (d) What types of curing agents are used with these epoxy resins? (e) Cite a common use for epoxy resins. (f) List your source(s) of information for this question.

Chapter 3

Melt Processes

3.1 INTRODUCTION

Melt processes convert a liquid melt into a solid article with a defined structure and shape. The physical and chemical characteristics of liquid melts are as diverse as those of their solid counterparts. For example, molten metals have low viscosities and can be poured like water, while molten thermoplastic polymers are so viscous that they have to be forced with pressure to flow at significant rates. On cooling, some melts solidify by crystallization, others by glass transition. In spite of this variety, materials processing operations that employ these melts follow the same general sequence of steps and are built on a common set of fundamental science and engineering concepts.

There are three steps in every melt-based process. The first step is *flow*—the melt flows either due to gravity or an external pressure. The second step, which usually overlaps the first, is the *shape definition*. The melt flows into a mold to create a three-dimensional (3D) shape, through a die or orifice to form a shape with a constant 2D cross-sectional geometry (e.g., tube), or onto a surface to form a slab or sheet. Melts can also be digitally directed to platform to create 3D objects by additive processes. In some processes, further deformation or shaping takes place after some initial shaping step. For example, in extrusion blow molding, an extruded tube of polymer is clamped into a mold and then inflated against the mold to form a hollow shape. The third step, *shape retention*, occurs by cooling the melt so that it solidifies by crystallization or glass transition into a solid. For reactive polymer melts, shapes are retained by cross-linking reactions. To design and control a melt process, the engineer must understand the flow properties and solidification behavior of the melt, and the fundamentals of heat transfer that govern the temperature of the material throughout the process.

There are distinct advantages to working with melts. Molten materials can adopt nearly any shape that can be engineered into a mold or a die. Therefore, melt-based processes are often used for complex 3D shapes. Also, diffusion is faster in the liquid state than in the solid state and therefore compositional changes can be achieved more quickly in melts. For example, alloying elements can be easily added to metal melts to tune composition and properties, but making the same compositional changes in the solid state would require a much longer time.

Most metals are cast from the melt at some stage in their processing. Casting is routinely used as the last step in the extractive metallurgy process that is used to create the alloy starting material itself, as described in Chapter 2. These *primary* casting processes include ingot casting and continuous casting. In this chapter, we focus on the use of melt casting to create a final shape. Melt is poured or forced under pressure into a mold made of sand or metal, and then solidified there by cooling. There are a variety of melt casting processes, typically classified by the type of mold and in some cases by the method of delivering the melt to the mold. After cooling and removal from the mold, the metal part usually requires some machining and surface finishing. Depending on the alloy, heat treatment is also commonly used to adjust microstructure and properties.

Ceramics by virtue of their strong bonds require high temperatures for melting. This fact alone does not prevent the adoption of a melt-based process. Other complications interfere with the use of melt processes for forming polycrystalline ceramics. For example, some metal carbide and metal nitride ceramics require extremely high temperatures (i.e., $>2500°C$) or decompose rather than melt on heating. Many ceramic oxides can be formed into stable melts, but these melts cannot be solidified into fine-grained, defect-free polycrystalline microstructures that are required for most applications. Ceramic glasses, on the other hand, are almost always formed from the molten state. Unlike polycrystalline ceramics, glasses solidify gradually by a glass transition to an amorphous solid state. At high temperature, the melt viscosity is low enough for pouring; on cooling, the viscosity increases to a point at which the melt can be shaped into an object easily and, finally, with more cooling, the viscosity is so high that the glass is a solid and shape is retained. Ceramic glasses are formed into flat, continuous sheets by casting processes, including the float glass process, and into hollow shapes (e.g., bottles) by a process known as blow molding. Thermal annealing and tempering are two common post-processing operations for glasses.

Polymers are also commonly formed from melts. Like ceramic glasses, thermoplastic polymers have temperature ranges for flow, shape formation, and solidification. The two most common processes for forming thermoplastic polymers, extrusion and injection molding, are melt processes. Due to their high viscosities, polymer melts require high pressures to cause the melt to flow through a die or into a mold. The polymer shapes solidify by glass transition or by a combination of glass transition and crystallization. Extrusion and injection molding also create preforms that are further processed by blow molding to make container shapes. Additionally, one of the more common additive manufacturing technologies, fused deposition modeling (FDM), is a melt-based process. Thermoset polymers can be formed by similar injection molding and extrusion processes; however, prepolymer liquids are used and solidification to an amorphous solid occurs by reaction, rather than cooling. A great advantage of these polymer melt-based processes is that the part is ordinarily finished after solidification—no post-processing operations are needed.

The engineering of melt processes involves selecting starting material compositions and conditions for the different stages in the process. Invariably, the processes take place in a series of steps: melting, flow, and solidification, the slowest of which is rate controlling. Predicting the rates of the various steps and choosing the correct conditions for adjusting the overall process as well as the structure and properties of the final part is key.

This chapter begins with the fundamentals of melt processes and then moves on to the major categories of melt-based processes. The fundamentals include characteristics of melts, rheology, heat transfer, and solidification. These fundamental topics are necessary for understanding the processes that follow: shape casting, casting of flat sheets, extrusion, injection molding, blow molding, and additive processes. Post-processing operations are introduced with the forming operations.

3.2 FUNDAMENTALS

3.2.1 Melt Structure and Surface Tension

Melts are engineering materials in the liquid state. The properties of liquid melts are impacted by the size, shape, electronic structure, and bonding of the constituent molecules, ions, or atoms. In this section, the structural features of liquid melts are introduced, followed by discussions of several of the basic characteristics of these melts, including surface tension, density and melting temperature. Melt rheology is treated separately in the next section.

Melt Structure. Figure 3.1 shows cartoons of the structures of melts that are frequently used in melt-based processes: metals, ceramic glasses, and thermoplastic polymers. These schematics provide a general understanding of the structural features that are useful as the characteristics of the melts are introduced.

At their simplest, molten metals are composed of a disorganized array of spherical atoms in thermal motion. There are both attractive and repulsive interactions between adjacent atoms, as in solids, but interatomic separations are larger than in solids and the interactions weaker. The attractive interactions that hold the liquid together are mainly electrostatic, similar to those in solid metals. The ion cores of metallic atoms are attracted to the surrounding sea of electrons. Metal alloys, which are composed of more than one type of metal atom, have more complex structures and interatomic bonds. Clusters of atoms may form in the liquid state.

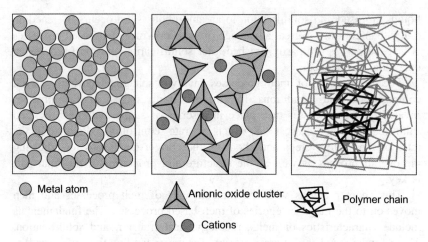

FIGURE 3.1 Schematic structures of liquids; metal (left), ceramic glass (center), and thermoplastic polymer (right). The schematics are not to-scale and include more free space than would be present in the real liquids.

Molten metal structure and properties change with temperature. As the temperature is raised, atomic vibrations and motions increase, atoms are driven further apart and interatomic bonds weaken. The density of the melt drops as a consequence. In contrast, as the melt temperature decreases, the reverse occurs until the melt crystallizes and the atoms become locked into the regular crystal structure of the solid. Some complex alloys can be cooled without crystallization, but these metallic glasses are the exception.

Ceramic glass melts consist of a disorganized array of small anionic oxide clusters interspersed with other cations, known as modifiers. Figure 3.1 depicts the most technologically relevant silicate glass structure, in which the anionic clusters are $[SiO_4]^{-4}$ tetrahedral units, represented in the cartoons as triangles (as if viewed from above) and the other cations, which modify the network created by the anionic units, would typically be alkalis, such as Na^+, and alkaline earths, such as Ca^{2+}. The short-range order of the tetrahedral unit persists in the melt and linkages between these units form dimers, chains, and other multiunit structures. However, the bonding is dynamic; bonds between the clusters continuously break and reform. On cooling, the dynamic melt becomes a glassy solid as the bonds gradually become more persistent. The solid glass consists of an interconnected silicate network with other cations interspersed and held in place by ionic bonds to the network.

Polymer melts are comprised of a disorganized and typically entangled conglomeration of polymer chains. The covalent bonding along the backbone of the chain is relatively stable with only bond rotation affected by temperature changes. The van der Waals bonds between adjacent chains also respond. As temperature increases, bonds weaken and allow more rotation and relative chain motion, leading to easier melt flow. On cooling, the weak bonds between chains strengthen and bond rotation in the backbone becomes less likely, eventually leading to a structural freezing, a transition to a glassy or partially crystalline state. For the partly crystalline materials, the solid structure consists of regions of crystalline order surrounded by glassy, random areas.

Surface Tension. The structure descriptions above paint a general picture of liquid melt structure as dynamic assemblies of atoms or molecules. Now, consider the surface of the liquid. Here, an abrupt change takes place from the liquid to the adjacent vapor phase. In the vapor, atoms and molecules are separated to a much greater degree than they are in the liquid. Therefore, the layer of molecules at the liquid surface is bonded on one side to molecules in the liquid and on the other side to far fewer molecules in the vapor. To compensate, the molecules in the surface layer become more strongly attracted to their neighbors within the liquid surface layer as compared to the molecules beneath, in the bulk of the liquid. The consequence of this imbalance is *surface tension*. See Figure 3.2a.

The mechanical nature of surface tension is best understood with the example of a liquid trapped on a metal frame. Surface tension causes the surface to contract whenever possible; therefore, a force must be placed on such a frame to prevent the liquid from retracting. The force needed to keep the wire of length l in Figure 3.2b in place is:

$$F = \gamma(2l) \tag{3.1}$$

where γ is the surface tension of the liquid with units of force per length (e. g., N/m). The factor of two accounts for the two surfaces of the trapped liquid.

This mechanical approach to understanding surface tension can be connected to an energetic or thermodynamic one. The work done to move the wire by an amount dx is given by:

$$dW = Fdx = \gamma(2l)dx = \gamma dA \tag{3.2}$$

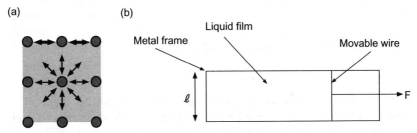

FIGURE 3.2 (a) Cartoon of liquid structure and interactions at the surface as compared with those in the bulk of the liquid. (b) Illustration of the effect of surface tension. To keep the liquid film captured on the metal frame, a force must be exerted.

Therefore, $\gamma = dW/dA$, and for small reversible changes, $dW = dG$, where G is Gibbs free energy and W is the work done. Therefore,

$$\gamma = \left(\frac{\partial G}{\partial A}\right)_{P,\,T,\,\text{composition}} \tag{3.3}$$

Hence, the surface tension is equivalent to a surface energy, which is defined as the change in Gibbs free energy due to a change in surface area, A. The subscripts indicate that this definition is taken at constant pressure, temperature, and composition. To be more precise, the surface energy is actually an interfacial energy. That is, it is the energy of the interface between the liquid and the vapor phase. This distinction is noted by subscripts; γ_{LV} is the interfacial energy between the liquid and the vapor. In this chapter, the subscripts help to distinguish surface energy from shear strain, which is also given the symbol γ.

Therefore, liquid surface tension has a dual nature. On the one hand, it is mechanical in nature with units of force/length. This definition is relevant to understanding how surface tension causes liquids to change shape. On the other hand, surface tension is equivalent to surface energy with units of energy/area. The surface energy represents the energetic penalty for creating more surface, i.e., moving atoms from inside the bulk of the liquid to the surface. The energetic definition is useful in understanding why liquids reconfigure themselves. For example, an irregularly shaped liquid drop suspended in a vapor reconfigures to a sphere to minimize the number of atoms on its surface and therefore minimize its Gibbs free energy. The surface tension force from the mechanical definition is responsible for the movement that takes place in this reconfiguration.

Unlike liquids, solids are not able to reconfigure themselves to minimize the surface contributions to their Gibbs free energy. Therefore a solid does not have a surface tension, but it does have a surface energy, as defined in Eq. (3.3). More properly, the surface energy of a solid is an interfacial energy between the solid and the surrounding vapor, γ_{SV}.

When a liquid drop comes into contact with a solid surface, it adopts a configuration based on the interfacial energies. The vapor, liquid, and solid meet at the "contact line" and at equilibrium, there is a characteristic contact angle, θ, as shown in Figure 3.3. The Young equation gives the relationship between the contact angle and the interfacial energies.

$$\gamma_{SV} = \gamma_{LV}\cos\theta + \gamma_{SL} \tag{3.4}$$

where γ_{SL} is the interfacial energy between the solid and the liquid. This equation is derived from a force balance that assumes all the interfacial energies can be expressed as interfacial tensions. While this assumption is not strictly correct, the equation describes the behavior of liquids on solids accurately and has also been derived by alternative, more rigorous approaches.

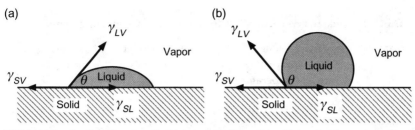

FIGURE 3.3 Configurations for a liquid drop on a solid surface with interfacial energies and contact angles shown: (a) a wetting condition with $\theta < 90$ and (b) a nonwetting condition with $\theta > 90$.

The magnitude of the contact angle defines wetting conditions. If the contact angle is less than 90° (Figure 3.3a), then the liquid spreads on the substrate and is said to "wet" the substrate. Strongly wetting liquids have contact angles close to zero. Nonwetting, on the other hand, is characterized by a contact angle greater than 90° and the liquid beading up (Figure 3.3b).

Surface tensions of common melts are given in Table 3.1, along with density and melting temperature. Molten metals have the highest surface tensions of the melts listed. Ceramic glasses are intermediate and polymers have lowest surface tensions. Understanding the origin of these differences in detail is no small task. Among the important factors are the chemical constitution, density of the material, and bonding strength. Within any given class of melts, trends in surface tension typically correlate well with bonding strength in the melt. For example, the surface tension of elemental metals at temperatures just above their melting points increases with the melting point of the metal. Since the melting point is an indicator of bonding strength in the solid, one can assume that the bonding in the liquid metal follows a similar trend. Stronger bonds in the liquid lead to higher surface tension. There are other factors. In silicate glasses, chemical composition of the glass strongly influences the surface tension. In polymer melts, the surface tension depends on density, molecular weight, and chemical composition.

For most liquids, surface tension drops as temperature increases. See Figure 3.4a for examples. This general change follows the expectation based on a correlation between bonding strength and surface tension. As temperature rises and thermal expansion takes place, the distance between atoms or molecules in the liquid increases, density decreases and the bond strength decreases, and therefore surface tension also decreases. There are exceptions to this general trend. Namely, some glass forming oxide melts, such as SiO_2 and B_2O_3, have positive temperature coefficients for surface tension, while common, more complex glass compositions, such as soda lime glass, have the usual negative coefficient.

Surface tension is also sensitive to also small changes in the chemical composition of the melt. Impurities and dopants frequently segregate to surfaces and change the local bonding. For example, sulfur impurities in steel

TABLE 3.1 Characteristics of Melts for a Variety of Engineering Materials

Material	Melting or processing temperature (°C)[a]	Surface tension of melt (mN/m)[b]	Density of melt (g/cm³)[b]
Metals[c]			
Aluminum	660	914	2.38
Copper	1083	1285	8.00
Iron	1536	1872	7.02
Tungsten	3400	2500	17.6
Ceramic Glasses[d]			
B_2O_3	700	83 (@1000°C)	NA
Soda-Lime-Silica Glass	1250	304 (@1300°C)	2.31 (@1400°C)
"E glass"	1600	320 (@1400°C)	2.47 (@1400°C)
SiO_2	2000	307 (@1800°C)	2.08 (@2000°C)
Thermoplastic Polymers[e]			
High Density Polyethylene	160–320	25.5 (@200°C)	0.94–1.05
Polypropylene	90–320	19.3 (@200°C)	0.89–1.70
Polycarbonate	220–330	32.1 (@200°C)	0.96–1.02
Polystyrene	190–225	27.8 (@200°C)	0.94–1.00

[a] Ceramic glass melting points taken as the temperature at which viscosity is ~100 Pa•s using data tabulated in Martlew (2005). Processing temperature ranges are given for polymers based on information on datasheets found on www.matweb.com.
[b] Values at temperatures close to melting or process temperature unless otherwise specified.
[c] Data tabulated in Brandes (1983).
[d] Surface tension data from Kingery (1959), Badger et al. (1938), and Clare et al. (2003) and density data tabulated in Mazurin (2005).
[e] Surface tension data tabulated in Brandrup et al. (1999) and density data from data sheets found on www.matweb.com (NA = not available).

melts lower the surface tension and change the temperature dependence of surface tension. See Figure 3.4b for an example.

Surface tension leads to pressure differences across curved liquid surfaces. The simplest example is a bubble. Consider a bubble of radius R in a melt or liquid. See Figure 3.5. Surface tension tends to act to decrease the size of the bubble, but as this occurs the pressure in the bubble rises. So the pressure on the inside of the bubble is higher than on the outside; $\Delta P = P_1 - P_2$. Qualitatively, the pressure difference should increase with the surface tension

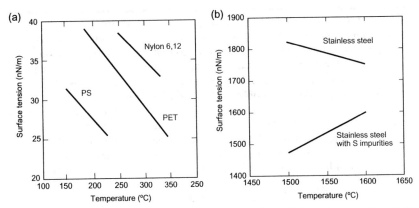

FIGURE 3.4 Effect of temperature on the surface tensions of (a) polymer melts and (b) stainless steel melts, showing effect of sulfur impurities. *Plotted using fitting parameters from (a) Sauer and Dee (2002) and (b) Li et al. (2005).*

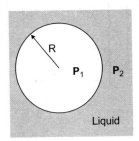

FIGURE 3.5 Cartoon of a gas bubble in a melt or liquid.

of the liquid. A quantitative relationship can be found by examining equilibrium where the work of expansion (or contraction) of the gas in the bubble is balanced with the work needed to increase (or decrease) the surface area:

$$\Delta P dV = \gamma_{LV} dA$$

$$\Delta P(4\pi R^2 dR) = \gamma_{LV}(8\pi R dR)$$

$$\Delta P = \frac{2\gamma_{LV}}{R} \tag{3.5}$$

Smaller bubbles have higher internal pressures than larger bubbles. Equation 3.5 is the basis for one method to measure surface tension (Example 3.1). In a more general sense, a surface with two principle radii of curvature, R_1 and R_2, experiences a pressure difference of:

$$\Delta P = \gamma_{LV}\left(\frac{1}{R_1} + \frac{1}{R_2}\right) \tag{3.6}$$

EXAMPLE 3.1 One method used to measure surface tension is the maximum bubble pressure technique. A thin capillary of radius R, which is composed of a nonreactive material, such as platinum, is immersed into a melt so that the tip is at a known distance, h, below the surface. The gas pressure is measured as a bubble is created. The pressure reaches a maximum, P_{max}, when the bubble at the end of the capillary is a hemisphere, and essentially breaks loose from the capillary. (a) Derive an expression for the determination of the surface tension. (b) Find the P_{max} required to measure the surface tension in a soda-lime-silica melt at 1300°C and that required for an iron melt at 1540°C, assuming the capillary is radius is 1 mm and it is immersed 1 cm below the surface.

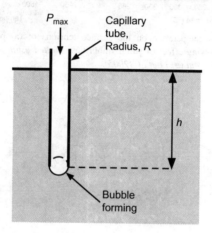

FIGURE E3.1

a. To create the hemisphere/half bubble at the end of the capillary, the pressure must first drive the gas beneath the surface of the liquid. When the gas just reaches the distance h beneath the surface and the interface is flat, the pressure in the gas is equal to the hydrostatic pressure in the liquid at that depth: $\rho g h$ where ρ is the density of the liquid and g is the acceleration due to gravity (9.8 m/s²). Next pressure is supplied to create the curved surface, as in Eq. (3.5).

$$P_{max} = \rho g h + \frac{2\gamma_{LV}}{R}$$

b. Using the data in Table 3.1,
   Glass:

$$P_{max} = (2310 \text{ kg/m}^3)(9.8 \text{ m/s}^2)(0.01 \text{ m}) + \frac{2(0.304 \text{ N/m})}{(0.001 \text{ m})} = 226 + 608 = 834 \text{ Pa}$$

Iron:

$$P_{max} = (7015 \text{ kg/m}^3)(9.8 \text{ m/s}^2)(0.01 \text{ m}) + \frac{2(1.872 \text{ N/m})}{(0.001 \text{ m})} = 687 + 3744 = 4431 \text{ Pa}$$

3.2.2 Melt Rheology

The term "rheology," coined by Bingham in the early 1900s, is defined as the study of the deformation and flow of matter. This definition is broad. In this chapter, the focus is on the rheology of engineering materials in the liquid or molten state.

Shear Viscosity. One fundamental distinction between a solid and a liquid is the ability to withstand a shear stress. In Figure 3.6a, a shear stress is applied to a solid element. We can physically picture this action as a force, F, in the x direction acting on the top face of the solid element, which has an area A. The resulting shear stress is given by $\tau = F/A$.

The solid resists the shear stress, deforming elastically by a small amount and then stopping. The deformation is characterized by shear strain, γ, which is given by:

$$\gamma = \tan \Delta\phi \cong \Delta\phi \quad (3.7)$$

where $\Delta\phi$ is the angle defined in Figure 3.6. The approximation in Eq. (3.7) holds for small deformations. For elastic solids, the shear stain is proportional to the shear stress, τ:

$$\tau = G\gamma \quad (3.8)$$

This relationship is Hooke's law in shear and the constant of proportionality is the shear modulus, G, a property of the solid. The shear modulus gives a measure of the solid's ability to resist shear stress. When the force is

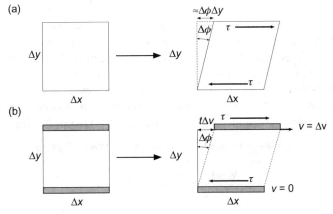

FIGURE 3.6 The action of a shear stress imposed by a moving plate on (a) an elastic solid and (b) a liquid. The solid resists the shear stress and deforms a small amount and then stops. The liquid does not resist the shear stress and instead continues deforming.

removed, the elastic solid element returns to its original shape. Work is done on the element in the amount given by:

$$\text{Work} = (\text{Force})(\text{Displacement}) = (\tau A)(\gamma \Delta y)$$

$$\frac{\text{Work}}{\text{Volume}} = \frac{(\tau A)(\gamma \Delta y)}{A \Delta y} = \tau \gamma = G\gamma^2 \tag{3.9}$$

The importance of the relationship between deformation and work for solids is explored in Chapter 4. Here, we introduce the concept to note the analogies to the viscous liquid case.

Liquids cannot resist shear stresses. When a shear stress is applied, the liquid in contact with the moving plate travels with the same velocity as the plate (Δv), while the liquid in contact with the stationary plate is also stationary. See Figure 3.6b. The deformation can also be represented by a shear stain, $\gamma = \tan \Delta \phi$, as above, but this shear strain increases with time and the shape of the liquid does not return to its original shape after the stress is removed. In the limit of a small time t, the top plate moves $t\Delta v$ and the shear strain is:

$$\gamma = \tan \Delta \phi = \frac{t \Delta v}{\Delta y} \tag{3.10}$$

The shear strain changes with time. The shear strain rate, or more simply, the "shear rate" is given by:

$$\dot{\gamma} = \frac{d\gamma}{dt} = \frac{dv}{dy} \tag{3.11}$$

Hence, the shear rate is equivalent to the velocity gradient in the liquid, which is a constant, in this case. The liquid between plates can be considered as a series of liquid layers, one atop the other, with the uppermost layer moving at the same velocity as the plate and the lowermost layer stationary. An analysis of the momentum transfer between layers also reveals that the velocity changes linearly with position. See Section 3.2.3. Shear rates encountered in melt-based processes are listed in Table 3.2.

Work is done on the liquid as it is sheared. However, since the liquid is flowing, the *rate* that work is done is important in understanding some aspects of liquid flow in melt processes, particularly viscous heating.

$$\text{Work} = (\text{Force})(\text{Displacement}) = (\tau A)(\gamma \Delta y)$$

$$\frac{\text{Work}}{\text{Volume}} = \frac{(\tau A)(\gamma \Delta y)}{A \Delta y} = \tau \gamma$$

$$\frac{d(\text{Work}/\text{Volume})}{dt} = \tau \dot{\gamma} \tag{3.12}$$

TABLE 3.2 Approximate Characteristic Shear Rates Associated with Melt-Based Processes

Process	Shear rate (s^{-1})
Pouring[a]	1–10
Extrusion[b]	50–1000
Injection Molding[c] and Die Casting	10^3–10^5
Fused Deposition Modeling	10–100
Blow Molding	10–1000

[a]Reed (1995).
[b]Giles et al. (2005).
[c]Morton-Jones (1989).

This equation shows that the shear rate is a key parameter in the describing the work that is done on the liquid as it is sheared. According to the first law of thermodynamics, if there is no heat exchange with the surroundings. the internal energy of the liquid increases as work is done on the liquid (i.e., as it is sheared), and hence its temperature increases. The importance of this "viscous heating" is explored later in the chapter.

Newton postulated that the shear rate, or velocity gradient developed in the liquid, is proportional to the shear stress imposed.

$$\tau = \eta \dot{\gamma} \qquad (3.13)$$

This relationship is known as Newton's law of viscosity, and the constant of proportionality is the shear viscosity, η, a property of the liquid. This measure of viscosity is important to most melt processes and is referred to here simply as viscosity. A liquid with a high viscosity requires larger shear stresses to create the same velocity gradient as compared with a less viscous or lower viscosity liquid. That is, moving the layers of liquid relative to one another requires more force. In common practice, more force is required to stir higher viscosity liquids at equivalent rates as compared with the force needed to stir low viscosity liquids.

Table 3.3 shows examples of viscosities of some common melts. Analysis of Eq. (3.13) shows that the standard units of viscosity are Pa•s. Another unit that is also used for viscosity is a poise, P = 1 g/(cm•s), a unit named after a French scientist Poiseuille. The conversion from P to Pa•s is straightforward: 1 P = 0.1 Pa•s or 1 cP = 1 mPa•s. As a reference point, the viscosity of water at room temperature is ~1 mPa•s. The role of the structure and composition of the melt on viscosity and rheological behavior is discussed later in this section.

TABLE 3.3 Examples of Melt Viscosities

Liquid	Conditions of measurement	Viscosity
Metals[a]		
Aluminum	700°C	1.15 mPa·s
	800°C	0.95 mPa·s
	900°C	0.81 mPa·s
Copper	1100°C	4.35 mPa·s
Iron	1650°C	4.79 mPa·s
Ceramic Glasses[b]		
B_2O_3	700°C	93 Pa·s
Soda-Lime-Silica	800°C	70,000 Pa·s
	1000°C	756 Pa·s
	1200°C	54 Pa·s
Aluminosilicate	1200°C	1066 Pa·s
Thermoplastic Polymers[c]		
Polypropylene	1 s^{-1}, 210°C	1000 Pa·s
	1 s^{-1}, 180°C	6000 Pa·s
	100 s^{-1}, 180°C	400 Pa·s
Polycarbonate	100 s^{-1}, 250°C	20,000 Pa·s
Nylon	100 s^{-1}, 260°C	2000 Pa·s
Low Density Polyethylene	100 s^{-1}, 190°C	1000 Pa·s

[a]From information tabulated in Brandes (1983).
[b]From information tabulated in Martlew (2005).
[c]From information tabulated in Osswald (1998) and Wilczynski and White (2003). Polymer melt viscosities depend on molecular weight.

Viscosity decreases with increasing temperature. A few examples are given in Table 3.3. The temperature dependence of viscosity for many liquids follows the general form:

$$\eta = \eta_o \exp\left(\frac{E}{RT}\right) \tag{3.14}$$

where η_o is a constant, R is the gas constant, and E is the activation energy for viscous flow. That is, the rate of liquid flow (i.e., in m^3/s) due to gravity or a pressure gradient, for example, varies inversely with the viscosity and follows the Arrhenius form:

$$\text{Flow rate} \propto \frac{1}{\eta} = \frac{1}{\eta_o}\exp\left(\frac{-E}{RT}\right) \tag{3.15}$$

The origin of this temperature effect is related to molecular flow processes and bonding in the liquid. Molecules in the liquid must squeeze by each other or move into small spaces between molecules, and for motions such as these, there is an energetic barrier. As the temperature increases, the fraction of molecules with enough thermal energy to overcome the barrier increases and the viscosity decreases. Decreasing viscosity with increasing temperature is characteristic of all liquids. However, more accurate empirical relationships for the temperature dependence have been developed for some melts, as noted below.

For some melts, viscosity also varies with the shear rate. These liquids are called non-Newtonian. One general way to describe the rheology of non-Newtonian liquids is to use a power law expression for the relationship between shear stress and shear rate:

$$\tau = K(\dot{\gamma})^n \tag{3.16}$$

where K and n are properties of the fluid, referred to, respectively, as the consistency index and the power law index. The consistency index has units of $Pa \bullet (s)^n$, while the power law index is unitless. Using this general relationship and Eq. (3.13), the viscosity is given by:

$$\eta = \frac{\tau}{\dot{\gamma}} = K(\dot{\gamma})^{n-1} \tag{3.17}$$

A *Newtonian* liquid has a power law index, n, of one. In other words, viscosity is independent of shear rate. When n is greater than one, the viscosity increases as shear rate increases, a phenomena known as shear thickening. When n is less than one, the viscosity drops with shear rate, a behavior known as shear thinning. The latter behavior is particularly common in polymer melts. The former is not typically found in melts. Figure 3.7 illustrates the difference between Newtonian and shear thinning behaviors.

Viscoelasticity. Real melts not only flow like viscous liquids, but they can also respond like an elastic solids under some conditions. The "real" melt, therefore, is classified as *viscoelastic*. While we could classify all melts as viscoelastic, in practice, the inclusion of the elastic response is only important in understanding the rheology of polymer melts. A simple way to interpret viscoelasticity is through mechanical models comprised of springs, representing the elastic response, and dashpots, representing the viscous response. The Maxwell and Kelvin models are shown in Figure 3.8. For melts, the Maxwell model is arguably more suitable because the dashpot in series with the spring allows continuous deformation.

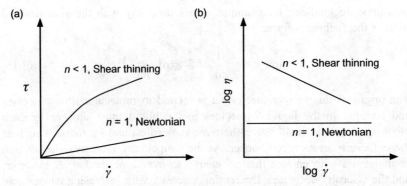

FIGURE 3.7 Relationships between (a) shear stress and shear rate and (b) viscosity and shear rate for: Newtonian and non-Newtonian (shear thinning) melts.

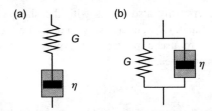

FIGURE 3.8 Mechanical models for viscoelastic materials: (a) Maxwell model and (b) Kelvin model.

The response of the Maxwell model can be predicted and the concept of relaxation time can be introduced using linear viscoelasticity. In this approach, the response of the combination of elements is found by summing the contributions of the individual elements. When a load is placed on the Maxwell model, the shear rate is the sum of the shear rates of each element:

$$\dot{\gamma} = \dot{\gamma}_{\text{Elastic}} + \dot{\gamma}_{\text{Viscous}} \tag{3.18}$$

$$\dot{\gamma} = \frac{d(\tau/G)}{dt} + \frac{\tau}{\eta} = \frac{1}{G}\frac{d\tau}{dt} + \frac{\tau}{\eta}$$

Rearranging this equation,

$$\tau + \frac{\eta}{G}\frac{d\tau}{dt} = \eta\dot{\gamma} \tag{3.19}$$

The term η/G has units of time, and is called the relaxation time. It represents the time needed for polymer molecules to relax back to an equilibrium configuration after being perturbed. When the relaxation time is short, the Maxwell model responds like a viscous liquid:

$$\text{Short relaxation time:} \frac{\eta}{G} \to 0, \quad \tau = \eta\dot{\gamma} \quad \text{(viscous liquid)} \tag{3.20}$$

Likewise, it can be shown that when the relaxation time approaches infinity, the Maxwell model behaves like an elastic solid.

Metal Melts. Molten metals are Newtonian and typically have viscosities in the range of ~1—5 mPa•s. The viscosities are therefore of the same order of magnitude as water and other simple liquids. Recalling the cartoon of the liquid metal structure, the structural length scale in the liquid is quite small, on order of atomic dimensions, and therefore, there is little resistance to the motion of liquid layers relative to each other and the viscosity is low. There are differences between metals, as shown in Figure 3.9. Among elemental metals, transition metals have higher viscosities than alkali metals, for example. These differences indicate a structure—viscosity connection. There are theories to account for these differences. (See the suggested reading list at the end of the chapter.) For example, one theory assumes that momentum transfer between the liquid layers is accomplished by atomic vibrations, leading to the following relationship for the viscosity of an elemental metal at its melting point:

$$\eta_{mp} = C \frac{(T_{mp} A)^{1/2}}{V_m^{2/3}} \qquad (3.21)$$

where C is a constant (1.655×10^{-7} $J^{1/2}$ $K^{-1/2}$ $mol^{-1/6}$), A is the atomic weight, and V_m is the molar volume at T_{mp}, the melting point in Kelvin. There is also a general connection between the viscosity at the melting point of the metal and its activation energy for viscous flow. This trend relates the

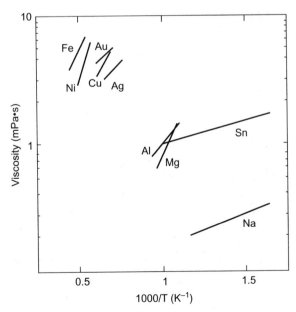

FIGURE 3.9 The effect of temperature on the viscosity of elemental metals. *Plotted with parameters compiled in Battezzati and Greer (1989).*

bonding in the metal melt to the energy needed for viscous flow. See Example 3.2.

EXAMPLE 3.2 Determine the activation energy for viscous flow for sodium, aluminum, copper, and iron from the data provided in Figure 3.9. Plot the activation energy as a function of the melting point of the metal in K.

From Eq. (3.14),

$$\eta = \eta_o \exp\left(\frac{E}{RT}\right) \rightarrow \log \eta = \log \eta_o + 2.303\left(\frac{E}{RT}\right)$$

The slope of a plot of $\log \eta$ versus $1000/T$ is therefore $2303E/R$. Larger slopes indicate higher activation energies. By inspection, Fe has the highest activation energy and Na, the least in the set of metals listed. To determine the activation energy values, two points are chosen on each line and the activation energy, E, using the equation above:

Na: $E = 5.24$ kJ/mol and $T_{mp} = 471$ K
Al: $E = 16.5$ kJ/mol and $T_{mp} = 933$ K
Cu: $E = 30.5$ kJ/mol and $T_{mp} = 1356$ K
Fe: $E = 41.4$ kJ/mol and $T_{mp} = 1809$ K

This data, along with the others from Figure 3.9, is plotted below. Notice the general trend is toward higher activation energy for viscous flow, as the melting point increases. The melting point correlates well with bonding strength. Materials with stronger interatomic bonding have higher melting points.

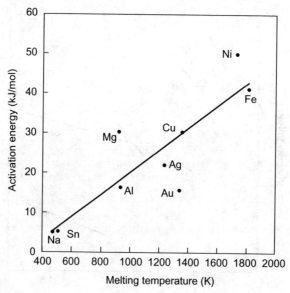

FIGURE E3.2

Alloy viscosities are more varied than elemental metal viscosities and the compositional dependence is complex. In binary alloy systems with complete solid solubility and no intermediate compounds, the viscosity of the alloy typically follows a smooth, sometimes linear variation between the viscosities of the two end members. However, in more complex systems, the viscosity is found to vary with composition near eutectics and intermediate compounds. Considering just the viscosity at the temperature near the liquidus, or solidification point, the viscosity at the eutectic composition is higher than those of surrounding compositions, indicating that some short range structure develops in these melts. However, if one measures the viscosity of melts in a binary system at some constant temperature above the liquidus temperature for all compositions, then the viscosity of the melt at the eutectic composition is lower than the others. In this comparison, the low viscosity of the eutectic composition is due to the fact that the eutectic liquid has been heated further above its melting point to reach the temperature of comparison.

Measuring the viscosity of liquid metals is challenging due to the high temperature of the melts and the potential for chemical reactions between the melt and the atmosphere, container, or measurement device. So, characterization is not routinely performed. For some melt casting processes, a practical measure, called fluidity, is determined. Fluidity measurements determine the length a melt travels through a standard spiral channel in a mold under a certain set of conditions that mimic the casting operation. Therefore, fluidity is not only affected by the viscosity of the melt, but also the heat transfer into the mold, which causes solidification and ceases the flow.

Ceramic Glass Melts. Like metal melts, glass melts tend to have Newtonian character, but their viscosities are much higher than those of metals, as shown in Table 3.3. The higher viscosity is due to the more complex melt structure; the oxide glass structure contains interconnected silicate anionic units.

The viscosity of oxide glass melts is a strong function of composition. The most common glass chemistries contain "network formers," such as SiO_2, which are the basis for the interconnected structure of oxide anionic units, and several "network modifiers," such as Na_2O and CaO. As the network modifier oxides dissolve into a silicate melt, the oxygen from the modifiers becomes part of the anionic oxide network and hence changes the ratio of O/Si from two, for a completely interconnected network of SiO_2 with each O bonded to two Si's, to a number greater than two, which indicates extra oxygen. Hence, glass structure in the composition with modifiers some oxygen ions are bound to only one Si. These so-called nonbridging oxygens break up the network, lowering the melting point and the viscosity at any given temperature. In technologically relevant silicate glasses, the addition of alkali oxides is particularly effective in breaking up the silicate network and lowering the viscosity. This effect is demonstrated in Figure 3.10. For

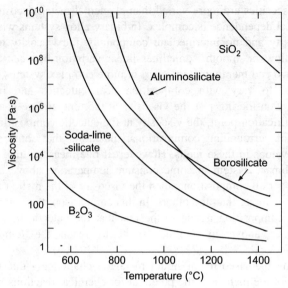

FIGURE 3.10 Viscosity as a function of temperature for several oxide glasses. *Plotted with data from Martlew (2005).*

example, the viscosity of pure silica is orders of magnitude higher than the soda-lime-silica glass at any given temperature.

The viscosity of glass melts drops with temperature. For many oxide glasses, a three parameter relationship, known as the Volgel–Fulcher–Tammann (VFT) equation provides a good fit. The VFT equation is given by

$$\eta = \eta_o \exp\left(\frac{B}{T - T_o}\right) \tag{3.22}$$

where η_o is a constant with the same units as η, and T_o and B are constants with the same units as T (typically, K). Notice that B is similar to E/R in Eq. (3.14), and represents the activation energy. In this expression, however, the activation energy depends on temperature (Example 3.3).

Thermoplastic Polymer Melts. Thermoplastic polymer melts are composed of long-chain molecules that are typically entangled with one another. Polymer melt viscosities are much higher than those of the melts considered so far. The relatively large and complex liquid structure is responsible for the higher viscosity. Polymer melt viscosity depends critically on the structure of the polymer molecule. In particular, the viscosity increases with the "stiffness" of the chain or, in other words, the molecule's ability to rotate and reconfigure itself. Chain stiffness increases when the polymer backbone contains rigid groups, such as aromatic rings, or bulky side groups. For any given polymer chemistry, the molecular weight is an important factor in

EXAMPLE 3.3 An extensive list of VFT parameters is given in Martlew (2005). In this compilation, the VFT equation is given in the form:

$$\log \eta = A + \frac{B}{T - T_o}$$

where viscosity is in Poise (P) and temperature is in °C. Use the equation to generate a data set of viscosity (Pa•s) and temperature in which there is a data point every 10°C. Using this data set, attempt to fit the data to a simple Arrhenius equation. Comment on the fit.

	A	B	T_o	T_{min} (°C)	T_{max} (°C)
Aluminosilicate	−2.5015	5348	381	695	1400

The VFT equation fits a straight line to data plotted as $\log \eta$ versus $1/(T-T_o)$

To fit to the simpler two parameter model (Eq. (3.14)) requires a plot of $\ln \eta$ versus $1/T$ to be linear. That is,

$$\eta = \eta_o \exp\left(\frac{E}{RT}\right) \rightarrow \ln \eta = \ln \eta_o + \frac{E}{RT}$$

Plotting the data in this way and carrying out a linear fit gives the following results. There is a clear difference between the VFT approach and the Arrhenius approach.

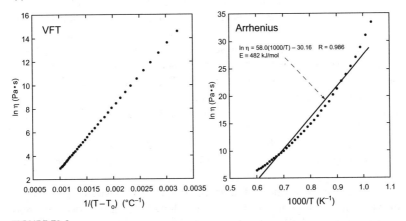

FIGURE E3.3

determining viscosity. As discussed more below, as the molecular weight increases, the polymer molecules become entangled. In other words, molecules intertwine with one another, which further impedes flow.

The rheology of polymer melts is affected by shear rate and temperature. See examples in Table 3.3. Figure 3.11a shows the effect of shear rate on a thermoplastic polymer melt in general. At very low shear rates the melt is

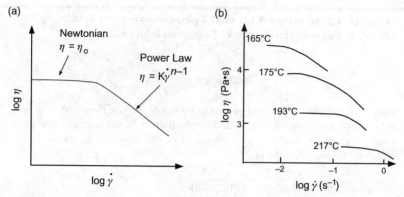

FIGURE 3.11 Effect of shear rate and temperature on the shear viscosity: (a) general shear rate dependence and (b) behavior of a nearly monodisperse polystyrene melt with a weight average molecular weight of 411,000 g/mol and polydispersity index of 1.06. *Data from Penwell and Graessley (1974).*

Newtonian; the viscosity is high and independent of shear rate. This low shear, plateau viscosity is called the zero shear viscosity, η_o. In this regime, the polymer molecules rearrange and reconfigure at a faster rate than the shearing rate. That is, the characteristic time for rearrangement or structural relaxation is shorter than the time scale of shearing, $1/\dot{\gamma}$. When a melt contains entanglements, the entanglement density remains constant at low shear rates. However, as the shear rate increases, the time scale for the shearing decreases. Eventually, it is shorter than the relaxation time (η/G). Under these conditions, the polymer melt viscosity drops as shear rate increases. The polymer molecules, unable to relax back to equilibrium random coil shapes, become distorted by the velocity gradient and present less resistance to flow. Additionally, the molecules become disentangled under the shearing action when they are not able to relax back or rearrange again fast enough. The disentanglement leads to a drop in the viscosity. The power law relationship (Eq. (3.16)) can be used over some span of shear rates as a good approximation.

The effect of shear rate and temperature on the viscosity of a polystyrene melt is shown in Figure 3.11b. The data show that the viscosity decreases with temperature, similar to other melts. But this figure also demonstrates how the competition between structural relaxation and distortion and disentanglement on shearing gives rise to shear thinning. As the temperature increases, the viscosity drops and structural relaxation becomes faster (i.e., relaxation time decreases). The data show that as temperature increases, the Newtonian plateau extends to higher shear rates. That is, at higher temperatures the melt is able to respond and reconfigure itself more effectively, remaining Newtonian at higher shear rates (i.e., shorter characteristic shear times). Similar trends are noted in polymer melts with broader molecular weight distributions (see Figure 3.12). However, since relaxation time

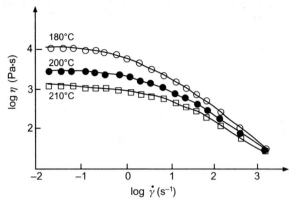

FIGURE 3.12 Effect of shear rate and temperature on the viscosity of a polypropylene melt. *Reproduced with permission of John Wiley & Sons from Wilczynski and White (2003); ©2003 Society of Plastics Engineers.*

TABLE 3.4 Rheological Parameters for Thermoplastics[a]

Polymer	K_o (Pa·s^n)	n	a (1/°C)	T_o (°C)
Polystyrene	28,000	0.28	0.025	170
High Density Polyethylene	20,000	0.41	0.002	180
Low Density Polyethylene	6000	0.39	0.013	160
Polypropylene	7500	0.38	0.004	200
Polyvinyl chloride	17,000	0.26	0.019	180

[a] *Examples from Osswald (1998).*

depends on molecular weight, the effect of temperature on the onset of non-Newtonian behavior is not as distinct.

The effects of temperature and shear rate on viscosity can be represented in a modified power law expression:

$$\eta = K(T)(\dot{\gamma})^{n-1} \qquad (3.23)$$

where the consistency index contains the temperature dependence in a three parameter empirical equation.

$$K(T) = K_o \exp(-a(T - T_o)) \qquad (3.24)$$

Values for various polymers are given in Table 3.4. Notice that this equation has the same form as the VFT equation used to describe the viscosity of glasses. The power law index for polymer melts indicates shear thinning behavior (Example 3.4).

EXAMPLE 3.4 The viscosity of a polymer melt is shown below. (a) Find the zero shear viscosity. (b) Write a power law expression for viscosity as a function of shear rate and specify applicability. (c) Predict the viscosity at a shear rate of 10^4 s^{-1}.

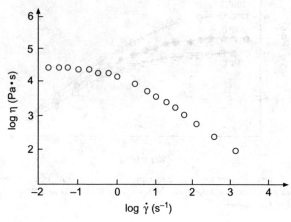

FIGURE E3.4A

a. The zero shear viscosity is the limit of viscosity as the shear rate goes to zero.

$$\log \eta_o = 4.4 \rightarrow \eta_o \cong 25{,}000 \text{ Pa} \cdot \text{s}$$

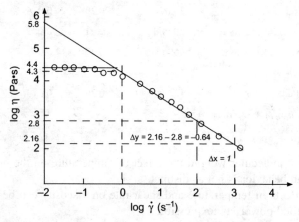

FIGURE E3.4B

b. The analysis of the data at higher shear rates requires an estimate of the linear regime from a shear rate of about 1 s^{-1} and higher for this melt.

$$\eta = K(\dot{\gamma})^{n-1} \rightarrow \log \eta = \log K + (n-1)\log(\dot{\gamma})$$

$$(n-1) = \frac{\Delta y}{\Delta x} = \frac{-0.64}{1} = -0.64 \rightarrow n = 0.36$$

$$\log K \cong 4.3 \rightarrow K \cong 20{,}000 \text{ Pa} \cdot \text{s}^{0.36}$$

$\eta = 20{,}000(\dot{\gamma})^{-0.64}$, where η has units of Pa·s and shear rate has units of s^{-1}.

c. $\eta = 20{,}000(10{,}000)^{-0.64} = 55.1$ Pa·s ($\log \eta = 1.7$)

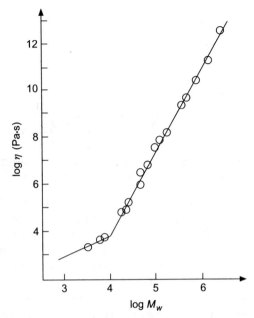

FIGURE 3.13 Effect of weight average molecular weight on the zero shear viscosity of nearly monodisperse polyisoprene at $-55°C$. *Data from Nemoto et al. (1971) and (1972).*

Molecular weight plays a key role in the viscosity of polymer melts, as shown in Figure 3.13. For monodisperse polymers, the zero shear viscosity increases proportionally with weight average molecular weight at low molecular weights, but at a critical molecular weight, the dependence becomes steeper to $\eta_o \propto M^{3.4}$. The transition represents the point at which the chains are long enough for entanglements to develop. After this transition, the polymer molecules move less freely through a process called reptation. In practice, polymer melts have a distribution of molecular weight, and while the molecular weight distribution determines the response, in general the viscosity increases with the weight average molecular weight.

3.2.3 Flow Fundamentals

A critical step in melt processing is the transport, or flow, of the melt. For example, molten metal flows, under gravity, into a mold during metal shape casting, while thermoplastic polymer melt flows, due to an applied pressure, in injection molding. Understanding the factors that control the rate of flow is essential for controlling and engineering a melt-based forming process. In this section, some of the important fundamentals of flow are introduced. The goal is to provide the essentials for engineering of melt-based processes.

There are three main principles governing liquid flow: mass conservation, energy conservation, and momentum conservation. The mass conservation

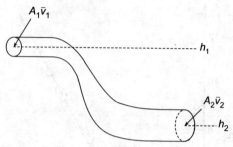

FIGURE 3.14 Schematic of a conduit with changing diameter and height. The cross-sectional areas of the conduit, A, average velocities, $\bar{v}$, and heights, h, at the two location are noted.

principle helps us understand flow rates for a liquid passing from one stage in a process to another. For example, the melt may flow through a series of passages with varying geometry as it makes its way to a mold cavity. Mass conservation is used to relate the flow velocity in different portions of the flow path. Energy conservation is also important in understanding flow. This principle accounts for heating and cooling of the material; however, in this section, we consider an isothermal flow, in which the energy conservation takes into account only kinetic and potential energy. Momentum conservation is central to quantifying the effects of viscosity, driving forces, and physical constraints on flow rate and velocity distribution.

Mass and Mechanical Energy Conservation. Some basic definitions can be made by considering the flow of a liquid through a cylindrical conduit, as shown in Figure 3.14. The mass flow rate, Q_m, is the mass of the liquid passing through the pipe per unit time (e.g., kg/s). Likewise, the volumetric flow rate, Q, is the volume of liquid passing through the tube per unit time (e.g., m³/s), where $Q = Q_m/\rho$. The velocity of the liquid ordinarily depends on the position in the tube, but the average velocity, $\bar{v}$, of the liquid is related to the volumetric flow rate and the cross-sectional area, A:

$$\bar{v} = \frac{Q}{A} \qquad (3.25)$$

The flow rate and average velocity depend on the liquid properties, such as viscosity and density, the magnitude of the force causing the flow, and the geometry of the conduit through which the liquid flows.

Figure 3.14 also illustrates the continuity principle that follows directly from mass conservation. Mass conservation requires that, in steady state, the mass of liquid entering and exiting a conduit in a given amount of time is the same. The continuity principle states that the volumetric flow rate of a liquid (i.e., incompressible fluid) in a series of interconnected conduits or passages is the same everywhere.

$$Q_1 = Q_2$$
$$\bar{v}_1 A_1 = \bar{v}_2 A_2 \qquad (3.26)$$

Energy conservation is essentially an application of the first law of thermodynamics to a flowing system. In an isothermal flowing system, energy conservation is restricted to a mechanical energy balance that includes the kinetic and potential energy as well as the work done by the flowing liquid and energy dissipation. Considering the system sketched in Figure 3.14, the change in potential energy (PE), kinetic energy (KE), and work done by the system are given by the following on a per unit mass basis:

$$\Delta PE = g(h_2 - h_1) \quad (3.27)$$

$$\Delta KE = \frac{1}{2\beta}(\bar{v}_2^2 - \bar{v}_1^2) \quad (3.28)$$

$$\text{Work done} = \frac{1}{\rho}(P_2 - P_1) \quad (3.29)$$

In the expression for kinetic energy, the factor β is included to account for the velocity profile in the cross-section. $\beta = 0.5$ for laminar flow and 1.0 for turbulent flow. In the expression for work done, P_2 and P_1 are the pressures in the melt at points 2 and 1, respectively. In addition, friction losses, represented as E_f, are included in the total energy conservation equation:

$$g(h_2 - h_1) + \frac{1}{2\beta}(\bar{v}_2^2 - \bar{v}_1^2) + \frac{1}{\rho}(P_2 - P_1) + E_f = 0 \quad (3.30)$$

This equation is known as Bernoulli's equation. It is helpful in the analysis of flow through a series of passages or conduits.

Flow Behaviors. Flowing liquid can be characterized as being laminar or turbulent. In laminar flow, layers of liquid move relative to one another such that there is no mixing of liquid, as shown in Figure 3.15. In contrast, turbulent flow is chaotic and irregular, and promotes mixing. For a given liquid and conduit geometry, laminar flow is found at low flow rates and turbulent flow at higher rates. In general, the transition from laminar to turbulent depends not only on the velocity, but also the liquid properties and conduit geometry. The transition occurs at a critical value of the Reynolds number, Re. The Reynolds number is a ratio of inertial force, which is based on the mass and acceleration of the liquid, to viscous force, which is based on the force between adjacent layers of liquid that are flowing at different velocities. The general form of the Reynolds number is:

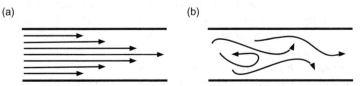

FIGURE 3.15 Cartoon showing flow streamlines in a conduit: (a) laminar and (b) turbulent.

$$Re = \frac{d\bar{v}\rho}{\eta} \qquad (3.31)$$

where d is a characteristic length scale for the flow (e.g., diameter of tube) and ρ is the density of the liquid. The cutoff for laminar flow is at a Re of about 2100. If the Reynolds number is less than 2100, the flow is laminar. Laminar flow is preferred for delivery of melt into a mold or through a die, for example. In contrast, turbulent flow is ideal for mixing.

In the next section, we examine two common liquid flow situations encountered in melt processing: drag flow and pressure-driven flow. For more details, the texts by Geiger and Poirier (1973) and Gaskell (1992) are excellent resources. The discussion here is restricted to the case of laminar flow under steady-state conditions (not changing with time). Several factors influence the flow: the liquid properties, the geometrical constrictions, such as the width of a flow passage, and the strength of the driving force of the flow. The principle of conservation of momentum is used to develop equations for the velocity of the liquid as a function of position in a flow configuration and the volumetric flow rate.

The moving liquid has momentum. When there is a velocity gradient in the liquid, as in the case of the simple shear flow used to define viscosity, there is momentum transfer or transport in the direction down the velocity gradient. That is, momentum is transported from a faster moving, higher momentum layer, to an adjacent, slower moving, lower momentum layer. The rate of momentum transport is equivalent to a force (e.g., $d(mv)/dt$ has units of $kg \cdot m \cdot s^{-2}$). The rate of momentum transport per unit area, or the momentum flux, is equivalent to a stress (e.g., $d(mv/A)/dt$ has units of $kg \cdot m^{-1} \cdot s^{-2}$), the shear stress between the layers. So, by Newton's law of viscosity, the momentum flux (i.e., the shear stress) is proportional to the velocity gradient and the constant of proportionality is the viscosity. To show the direction of momentum flux down the velocity gradient, a negative sign is added to Newton's law of viscosity:

$$\tau = -\eta \frac{dv}{dy} \qquad (3.32)$$

More important for the analysis below, however, is the conservation, or balancing, of the rate of momentum transport—in other words, the force balance.

Drag Flow. In drag or Couette flow, a liquid is sandwiched between a moving object and a stationary object. A simple case, shown in Figure 3.16, involves a liquid layer confined between two plates, not unlike the situation used to develop a definition of viscosity (Figure 3.6). One plate moves with velocity, v_{plate}, while the other is stationary. The plates are separated by a gap, H, and they have a width, W. We assume that the width is much larger than the gap, such that effects at the extremities can be ignored, and a steady-state condition. The moving plate sets into motion the liquid adjacent to it.

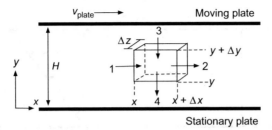

FIGURE 3.16 Coordinate system and control volume for analysis of drag flow between two parallel plates.

That is, momentum from the moving plate is transferred to the liquid layer adjacent to the plate and this liquid layer, in turn, sets into motion the layer of liquid beneath it, transferring momentum to it, and so on. This sort of momentum transfer is called viscous momentum transfer.

To determine the velocity profile in the liquid in steady state, the forces on the volume element are balanced. Given that there are no external forces in this case, the force balance requires a balance of rate of momentum transport in and out of the volume element. Let us start with the momentum that might be transferred in the x direction. Since the liquid is moving in the x direction, there is momentum due to the flow itself. For steady state conditions, however, the velocity of the liquid in the x direction (v_x) at any position y between the two plates is the same at the point x (face 1) and the point $x + \Delta x$ (face 2). Hence it is only the rate of viscous momentum transfer in the y direction that must be balanced. For the control volume, viscous momentum is transported into face 3 and out of face 4.

$$\tau_{yx}|_{y+\Delta y} \Delta z \Delta x - \tau_{yx}|_y \Delta z \Delta x = 0 \tag{3.33}$$

$$\frac{\tau_{yx}|_{y+\Delta y} - \tau_{yx}|_y}{\Delta y} = 0 \tag{3.34}$$

$$\frac{d\tau_{yx}}{dy} = 0 \tag{3.35}$$

Therefore, the shear stress, τ_{yx}, is constant with position y. The relationship between the shear stress and the velocity is found in Newton's law of viscosity. Since the momentum is transferred down the velocity gradient, a negative sign is added, as mentioned above.

$$\tau_{yx} = -\eta \frac{dv_x}{dy} = \text{constant} \tag{3.36}$$

$$\frac{dv_x}{dy} = \text{constant} \tag{3.37}$$

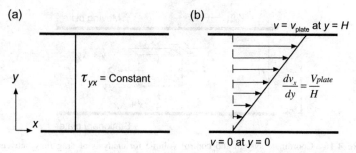

FIGURE 3.17 (a) Shear stress and (b) velocity profiles for drag flow between two parallel plates.

The velocity profile has a constant slope, proving the linear relationship noted earlier, and the shear stress on the liquid is constant throughout the gap. See Figure 3.17.

Using the boundary conditions ($v_x = 0$ at $y = 0$ and $v_x = v_{plate}$ at $y = H$),

$$v_x = \left(\frac{v_{plate}}{H}\right)y \qquad (3.38)$$

The same velocity would be found for a non-Newtonian liquid following the power law (Eq. (3.16)) as well. The average velocity of the liquid flowing in the channel is $\tfrac{1}{2}v_{plate}$; therefore, the volumetric flow rate is

$$Q = \bar{v}_x(\text{cross-sectional area of channel}) = \frac{1}{2}v_{plate}HW \qquad (3.39)$$

Pressure-Driven Flow. Consider again the liquid sandwiched between two plates, but now both plates are stationary and a pressure gradient drives the flow. Physically, a piston or a pump could impose the higher pressure on the left side. Here, we analyze the steady-state, well-developed flow profile that occurs far enough away from the point where the piston or pump is acting so that the velocity profile does not vary with time or position x. The coordinate system is chosen with the center of the gap defined as $y = 0$. See Figure 3.18. This choice is convenient since the liquid adjacent to either of the stationary plates has a velocity of 0, and the center is a symmetry plane.

The force balance on the control volume includes the external forces responsible for the pressure gradient as well as the viscous momentum transport. Again, the momentum transport in the x direction is not important because the flow is steady state. However, there is a force balance in the x direction. For flow in the positive x direction, the pressure at point x, P_x, is higher than the pressure at point $x + \Delta x$, $P_{x+\Delta x}$. The force balance includes these pressure-based forces acting on faces 1 and 2 of the control volume. Given the fact that the velocity of the liquid is zero at the plates, the

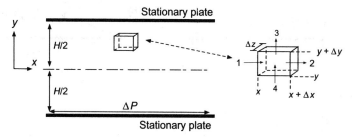

FIGURE 3.18 Coordinate system and control volume for analysis of pressure-driven flow between two parallel plates.

direction of viscous momentum transport is from the center towards the plates and therefore, the force balance includes rate of viscous momentum transport into face 4 and out of face 3 of the control volume.

$$P|_x \Delta z \Delta y - P|_{x+\Delta x} \Delta z \Delta y + \tau_{yx}|_y \Delta z \Delta x - \tau_{yx}|_{y+\Delta y} \Delta z \Delta x = 0 \quad (3.40)$$

$$\frac{d\tau_{yx}}{dy} = -\frac{dP}{dx} \quad (3.41)$$

In well developed flow, the pressure varies linearly with x such that:

$$-\frac{dP}{dx} = \frac{\Delta P}{L} \quad (3.42)$$

where ΔP is the pressure drop over a length L in the $+x$ direction. Now, we can integrate equation.

$$d\tau_{yx} = \frac{\Delta P}{L} dy \quad (3.43)$$

$$\tau_{yx} = \frac{\Delta P}{L} y + A \quad (3.44)$$

where A is an integration constant. Since there is a symmetry plane at $y = 0$, the velocity gradient is zero and $\tau_{yx} = 0$ at $y = 0$. Hence, $A = 0$. This conclusion can also be reached using the velocity boundary conditions as well.

$$\tau_{yx} = \frac{\Delta P}{L} y \quad (3.45)$$

The shear stress is a linear function of y and reaches a maximum at the walls of the parallel plate channel. See Figure 3.19.

To find the velocity profile, a constitutive equation is needed. For a Newtonian liquid, Eq. (3.32) is substituted into Eq. (3.45):

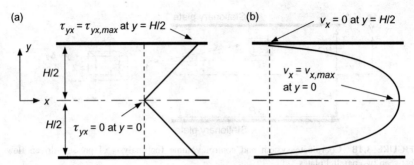

FIGURE 3.19 (a) Shear stress and (b) velocity profiles for pressure-driven flow between two parallel plates.

$$-\eta \frac{dv_x}{dy} = \frac{\Delta P}{L} y \tag{3.46}$$

Rearranging and integrating this expression,

$$v_x = -\frac{\Delta P}{2\eta L} y^2 + B \tag{3.47}$$

The integration constant B is found using this boundary condition, $v_x = 0$ at $y = H/2$, and the final expression for the velocity profile is

$$v_x = \frac{\Delta P}{2\eta L}\left(\frac{H^2}{4} - y^2\right) \tag{3.48}$$

The subscript x on v_x reminds us that the direction of the velocity is in the x direction and varies as a function of y, the position in the channel. The velocity follows a parabolic behavior as shown in Figure 3.19.

To find the volumetric output, Q, an expression for the average velocity is needed. This expression is found by integration:

$$\bar{v}_x = \frac{1}{H/2}\int_0^{H/2} v_x dy = \frac{1}{H/2}\int_0^{H/2} \frac{\Delta P}{2\eta L}\left(\frac{H^2}{4} - y^2\right) dy = \frac{H^2 \Delta P}{12\eta L} \tag{3.49}$$

The volumetric flow rate is therefore this average velocity multiplied by the cross-sectional area of the channel

$$Q = \bar{v}_x HW = \frac{H^3 W \Delta P}{12\eta L} \tag{3.50}$$

The above analysis can be carried out for non-Newtonian liquids as well. The appropriate constitutive equation, such as the power law equation, is used in place of Newton's law and the solution proceeds in the same way. The integration is a bit more time consuming but the results are easily found. See Table 3.5 for the solutions and the cylindrical or pipe geometry, which is also frequently encountered in processing (Examples 3.5 and 3.6).

TABLE 3.5 Pressure-Driven Flow Equations

Configuration	Newtonian	Non-Newtonian, power law liquid
Pressure-driven flow through a rectangular channel of width W	$v_x = \dfrac{\Delta P}{2\eta L}\left(\dfrac{H^2}{4} - y^2\right)$	$v_x = \dfrac{H}{2(s+1)}\left(\dfrac{H\Delta P}{2KL}\right)^s \times \left[1 - \left(\dfrac{2y}{H}\right)^{s+1}\right]$
	$Q = \dfrac{H^3 W \Delta P}{12\eta L}$	$Q = \dfrac{WH^2}{2(s+2)}\left(\dfrac{H\Delta P}{2KL}\right)^s \quad s = \dfrac{1}{n}$
Pressure-driven flow through a tube	$v_x = \dfrac{\Delta P}{4\eta L}(R^2 - r^2)$	$v_x = \dfrac{R}{s+1}\left(\dfrac{R\Delta P}{2KL}\right)^s\left[1 - \left(\dfrac{r}{R}\right)^{s+1}\right]$
	$Q = \dfrac{\pi R^4 \Delta P}{8\eta L}$	$Q = \dfrac{\pi R^3}{s+3}\left(\dfrac{R\Delta P}{2KL}\right)^s \quad s = \dfrac{1}{n}$

EXAMPLE 3.5 Consider the pressure-driven flow of HDPE polymer melt with a viscosity of 20,000 Pa·s and an aluminum melt with a viscosity of 1 mPa·s through a slit of gap 1 mm and length of 5 cm. (a) What is the maximum average velocity for laminar flow of each melt? Assume Newtonian behavior. (b) Find the pressure needed to develop that velocity.

a. For laminar flow, the Reynolds number must be less than 2100. So, we can use the definition of the Re to find the maximum average velocities for the melts at this cutoff.

$$Re = \frac{d\bar{v}\rho}{\eta}$$

HDPE melt:
$\eta = 20{,}000$ Pa·s (problem statement)
$\rho = 1000$ kg/m^3 (Table 3.1)

$$Re = 2100 = \frac{(0.001 \text{ m})\bar{v}(1000 \text{ kg/m}^3)}{20{,}000 \text{ Pa·s}} \Rightarrow \bar{v} = 4.2 \times 10^7 \text{ m/s}$$

This calculation shows that it is virtually impossible for the HDPE melt to enter the turbulent regime. The narrow constriction is one factor that inhibits turbulence, but more important is the extremely high viscosity of the HDPE melt.

Al melt:
$\eta = 1$ mPa•s (problem statement)
$\rho = 2380$ kg/m³ (Table 3.1)

$$Re = 2100 = \frac{(0.001 \text{ m})\bar{v}(2380 \text{ kg/m}^3)}{0.001 \text{ Pa}\cdot\text{s}} \Rightarrow \bar{v} = 0.88 \text{ m/s}$$

Therefore it is quite possible that the aluminum melt will enter the turbulent regime.

b. To find the pressure needed to drive the flows found in (a) we use Eq. (3.49).

$$\bar{v}_x = \frac{H^2 \Delta P}{12\eta L}$$

HDPE melt:

$$4.2 \times 10^7 \text{ m/s} = \frac{(0.001 \text{ m})^2 \Delta P}{12(20,000 \text{ Pa}\cdot\text{s})(0.05 \text{ m})} \Rightarrow \Delta P = 5 \times 10^{17} \text{ Pa}$$

As we might have expected from the answer to (a), this pressure is extremely large and in fact not accessible. So, it is impossible to enter the turbulent regime.

Al melt:

$$0.88 \text{ m/s} = \frac{(0.001 \text{ m})^2 \Delta P}{12(0.001 \text{ Pa}\cdot\text{s})(0.05 \text{ m})} \Rightarrow \Delta P = 528 \text{ Pa}$$

EXAMPLE 3.6 Plot the velocity profile for flow of a LDPE melt at 160°C through a 2 cm diameter tube. Compare to the shape expected for a Newtonian fluid. Assume that the pressure driving flow is 1 MPa and that the tube length is 10 cm.

$R = 1$ cm, $L = 10$ cm, $\Delta P = 10^6$ Pa

From Table 3.4: $n = 0.39$ and $K = 6000$ Pa•s$^{0.39}$ $s = 1/n = 2.56$

For pressure-driven flow of a power law liquid through a tube of radius R:

$$v_x(r) = \frac{R}{1+s}\left(\frac{R\Delta P}{2KL}\right)^s \left[1 - \left(\frac{r}{R}\right)^{s+1}\right]$$

$$= \frac{1 \text{ cm}}{2.56}\left(\frac{(1 \text{ cm})(10^6 \text{ Pa})}{2(6000 \text{ Pa}\cdot\text{s}^{0.39})(10 \text{ cm})}\right)^{2.56}\left[1 - \left(\frac{r}{1 \text{ cm}}\right)^{2.56}\right]$$

$$= 89[1 - r^{2.56}]$$

where the velocity has units of cm/s.

For pressure-driven flow of a Newtonian liquid with same maximum velocity:

$$v_x(r) = 89[1 - r^2]$$

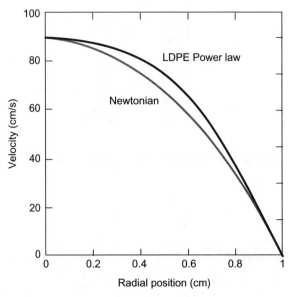

FIGURE E3.6

3.2.4 Heat Transfer Fundamentals

Heat transfer is essential not only for achieving the high temperature molten state, but also for solidifying the melt into a solid part. The speed of the latter has greater importance to the shaping process and structure of the final part. Usually, we are interested predicting the time needed to solidify a part from the conditions of cooling as well as the geometry and properties of the material itself. Two primary heat transfer mechanisms come into play: conduction and convection.

Conduction is transfer of heat by motion or vibration in a material or between two materials in contact. In liquids, conduction occurs by molecular collisions, while in solids, the mechanism involves free electron drift (metals) or lattice vibrations called phonons (nonmetals). Fourier's law of heat conduction gives the relationship between the heat flux, q, in J/s, and the temperature gradient, dT/dx:

$$q_x = -kA_x \frac{dT}{dx} \qquad (3.51)$$

where k is the thermal conductivity of the material and A_x is the area perpendicular to the heat flux, as shown in Figure 3.20. The thermal conductivity is a materials property and ranges over several orders of magnitude, as shown in Table 3.6. Thermal conductivities can vary with temperature and so frequently the temperature range over which the value holds is specified.

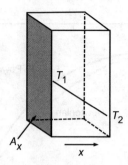

FIGURE 3.20 Coordinate system and conditions for Fourier's law of conduction.

TABLE 3.6 Room Temperature Thermal Properties of Solid Engineering Materials[a]

Material	Thermal conductivity (W/(m·K))	Specific heat capacity (J/(kg·K))	Thermal expansion coefficient $\times 10^{-6}$ (°C^{-1})
Metals			
Aluminum (alloy 1100)	222	904	23.6
Copper (C11000)	388	385	17.0
Plain carbon steel (1020)	51.9	486	11.7
Stainless Steel (316)	16.2	500	15.9
Ceramics			
Soda-Lime-Silica Glass	1.7	840	9.0
Fused Silica Glass	1.4	740	0.4
Alumina	39	775	7.4
Silicon Carbide	80	670	4.6
Thermoplastic Polymers			
Polypropylene	0.12	1925	146–180
Polyethylene (HDPE)	0.48	1850	106–198
Polycarbonate	0.20	840	122
Nylon 6,6	0.24	1670	81–117

[a] Data tabulated in Callister (2007).

Figure 3.20 shows a steady state in which the temperature gradient is stable with time; however, cooling is inherently transient, not steady state. To predict the temperature as a function of position and time, the "heat equation" is solved with a specified set of boundary conditions.

$$\frac{\partial T}{\partial t} = \alpha \frac{\partial^2 T}{\partial x^2} \quad (3.52)$$

In this equation, α is the thermal diffusivity, which itself depends on the specific heat or heat capacity per unit mass, C_p, density, ρ, and thermal conductivity,

$$\alpha = \frac{k}{\rho C_p} \quad (3.53)$$

The thermal diffusivity has units of m²/s. The heat equation is analogous to Fick's second law that governs mass diffusion, and the thermal diffusivity is analogous to the diffusion coefficient. Solutions take forms familiar to mass diffusion; some examples follow after convection is introduced.

Convection involves transfer of heat by flow or movement of a gas or liquid from one part of a system to another. Convection is not possible in solids. Figure 3.21 shows a situation in which convection dominates: the cooling of a hot object. For simplicity, this situation, like the one illustrating conduction, is one-dimensional. In this special case, the rate of cooling of the object is limited by the convective transport of the heat away from the surface and conduction to the surface is fast by comparison. The temperature of the object remains virtually constant spatially as it cools. This situation is called "Newtonian cooling." Newton's law of cooling shows the factors that influence the heat flux for convective heat transfer:

$$q_x = hA_s(T_s - T_\infty) \quad (3.54)$$

where h is the heat transfer coefficient, A_s is the surface area perpendicular to the direction of heat transfer, and the other terms are defined in Figure 3.21.

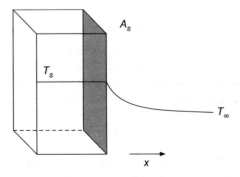

FIGURE 3.21 Coordinate system and conditions for Newtonian cooling by convection.

The heat transfer coefficient is not a physical property but rather a parameter that depends on the conditions of gas or liquid flow around the material. For example, the heat transfer coefficient is larger for a fan blowing as compared to conditions without the fan (i.e., free convection vs. forced convection).

Under Newtonian cooling conditions, the temperature of a part remains constant throughout the part and simply decreases with time. The time needed to cool an object to a specified temperature can be found by equating the heat flux away from the surface to the rate of decrease in the thermal energy of the object:

$$hA_s(T_s - T_\infty) = -m\frac{dH}{dt} = -\rho V C_p \frac{dT}{dt} \tag{3.55}$$

where H, m, V and ρ are the enthalpy, mass, volume and the density of the object, respectively. Note that dT/dt is a negative number for a situation of cooling ($T_s > T_\infty$). This differential equation can be solved to find the temperature of the part as a function of time.

$$\frac{T - T_\infty}{T_i - T_\infty} = \exp\left(\frac{-hA_s}{\rho V C_p}t\right) \tag{3.56}$$

where T and T_i are the temperature of the material at time t and $t = 0$, respectively. Considering a slab of thickness $2L$ cooled from two faces of area A_s, the half thickness of the slab, L, is a characteristic length, $L_c = V/A_s$.

Newtonian cooling requires that the resistance to conductive heat transfer ($1/k$) is less than the resistance to convective heat transfer ($1/hL_c$); the ratio of these two resistances defines a dimensionless number called the Biot number, Bi:

$$Bi = \frac{hL_c}{k} \tag{3.57}$$

When Bi is low (i.e., $Bi < 0.1$), then the conditions are those of Newtonian cooling. The Fourier number, Fo, is characteristic dimensionless number for heat conduction:

$$Fo = \frac{\alpha t}{L_c^2} \tag{3.58}$$

It can be thought of as a dimensionless time related to conduction by noticing that a small Fo represents a situation in which the length scale for conduction ($\sqrt{\alpha t}$), which scales with time, is small relative the characteristic length. With these definitions, Eq. (3.56) can be rewritten as:

$$\frac{T - T_\infty}{T_i - T_\infty} = \exp(-Bi \cdot Fo) \tag{3.59}$$

This equation only holds for low Bi. Here, the temperature inside the material remains fairly constant during cooling. That is, there are no spatial gradients in temperature. However, if the conductivity of the material is sufficiently low or the length scale of the object is sufficiently large, then temperature is a function of both position and time.

To determine the temperature of a material as a function of position and time, the heat equation is solved with appropriate boundary conditions. Let's consider two general situations. In the first, the heat transfer takes place by conduction from a "semi-infinite" material to a heat sink. The semi-infinite condition requires that the temperature in the material some distance away from the interface remains at the initial condition. That is, the physical length scale of a semi-infinite material is larger than the length scale over which the temperature changes take place. See Figure 3.22a. If the surface temperature remains T_o as time goes on, then the heat equation can be solved with these boundary conditions and the temperature in the material at any position x away from the interface with the heat sink at time t is given by:

$$\frac{T - T_o}{T_i - T_o} = erf\left(\frac{x}{2\sqrt{\alpha t}}\right) \quad \text{or} \quad \frac{T - T_i}{T_o - T_i} = erfc\left(\frac{x}{2\sqrt{\alpha t}}\right) \qquad (3.60)$$

In this solution, *erf* is the error function and *erfc* is the complementary error function.

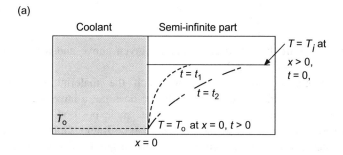

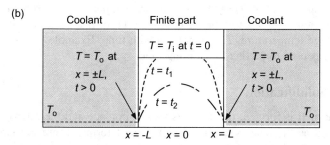

FIGURE 3.22 Transient temperature distributions for the situation with (a) heat transfer from a semi-infinite part to a coolant or heat sink under conditions with constant surface concentration and (b) heat transfer from a finite part to heat sinks.

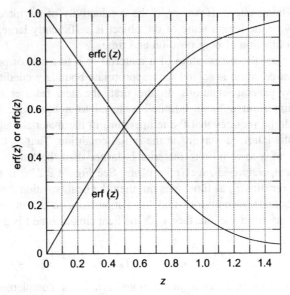

FIGURE 3.23 Error function and complementary error function.

$$erf(z) = \frac{2}{\sqrt{\pi}} \int_0^z e^{-z^2} dz \quad \text{and} \quad erfc(z) = \frac{2}{\sqrt{\pi}} \int_z^\infty e^{-z^2} dz \quad (3.61)$$

The definitions of these two functions are given above and a plot showing the functions is in Figure 3.23.

The second situation is one in which the material is finite. See Figure 3.22b. For this situation, the solution does not a have a closed form, but is an infinite series. The temperature at some position in the material or the average temperature of the material can be found. Graphical solutions are common. For example, Figure 3.24 shows a graphical solution for cooling of a material under the conditions described in Figure 3.22b. That is, the initial temperature is T_i and the surface temperature is forced to T_o, the temperature of the surrounding cold material or heat sink, at time $t > 0$. The average temperature, T_{ave}, increases with time. Example 3.7 shows how the graph is used to predict the cooling time.

3.2.5 Solidification

After the melt is formed, the shape is locked in place and the microstructure established as the heat is extracted and a solid forms. There are two main modes of solidification: crystallization and glass transition. Understanding and controlling solidification is necessary to establish the desired microstructure and properties of the solid, and prevent damaging defects.

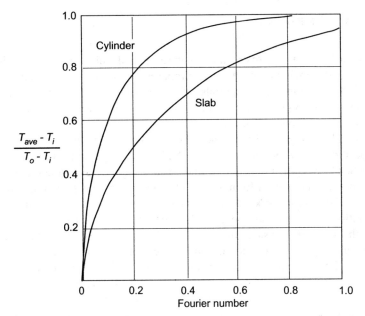

FIGURE 3.24 Graphical solutions to the heat equation for a finite medium. The initial temperature of the material is T_i, the surrounding coolant has a temperature of T_o, and the average temperature of the material after time t is T_{ave}. The solution for the slab has thickness $2L$ ($L_c = L$) and for the cylinder of radius R ($L_c = R$; note that this is a different convention than Table 3.8). *Adapted from Carslaw and Jaeger (1959).*

EXAMPLE 3.7 Polycarbonate melt at 300°C is injected into a steel mold at 30°C. Estimate the time for the polymer to reach the heat distortion temperature (150°C) for (a) 2 mm thick slab and (b) 10 mm radius cylinder.

So, the goal is to achieve an average temperature of 150°C. $T_o = 30°C$ and $T_i = 300°C$.

$$\frac{T_{ave} - T_i}{T_o - T_i} = \frac{150°C - 300°C}{30°C - 300°C} = \frac{-150°C}{-270°C} = 0.56$$

Therefore according to Figure 3.24, $Fo = 0.27$ for the slab geometry and $Fo = 0.09$ for the cylinder geometry.

First, we need to find the thermal diffusivity. The density of polycarbonate is ~ 1200 kg/m^3. The other properties of polycarbonate are in Table 3.6.

$$\alpha = \frac{k}{\rho C_p} = \frac{0.20 \ W/(m \cdot K)}{(1200 \ kg/m^3)(1250 \ J/(kg \cdot K))} = 1.3 \times 10^{-7} \ m^2/s$$

(a) $L_c = 1$ mm

$$Fo = \frac{\alpha t}{L_c^2} = \frac{(1.3 \times 10^{-7} m^2/s)t}{(0.001 \ m)^2} = 0.13t$$

$0.27 = 0.13t$
$t = 2.08$ s

(b) $L_c = 10$ mm

$$Fo = \frac{\alpha t}{L_c^2} = \frac{(1.3 \times 10^{-7} m^2/s)t}{(0.01 \ m)^2} = 0.0013t$$

$0.09 = 0.0013t$
$t = 69.2$ s

Crystallization. Microstructural features that develop during crystallization from the melt influence many properties of polycrystalline and semicrystalline engineering materials. Conceptually simple, crystallization consists of two steps: nucleation, or the formation of small regions of crystalline order, and growth of these small regions into crystals. Crystallization is also complex, influenced profoundly by chemical composition, surfaces, and compositional and thermal gradients. Additionally, crystallization, or rather the lack of crystallization, is essential to forming a glass, on cooling. And, controlled crystallization from a solid glass creates a "glass-ceramic." In this section, the groundwork for understanding crystallization is laid with a description of nucleation and crystal growth of a pure material as it cools from the melt in a system free of thermal gradients. Later in the chapter, these concepts are elaborated on in the context of melt-based processes.

As a liquid melt is cooled below its equilibrium melting point, T_{mp}, the solid, crystal phase becomes thermodynamically stable, and the liquid phase becomes unstable and is considered "supercooled." That is, the Gibbs free energy per unit volume of the solid G_s is less than that of the liquid, G_l. The change in the overall Gibbs free energy on crystallization is:

$$\Delta G_v = G_s - G_l < 0 \qquad (3.62)$$

As shown in Figure 3.25, this difference, ΔG_v, becomes a larger and larger negative number as the temperature drops further below T_{mp}. Hence the thermodynamic driving force for the liquid to solid transformation increases as the temperature drops. Once this driving force appears, the first step in crystallization, nucleation, can take place.

Homogeneous nucleation involves the formation of nuclei at random in the bulk of the supercooled liquid melt. The energetic benefit is clear—the solid nucleus has lower free energy than the liquid. However, there is also an

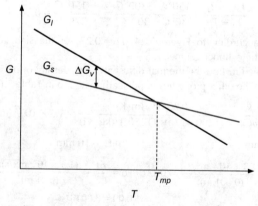

FIGURE 3.25 Change in Gibbs free energy with temperature for a liquid and solid of the same composition.

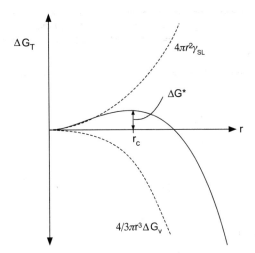

FIGURE 3.26 Change in Gibbs free energy with a nucleus radius.

energetic penalty—an interface with a higher free energy than the bulk material is formed. Hence the total change in the Gibbs free energy on forming a spherical nucleus of radius r is:

$$\Delta G_T = \frac{4}{3}\pi r^3 \Delta G_v + 4\pi r^2 \gamma_{SL} \quad (3.63)$$

where γ_{SL} is the interfacial energy between the solid nucleus and the surrounding, supercooled liquid. Since ΔG_v is negative, the first term in Eq. (3.63) becomes more negative as r increases. The second term, however, becomes more positive as the size of the nucleus increases. As shown in Figure 3.26, ΔG_T is dominated by the interfacial effect at low nucleus size and the volume free energy effect at larger sizes. An energy barrier, ΔG^*, appears, representing the free energy of formation of a critically sized nucleus, r_c. Nuclei or embryos smaller than this critical size are not stable, tending to shrink in order to lower the energy of the system. However, a nucleus that is of critical size grows. Analysis of Eq. (3.63) shows that both the critical nucleus and the barrier decrease as the temperature is lowered below T_{mp}. See Example 3.8. Therefore, as temperature decreases, the likelihood that nucleation occurs increases. That is, the rate of nucleation is a strong function of temperature.

The rate of nucleation, I, is defined as the number of nuclei formed per volume per time. Considering that the system has a statistical distribution of nuclei sizes, the concentration of those nuclei of critical size in number per volume is given by:

$$n_c = n_o \exp\left(\frac{-\Delta G^*}{kT}\right) \quad (3.64)$$

EXAMPLE 3.8 Derive expressions for the critical radius, r_c, and energy barrier, ΔG^*, and show that both become smaller as the temperature drops further below T_{mp}.

To find r_c, we need to take the derivative of the expression for ΔG_T and set it equal to 0.

$$\frac{d\Delta G_T}{dr} = 0 \text{ at } r = r_c$$

$$\frac{d\Delta G_T}{dr} = 4\pi r_c^2 \Delta G_v + 8\pi r_c \gamma_{SL} = 0$$

$$r_c = \frac{-2\gamma_{SL}}{\Delta G_v}$$

As temperature decreases, ΔG_v becomes a larger negative number and therefore r_c becomes smaller.

$$\Delta G_T = \Delta G^* \text{ when } r = r_c$$

$$\Delta G_T = \Delta G^* = \frac{4}{3}\pi r_c^3 \Delta G_v + 4\pi r_c^2 \gamma_{SL}$$

$$\Delta G^* = \frac{4}{3}\pi \left(\frac{-2\gamma_{SL}}{\Delta G_v}\right)^3 \Delta G_v + 4\pi \left(\frac{-2\gamma_{SL}}{\Delta G_v}\right)^2 \gamma_{SL}$$

$$= \frac{-32\pi \gamma_{SL}^3}{3(\Delta G_v)^2} + \frac{16\pi \gamma_{SL}^3}{(\Delta G_v)^2} = \frac{-32\pi \gamma_{SL}^3}{3(\Delta G_v)^2} + \frac{48\pi \gamma_{SL}^3}{3(\Delta G_v)^2}$$

$$\Delta G^* = \frac{16\pi \gamma_{SL}^3}{3(\Delta G_v)^2}$$

As temperature decreases, ΔG_v becomes a larger negative number and therefore, the barrier also becomes smaller.

---

where n_o is the concentration of atoms or molecules overall. The nucleation rate is the product of n_c and the frequency at which atoms or molecules are added to the critically sized nucleus, ν, making it a stable nucleus that grows.

$$I = n_c \nu \tag{3.65}$$

The frequency of atom addition is related to the diffusion of atoms or molecules in the melt and takes the form:

$$\nu = \nu_o \exp\left(\frac{-Q_d}{kT}\right) \tag{3.66}$$

where Q_d is the activation energy for diffusion and ν_o is a fundamental vibrational frequency. Therefore, the nucleation rate has two important terms, one representing the number of potential nuclei per unit volume and the other, the frequency at which these potential nuclei become stable.

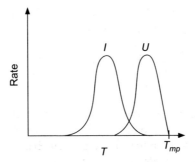

FIGURE 3.27 Schematic diagram showing the effect of temperature on the homogeneous nucleation rate (I) and crystal growth rate (U).

Lumping the pre-exponential terms from Eqs. (3.64) and (3.66) into a single term A,

$$I = A \exp\left(\frac{-\Delta G^*}{kT}\right) \exp\left(\frac{-Q_d}{kT}\right) \qquad (3.67)$$

The temperature dependence of nucleation rate is shown schematically in Figure 3.27. At the melting temperature, there is no thermodynamic driving force for nucleation (i.e., $\Delta G_v = 0$ $\Delta G^* = \infty$) and hence the nucleation rate is zero. As the temperature falls, ΔG_v becomes a larger and larger negative number and ΔG^* drops so that the concentration of critically sized nuclei increases. At the same time, the frequency of atom or molecule addition to the critical nucleus drops. These competing effects result in a peak in nucleation rate at some temperature below T_{mp}.

Heterogeneous nucleation is a similar process, but it involves the formation of the nucleus on a solid surface, such as a container wall or an impurity particle in the melt. Heterogeneous nucleation takes place on favorable surfaces, those with low interfacial energy with the crystal nucleus, and hence the energetic penalty for the formation of the new interface drops. The barrier, ΔG^*, for heterogeneous nucleation is therefore considerably lower and the peak in nucleation rate shifts closer to the melting point.

Crystal growth follows nucleation. Growth is basically the process of atom or molecule addition to a crystal and the growth rate U, is given as a linear velocity of the crystal/melt interface. While the structure and chemistry of the crystallizing material determines the details of the growth process, the crystal growth rate, in general, follows the schematic representation shown in Figure 3.27. The rate increases steeply as temperature drops below T_{mp}, reflecting the thermodynamic driving force for the phase change, reaches a peak, and then falls as temperature drops further.

This decreasing growth is due to slower diffusion at lower temperatures. A general expression for growth rate is:

$$U = \lambda v \left[1 - \exp\left(\frac{-V\Delta G_v}{kT}\right)\right] = \lambda v_o \exp\left(\frac{-Q_d}{kT}\right)\left[1 - \exp\left(\frac{-V\Delta G_v}{kT}\right)\right]$$

(3.68)

where λ is the width of the atomic or molecular layer on the crystal, and V is the volume per atom or molecule. Since ΔG_v increases dramatically as T drops below T_{mp}, the term in the square brackets becomes approximately one after a certain amount of undercooling.

In systems designed to form glasses on cooling, preventing nucleation and growth is necessary. Melts that form a glass easily, those with sluggish crystallization, tend to have high viscosity. The atom or molecule addition requires motion through the melt, as specified in the rate expressions by a frequency term, v. This frequency is expected to drop as the melt viscosity increases. Hence, the good glass forming abilities of melts with high viscosities is linked to decreased I and U for these melts.

Crystallization is the combined process of nucleation and growth that converts a liquid into a crystalline solid. Experimentally, it is easier to track the overall change as a fraction crystallized than it is to determine the individual rates of nucleation and growth. The Johnson–Mehl–Avarami equation can be used to model the isothermal change in the fraction crystallized, F, as a function of time:

$$F = 1 - \exp(-kt^n)$$

(3.69)

The rate constant, k, and the constant, n, depend on the nucleation and growth processes. For spherical growth from random, homogeneous nucleation:

$$F = 1 - \exp\left(-\frac{\pi}{3}U^3 I t^4\right)$$

(3.70)

The fraction crystallized as a function of time during isothermal holds at several temperatures is shown in Figure 3.28a. From this data, one can construct a time-temperature-transformation (TTT) plot. See Figure 3.28b. The TTT plot consists of lines of equal extents of crystallization. TTT diagrams are frequently used to plot the course of solid state transformations in steel alloys and tailor microstructures during heat treatment. In the context of melt-based processes, one important application is in the determination of conditions that prevent crystallization entirely. See Example 3.9.

Glass Transition. As a glass-forming melt cools, it undergoes a gradual transition from liquid to solid. As described earlier in this chapter, the viscosity of a liquid melt increases as temperature falls. If the melt crystallizes on cooling then its viscosity increases suddenly, essentially to infinity, at the

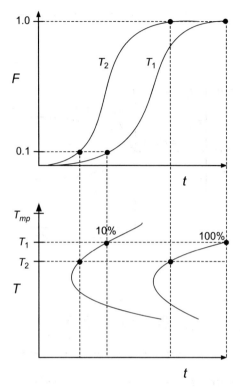

FIGURE 3.28 Schematic diagrams showing (a) the fraction crystallized as a function of time of isothermal holding at two temperatures and (b) time-temperature-transformation (TTT) diagram.

EXAMPLE 3.9 The time–temperature–transformation diagram for a glass is shown below. The line marks the beginning of crystallization. Find the cooling rate necessary to avoid crystallization for a melt at an initial temperature of 1000°C.

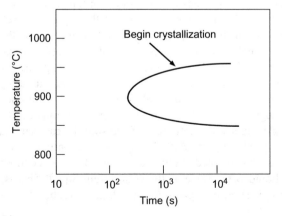

FIGURE E3.9A

To find the critical cooling rate, the time at the "nose" of the transformation is found. Notice that the scale on the x-axis is logarithmic. The paths for a constant cooling rate would be curved on this diagram.

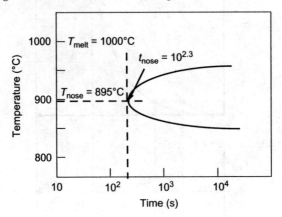

FIGURE E3.9B

We can estimate the cooling rate needed to just avoid the nose.

$$\text{Critical Cooling Rate} = \frac{T_{melt} - T_{nose}}{t_{nose}} = \frac{1000°C - 895°C}{200\ s} = 0.53°C/s$$

---

crystallization temperature. For glass forming materials, however, viscosity versus temperature data shows no discontinuous change dividing solid and liquid. If instead the volume or density of the material is monitored as a function of temperature, the difference between liquid and solid is apparent, and a characteristic temperature, the glass transition temperature, T_g, can be assigned. Likewise, experimental measurements of the heat capacity and dynamic mechanical behavior also show the glass transition.

Figure 3.29 shows a schematic of the specific volume of a material (volume per gram, inverse of density) as a function temperature during constant rate cooling. Following first the path ABCD, the liquid contracts as it cools until the equilibrium melting point is reached. The slope of the line AB is the volume thermal expansion coefficient of the liquid. Then a dramatic change takes place just below the melting point—the specific volume drops discontinuously as the material crystallizes into a solid with a higher density. Continued contraction of the solid takes place as the solid cools. Note that the slope of segment EF is less than the slope of segment AB. That is, the volume thermal expansion coefficient of the solid is less than that of the liquid. The noncrystalline path (ABEF) shows no discontinuity at T_{mp} on cooling; rather, a supercooled liquid is retained. The contraction of the supercooled liquid is the same as the equilibrium liquid. That is, the volume relaxation time is short enough so that the liquid structure has time to adjust

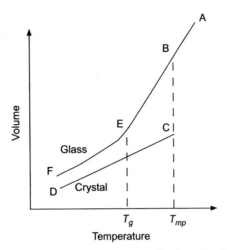

FIGURE 3.29 Comparison in the volume change on cooling for a glass forming material and a material that crystallizes.

as the temperature changes. This relaxation time scales with the viscosity so it becomes longer and longer as the temperature falls. At E, the relaxation time is long enough that the structure cannot change as a liquid would, but instead it is "frozen in place" and then from E to F contracts in the approximately the same way as the crystalline solid. This change takes place at the glass transition temperature, T_g. Not a sharply defined point, T_g is found by extrapolating the liquid and solid segments. This temperature depends on the cooling (or heating) rate.

The glass transition temperature depends on cooling or heating rate. Fast cooling results in higher T_g because the supercooled liquid structure does not have time to relax. The glass structure is frozen in at a higher temperature and it has a higher specific volume–lower density.

3.3 SHAPE CASTING

3.3.1 Process Overview

Melt shape casting processes involve pouring a melt into a mold or onto a surface where it solidifies into a shape. A general flow diagram and schematic for shape casting of metal parts is shown in Figure 3.30. The first step in the process is fabrication of the mold. The mold is designed with features to control the flow of the metal melt and define the final shape of the part, which is known as a "casting." Then, the main casting process begins with the preparation of a melt with controlled composition and temperature. The melt is transferred to the mold. For example, a ladle, filled with melt from the melting furnace, is used to pour the melt into the mold. The melt flows through a series

154 Materials Processing

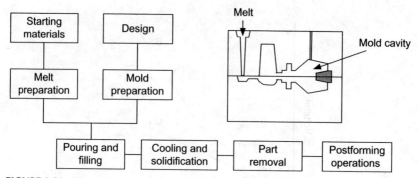

FIGURE 3.30 Flow diagram and schematic drawings for shape casting by the sand casting process as an example.

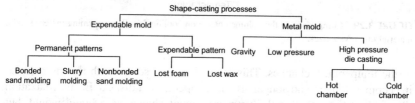

FIGURE 3.31 Shape casting categories and methods. *Adapted from Schey (2000).*

of internal mold features and into the mold cavity. Heat is extracted and the melt solidifies. After solidification and further cooling, post-forming operations are needed, including removal of melt solidified in the channels leading to the main cavity, surface finishing, and heat treatment.

For shape casting processes, the melts have low enough viscosities for pouring or flow under gravity. Polymer melts are not cast, but rather injection molded using pressurized flow and solidification. Casting processes are used to make some glass shapes, but the majority are formed into shapes by blow molding. Therefore, the emphasis in this section is on metals, which have low enough viscosities for efficient gravity induced flow. In addition to gravity casting, metal casting processes that involve pressure assistance are also included for comparison.

Casting is an important process for producing complex 3D metal parts. As such, there are numerous variations in the general casting process. Each variation addresses a specific need in terms of complexity of shape, speed of process, or control of microstructure. The myriad of processes is typically categorized first by the type of mold: expendable or permanent. In expendable mold processes, the mold is made, used, and then destroyed as the metal part is extracted. Permanent mold processes, on the other hand, involve a mold that is used repeatedly. Within each category, lie various subcategories, as shown in Figure 3.31. This section begins with melt preparation, a step that is largely the same for the different casting processes, and then continues with the fundamentals of sand casting, including mold preparation, mold filling, and solidification. Then, we explore faster alternatives to sand casting—permanent mold casting and die casting. A discussion of post-processing operations completes this section.

3.3.2 Metal Melt Preparation

The aim of the first step in the shape casting process is to create a uniform metal melt with controlled chemical composition and temperature. Several furnace types are available to create the melt, including electric arc, induction, and crucible furnaces. The "charge" for the furnace consists of the individual components in the alloy or prealloyed bars, scrap, and additives. Additives include fluxes, which help lower the gas or impurity content in the melt, and additives, which help form a slag. The slag traps impurities and helps to prevent continued reaction of the molten metal with the gas atmosphere. Key challenges in the melting stage include achieving the correct chemistry, removing solubilized gases, and limiting reactions with the container material and the atmosphere.

Reaction between metal melts and gases is a challenge and sometimes a significant safety risk. Gases from the atmosphere and the combustion process (if gas firing is used to heat the furnace) can react with and dissolve into alloy melts. Surface films or slags typically form on the molten metal as a result. Because incorporation of these reaction products into the final casting would degrade its properties, careful pouring and filling are needed, and filters are used. Also, gases that dissolve into the melt are a problem as they can generate defects on solidification.

The important melt-gas interactions and practices to minimize their effects vary from alloy to alloy. For example in steel alloys, control of oxygen is particularly important. Oxygen dissolved in the steel melt reacts with carbon, forming carbon monoxide and depleting the metal of carbon. Deoxidants that react readily with oxygen (i.e., Si) can be added to mitigate this problem. By contrast, aluminum alloys form an oxide film rapidly, but this film is protective and blocks continued reaction between the gas and melt beneath it. Pouring practices are designed to prevent the oxide film from entering the casting. Hydrogen is perhaps more damaging to Al alloys. Hydrogen forms by reaction between the melt and water vapor and can dissolve into the Al melt. The solubility of hydrogen in molten aluminum is much higher than in solid aluminum. Because of the vast difference in solubility, pores can form in the metal as hydrogen comes out of solution during solidification. Magnesium alloys also react rapidly with oxygen, but their surface films are not protective; therefore, special precautions, including inert gas blanketing and addition of fluxes, are necessary.

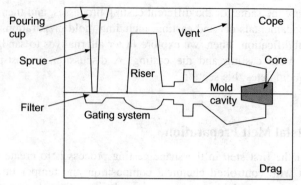

FIGURE 3.32 Schematic diagram of a sand casting mold, showing the upper half (cope), lower half (drag), and core in place.

3.3.3 Sand Casting

Mold Features. A schematic diagram of a sand mold for is shown in Figure 3.32. While used for only one casting, this type of mold has extreme flexibility in terms of its size and complexity. The importance of the various features to flow and solidification is described more below, but for now, let's follow the path of the melt and relate it to the general sequence of events in the casting process. The melt is poured into the pouring basin (or pouring cup) and from there, it travels down the sprue, through the gating system with its gates and runners, and into the riser and mold cavity, which contains a core to define an internal feature. During filling, vents allow the air originally occupying the features to exit effectively, and a filter captures inclusions in the melt. As solidification takes place, the riser feeds the mold cavity to combat solidification shrinkage.

A sequence of steps is used to create the sand mold, as described in Figure 3.33. First, the mold features are carefully designed and a pattern that details most of the internal features of the mold, including the mold cavity, is fabricated, typically in metal or plastic. The pattern can be a simple, single piece for each half of the mold, or a more complex structure with more pieces. The core box must also be fabricated and then used to create the core in two sand pieces. The mold and core are made of a sand mix, consisting of silica sand mixed with a binder (e.g., clay, organic resin). The sand mix is compacted in the flask around the cope pattern, which is typically augmented with inserts for the sprue, pouring cup, and sometimes a riser. Then the pattern and inserts are removed. A similar process is used for the lower half of the mold, the drag. The core is carefully put in place and the two halves of the mold are assembled.

Process Steps. In the sand casting process, the melt is traditionally transferred to the mold in a ladle or a series of ladles or specialty containers.

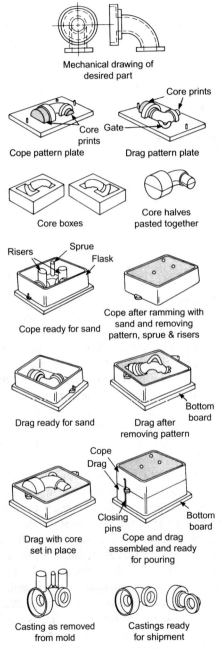

FIGURE 3.33 Preparation of a green sand casting mold. *From Blair and Stevens (1995), reproduced by permission of the Steel Founders' Society of America.*

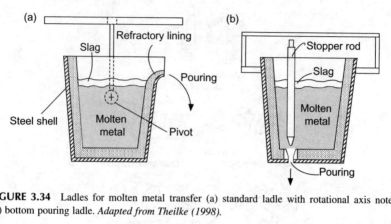

FIGURE 3.34 Ladles for molten metal transfer (a) standard ladle with rotational axis noted (b) bottom pouring ladle. *Adapted from Theilke (1998).*

Final adjustments in the melt temperatures, and sometimes composition, are carried out in the ladle. These transfers can be automated. The ladles themselves have features that help to protect the melt and create conditions for effective pouring. Figure 3.34 shows ladle configurations. The bottom pouring ladle has the advantage of pulling metal from beneath the surface of the melt and therefore, inclusions from the slag are less likely to enter the mold. The melt stream, as it exits the ladle, can be shrouded to limit exposure to oxygen in the air. Various techniques have been developed to protect the metal, especially for highly reactive metals like magnesium alloys.

In the mold filling step, the melt flows under gravity through a series of features designed to effectively direct the melt to the mold cavity. The melt enters the mold at the pouring cup or basin, which is designed as a rectangle or other asymmetric shape so that a vortex does not form and draw in air. From the basin, the melt flows down the sprue to the gating system. As the melt descends, it accelerates due to gravity. The higher velocity could cause some problems, including mold erosion. Additionally, the higher velocity causes the melt stream to contract according to the continuity principle (Eq. (3.26)). The sprue taper, however, is designed accommodate this contraction, effectively preventing a pressure difference that would otherwise result. Such a pressure difference would cause aspiration, the drawing in of air from the mold. The taper also minimizes turbulence because the diameter decreases, which lowers the Reynolds number, Re. However, turbulence is typically unavoidable, as we might anticipate from Example 3.5. Another critical point is the onset of surface turbulence, which takes place at Re of 20,000.

Bernoulli's equation is used to design sprue dimensions that prevent aspiration. See Figure 3.35. Assuming turbulent conditions and ignoring friction losses, the mechanical energy balance applied to positions 1 and 3 is:

$$g(h_3 - h_1) + \frac{1}{2}(\bar{v}_3^2 - \bar{v}_1^2) + \frac{1}{\rho}(P_3 - P_1) = 0 \qquad (3.71)$$

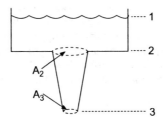

FIGURE 3.35 Schematic diagram of a pouring basin and sprue, showing the taper. See text for analysis.

The velocity of the melt is zero at position 1, the top of the pouring cup. To prevent aspiration, the pressure at point 3 must be the same as the pressure at point 1 (i.e., atmospheric pressure). Therefore, the velocity at position 3 can be found:

$$\bar{v}_3 = \sqrt{2g(h_1 - h_3)} \tag{3.72}$$

Likewise, applying the mechanical energy balance to positions 1 and 2 results in:

$$\bar{v}_2 = \sqrt{2g(h_1 - h_2)} \tag{3.73}$$

Applying the continuity principle ($\bar{v}_2 A_2 = \bar{v}_3 A_3$) results in the geometric constraint on the sprue design:

$$\frac{A_2}{A_3} = \sqrt{\frac{(h_1 - h_3)}{(h_1 - h_2)}} \tag{3.74}$$

The melt is then directed through the gating system. Here again, the goal is to minimize turbulence and achieve fast, uniform filling of the riser and mold cavity. See Example 3.10. The mold filling time is estimated by multiplying the volumetric flow rate, anywhere in the system, by the total volume of the internal spaces (i.e., sprue, runners, riser, cavity, etc). It typically takes seconds to fill a mold, depending on the size. Importantly, the mold filling time is much shorter than solidification time.

Reaction with the atmosphere continues to be an issue during pouring and mold filling. The high surface to volume ratio of the melt stream renders it susceptible to oxidation. Oxide layers can become incorporated into the melt. Filters installed in the passages prevent oxide from entering the casting itself. The melt at the leading edge of the filling liquid is most likely to have oxide incorporation and hence another design strategy is to include dead ends in the runners leading to the mold cavity. The leading edge melt goes first to the dead end before the melt following enters into the mold cavity.

At some point in the melt path to the mold cavity, the melt enters and fills a riser. The riser feeds melt to the mold cavity to compensate for the

EXAMPLE 3.10 Consider a sprue in a sand casting mold used to make Al castings. The sprue length is 6 cm and the upper opening is circular with a diameter of 1 cm. The pouring basin has a depth of 1 cm. (a) Find the diameter of the sprue at its base. (b) Calculate the volumetric filling rate. (c) Find the Reynolds number in the sprue. State assumptions.

a. Referring to Figure 3.35, let $h_3 = 0$, then $h_2 = 6$ cm and $h_1 = 7$ cm. Plugging into Eq. (3.74),

$$\frac{A_2}{A_3} = \sqrt{\frac{(h_1 - h_3)}{(h_1 - h_2)}} \Rightarrow \frac{\pi(\frac{1}{2}\text{cm})^2}{\pi(\frac{d_3}{2}\text{cm})^2} = \sqrt{\frac{(7\text{ cm} - 0\text{ cm})}{(7\text{ cm} - 6\text{ cm})}} \Rightarrow d_3 = 0.61 \text{ cm}$$

b. The volumetric filling rate can be found by finding the average velocity at the base of the sprue.

$$\bar{v}_3 = \sqrt{2g(h_1 - h_3)} = \sqrt{2(9.8 \text{ m/s}^2)(0.07 \text{ m} - 0 \text{ m})} = 1.17 \text{ m/s}$$

$$Q = A_3\bar{v}_3 = \pi\left(\frac{0.61 \text{ cm}}{2}\right)^2 (117 \text{ cm/s}) = 34.2 \text{ cm}^3/\text{s}$$

c. The Reynolds number is given by: $Re = \frac{d\bar{v}\rho}{\eta}$. Assume the density of the Al melt is 2380 kg/m³ and the viscosity is 1 mPa·s. Using the velocity and diameter above to find Re.

$$Re = \frac{d\bar{v}\rho}{\eta} = \frac{(0.0061 \text{ m})(1.17 \text{ m/s})(2380 \text{ kg/m}^3)}{0.001 \text{ Pa·s}} = 16,986$$

This Re is high enough for turbulence (>2100), but less than the Re for value noted for surface turbulence (20,000). Therefore, the flow will likely not cause problems with erosion of the mold. It is impossible to fill at a reasonable rate without surpassing 2100.

---

shrinkage there. Shrinkage occurs due to liquid metal thermal contraction (from pouring temperature to liquidus temperature) and more importantly, from the shrinkage that accompanies the change from liquid to solid. This "solidification shrinkage" varies with composition and can be substantial, as shown in Table 3.7. The riser can be located, as in Figure 3.32, at the side of the mold cavity or alternatively, it may be located in the melt path after the mold cavity and is sometimes even open to the top of the sand mold. The riser, regardless of location, is designed to solidify after the mold cavity. The riser must continue to feed melt to the mold cavity as the part solidifies. The rule of thumb is solidification time for the riser is about 1.25 times greater than the solidification time for the part itself. Inserts, known as "chills," can be added to the sand to enhance the heat transfer to the mold and locally increase solidification rate. Chills help direct the progress of solidification so that the riser can effectively feed the mold cavity.

TABLE 3.7 Shrinkage Characteristics of Metals[a]

Metal or alloy	Solidification contraction (vol%)
Aluminum	6.6
Magnesium	4.2
Cast Iron	expansion 2.5
Steel (1% Carbon)	4
Copper	4.9
Zinc	6.5
70% Cu − 30% Zn	4.5

[a]Data tabulated in Kalpakjian (1997) from original source Flinn (1963).

Likewise, the mold cavity design depends heavily on the solidification behavior. The size of the mold cavity must account for the contraction of the solid from the solidification temperature to room temperature. The size must also account for the surface finishing that is necessary to smooth surfaces of the sand casting (Example 3.11).

Solidification. Predicting solidification time is critical to optimizing production. In melt casting, solidification is linked to the heat extracted from the melt. For sand casting, a simple analysis makes use of the fact that the molten metal has a much higher thermal conductivity than the sand, and therefore, the removal of heat is limited by the conduction through the sand. Additionally, some other assumptions are made. First, we assume that the metal is pure so that it solidifies at a single temperature and as it does, latent heat is evolved. Also, the melt is assumed to be at the melting point when it comes into contact with the mold as solidification occurs; therefore, the heat transferred from the metal to the mold is entirely due to the latent heat evolution. The temperature of the metal does not decrease during solidification. The heat transfer is one-dimensional. As a consequence of these assumptions, the temperature of the casting remains constant while a changing temperature gradient develops in the adjacent sand. The situation is summarized in Figure 3.36.

The temperature distribution in the sand can be found by solving the heat equation with the appropriate boundary conditions.

$$\frac{\partial T}{\partial t} = \alpha_{\text{sand}} \frac{\partial^2 T}{\partial x^2} \qquad (3.75)$$

Here, α_{sand} is the thermal diffusivity of the sand. Recall that the thermal diffusivity is found from the thermal conductivity, specific heat capacity, and density of the material, $\alpha = k/(\rho C_p)$. Sand has a thermal conductivity of approximately 0.3 W/(m•K), a specific heat of 800 J/(kg•K), and density of

EXAMPLE 3.11 A sand mold is designed to make an aluminum casting that is $10 \times 10 \times 2$ cm. The casting is made by pouring the Al melt at 725°C. (a) Specify the mold dimensions necessary to make up for contraction of the solid on cooling, assuming that the riser effectively compensates for the shrinkage due to liquid contraction and solidification. (b) Find the volume of Al melt that should be supplied by the riser. Assume the linear thermal expansion coefficient of Al melt is 30×10^{-6} (°C)$^{-1}$.

a. According to Table 3.6, the thermal expansion coefficient of Al metal is 23.6×10^{-6} (°C)$^{-1}$. The solid contracts from 660°C (the solidification point) to 20°C ($\Delta T = 640$°C).

$$\text{Linear thermal expansion coefficient} = 23.6 \times 10^{-6} \, °C^{-1} = \frac{1}{L_o} \frac{\Delta L}{\Delta T}$$

$$\frac{\Delta L}{L_o} = 0.015 \Rightarrow \Delta L = L - L_o = 0.015 L_o$$

$$L = 1.015 L_o$$

The dimensions of the mold cavity should be $10.15 \times 10.15 \times 2.03$ cm.

b. The thermal contraction of the liquid from 725°C to 660°C are:

$$\text{Linear thermal expansion coefficient} = 30 \times 10^{-6} \, C^{-1} = \frac{1}{L_o} \frac{\Delta L}{\Delta T}$$

$$\text{Volume thermal expansion coefficient} = 90 \times 10^{-6} \, C^{-1} = \frac{1}{V_o} \frac{\Delta V}{\Delta T}$$

$$\frac{\Delta V}{V_o} = 0.006 \, (0.6\%)$$

The solidification shrinkage is 6.6% by volume. So the total amount of liquid that needs to be supplied by the riser is 7.2% the cavity volume (209 cm^3) or 15.0 cm^3. (Note: For isotropic expansion/contraction, the volume thermal expansion coefficient is three times the linear thermal expansion coefficient.)

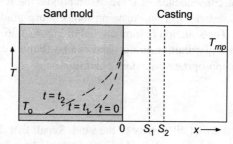

FIGURE 3.36 Temperature distribution and solidification front in sand casting at various times ($t = 0$, t_1, and t_2 where $t_2 > t_1 > 0$. Solidification front position S_1 corresponds to time t_1 and position S_2 corresponds to time t_2.

1.5 g/cm^3. Hence, $\alpha_{sand} = 2.5 \times 10^{-7}$ m^2/s. The boundary conditions call for the initial sand temperature to be T_o at $x \leq 0$, $t = 0$, and the sand temperature at the mold surface to be at the melt temperature, which is at the melting point, T_{mp} at $x = 0$, $t \geq 0$. See Figure 3.36. The solution to the heat equation, for this case, gives the temperature in the sand mold, T, as a function of position and time:

$$\frac{T(x,t) - T_o}{T_{mp} - T_o} = erfc\left(\frac{-x}{2\sqrt{\alpha_{sand}t}}\right) \quad (3.76)$$

The solution is related to that given in Eq. (3.60); the negative sign accounts for the difference in the boundary conditions. The result shows that the temperature distribution broadens with time. More importantly, the quantity of heat extracted from the casting can be calculated and used to find the position of the solid/liquid interface, the solidification front.

The position of the solidification front is found by a heat balance. The rate of evolution of latent heat as solidification occurs is given by:

$$\text{Rate of heat evolution} = -\rho_{metal}\frac{dV_s}{dt}\Delta H_{f,metal} \quad (3.77)$$

where V_s is the volume of the solidified metal, ρ_{metal} is the density of the solid metal, and $\Delta H_{f,metal}$ is the latent heat of solidification in units of J/kg. The negative sign indicates that the heat is exiting the solidifying casting. The volume rate of solidification, dV_s/dt, can be converted to the rate of thickness increase:

$$\frac{dV_s}{dt} = A\frac{dS}{dt} \quad (3.78)$$

The rate of heat extraction from the melt into the mold is determined by applying Fourier's Law (Eq. (3.51)):

$$\text{Rate of heat transfer from casting} = q_x = -k_{sand}A\frac{\partial T}{\partial x}\bigg|_{x=0} \quad (3.79)$$

where A is the area of the interface between the mold and the casting. Since the temperature gradient changes with time, so does the rate of heat extraction. The negative sign indicates that heat is extracted in the negative x direction. It is left to an exercise to show that:

$$\frac{\partial T}{\partial x}\bigg|_{x=0} = \frac{T_{mp} - T_o}{\sqrt{\pi \alpha_{sand} t}} \quad (3.80)$$

Therefore, the heat balance equates the evolution of the latent heat to the heat transfer to the mold:

$$-\rho_{metal}A\frac{dS}{dt}\Delta H_{f,metal} = -k_{sand}A\left[\frac{T_{mp} - T_o}{\sqrt{\pi \alpha_{sand} t}}\right] \quad (3.81)$$

By integration, the thickness of the solidified layer is given by:

$$S = \frac{2}{\sqrt{\pi}} \left(\frac{T_{mp} - T_o}{\rho_{\text{metal}} \Delta H_{f,\text{metal}}} \right) \sqrt{\frac{k^2_{\text{sand}} t}{\alpha_{\text{sand}}}} \tag{3.82}$$

The thickness of the solidified metal grows with the square root of time at a rate that depends on the properties of the metal being cast (i.e., melting point, density, heat of fusion) as well as the sand properties (i.e., thermal conductivity, specific heat, density).

The solidification time is the time needed for the casting to be entirely solidified. Consider first a simple slab with a thickness of $2L$ bounded on both sides by sand mold. See Example 3.12. There are two solidification fronts that move toward each other. When $S = L$, then the slab is entirely solid and $t = t_s$ (solidification time). So,

$$t_s = \frac{\pi L^2}{4 k_{\text{sand}} C_{p,\text{sand}} \rho_{\text{sand}}} \left(\frac{\rho_{\text{metal}} \Delta H_{f,\text{metal}}}{T_{mp} - T_o} \right)^2 \tag{3.83}$$

Notice that for a given metal and mold, most terms are constants that can be lumped into a constant, C.

$$t_s = CL^2 \quad C = \frac{\pi}{4 k_{\text{sand}} C_{p,\text{sand}} \rho_{\text{sand}}} \left(\frac{\rho_{\text{metal}} \Delta H_{f,\text{metal}}}{T_{mp} - T_o} \right)^2 \tag{3.84}$$

The results above can be generalized to other shapes by defining a characteristic length scale as:

$$L_c = \frac{\text{volume of mold cavity}}{\text{surface area through which the heat is extracted}} \tag{3.85}$$

The characteristic lengths for standard shapes are shown in Table 3.8. The characteristic length of the slab is half the thickness. So the solidification time can be stated in a general rule known as Chvorinov's rule:

$$t_s = C L_c^2 \tag{3.86}$$

where C is given in Eq. (3.84). The characteristic length can be estimated using the general definition and common sense. It also is helpful in designing riser dimensions. The rule gives a good general sense for the solidification time, but it is limited because the three-dimensional nature of the mold is not included. For example, the effect of curvature on the heat transfer is not taken into account. Finite element-based software is used for precise descriptions of solidification.

Microstructure Development. The polycrystalline microstructure is established during solidification. There are up to three zones in the microstructure of a typical sand casting, as shown in Figure 3.37. The chill zone consists of small, equiaxed, randomly oriented crystals near the mold. The

EXAMPLE 3.12 An iron melt at its melting temperature of 1536°C is cast into a sand mold to form a slab of thickness 0.05 m. (a) How long will it take to solidify the slab completely? (b) How thick must the sand be in order for the sand to be considered semi-infinite?

The sand is initially at 25°C. It has a thermal conductivity of 0.3 Wm^{-1}K^{-1} and a thermal diffusivity is 2.5×10^{-7} m^2/s. Iron has a density of 7265 kg/m^3 (solid near melting point) and latent heat of melting is 2.8×10^5 J kg^{-1}.

a. For a slab shape, heat is extracted out of the two large faces and solidification fronts move in from the these two faces.

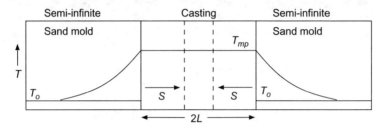

FIGURE E3.12

So, when the position of the solidification front, S, reaches half the thickness of the slab, the part is solidified. For this problem solidification takes place when S = 0.025 m

$$S = \frac{2}{\sqrt{\pi}} \left(\frac{T_{mp} - T_o}{\rho_{metal} \Delta H_{f,metal}} \right) \sqrt{\frac{k^2_{sand} t}{\alpha_{sand}}}$$

$$= \frac{2}{\sqrt{\pi}} \left(\frac{1809 \text{ K} - 298 \text{ K}}{(7265 \text{ kg/m}^3)(2.8 \times 10^5 \text{ J kg}^{-1})} \right) \sqrt{\frac{(0.3 \text{ W}/(\text{m} \cdot \text{K}))^2 t}{2.5 \times 10^{-7} \text{ m}^2/\text{s}}} = 0.025 \text{ m}$$

$t = 2471 \text{ s} = 41 \text{ min}$

b. To determine the thickness of sand required for the semi-infinite approximation to hold, one approach is to plot the temperature distribution into the sand at the solidification time. The temperature distribution is given by:

$$\frac{T(x,t) - T_o}{T_{mp} - T_o} = \text{erfc}\left(\frac{-x}{2\sqrt{\alpha_{sand} t}} \right)$$

$$\sqrt{\alpha_{sand} t} = \sqrt{(2.5 \times 10^{-7} \text{ m}^2/\text{s})(2471 \text{ s})} = 0.0249 \text{ m} \approx 2.5 \text{ cm}$$

Notice that the quantity $(\alpha_{sand} t)^{1/2}$ has units of length; it is called the thermal diffusion length. For this example, the thermal diffusion length is 2.5 cm. Therefore, the sand should be on the order of 4 to 5 thermal diffusion lengths in dimension ($\sim$10–12 cm).

TABLE 3.8 Geometrical Factors Used for Calculating Solidification Time

Shape	Geometry	Characteristic length
Slab		$L_C = \dfrac{2LA}{2A} = L$
Cylinder		$L_C = \dfrac{\pi R^2 L}{2\pi RL} = \dfrac{R}{2}$
Sphere		$L_C = \dfrac{\frac{4}{3}\pi R^3}{4\pi R^2} = \dfrac{R}{3}$

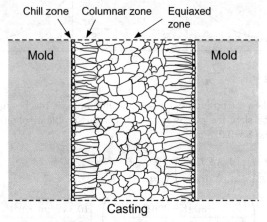

FIGURE 3.37 Schematic of the microstructure features in a section of a sand casting, showing microstructural zones. *Adapted from Beddoes and Bibby (1999).*

next zone—the columnar zone—results from the selective growth of favorably oriented chill grains. The elongated crystals grow parallel to the direction of heat flow and may contain morphological features known as dendrites, as described more below. The equiaxed zone, the innermost zone, forms away from the solidification front and consists of equiaxed crystals. All three zones may not be present. In particular, the relative proportions of the columnar and equiaxed zones depend on variables such as the alloy composition, mold geometry, and pouring temperature.

The crystallization process is coupled to heat transfer. In particular, the rate of heat extraction from the casting and the temperature gradient play roles. Crystallization of a pure metal in the absence of a temperature gradient is discussed in Section 3.2.5. Metal casting, however, is non-isothermal; the hot metal is poured into a cold mold. Since the mold is cold, nucleation occurs preferentially on the mold surface. Furthermore, the mold surface serves as a location for heterogeneous nucleation. The rapid nucleation is responsible for the fine, chill zone morphology. The details of the structure development in the columnar zone depend on the alloy system, the temperature gradient, and the rate of solidification.

Let's consider first the case of solidification of a pure metallic element (e.g., Al). In the heat transfer analysis, simplifying assumptions led to a uniform temperature in the metal melt and casting. More realistically, the melt is poured at a higher temperature than the equilibrium melting point. This temperature excess is called superheat. Superheat results in a temperature gradient in the solidifying melt. See Figure 3.38a. The temperature at the melt–solid interface is at the equilibrium melting temperature.

The mold is at a lower temperature than the melt and so nucleation occurs preferentially at the mold surface. As heat is extracted into the mold, the metal cools progressively away from the mold surface, and hence, crystals grow away from the mold. The morphology of this "solidification front" depends on the temperature gradient. Consider first the typical temperature gradient in which the melt ahead of the solidification front is hotter than the solid metal. In this case, the crystal growth occurs in a uniform, planar

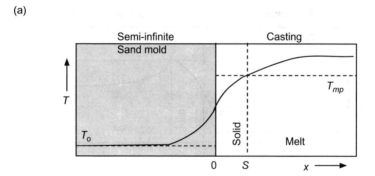

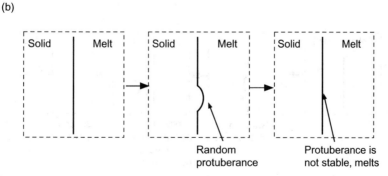

FIGURE 3.38 (a) A more realistic temperature gradient in sand casting and (b) resulting progression of planar growth.

solidification mode. Small protuberances to the front are not stable, as shown in Figure 3.38b. These random, solid projections melt in the surrounding hot liquid and growth continues as a plane.

In some instances, the melt ahead of the solidification front becomes colder than the adjacent solidified metal. That is, the melt is supercooled; it is at a temperature below its equilibrium solidification temperature. Such a temperature profile might occur if the mold cavity is open so that heat is removed out of a free surface. See Figure 3.39a. If the melt is at a lower temperature than the solid, then small protuberances can grow because latent heat is extracted through the melt, which is at lower temperature. The solidification front becomes unstable. For lower gradients or slower rates of growth, the instability leads to a mode of growth called cellular growth; uniformly spaced projections, or cells, form and grow outward in the direction of heat extraction. For steeper gradients or faster growth rates, the mode of growth becomes dendritic. See Figure 3.39b. Protuberances form and extend into the melt. Then an instability again occurs at the new solidification fronts and additional

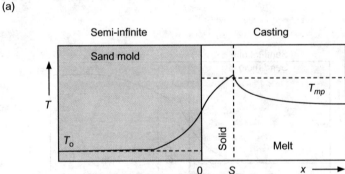

FIGURE 3.39 (a) An inverse temperature gradient in sand casting and (b) resulting progression of dendritic growth.

protuberances form. Eventually, a tree-like structure called a dendrite forms. Dendrite formation in pure metals requires this inverse temperature gradient and therefore, it is relatively rare. But in alloys, dendrite growth is more common because of a composition-based or "constitutional" supercooling.

To understand constitutional undercooling, consider a simple alloy of metal A with additive B as shown in the phase diagram of Figure 3.40. When a melt of composition C_o is cooled, it begins to crystallize at the liquidus temperature, T_L, into a solid that has a lower amount of B than C_o. As cooling continues, the "partitioning" of the solute B between the solid and liquid phases continues. The partition coefficient, k, is the ratio of the composition of the solid to that of the liquid, $k = C_{solid}/C_{liquid}$. If the liquidus and solidus lines are approximately straight, then k is a constant. The figure depicts an alloy with $k < 1$, but it is also possible to have $k > 1$. As k approaches 1, the spacing between the solidus and liquidus lines decreases. If equilibrium is maintained, the composition of the liquid and the composition of the solid change continuously during cooling until the solidus temperature, below which the material is entirely solid with the same composition, C_o, as the original liquid. Solidification during casting, however, is nonequilibrium. There are thermal gradients and diffusion in the solid and the liquid is not fast enough to erase the concentration gradients that develop during crystallization. As a consequence, solute that is rejected to the liquid accumulates at the solid/liquid boundary.

The accumulation of solute in the liquid in advance of the solidification front results in a chemical gradient in the liquid, and this chemical gradient leads to a gradient in the liquidus temperature. Figure 3.41a depicts a steady state compositional profile in the liquid adjacent to the solidification front. The local liquidus temperature is depressed near the solidification front where the concentration of B is higher. Far from the solidification front, the liquid composition is C_o and the liquidus temperature is higher. Constitutional undercooling occurs if the actual temperature is lower than

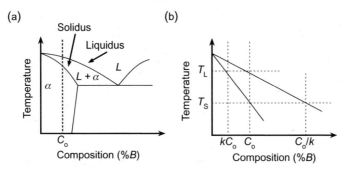

FIGURE 3.40 Example phase diagram of the A-B binary alloy system: (a) overview and (b) close up of solidification region for composition C_o.

the liquidus temperature as in Figure 3.41b. The constitutionally undercooled melt is subject to the same sort of instability in growth experienced by the pure metal melt due to an adverse temperature gradient. If, however, the temperature gradient is steep enough, as in Figure 3.41c, the temperature of the liquid remains higher than the liquidus temperature, constitutional undercooling does not occur, and the growth is planar.

Dendrites have several consequences. First, the formation of the dendrites itself is a consequence of compositional heterogeneity so this heterogeneity may impact the performance of the solidified metal part if it is not reversed by a post-forming heat treatment. More serious is the possibility of microporosity. The dendrite morphology can inhibit the flow of melt to the now complicated solidification front, preventing the compensation of solidification shrinkage. Lastly, the strength of the alloy may suffer if the dendrites are large. Fast solidification is preferred as the increased speed leads to smaller dendrite arm spacings and higher strengths, the strengthening effect being akin to the enhancement of strength that accompanies grain size reduction.

Casting microstructure is influenced by many variables, including alloy composition, additives, solidification rate, pouring temperature, and mold temperatures. Procedures and recommended practices have been developed for the different alloy systems and in many cases, specific alloy chemistries have been designed in order to achieve microstructure and property control. Control over thermal conditions during pouring and in the mold itself improves results. Of particular interest is encouraging the transition from the columnar zone to the central equiaxed zone. The columnar zone can be a site of weakness due to the directional microstructure. The central equiaxed zone forms by nucleation away from the solidification front that moves in from the mold surfaces. The melt needs to be supercooled away from the

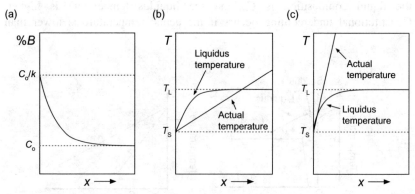

FIGURE 3.41 (a) Steady state compositional profile in the liquid adjacent to the solidification front for the alloy system depicted in Figure 3.40. Gradients in liquidus temperature and actual temperature adjacent to the solidification front for conditions of (b) constitutional undercooling and dendritic growth and (c) no constitutional undercooling and planar growth.

solidification front for this zone to form. Its formation is also encouraged by additives, known as grain refiners or inoculants, or by broken off dendrites. These serve as sites for heterogeneous nucleation away from the main growth front.

Advantages and Disadvantages. Sand casting has a number of advantages for metal shape casting. Perhaps one of the most important is that the size and shape of the casting are virtually without bounds. Extremely large castings, such as large anchors for cruise ships, and complicated shapes, such as engine blocks can be produced. Additionally, nearly all metals can be sand cast. The only exceptions are those that cannot be easily melted. Sand casting is inexpensive and suitable for short runs of only a few castings as well as long production runs.

The downsides of sand casting are related to the use of sand as a casting medium. First, the production speed is slow due to the time needed to prepare each mold and for the solidification itself. The interface between the casting and the sand is rough and therefore, surface finishing is needed and the dimensional tolerances are not that high. In particular, small precision parts cannot be made by sand casting. These disadvantages are overcome by faster, more precise metal casting methods, such as permanent mold casting and die casting.

3.3.4 Permanent Mold Casting

Molds and Process Steps. Permanent mold casting is a metal shape casting process that is similar to sand casting, but this process uses permanent, high thermal conductivity molds. For example, Figure 3.42 shows an example of

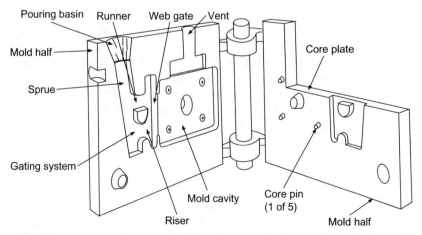

FIGURE 3.42 Example of a permanent mold for metal casting. *From West and Grubach (1998), reproduced by permission of ASM International.*

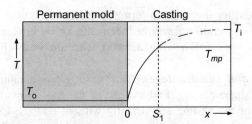

FIGURE 3.43 Temperature distribution and solidification front position in permanent mold casting at time $t = t_1$. At time $t = 0$, the casting has a $T = T_{mp}$ at $x = 0$. See text for explanation.

a permanent mold design. The features bear some similarity to those in sand molds, but the mold itself must be made of a durable material capable of withstanding repeated use, especially thermal cycling. Common materials for permanent molds include steel, iron, and bronze. The metal molds themselves are frequently called "dies." Coatings on the interior mold surfaces are applied to prevent adhesion and assist part removal. The features of the permanent mold must allow for the part to be removed as the two halves are separated. "Draft," or angled, surfaces are designed into all internal features so that the solidified metal can be removed. By contrast, a sand mold is destroyed as the part is extracted. Cores made of metal or ceramic are also used, keeping in mind the restrictions of opening and closing the mold. Permanent molds are made by machining. Vents are more important in permanent molds as the mold material is not permeable. Ejector pins are also often present to assist in part removal.

The higher thermal conductivity of permanent molds greatly increases the rate of heat transfer and reduces the solidification time. For a quantitative assessment, consider a metal mold that remains at a constant temperature by water. To the first approximation, the conduction of heat through the metal casting, not the mold, limits the solidification rate. Figure 3.43 shows the boundary conditions for the solution of the heat equation and solidification time for permanent mold casting. At $t = 0$, the metal melt is assumed to have a constant temperature of T_{mp} and the mold has a constant temperature of T_o. Because the mold stays at T_o, the metal in contact with the mold is forced to T_o and a temperature gradient develops into the casting at $t > 0$. Like the analysis of sand casting, we assume that the solidification is coincident with the removal of latent heat.

Solidification. The procedure to find the solidification position as a function of time is similar to that used for sand molds. However, the metal in the casting is assumed to be semi-infinite and the temperature distribution in the casting is determined rather than the temperature distribution in the mold. Therefore, the thermal diffusivity in the heat equation is that of the solid metal casting (α_{metal}) and solving for the temperature distribution requires an approximation. The temperature distribution is found by assuming that the metal casting is semi-infinite, reaching some constant temperature, T_i, at some distance away from the mold-casting interface. Equation (3.60) gives the solution to the heat equation for the semi-infinite boundary conditions. However, the metal is not really semi-infinite; therefore, an approximation is made by introducing another restriction. The temperature at the solidification front, $x = S$, is T_{mp}, a condition which effectively truncates the temperature distribution. With this condition, an expression for T_i is found and the solution becomes:

$$\frac{T - T_o}{T_{mp} - T_o} = \frac{erf\left(\frac{x}{2\sqrt{\alpha_{metal}t}}\right)}{erf\left(\frac{S}{2\sqrt{\alpha_{metal}t}}\right)} \quad \text{for } 0 < x \leq S \tag{3.87}$$

To simplify the expression, let

$$\phi = \frac{S}{2\sqrt{\alpha_{metal}t}} \tag{3.88}$$

For any given time, ϕ is a constant. Therefore, the position of the solidification front, S, varies with the square root of time, the same dependence as sand casting.

Once again, to find S we need to balance the latent heat evolved during solidification with the rate of heat removal. Now, the rate of heat removal is determined by finding the heat flux per unit area at the solidification front.

$$-k_{metal}\frac{\partial T}{\partial x}\bigg|_{x=S} = -k\left[\frac{T_{mp} - T_o}{erf\phi\sqrt{\pi\alpha_{metal}t}}e^{-\phi^2}\right] \tag{3.89}$$

The latent heat removed is the same as that derived earlier. Converting this expression to a per unit area basis, in a similar way as the analysis of sand casting,

$$-\rho_{metal}\frac{dS}{dt}\Delta H_{f,metal} = -k\left[\frac{T_{mp} - T_o}{erf\phi\sqrt{\pi\alpha_{metal}t}}e^{-\phi^2}\right] \tag{3.90}$$

Given the definition of ϕ:

$$\frac{dS}{dt} = \phi\sqrt{\frac{\alpha_{metal}}{t}} \tag{3.91}$$

Grouping all the terms with ϕ on one side and using the definition of thermal diffusivity:

$$\phi e^{\phi^2} erf\phi = \frac{C_{p,metal}}{\Delta H_{f,metal}}\left[\frac{T_{mp} - T_o}{\sqrt{\pi}}\right] \tag{3.92}$$

To find solidification time, the information about the metal and conditions is put into the right side of the equation and then the value of ϕ is found either graphically or with a simple spreadsheet or computer program (Example 3.13).

EXAMPLE 3.13 A copper melt at its freezing temperature of 1084°C is cast into a chilled steel mold to form a slab of thickness 0.05 m. How long will it take to solidify the slab completely? (b) The copper melt has a thermal diffusivity is 3.77×10^{-5} m²/s, latent heat of melting is 2.054×10^5 J kg^{-1}, and specific heat of 531 J kg^{-1}K^{-1}. The metal mold is kept at 20°C.

$$\phi e^{\phi^2} \operatorname{erf} \phi = \frac{C_{p,\text{metal}}}{\Delta H_{f,\text{metal}}} \left[\frac{T_{mp} - T_o}{\sqrt{\pi}} \right] = \frac{531 \text{ J/(kg} \cdot \text{K)}}{205{,}400 \text{ J/kg}} \left[\frac{1357 \text{ K} - 293 \text{ K}}{\sqrt{\pi}} \right] = 1.552$$

Now we can find ϕ using a graph or a spreadsheet. From this analysis, we find that $\phi = 0.89$.

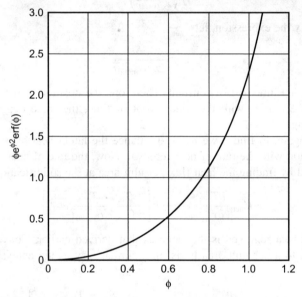

FIGURE E3.13

Solidification takes place at $S = 0.025$ m

$$\phi = \frac{S}{2\sqrt{\alpha_{\text{metal}} t}} = \frac{0.025 \text{ m}}{2\sqrt{(3.77 \times 10^{-5} \text{ m}^2/\text{s})t}} = 0.89$$

$t = 5.2$ s

It seems amazing, but the solidification is very fast when the thermal conductivity of the mold is high.

The rapid rate of solidification in permanent mold casting leads to the use of different methods to enhance the filling rate and ensure uniformity of fill. Simple gravity pours (i.e., static pours) are used, but often a tilted pour strategy is employed. In the tilted pour, the mold is oriented with its parting line horizontal, a pouring basin is filled, and then the mold is tilted so that

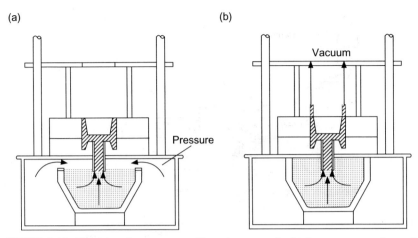

FIGURE 3.44 Examples of permanent mold casting setups for (a) low pressure casting and (b) vacuum casting. *From West and Grubach (1998), reproduced by permission of ASM International.*

the parting line is vertical. During the tilt, melt flows down the runners into the riser and mold cavity. Turbulence is decreased and the vertical orientation for parting is best for ejection of the part. Vacuum and low pressure casting methods employ pressure assist to the melt flow. In vacuum casting, a vacuum is pulled on the mold to draw in melt from a reservoir, which is located below the mold. Similarly, in low pressure casting gas pressure is applied to the metal melt in the reservoir, forcing it upward into the mold. Figure 3.44 shows examples. Finally, controlling the progression of solidification in the mold is essential so that the riser can adequately feed into the mold cavity. Risers feed the thicker sections of the mold cavity.

Advantages and Disadvantages. Permanent mold casting is typically selected for higher volume production runs or when superior surface finish, microstructure, or dimensional accuracy are needed. Permanent mold cost is considerable, scaling with the complexity and size of the mold. The cost of the mold must be justified based on the increased number of units sold. The rapid solidification tends to create fine microstructures, often providing superior properties compared with sand casting. However, the size of the cast part is limited in permanent mold casting, as are the alloy selections. Cast parts are typically no more that 100 lbs or so. As mentioned, the parting of the mold halves leads to shape limitations. And, typically only non-ferrous alloys are cast by these methods.

3.3.5 Die Casting

In die casting, the metal melt is driven under high pressure into a permanent metal mold or "die." The pressure ensures a rapid fill. Also, pressure, rather

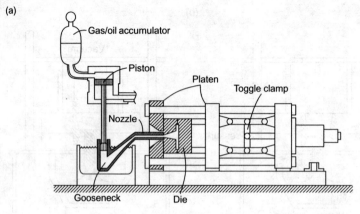

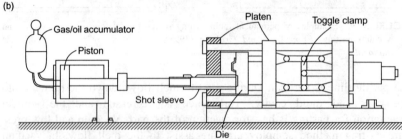

FIGURE 3.45 (a) Hot chamber die casting and (b) cold chamber die casting. *From Piwonka (1998), reproduced by permission of ASM International.*

than a riser, is used to compensate for shrinkage. That is, a pressurized melt feeds the mold cavity as solidification occurs. Like permanent mold casting, die casting is limited to lower melting point, nonferrous alloys. Aluminum, magnesium, copper, and zinc alloys are the most common.

There are two types of die casting processes, based on the handling of the metal melt. The hot chamber die casting process incorporates a reservoir for the melt. See Figure 3.45a. A "shot" of metal melt fills a gooseneck and is then ready for injection into the die via a nozzle. In cold chamber die casting, the melt is prepared in a separate furnace, and then a shot of melt is transferred via an automated ladle to the shot sleeve, which is connected to the die. See Figure 3.45b. In the cold chamber process, the molten metal is exposed to the piston apparatus for a short amount of time; therefore, the cold chamber process is preferred for higher temperature alloys, such as Al alloys. In both apparatuses, there is a hydraulic actuating system with a piston, a metal die that contains the mold cavity as well as various channels to direct the melt, as described below, and a clamping system to hold the die closed during injection and to release the part at the end of the casting cycle.

The metal die is comprised of two halves that are precisely manufactured out of a high strength alloy steel. In one half, the die connects with the nozzle or shot sleeve, depending on whether the die is designed for cold or hot chamber operation. From this entry point, the melt travels down the sprue and through runners to the gates and eventually into the mold cavity or mold cavities, in multipart molds. These multipart molds are similar to family molds that are described with injection molding in the next section. No riser is needed. Vents are machined into the die so that there is a path for air escape. Simple cores can be used to create some internal features, but like permanent mold casting, the shapes are limited. Mold temperature is controlled by flowing liquid, such as chilled water, through channels that are machined into the mold. The mechanism for ejecting the solidified part is included in the other half of the die.

The die casting process is executed in a series of steps. In the first step, the shot is prepared either by moving the piston in the hot chamber machine to fill the gooseneck or by careful addition of a shot to the shot sleeve of a cold chamber machine. Second, the mold is prepared with a coating, closed, and clamped in place. The third step is injection; the piston moves forward with a controlled motion. The piston velocity is controlled so that air can be effectively expelled through vents initially, and then the speed can be fast enough to fill the cavity before solidification. When the cavity fill is complete or nearly complete, the piston slows but the force on the piston increases to keep the pressure on the melt high as the fourth step, solidification, takes place. Lastly, the part is ejected as the mold is opened, and ejector pins push out the part. The process is typically highly automated and can produce hundreds of parts per hour.

Die casting involves rapid injection velocities and high pressures. The liquid metal enters the die rapidly, up to 80 m/s. Air can become entrapped and surface oxide films incorporated into the casting due to this rapid injection. Ramping the velocity of the piston can help with this initial step. 3D flow and modeling can also be employed to optimize the die design and conditions for casting. Maintaining pressure on the melt during solidification combats solidification shrinkage. Up to ~200 MPa pressure is used and hence, large forces are needed to keep the mold closed. The clamping force can be estimated by multiplying the cumulative projected area of the mold cavities and gate/runner system by the pressure of the metal melt. The projected area is the area of the features of the mold projected onto a plane that is perpendicular to the direction of injection. Hence, larger parts require larger clamping forces. Die casting machines are rated by their clamping force, which typically ranges from 450 kN to 35 MN.

Advantages and Disadvantages. Die casting's chief advantages are high production rate, dimensional accuracy, and smooth surface finish. After removal of the extraneous solidified metal (i.e., sprue and runners), the die cast part is usually nearly finished. Some machining may be necessary, depending on the final product requirements, but typically these steps are much less extensive than other casting processes, especially sand casting. Because steel molds are required, the size of the die cast part and the alloys that can be used are limited, similar to permanent mold casting. The part design is centered on thin sections, similar to injection molding polymers.

3.3.6 Post-Processing of Cast Metal Parts

After the casting is cooled, there are several post-processing (i.e., post-forming) operations that may be done, depending on the specific casting process, alloy composition, and final product geometry and property requirements. All castings are solidified along with extraneous metal in the form of risers, runners, sprue, etc. These extra metal pieces are removed by fracture or a cutting process with a saw or torch. This step can be automated and the extra metal recycled back into the process. Following this step, there are three main operations that are carried out: surface cleaning and finishing, machining, and heat treatment.

Surface cleaning and finishing are typically accomplished by blasting some form of steel shot or refractory grit. For most castings, the removal of extraneous metal leaves behind ragged edges that must be smoothed. Further, the surfaces of sand cast pieces frequently contain embedded sand and are naturally rough; therefore, the entire surface must be smoothed. The blasting process is carried out in chambers to contain the grit. Some alloys are also subjected to chemical treatments for surface cleaning.

Machining of a casting may be necessary to add in some extra details or features that are not easily created by the casting process. For example, holes may be drilled in some locations to provide points for an assembly process. Some machining may also be performed to increase the dimensional accuracy of critical portions of the casting. There are a wide variety of machining operations, most automated with tight control of tool positions and final dimensions.

Heat Treatment. Melt cast alloy products are frequently heat treated to either remove thermal stresses, enhance chemical homogeneity, or modify the microstructure and properties. The time–temperature profiles required for these heat treatments vary with alloy composition as well as the size of the casting. Heat treatment is not used for die cast parts as these may have considerable porosity. Gas encased in the pores expands and causes distortion or cracks. The fundamentals of heat transfer can be used to predict the time necessary to achieve a desired temperature or temperature profile. Additionally, phase diagrams and considerations of the kinetics of solid-solid phase transformations are needed to understand the heat treating process. In this section, heat treatments are discussed for three examples of cast alloy systems: medium plain carbon steel, ductile cast iron, and an aluminum alloy.

Heat treatment for thermal stress relief follows the same general procedure regardless of alloy chemistry. Thermal stresses may result from nonuniform cooling and are typically found in castings with complex geometric features or sections of varying thickness. Essentially, the equilibrium thermal contraction of one part of a casting is restricted by an adjoining part that is cooling and contracting at a different rate. Thermal stress relief is accomplished by reheating the parts slowly to a temperature high enough to allow atomic diffusion and dislocation motion to occur at an appreciable rate, but low enough so that microstructure and phase content are not changed. After being held at this temperature, the part is cooled very slowly. During this thermal treatment, stress is relieved by mechanisms similar to creep. It should be noted that if a casting is to receive a heat treatment for another purpose, such as strengthening, then thermal stress relief does not need to be accomplished in a specific step.

The Fe–Fe$_3$C phase diagram shown in Figure 3.46 is an essential reference for the heat treating of steel alloys. Steels have relatively low carbon contents, below ~1.5 wt%; therefore, the left side of the phase diagram is analyzed for these materials. In particular, the heat treatment process revolves around the transformation of a high temperature alloy, austenite (γ-Fe), to a lower temperature structure, which is either a combination of ferrite (α-Fe), and cementite (Fe$_3$C), as one might expect from the phase diagram, or a metastable phase called marstenite, which is not represented on the phase diagram. The temperature to achieve the austenite structure is given by the line marked A_3, for hypoeutectiod alloys (<0.77 wt% C), or A_{cm} for hypereutectoid alloys (>0.77 wt% C). Medium carbon steels are commonly used in casting. They are hypoeutectiod alloys with carbon contents between 0.2 and 0.5 wt%. The temperature at which the transformation to ferrite and cementite begins on cooling is marked with A_1. These temperatures change as alloying elements, such as Cr and Si, are added, as does the carbon content at the eutectoid. Below, the heat treatments used for medium carbon steel are discussed. Similar treatments are used for low and high carbon steels as well as alloy steels.

There are several treatments designed to either prepare the steel casting for an additional post-forming operation or enhance its properties: full annealing, normalizing, quenching/hardening, and tempering. All heat treatments begin with heating the casting to about 50°C above A_3 and transforming the structure to austenite. Austenite is a single phase material with a uniform chemical composition. For casting microstructures that contain dendrites, heating to this high temperature and holding also enhances the chemical homogeneity.

For a full anneal, the casting is cooled slowly to a temperature about several hundred degrees below A_1 (e.g., 425°C). Typically, the heat treating furnace is turned off and the atmosphere is still and usually set to be nonoxidizing. On cooling below A_3, ferrite precipitates in the austenite microstructure, and the composition of the austenite becomes richer in carbon. At A_1, the austenite has the eutectoid composition. On further cooling, the austenite transforms into a two phase "microcontituent" called pearlite.

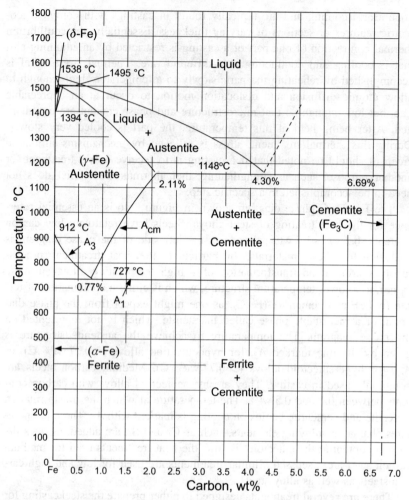

FIGURE 3.46 Fe−Fe$_3$C phase diagram. *Adapted from Hawkins and Hultgren (1974).*

Pearlite contains alternating lamellae or sheets of α-Fe and Fe$_3$C. Austenite has a uniform carbon content, but the two phases that form in pearlite have very different carbon contents; the lamellae structures minimize the diffusion distance. However, there is an energetic penalty for interfaces and hence if the cooling is slow, then these layers have time to become thicker; this material is known as coarse pearlite. After reaching a temperature at which further changes do not take place at an appreciable rate (e.g., 425°C), the casting is removed and cooled more quickly to room temperature. The fully annealed material has high ductility due to the microstructure, which contains larger regions of the relatively ductile ferrite and good machinability.

Normalizing produces a ductile, machinable casting in less time than annealing. A normalizing treatment begins in the same way as annealing—heating above A_3. Then the cooling process occurs a bit more quickly by cooling the casting in a freely circulating air. The faster cooling results in a pearlite structure with thinner layers. The properties are typically still adequate for achieving the softness and machinability required for some castings.

Quenching results in strengthening and hardening, the opposite effect of that provided by annealing and normalizing. In this heat treatment, the casting again begins at high temperature as austenite and then it is cooled rapidly by quenching it into a liquid, such as oil, or gas. To understand this more dynamic, nonequilibrium process, continuous cooling transformation diagrams are used. Considering simple carbon steels, if the cooling rate is fast enough, austenite transforms completely to martensite, a metastable, high hardness structure. At slower rates, a mixture of pearlite and martensite results. The cooling rate is determined by the quenching conditions as well as the size of the casting. The surface cools fastest, while the interior metal in thicker sections cools more slowly and therefore may end up with a different microstructure and properties. Modeling this heat transfer is more complicated than the solidification examples described earlier in this chapter. When the hot metal is immersed in a liquid, local boiling, vapor barriers, and convection influence the heat transfer. Tempering typically follows quenching.

Tempering is a heat treatment carried out after quenching and sometimes after normalizing. The primary functions of tempering are relieving thermal stress imparted by quenching and modifying the structure and properties of martensite. The tempering treatment is carried out below A_1. Fairly wide ranges of temperature (175−705°C) and time (30 min−4 h) are used with either longer treatment times at lower temperatures or shorter times at higher temperatures. On heating, stresses are relieved and carbon diffuses out of the metastable martensite, leading to the formation of ferrite and cementite. The microstructure is different than pearlite; very fine cementite particles are completely surrounded by a ferrite matrix. This change leads to an increase in ductility, but with some sacrifice of the strength and hardness.

Similar heat treatments are used for ductile cast iron. Cast irons have higher carbon concentrations than steels, and their microstructures contain various forms of graphite as a consequence. Ductile iron has about 3−4.2 wt% C and 1.5−2.8% Si. As-cast, ductile iron has a ductile ferrite or pearlite matrix with dispersed, rounded graphite particles. Heat treatments are used to relieve stress as well as alter the microstructure and properties. All of the latter heat treatments begin with austenitizing, heating the casting to a temperature above A_1, and holding it there until all pearlite is removed and an austenite with graphite results. To produce annealed ductile iron, the casting is cooled very slowly such that ferrite and nodular graphite result. In normalizing, a faster cooling rate is using, similar to the treatment described for steel, and the result is the formation of a fine pearlite matrix containing some ferrite, with graphite dispersed. Quenching and tempering also follow analogous procedures as in steel, resulting in martensite with dispersed graphite and tempered martensite with dispersed graphite, respectively. These treatments increase the strength of the ductile iron. Lastly, austempering is a heat treatment commonly used on cast irons. After austenitizing, the casting is quenched to intermediate temperatures higher than the martensite transformation temperature. The casting is held at that temperature and the austenite is partly converted to bainite; the result is high strengths and ductility.

For aluminum alloys, heat treatments induce strengthening by a mechanism called precipitation hardening. The heat treatment consists of three steps. First, the casting is heated to a temperature at which the alloy can reach a uniform, single phase, solid solution microstructure. As an example, consider an alloy with copper as the primary additive. According to the phase diagram in Figure 3.47, an alloy with 4 wt% Cu would require heating to about 500°C. Higher temperatures within the single phase region require less time for the homogenization process to be complete. This step is called the "solution heat treatment." The second step is quenching to retain the solid solution structure in a metastable state. Lastly, the casting is reheated to precipitate the $CuAl_2$ in the A_l solid solution matrix. This step is called "precipitation heat treating." The result is an increase in strength, so long as the precipitate has the desired coherent interface with the aluminum matrix.

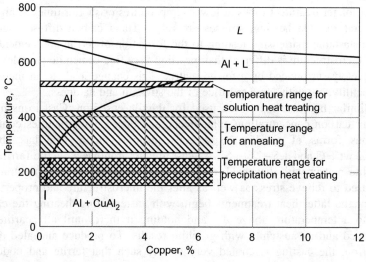

FIGURE 3.47 Al-rich end of the Al-Cu phase diagram. *From Hatch (1984), reproduced by permission of ASM International.*

3.4 CASTING OF FLAT SHEETS

3.4.1 Process Overview

Casting processes involving flow under gravity are also used to produce flat sheets. Again, low viscosity melts are needed for this casting process. In this case, using casting to make a flat sheet of metal is not practical. Much more efficient and controllable solid deformation rolling processes are used to create thin sheets of metal. Instead, the application of casting to sheet fabrication is limited to ceramic glasses.

The two main methods for creating flat glass sheets, the float glass process and the fusion downdraw process, are melt casting processes. These two processes account for the majority of flat glass products, including the glass used for architectural and automotive applications, and glasses for electronic displays. In this section, the glass melt preparation is briefly described, followed by descriptions of the two flat glass forming processes and a summary of post-forming operations for glass sheets.

3.4.2 Glass Melt Preparation

Large-scale glass manufacturing, which would be typical of the flat glass production methods described in this section, is carried out in integrated facilities. As described in Chapter 2, the raw materials that make up the glass batch enter on one end of the facility and finished products exit out the other. The glass melting furnace is the first stop in the procession. Horizontal glass melting furnaces consist of several sections. The furnace is fed continuously with premixed, raw, powdered batch materials, such as glass sand, feldspar, soda ash, limestone, and metal oxides, and glass fragments called cullet, which is recycled from elsewhere in the facility. As these solids are heated, they react with one another to form a melt. Reaction and dissolution into the melt requires time, depending on the raw material size and chemical composition. Thermal gradients lead to convective flows within the melt (i.e., temperature differences lead to density differences that drive the flow). Typically, these flows circulate from top to bottom and help to mix the melt. Additionally, the melt contains bubbles, mainly from gaseous products (e.g., carbon dioxide) formed during reactions. These bubbles must be removed before forming. Bubbles rise to the top of the glass melt; their speed increases as the size of the bubble increases or as the viscosity of the melt decreases. Additives known as fining agents are sometimes added to speed this process. Fining agents undergo redox reactions to form gases (e.g., $Sb_2O_5 = Sb_2O_3 + O_{2(g)}$). The additional gas causes the existing bubbles to grow and rise more quickly to the surface. As the melt proceeds down the length of the furnace, it reaches a conditioning section where chemical homogeneity is achieved.

184 Materials Processing

The melt is prepared to enter into the forming operation at far end of the furnace or in an adjacent tank called the forehearth. Here, the temperature is made uniform and adjusted for the needs of the forming operation. For a given glass composition, the temperature determines the viscosity of the glass and therefore, the rate or flow and the extent of cooling needed to solidify a glass piece.

3.4.3 Float Glass Process

The float glass process was invented in the 1950s in response to a pressing need for an economical method to create flat glass for automotive as well as architectural applications. Existing flat glass production methods created glass with irregular surfaces; extensive grinding and polishing was needed for many applications. The float glass process involves floating a glass ribbon on a bath of molten tin and creates a smooth surface naturally. Floating is possible because the density of a typical soda-lime-silica glass (~ 2.3 g/cm^3) is much less than that of tin (~ 6.5 g/cm^3) at the process temperature. After cooling and annealing, glass sheets with uniform thicknesses in the $\sim 1-25$ mm range and flat surfaces are produced. The float glass process is used to produce virtually all window glass as well as mirrors and other items that originate from flat glass. Since float glass is ordinarily soda-lime-silica, the reference temperatures and behavior of this glass are used in the discussion below.

Figure 3.48 shows the basic layout of the float glass line. The glass furnace is a horizontal type, as described above. For a float line, the glass furnace is typically on the order of ~ 150 ft long by 30 ft wide and holds around 1200 tons of glass. To achieve good chemical homogeneity, the glass is heated to $\sim 1550-1600°C$ in the furnace, but is then brought to about $1100-1200°C$ in the forehearth. From there, the glass flows through a channel over a refractory lipstone or spout onto the tin bath. As it flows, the glass has a temperature of about 1050°C and viscosity of about 1000 Pa•s. A device, called a tweel, meters the flow of the molten glass.

There are two basic designs for the float portion of the line. In the original Pilkington design (Figure 3.49a), the channel between the forehearth and

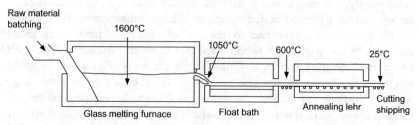

FIGURE 3.48 Schematic overview of a float glass process line from batching to product shipping.

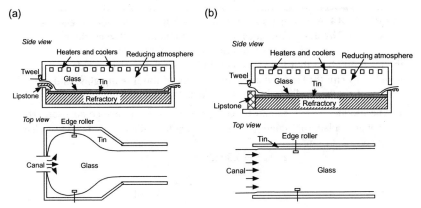

FIGURE 3.49 Schematics the tin bath sections of float glass production lines based on (a) the Pilkington version and (b) the PPG version. *Adapted from McCauley (1980).*

the tin bath is fairly narrow (~1 m). The glass flows over the lipstone, which is not in contact with the tin bath. As the glass flows onto the tin bath, it spreads unconstrained over the surface and progresses down the tin bath. By downstream processes, the glass ribbon's width and thickness are controlled. In the Pittsburgh Plate Glass (PPG) design (Figure 3.49b), the channel is as wide as the ribbon and the glass flows over a lipstone, which is in contact with the tin bath. Various methods are used to regulate the glass thickness and the production rate, as described below.

Tin is an ideal bath material because it has the right set of physical properties. Tin melts at 232°C, has relatively low volatility, and does not boil until over 2000°C. Molten tin is denser than molten glass and is not miscible or reactive with molten glass. The gas atmosphere is controlled so that tin does not oxidize at a fast rate. Any oxide that does form is collected in a dross container on the bath.

Regulating the flow of the glass is important at this stage, both from the entry point and the lateral flow. The glass flow onto the tin bath is regulated by a gate, called a tweel, which is located in the canal between the forehearth and spout. The glass flows down the spout or lipstone onto the tin surface. There is some pressure driving this flow through the gap of the tweel. See Example 3.14. As the glass flows onto the tin bath, the thickness of the glass sheet depends on how that flow is controlled laterally and along the length of the bath. The first step to understanding thickness control is to examine the equilibrium thickness.

Molten glass simply poured atop molten tin adopts an equilibrium thickness that is determined by a balance of gravity and surface tension. Consider a large pool of glass on tin, such that the effects of the curvature where the glass and tin meet can be ignored. Surface tension forces exert a net force on the glass inward toward the center of the glass pool, because the sum of the

EXAMPLE 3.14 A float glass plant produces 500 tons of soda-lime-silica glass a day. The design includes a 1 m wide channel between the glass tank and the tin bath. (a) At what position should the tweel be set above the lipstone given that the glass melt height on the upstream side of the tweel is 15 cm. (b) Find the average velocity of the flowing glass and comment on the likelihood that the flow is laminar.

a. The production rate should be converted to m³/s for easy comparison. The float glass plant runs continuously. The soda-lime-silica glass melt has a density of ~2.3 g/cm³.

500 tons/day → 5.25 kg/s → 0.00229 m³/s

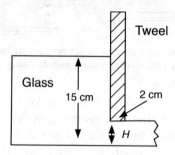

FIGURE E3.14

The pressure on the upstream side of the tweel is higher than on the downstream side due to the bath height. The pressure at a depth of 13.5 cm (midway in the channel) is

$$P = P_{atm} + \rho g h = P_{atm} + (2300 \text{ kg/m}^3)(9.8 \text{ m/s}^2)(0.135 \text{ m}) = P_{atm} + 3043 \text{ Pa}$$

Hence the pressure difference with the downstream side, which is at atmospheric pressure is 3043 Pa. The volumetric flow rate beneath the tweel is therefore:

$$Q_v = \frac{H^3 W \Delta P}{12 \eta L} = \frac{H^3 (1 \text{ m})(3043 \text{ Pa})}{12(1000 \text{ Pa} \cdot \text{s})(0.02 \text{ m})} = 0.00229 \text{ m}^3/\text{s}$$

$$H = 5.65 \text{ cm}$$

b. The average velocity is the volumetric flow rate divided by the cross-sectional area over which flow is taking place. In this calculation, we use the volumetric flow rate from the production.

$$\bar{v} = \frac{Q_v}{WH} = \frac{0.00229 \text{ m}^3/\text{s}}{(1 \text{ m})(0.0565 \text{ m})} = 0.041 \text{ m/s}$$

The Reynolds number is given by:

$$Re = \frac{d \bar{v} \rho}{\eta} = \frac{(0.0565 \text{ m})(0.041 \text{ m/s})(2300 \text{ kg/m}^3)}{1000 \text{ Pa} \cdot \text{s}} = 0.0053$$

Here, we chose the thickness as the characteristic length scale to find the Reynolds number; Re is very low and laminar flow is a good assumption.

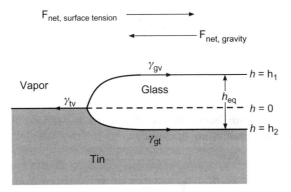

FIGURE 3.50 Cartoon of the edge of a glass pool floating on the molten tin bath. *Adapted from Narayanswany (1977).*

interfacial tension of the glass-tin interface (~0.5 mN/m) and the surface tension of the glass or glass-vapor interfacial tension (~0.35 mN/m), which both act inward, is greater than the surface tension of the tin (~0.5 mN/m), which acts outward. Gravity, however, results in a net force on the glass outward.

In the Figure 3.50, h_{eq} is the equilibrium film thickness; h_1 and h_2 are the positions of the glass surface and the interface between the glass and tin relative to the tin surface. These relative positions can be found, given that the pressure in the glass at the interface must be equal to the pressure in the tin bath at the same position.

$$P_{\text{glass}}|_{h=h_2} = P_{\text{tin}}|_{h=h_2}$$

$$P_{\text{atm}} + \rho_g g h_{eq} = P_{\text{atm}} + \rho_t g h_2 \quad (3.93)$$

$$h_2 = \frac{\rho_g}{\rho_t} h_{eq}$$

$$h_1 = h_{eq} - h_2 = \frac{h_{eq}(\rho_t - \rho_g)}{\rho_t} \quad (3.94)$$

The force equilibrium is analyzed in a plane cutting through the glass far from the edge where the glass contacts tin. The net surface tension force per unit length is given by:

$$F_{\text{net,surface tension}} = \gamma_{gv} + \gamma_{gt} - \gamma_{tv} \quad (3.95)$$

The gravitational force per unit length is found by integrating the hydrostatic pressure ($\rho g h$) with respect to position. The glass pushes outward over the entire cross-section of the glass sheet from its surface to a depth of h_{eq}, while the tin bath resists from its surface to a depth h_2.

$$F_{\text{net,gravity}} = \frac{1}{2}\rho_g g h_{eq}^2 - \frac{1}{2}\rho_t g h_2^2 \tag{3.96}$$

Substituting in for h_2 gives:

$$F_{\text{net,gravity}} = \frac{1}{2}g h_{eq}^2 \left[\frac{\rho_g(\rho_t - \rho_g)}{\rho_t}\right] \tag{3.97}$$

Lastly, the forces are balanced and the terms rearranged to provide an expression for the equilibrium thickness:

$$h_{eq} = \sqrt{\frac{2\rho_t(\gamma_{gv} + \gamma_{gt} - \gamma_{tv})}{g\rho_g(\rho_t - \rho_g)}} \tag{3.98}$$

Substitution of typical values results in the 7 mm equilibrium thickness. While this thickness is a useful benchmark, in practice the thickness of the glass is not determined by equilibrium, as described below.

As the floating glass ribbon traverses down the length of the tin bath, its properties change dramatically. The glass enters as a viscous liquid and exits virtually a solid at a temperature very close to its glass transition temperature. The details of how the temperature changes and the viscosity builds are complicated. On one side, the free surface of the glass is exposed the atmosphere; heat can leave this surface by radiation or convection. Cooling and heating apparatuses are stationed above the glass ribbon down the length of the bath to allow adjustment of the ribbon temperature. On the other side, the glass is in contact with the tin bath, which can absorb some of the heat and transport it away from the ribbon. The tin bath is in constant motion due to the moving glass above it as well as the thermal convection currents. Unfortunately, no simple approximations can be made to make the modeling of the heat transfer.

The thickness of the float glass sheet is adjusted by controlling flow onto the tin bath as well as by tension exerted along the length of the bath by rollers in the annealing lehr and sometimes by rollers in the bath unit itself. In the Pilkington design, the melt enters the bath and spreads out laterally to a thickness near the equilibrium value. If a sheet thicker than the equilibrium is required, then this spreading is constrained with physical barriers. If a sheet thinner than equilibrium is needed. then the glass ribbon is pulled in tension by rollers. In the PPG design, thickness is regulated by the tweel position and by tension from rollers in the lehr. The thermal profile allows the thinning deformation to take place effectively. A short distance away from the entry point, the temperature of the ribbon drops and the viscosity rises. Overhead coolers help this process. The glass viscosity is high enough so that knurled rollers contact the glass ribbon and pull it forward (and in some operations, laterally as well). Heaters are placed shortly downstream of these edge rollers to raise the temperature of the ribbon and create a

EXAMPLE 3.15 Find the velocity of float glass as it exits the tin bath on a line that produces 500 tons of soda-lime-silica glass a day. The flat glass produced has a thickness and width of 3 mm and 4 m, respectively, at room temperature. At exit, the glass is at about 600°C. Assume that the average thermal expansion coefficient over the range from room temperature to 9×10^{-6} °C^{-1}.

As shown in Example 3.14, 500 tons a day corresponds to a volumetric flow rate of 0.00229 m³/s. To find the exit velocity, we need to divide this volumetric flow rate by the cross-sectional area of the glass sheet at this point in the process.

The dimensions at 600°C can be found using the thermal expansion coefficient:

$$\alpha_t = \frac{1}{L_o}\frac{\Delta L}{\Delta T}$$

$$9 \times 10^{-6}\ °C^{-1} = \frac{1}{3\ mm}\frac{(h_{600} - 3\ mm)}{(600°C - 25°C)}$$

$$h_{600} = 3.016\ mm$$

Likewise, $W_{600} = 4.02$ m

Therefore, the velocity exiting the bath is:

$$\overline{v} = \frac{Q_v}{Width \cdot Thickness} = \frac{0.00229\ m^3/s}{(0.003016\ m)(4.02\ m)} = 0.189\ m/s$$

---

deformable zone. This zone is followed by coolers that again lower the temperature and raise the viscosity. At exit from the lehr, the ribbon is virtually solid. The main deformation is due to the rollers in the lehr, which pull on the glass ribbon from the lehr to the edge rollers; extension takes place in the deformation zone. Example 3.15 considers the exit velocity of glass from the process.

3.4.4 Fusion Downdraw Process

The fusion downdraw glass process was invented in the 1960s; like float glass, it is a process developed in pursuit of higher quality, flat glass. The fusion process is also innovative in its approach. Figure 3.51a shows a schematic diagram of the refractory trough that is central to the process. Molten glass pumped into one side of the trough fills the trough and then overflows on both sides. The glass then flows down over the sides of the trough and joins together at the root. The two layers of glass on either side of the trough "fuse" together to produce the glass sheet. The geometrical features of the taper in the channel of the trough produce a uniform overflow rate and eventually, a uniform glass thickness across the width. The features of the

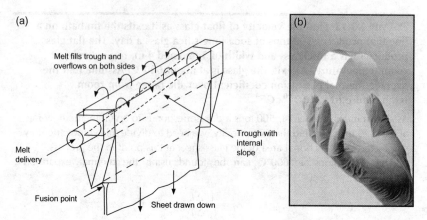

FIGURE 3.51 (a) Schematic of the fusion downdraw process. *Adapted from Dockerty (1967) and Hynd (1984).* (b) An image of flexible glass sheet, Corning® Willow™ Glass, made by the fusion downdraw process. *Courtesy of Corning, Inc., www.corning.com.*

keel-like, external surface of the trough are also important to controlling the flow. The glass entering the trough has a viscosity of about 4000 Pa•s, but by the time it reaches the root, it has cooled and reached a much higher viscosity ($\sim$50,000 Pa•s). Control of local temperature and hence viscosity is another necessity for a successful process. The glass is drawn downward by edge rollers. By adjusting the flow rate and tension, glass with variety of thicknesses can be produced. Thicknesses range from $\sim$100 μm to 15 mm. Thin glass with pristine surfaces is in particular demand for displays and other electronic applications (see Figure 3.51b).

3.4.5 Post-Processing Operations for Glass Sheets

The critical post-processing or post-forming operations for flat glass sheets include annealing to relieve thermal stress, coating, cutting, and strengthening treatments. Most of the post-forming is done directly on the process line. However, strengthening treatments are typically performed on cut glass after it has exited the process line.

Thermal annealing is a necessary post-forming operation for many glass products, including glass sheets. Thermal stresses develop in glass sheets that are cooled quickly. The cooler glass on the surface has a lower equilibrium volume than the warmer inner glass; strain arises from the difference in local temperature, with the average temperature dictating the actual volume. The exterior of the quickly cooled glass would be in tension and the interior in compression. Such stresses might be expected to be transient because the temperature gradients eventually equalize, but glass is viscoelastic and can relax stresses if its temperature is high enough. Hence different extents of stress relaxation produce persistent stresses in glass. Therefore, annealing is used to purposefully expose the glass to the proper thermal conditions so that the stress state at the end of cooling can be very low.

Stress relaxation in glasses can be understood with a modified Maxwell model, which consists of a spring and a dashpot connected in series. In this model, the viscous nature of the glass is represented by the dashpot with viscosity, η, and the elastic nature by a spring with a shear modulus, G. The characteristic relaxation time is given by that ratio of these two: $\tau = \eta/G$. Since viscosity is strongly temperature dependent, so is the relaxation time. At high temperatures, viscosities are low and relaxation times short, but as the glass cools, the relaxation time increases dramatically. Each glass composition has a characteristic annealing point, where the viscosity is $\sim 10^{12}$ Pa•s and stresses of a 10 mm thick sheet are relieved in ~ 15 minutes. For soda lime glass, the annealing point is $\sim 550°C$.

Annealing of flat glass occurs directly after the forming operation. The glass sheet exiting the tin bath or moving down from the fusion drawdown trough is sent directly to an annealing lehr. The glass is cooled slowly as it moves down the lehr so that at it has very low stresses as it exits the line at room temperature. The temperature of the first zone in the annealing lehr is at or just above the annealing point. As the glass travels through this zone, stresses present in the glass from the previous process steps are relaxed. Then, the temperature is slowly lowered as the glass continues down the lehr. The cooling rate is slow until a second critical point, the strain point at which the viscosity is $\sim 10^{13.5}$ Pa•s. After this temperature, the cooling rate can be increased safely.

The glass should be in an annealed, nearly stress-free state when it is cut into discrete glass panels. Thermal stresses interfere with the precision of the cutting process since residual tensile stress drives crack extension. Cutting is typically a mechanical operation, a scoring and fracture, but can also be accomplished be other means such as lasers.

Frequently, glass sheets are strengthened in post-forming treatments. Thermal tempering is one way to strengthen glass sheets. In a thermal tempering, glass sheets travel on rollers through tunnel-like unit that exposes the glass to a series of thermal steps. On exit, the glass has residual compression on its surface. Since the glass surfaces are most likely to become damaged and are therefore the site of flaws, surface compression combats the growth of these flaws under a mechanical stress and creates a stronger glass sheet. Tempering begins with heating the glass sheet to a temperature above the glass transition temperature; for soda-lime-silica glass temperatures of 620–640°C are used. Once at temperature, the glass traverses to the next stage in the process—controlled and rapid cooling of the surface. Air jets impinge on both sides of the glass sheet, cooling the surface to well below T_g, at which it is rigid. In the next stage, the glass, with its nonuniform structure and temperature, is cooled more slowly. By a combination of stress relaxation and temperature gradient-induced stress development, the final stress profile ends up parabolic with the surfaces in compression and the interior in tension.

Chemical tempering or strengthening is an alternative method to strengthen glass sheets. This method is used for thin glass sheets, such as those produced by the fusion downdraw process. Thermal tempering is less effective for thin sheets, because it is not possible to produce the thermal gradients through the thickness that are needed for high strength. In chemical strengthening, glasses are immersed in a heated ionic liquid, such as a molten salt. The chemistry of the salt is chosen so that it contains large cations, typically potassium, which then exchange with smaller sodium cations present in the glass. The exchange takes place selectively on the surface and results in compression, due to the size difference between the native ion in the glass and the ion entering from the molten salt bath. The result is high levels of surface compression and enhanced strength.

3.5 EXTRUSION

3.5.1 Process Overview

Extrusion is a process that has many forms. In this chapter, extrusion is described as a melt process; later in the text, extrusion processes that are based on solids and dispersions are discussed. Some of the principles are similar but more are different. Here, we are concerned with melts so extrusion, in that context, involves forcing a melt through a die to define shape and then solidification to retain the shape. Unlike casting, extrusion requires a high viscosity to work properly. As the melt exits the die, it must hold its shape long enough for heat transfer and solidification to take place. A metal melt, with its very low viscosity, would not hold its shape after passing through the die. Thermoplastic polymer melts, on the other hand, are ideally suited to this process. As they exit the die, they can retain their shape during solidification on cooling. Likewise, some reactive polymers can be engineered with the desired flow and solidification characteristics to be extruded.

In extrusion, a molten polymer is forced under pressure through a die to define a constant cross-sectional shape or profile after the exiting polymer is solidified, typically by cooling. Extrusion is used to make simple profiles such as flat sheets, rods and tubes, and more complicated profiles, such as honeycombs. The main focus of this section is extrusion with a single screw extruder, including the structural changes and flow that occur in the extruder, the operating diagram, die design, exit effects, and solidification. Along the way, twin screw extrusion is described briefly. Extruded products after solidification rarely require significant post-processing, with the exception of simply cutting lengths of the extruded product.

3.5.2 Melting and Flow in a Single Screw Extruder

Figure 3.52 is a schematic diagram of a single screw extruder. The starting material, polymer pellets or granules, is fed into the unit through a hopper.

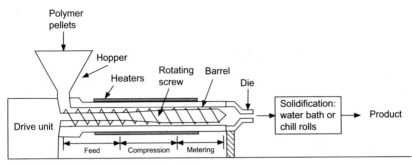

FIGURE 3.52 Schematic of a single screw extruder, showing the production flow from feeding of starting material (polymer pellets) into the hopper of the extruder, through the three stages of the extruder body (conversion from pellets to pressurized melt) and out the die to the solidification process and ending in a solid polymer product.

The pellets enter the body of the extruder and encounter the turning screw. The flights push the pellets forward. As the polymer pellets are pushed forward, they are compacted and heated, both by external heaters and by internal processes, including friction and viscous dissipation. Then, in the compression zone, the polymer pellets become molten and eventually fill the entire space in the screw channel. To accommodate the volume change, the channel depth drops. In the last section, the metering section, the melt becomes homogenized and pressurized. The pressure builds due to the constrictions at the end of the extruder and is used to push the molten polymer through the die.

The fundamental features of the screw geometry are shown in Figure 3.53. The screw is positioned inside a cylindrical barrel of diameter D_b. The screw flights nearly touch the barrel, separated by a small clearance δ, which is on order of a fraction of a millimeter. So for all practical purposes, the screw diameter, D_s, measured from flight tip to flight tip, is the same as the barrel diameter. Extruders are characterized by their L/D, where L is the axial length of the screw and D is the diameter (i.e., $D \approx D_b \approx D_s$). For a typical extruder, L/D is 20, but this value can vary from around 10 to 30. Actual screw diameters range from around 2 cm (or ¾") to 50 cm (or 20"). The polymer is enclosed in a channel of height, H, and width, W. While D_b and D_s are constant down the length of the extruder, H varies, as described below. The flights of the screw have a finite width, which is typically small enough to be ignored in some calculations, and make an angle ϕ with the vertical. The angle ϕ is known as the helix angle. Ignoring the flight width, the lead, L_s, is related to the diameter and the helix angle by $L_s = \pi D \tan\phi$. Commonly, the helix angle is 17.65°, which is the "square angle," or the angle needed for the distance between flights to equal one diameter.

The changes in structure down the length of the extruder can be illustrated by examining the material that occupies the helical channel as a function of position in the extruder. Experimentally, the extrusion process can be

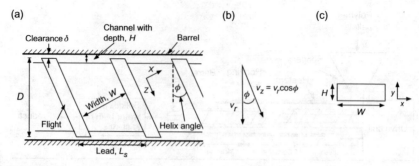

FIGURE 3.53 (a) Schematic showing the features of the extruder screw. The coordinate system shown defines z as the direction down the helical path of the channel and x as the cross-channel direction. The clearance depth and width of the flights are exaggerated in size and the complexity of the flight geometry is not show. The lead, L_s, is given by $L_s = \pi D \tan\phi$. (b) The relationship between the rotational velocity v_r and the velocity down the channel, v_z. (c) A view of the cross-section of the channel, showing the direction y through depth of the channel.

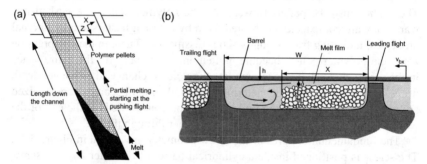

FIGURE 3.54 Schematics of (a) unwrapped material from extruder screw [*adapted from Osswald (1998)*] and (b) distribution of material in a partially melted zone [*reproduced by permission of Carl Hanser Verlag Munich from Osswald (1998)*].

halted and the contents of the extruder cooled, and then unwrapped to show the progression from a porous, packed solid composed of polymer granules to a melt, which would be a dense polymer after cooling. Figure 3.54 shows this transformation schematically. Notice that there are three main morphologies: polymer solids in the feed section, mixed solids and melt in the compression section, and melt in the metering section. The changes in structure and flow in each of these sections of the extruder are examined below.

Flow in the first section of the extruder, the feed zone, involves polymer pellets or granules filling the screw channel and moving forward by a pushing action from the screw flights. The particulate polymer starting material must enter freely into the feed zone from the hopper. To prevent sticking, the hopper is cooled. The polymer pellets falling into the channel do not pack well and occupy only about 30–40% of the space available in the screw channel. The pellets are large enough so that they only occupy the channel and are not susceptible to entering the narrow clearance between the channel and the barrel. To move forward, the pellets must stick to the barrel and slip on the screw. If this condition is met, then moving screw flights meet the pellets stuck on the barrel and push them forward. If the pellets stick on the screw, then they turn with the screw rather than moving forward. To enhance the frictional contrast between the screw and the barrel, the screw has a highly polished surface and the barrel in the feed section may be cooled or grooved. The polymer pellets, as they are pushed down the feed section, begin to compact; air from the spaces between them is expelled out the hopper, and the pressure starts to increase in this section. The volumetric flow rate in this section increases with the screw rotational speed and the space available in the channel of the screw. The main principle is drag flow, described later for the metering zone, but the details are a bit different.

The polymer begins to melt as it enters the compression zone. Here, the heating from the barrel heaters leads to a layer of melt at the interface with the barrel. The molten polymer forming on the barrel is swept into the channel and concurrently the air originally between the granules is expelled in the opposite direction out the hopper. Additionally, friction and viscous heating contribute to a rise in temperature. To understand viscous heating, we need to recall that work is being done on the polymer melt. The first law of thermodynamics states that the internal energy, U, of a system depends on the work done, w, and the heat exchanged with the surroundings, q:

$$\Delta U = q - w \tag{3.99}$$

where positive q refers to heat added to the system from the surroundings and positive w refers to work done *by the system* on the surroundings. So, if the heat exchange with the surroundings is slow ($q = 0$) and work is done *on the system*, then the internal energy increases and hence the temperature increases. Since polymers have low thermal conductivity, the heat flow may be limited, especially at short times, so we can use this approximation to develop an expression for viscous heating.

An expression for viscous heating is found by equating the rate of internal energy increase to the rate of work done on the polymer. The following expression, introduced earlier in the chapter, gives the *rate* of work done on the liquid during shear:

$$\frac{d\,(\text{Work/Volume})}{dt} = \tau \dot{\gamma} \tag{3.12}$$

Therefore, for a Newtonian liquid, substitution for the shear rate gives:

$$\frac{d\,(\text{Work/Volume})}{dt} = (\eta \dot{\gamma})\dot{\gamma} = \eta(\dot{\gamma})^2 \tag{3.100}$$

An alternative expression that assumes power law behavior is easily found. To find the rate of increase of internal energy, the starting point is the relationship between internal energy and heat capacity. The internal energy increase is related to the heat capacity at constant volume c_V by:

$$c_V = \left(\frac{dU}{dT}\right)_V \tag{3.101}$$

The heat capacity at constant pressure is a more readily available property, is approximately equal to c_v. Therefore, using the specific heat capacity at constant pressure [e.g., in J/(kg•K)], C_p, an expression for the rate of internal energy increase per unit volume is found:

$$\frac{d(U/\text{Volume})}{dt} = \rho C_p \frac{dT}{dt} \tag{3.102}$$

Therefore, if heat exchange is limited then:

$$\rho C_p \frac{dT}{dt} = \eta(\dot{\gamma})^2 \tag{3.103}$$

If we analyze a short period of time in which all the properties and the shear rate are constant, then the temperature rise is found by integration:

$$\int_{T_1}^{T_2} dT = \int_0^t \frac{\eta(\dot{\gamma})^2}{\rho C_p} dt$$

$$\Delta T = \frac{\eta(\dot{\gamma})^2}{\rho C_p} t \tag{3.104}$$

The temperature increase is therefore higher when higher shear rates are used and also when the polymer itself has a higher viscosity. This viscous heating under shear is also important for the flow through the die and in the injection molding process (Example 3.16).

Melting due to viscous heating and conduction of heat from the barrel gradually changes the solid pellets to melt. From Figure 3.54, the melt accumulates on the "pushing" or "trailing" flight, which is closer to the hopper. The melt also covers the interface with the extruder barrel at the top of the solid's bed and there is circulation in the melt pool. Again, the rotating screw drags the polymer forward. The fraction of melt increases down the length of the channel in the compression zone. Concurrently, a reduction in channel depth in engineered into the screw to accommodate the densification that occurs during the melting process. The length of the compression zone in extruder screws varies; relatively short compression zones are engineered into screws used for polymers that melt abruptly with temperature increase (e.g., PP) while longer compression zones are found in screws used to extrude for polymers that melt more gradually (e.g., PE). At the end of the zone, the polymer is molten and pressure is beginning to build.

EXAMPLE 3.16 Find an expression for the temperature rise in a polystyrene melt under shear for a short time, assuming the melt rheology follows a power law expression. Use parameters for polystyrene initially at 200°C to estimate the temperature rise in 10 s for a shear rate of 100 s^{-1}. Assume that the melt has a heat capacity of 2200 J kg^{-1} K^{-1}.

$$\frac{d\,(\text{Work/Volume})}{dt} = \tau\dot{\gamma} = [K(\dot{\gamma})^n](\dot{\gamma}) = K(\dot{\gamma})^{n+1}$$

$$\rho C_p \frac{dT}{dt} = K(\dot{\gamma})^{n+1}$$

$$\Delta T = \frac{K(\dot{\gamma})^{n+1}}{\rho C_p} t = \frac{(28{,}000 e^{-0.025(200°C - 170°C)} \text{Pa} \cdot \text{s}^{0.28})(100\ \text{s}^{-1})^{1.28}}{(960\ \text{kg/m}^3)(2200\ \text{J/(kg} \cdot \text{K)})}(10\ \text{s}) = 22.7°\text{C}$$

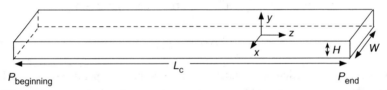

FIGURE 3.55 Schematic of unwrapped material from the channel of the extruder screw in the metering section. The top surface is in contact with the barrel and the bottom surface is in contact with the screw.

In the next zone, the metering zone, the polymer is prepared for shaping by the die. The melt reaches a uniform temperature and viscosity, and any stray solid pellets are melted. In this zone, the channel depth is constant and the pressure builds considerably due to the constriction from the die. In this zone, the flow is determined by the drag from the moving screw, which pushes the polymer forward, and the pressure gradient, which acts in the opposite direction. These two components of the flow can be incorporated in to a flow model for the metering section. The forward motion is represented by a drag flow; here, the fact that the screw is moving and the barrel is stationary brings to mind drag. And, the flow in the opposite direction is a pressure-driven. These two contributions are summed to define the total flow. In the development of this simple model, the following assumptions are made: the melt is Newtonian, the flow is steady state, the temperature is constant, the channel width is much greater than channel depth, and the leakage flow between the screw flights and the barrel is negligible. Additionally, the channel, which winds around the screw in a helical path, is viewed in its "unwrapped" state as a long rectangle, as shown in Figure 3.55.

The motion of the polymer forward is approximated as drag flow imposed by one moving surface along the channel. While this construction neglects the action of screw flights directly, the approach captures the important effects of screw rotation speed, screw geometry, and polymer rheology. The moving surface for the drag flow is taken to be the barrel for convenience and the stationary surface is the screw. The velocity of the moving surface is relative to a rotational velocity that is perpendicular to the axis of the screw:

$$v_r = \pi DN \tag{3.105}$$

where πD is the distance per rotation and N is the rotations per second. The velocity along the channel direction z is therefore:

$$v_z = \pi DN \cos \phi \tag{3.106}$$

According to the drag flow analysis earlier in the chapter, the velocity gradient through the channel thickness is linear and the volumetric flow rate due to drag is given by:

$$Q_d = \bar{v}_z(\text{area}) = \frac{1}{2} v_z WH = \frac{1}{2}(\pi DN \cos \phi)(\pi D \sin \phi)H$$

$$Q_d = \frac{1}{2}\pi^2 D^2 HN \cos \phi \sin \phi \tag{3.107}$$

Therefore, for a particular extruder screw, N is the only parameter that influences the forward drag flow. The effect of the screw parameters is also apparent. More output occurs in screws with larger diameters and channel depths as these factors determine the capacity of the screw channel. The helical angle, ϕ, also plays a role. As ϕ increases up to 45°, the output increases, reflecting the effects of the angle on both the channel width and velocity along the channel direction.

Pressure builds down the length of the helical channel due to the constriction from the die at the end. Therefore, the pressure at the end of the channel is greater than the pressure at the beginning and pressure-driven flow drives the polymer back toward the hopper. For the analysis of this flow component, the two surfaces—barrel and screw—are stationary and only the pressure drives flow. Hence the velocity profile is parabolic and Q_p, the volumetric flow rate in the extruder channel due to the pressure gradient is given by:

$$Q_p = \frac{H^3 W \Delta P}{12 \eta L_c} \tag{3.108}$$

where H and W are defined in Figure 3.55. L_c is the length of the channel as measured down the helical path; therefore, we can relate this length to L_m, the length of the metering section as measured down the axis of the extruder:

$$L_c = \frac{L_m}{\sin \phi} \tag{3.109}$$

and as defined before, ΔP is the pressure change down the length, in this case:

$$\Delta P = P_{beginning} - P_{end} \tag{3.110}$$

Substituting in for W and L_c:

$$Q_p = \frac{\pi D \, \sin^2 \phi H^3 \Delta P}{12 \eta L_m} \tag{3.111}$$

Q_p is negative because ΔP is negative (i.e., $P_{end} > P_{beginning}$). To the first approximation, the pressure at the end of the metering section is so much bigger than the pressure at the beginning that the following approximation is made:

$$\Delta P = -P_{end} = -P$$

$$Q_p = -\frac{\pi D \, \sin^2 \phi H^3 P}{12 \eta L_m} \tag{3.112}$$

In this expression, it is clear that aside from the screw geometry parameters that are not adjustable for a given extruder, the pressure and the viscosity are the variables that control the volumetric flow backwards. As will be shown, the pressure is not really an adjustable parameter, but rather set by the die.

The total extruder output is the sum of the drag flow and the pressure-driven flow components. Therefore,

$$Q_{EX} = Q_d + Q_p = \frac{1}{2}\pi^2 D^2 H N \, \cos\phi \, \sin\phi - \frac{\pi D \, \sin^2 \phi H^3 P}{12 \eta L_m} \tag{3.113}$$

Lumping all the constants and screw geometry factors into constants α and β, the equation becomes:

$$Q_{EX} = Q_d + Q_p = \alpha N - \frac{\beta P}{\eta} \tag{3.114}$$

The main adjustable parameter, N, has the effect of increasing the output at any particular level of pressure. It is also instructive to consider the velocity profile as this gives information on how the polymer flows relative to the channel features. See Example 3.17.

The flow analysis above only took into account the variations in the z-direction, down the length of the channel. One can imagine what would happen as the die becomes more constricted, limiting the output. In this case, the pressure backflow becomes stronger. Flow also takes place in the x-direction, or cross-channel. These cross-channel motions are mainly responsible for the mixing that takes place in the extruder. The analysis requires consideration of flow in two dimensions and is beyond the scope of this text. See the sources listed in the bibliography and recommended reading section at the end of the chapter for more information.

EXAMPLE 3.17 A lab scale extruder has a diameter of 2.0 cm, an L/D of 20, a channel depth of 0.25 cm in the metering section, and a helix angle of 17.65. The axial length of the metering zone is 5D. The extruder operates at 100 RPM, with a polymer that has a viscosity of 300 Pa·s. Plot the velocity profile through the thickness of the channel given that the extruder pressure is 1 MPa at the end of the metering zone.

Let $v_z(y)$ represent the velocity of the polymer melt through the channel where $y = -H/2$ is the stationary surface and $y = H/2$ is the surface moving at a velocity of v_z. We can find $v_z(y)$ for the drag flow and for the pressure-driven flow and add them together.

Pressure flow: $v_z(y) = \dfrac{\Delta P}{2\eta L_c}\left(\dfrac{H^2}{4} - y^2\right)$, $-\dfrac{H}{2} < y < \dfrac{H}{2}$

Drag flow: $v_z(y) = \dfrac{v_z}{H} y$, $0 < y < H$ We need to convert this to the same coordinate system as above

$$v_z(y) = \dfrac{v_z}{2} + \dfrac{v_z}{H} y, \quad -\dfrac{H}{2} < y < \dfrac{H}{2}$$

Combining the two:

$$v_z(y) = \dfrac{v_z}{2} + \dfrac{v_z}{H} y + \dfrac{\Delta P}{2\eta L_c}\left(\dfrac{H^2}{4} - y^2\right)$$

Rotational speed is 100 RPM. Geometrical parameters

$v_r = \dfrac{100 \text{ rot}}{\text{min}}\left(\dfrac{1 \text{ min}}{60 \text{ s}}\right)\left(\dfrac{\pi D}{\text{rot}}\right) = 10.47$ cm/s $H = 0.25$ cm $D = 2$ cm $\phi = 17.65°$

$v_z = v_r \cos\phi = 9.98$ cm/s $L_m = 5D = 10$ cm $\Rightarrow L_c = \dfrac{L_m}{\sin\phi} = 33$ cm

Substituting in from above and viscosity = 300 Pa·s.

$$v_z(y) = 4.99 + 40y - 50.5(0.0156 - y^2)$$

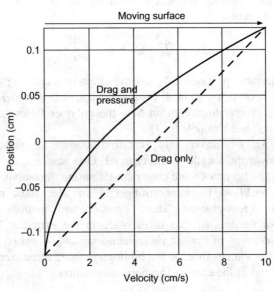

FIGURE E3.17

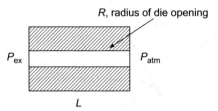

FIGURE 3.56 Schematic of a cylindrical die. Pressure develops at the end of the extruder, P_{ex}, and pushes the polymer melt thought the die opening.

3.5.3 Die Flow

The pressure built in the main body of the extruder is used to drive the flow through the die. There is no drag flow in the die; the polymer is conveyed purely by pressure-driven flow. Earlier in the chapter, we considered the pressure-driven flow through channels of various geometries. Let's consider the pressure-driven flow through a cylindrical channel again here, but now with the idea that this cylinder is the die, which defines the shape of the extruded product, a circular rod, in this case. Figure 3.56 shows the simple geometry: the die "land" has length L and the radius of the cylindrical channel is R. There is a pressure difference, ΔP, between the extruder pressure and atmospheric pressure at the exit. For a Newtonian liquid, the volumetric output is given by:

$$Q_{\text{DIE}} = \frac{\pi R^4 \Delta P}{8 \eta L} \tag{3.115}$$

Considering a die of fixed geometry, the volumetric flow rate depends only on the pressure built in the extruder, which we assume is much greater than atmospheric pressure (i.e., $\Delta P = P_{ex} - P_{atm} = P_{ex} = P$), and the polymer melt viscosity. The die constant, K_D, contains the geometry. In this particular case, $K_D = \pi R^4/8L$. For a slit die with width W, slit height H, and length L, $K_D = H^3 W/12L$. Likewise, other shapes of dies have different values of K_D. Hence, the general form for the volumetric flow rate of a Newtonian liquid through a die is:

$$Q_{\text{DIE}} = \frac{K_D P}{\eta} \tag{3.116}$$

To develop a higher output for a given die, either a larger pressure drop is needed or a lower viscosity polymer. More discussion about the features of dies and how they affect the extrudate is presented later in the section.

3.5.4 Single Screw Extruder Operating Diagram

The operating diagram for a single screw extruder takes into account the ability of the screw to deliver melt and the role of the die in controlling pressure and output. See Figure 3.57. The x-axis of the diagram is the pressure at the

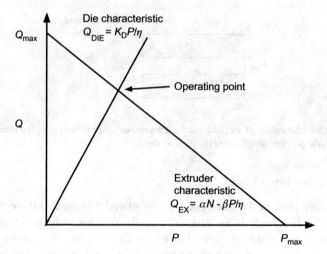

FIGURE 3.57 Operating diagram for a single screw extruder.

end of the metering section of the extruder. This pressure, simply noted as P here, controls the small backflow in the extruder screw, as shown in Eq. (3.114), and the output from the die, Eq. (3.116). The y-axis on the operating diagram is the output, Q. Two lines are placed on the diagram: one for the extruder output, Q_{EX}, and the other for the die, Q_{DIE}. Since the extruder screw and the die are in series, one feeds the other and the volumetric flow rates or outputs of those sections must be the same. The intersection of the two lines on the operating diagram is therefore the operating point. This point defines the output of the process and the pressure generated in the extruder. Also of interest are the extremities of the extruder characteristic line, although they represent fictional conditions that are not encountered in practice. On one end, Q_{max} is found when P is 0; this condition corresponds to an "open die," a state in which no constriction is placed at the end of the extruder screw. The other end, noted by P_{max}, represents a "closed die" that does not allow any output. Of course, operating at or near this condition is not recommended as excessive pressures can damage the extruder or cause it to explode.

The operating diagram is easily constructed (see Example 3.18) and provides general guidance on how conditions and parameters influence the operating point. For example, raising the rotational speed shifts the extruder characteristic line upward uniformly, which leads to an increase in both output and pressure at the operating point. See Figure 3.58. Making the die opening smaller and hence decreasing K_D, changes the slope on the die characteristic line and creates conditions with lower output and higher pressure. Likewise, the effects of channel depth, helix angle, and viscosity can be explored conceptually.

The flow model used to construct the lines on the diagram has some limitations, some of which can be addressed. First, the melt was assumed to be

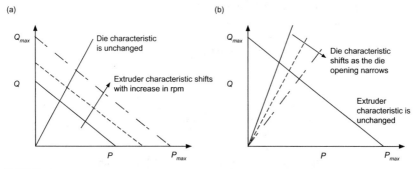

FIGURE 3.58 Operating diagrams for a single screw extruder, showing the effect of (a) increasing the rotational speed of the screw and (b) constricting the die (right).

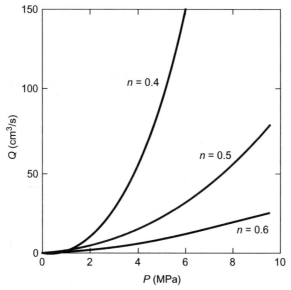

FIGURE 3.59 Die characteristic line for extrusion of a polymer with $K = 15{,}000 \text{ Pa} \cdot \text{s}^n$ and various power law indices, n, for a die with diameter of 5 mm and land length of 2 cm.

Newtonian, but most polymer melts are shear thinning. The effect of shear thinning on the die flow has already been presented, at least in equation form. The die characteristic line becomes nonlinear as expected from the equation in Table 3.5. An example is shown in Figure 3.59. The pressure-driven flow term of the extruder characteristic also changes, although the effect is not as dramatic as the die characteristic effect. Other factors, including temperature variations, friction, and more complicated screw geometries, are not as easily addressed.

EXAMPLE 3.18 Construct an operating diagram for the extruder and die described below and find the output at 60 RPM.

Extruder: $D = 6.35$ cm, $H = 3.23$ mm, $\varphi = 17.65°$
 Length of metering zone: 0.286 m
 Die: $R = 3$ mm, $L = 3$ cm
 Polymer: $\eta = 100$ Pa•s (assume Newtonian)
 Extruder characteristic:

$$Q_{EX} = Q_d + Q_p$$

$$Q_{EX} = \frac{1}{2}\pi^2 D^2 HN \cos\phi \sin\phi - \frac{\pi D \sin^2\phi H^3 P}{12\eta L_m}$$

$$Q_{EX} = \frac{1}{2}\pi^2 (0.0635 \text{ m})^2 (0.00323 \text{ m})\left(\frac{60 \text{ rot/min}}{60 \text{ s/min}}\right)\cos(17.65°)\sin(17.65°)$$

$$- \frac{\pi(0.0635 \text{ m})\sin^2(17.65°)(0.00323 \text{ m})^3 P}{12(100 \text{ Pa•s})(0.286 \text{ m})}$$

$$Q_{EX} = 1.86 \times 10^{-5} \text{ m}^3/\text{s} - 1.80 \times 10^{-12} P \text{ m}^3/\text{s}$$

$$Q_{max} = 1.86 \times 10^{-5} \text{m}^3/\text{s} = 18.6 \text{ cm}^3/\text{s}$$

$$P_{max} = \frac{1.86 \times 10^{-5} \text{m}^3/\text{s}}{1.80 \times 10^{-12} \text{m}^3/(\text{Pa•s})} = 1.03 \times 10^7 \text{ Pa} = 10.3 \text{ MPa}$$

Die characteristic: $Q_{DIE} = \dfrac{\pi R^4 \Delta P}{8\eta L} = \dfrac{\pi(0.003 \text{ m})^4}{3(100 \text{ Pa•s})(0.03 \text{ m})} P = 1.06 \times 10^{-11} P$

At operating point, $Q_{EX} = Q_{DIE}$

$$1.86 \times 10^{-5} - 1.80 \times 10^{-12} P = 1.06 \times 10^{-11} P$$
$$P = 1.50 \times 10^6 \text{ Pa} = 1.5 \text{ MPa}$$
$$Q = 1.59 \times 10^{-5} \text{ m}^3/\text{s} = 15.9 \text{ cm}^3/\text{s}$$

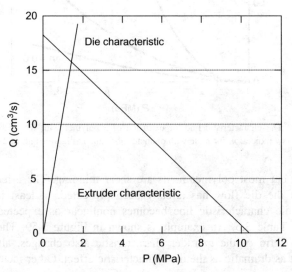

FIGURE E3.18

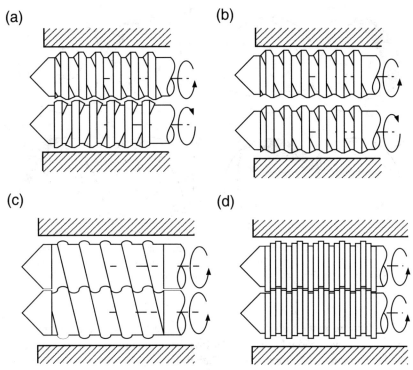

FIGURE 3.60 Twin screw extruders designs viewed from above: (a) counterrotating intermeshing elements, (b) counterrotating nonintermeshing elements, (c) corotating self-wiping intermeshing elements, and (d) corotating intermeshing kneading blocks. *Reproduced with permission of John Wiley & Sons from Baird and Collias (1998); Copyright ©1998 by John Wiley & Sons, Inc.*

3.5.5 Twin Screw Extrusion

As the name suggests, twin screw extruders employ a pair of screws that are rotated in tandem, either in the same direction (corotating) or in opposite directions (counterrotating). Figure 3.60 shows these two general options. The two screws are placed side-by-side in a barrel that closely accommodates the two screws with a tight clearance between the barrel and the flight tips. The design is typically based on intermeshing of the screws so that the flights of one screw penetrate into the channels of the other. Nonintermeshing, counterrotating screws are used in some circumstances, including reactive mixing. In this option, the operating principles are similar to single screw extrusion. One advantage of twin screw extruders is their modular designs that allow the incorporation of a series of different screw elements or sections down the length of the screw. Overall, the twin screw extruders are more complex; the flow mechanisms depend on the operation type (corotating vs. counterrotating) and specifics of the screw design.

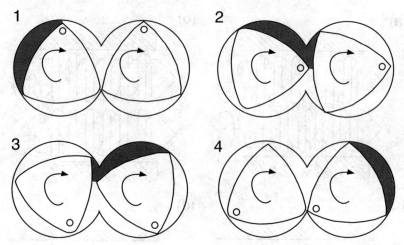

FIGURE 3.61 Melt transfer from one screw to the other with three lobed elements. In this figure, the view is down the length of the extruder. As the melt passes from one screw to the other, it is sheared extensively and propelled down the length of the extruder. *Adapted from Rauwendaal (2001)*.

The forward motion down the length of the extruder is different than that in the single screw version. Let's consider first the motion in a twin screw extruder with intermeshing corotating screws. Rather than the drag of a single screw, in this twin screw setup, the material is transferred back and forth between screws. As a result, the forward motion in the early stages depends less on friction; therefore, hard to feed polymers can be more effectively transported by the twin screw design as compared to the single screw extruder. As the polymer makes its way down the length of the extruder, the zones familiar from single screw extruders are not present, at least, not in the same way. The modular design allows for the incorporation of mixing blocks and features that aid in the compounding action. For example, the material motion and mixing action is illustrated in the sequence shown in Figure 3.61, where three lobed screw elements are used. The high shear in the corotating design makes it popular for compounding operations in which additives are incorporated or multiphase polymer blends created. Unlike the single screw extruder, the twin screw extruder is not typically operated full. Rather, only the zone nearest the die is filled so that sufficient pressure builds to push the polymer through the die.

The action in a twin screw extruder with intermeshed counterrotating screws is different and lends itself to extrusion of reactive and heat-sensitive polymers. The forward motion is similar to a gear pump. The material is enclosed in helical C-shaped pockets that are propelled forward by the turning of the screws. Since the two screws are rotating in opposite directions, there is a squeezing action that increases the pressure in the material as it is forced into the region between the screws. As a result, there is greater stress on the screws, which are forced apart, and operating conditions must be chosen to avoid high pressures that damage the screws. This design is most frequently chosen for extrusion of heat sensitive polymers such as PVC, due to lack of frictional heating.

3.5.6 Die Exit Effects

Extruder dies are designed to impart the desired cross-sectional shape to the extruded product and to prevent defects and problems on exit from the die. Importantly, laminar flow should be maintained through the die, and recirculations that might lead to polymer degradation should be avoided. Further, some features in the die, such as the taper, influence the stresses in the polymer and hence the defects that might occur on exit. The overall goal of die design is to create a die that produces an extruded product with the proper shape and size and a smooth surface. In this section, we consider the die flow in more detail, particularly addressing the flow and stress in the melt as it passes through the die. Then, the general features of the die are introduced and the two main exit effects—die swell and melt fracture—are explored.

Let's consider the flow of the melt from the extruder, which has a large diameter, through a simple cylindrical die, which has a much smaller diameter. See Figure 3.62. The pressure reaches a maximum at the end of the metering section or the point just beyond the screw where there may be a screen or breaker plate. Pressure drives the flow first through the larger diameter region and then through the taper and the narrower diameter die. Assuming the flow is steady state, the velocity profile is parabolic for a Newtonian liquid and a more flattened shape for a shear thinning liquid. See Example 3.6. The velocity is zero at the walls and maximum at the center. On the other hand, the shear rate and the shear stress are maximum at the walls. To keep the volumetric flow rate the same in the two nominal sections, the maximum velocity and the shear stress at the wall are markedly higher in the narrower section. Importantly, the melt experiences high stress

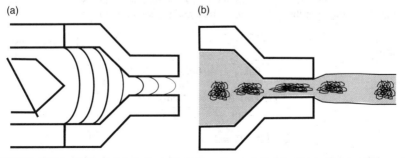

FIGURE 3.62 Schematic of the end of the extruder and the die: (a) velocity profiles and (b) polymer structure changes and die swell.

as it makes its way through the die—it is elongated due to the axial acceleration and compressed in the direction perpendicular to the axis due to the constraint of the die. Example 3.19 explores the flow in a die.

The taper and land in die are important to producing a high-quality extruded project. The taper allows a smooth laminar flow to be maintained as the diameter transitions from large to small and the polymer accelerates. The high viscosity of the polymer melt makes turbulent flow unlikely, but a discontinuity that occurs in an abrupt, nontapered transition leads to regions of circular motion called recirculations. Since the molten polymer is held in

---

EXAMPLE 3.19 Consider the end of the extruder and the die zone as two cylinders in series with a pressure of 5 MPa on the extruder end and atmospheric pressure (~ 0.1 MPa) at the exit. Assume steady state flow of a Newtonian liquid with viscosity of 1000 Pa·s and ignore the taper. Section 1: radius = 2 cm and length = 4 cm. Section 2: radius = 0.5 cm and length = 2 cm. Predict the linear output through the die, as well as the maximum velocity and max shear rate in both sections.

$$Q_1 = \frac{\pi R_1^4 \Delta P_1}{8\eta L_1} \quad \text{and} \quad Q_2 = \frac{\pi R_2^4 \Delta P_2}{8\eta L_2}$$

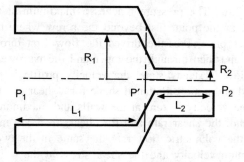

FIGURE E3.19

The two cylinders are in series and hence the output through section 1 (Q_1) is equal to the output through section 2 (Q_2) and the total output (Q_T).

$$Q_T = Q_1 = Q_2$$

$$\frac{\pi R_1^4 \Delta P_1}{8\eta L_1} = \frac{\pi R_2^4 \Delta P_2}{8\eta L_2}$$

$$\frac{(2 \text{ cm})^4 \Delta P_1}{4 \text{ cm}} = \frac{(0.5 \text{ cm})^4 \Delta P_2}{2 \text{ cm}}$$

$$128 \Delta P_1 = \Delta P_2$$

Most of the pressure drop is over section 2. Assuming some intermediate pressure P' between sections 1 and 2, the individual pressure drops can be found.

$\Delta P_1 = P_1 - P' = 5$ MPa-P' and $\Delta P_2 = P' - P_2 = P' - 0.1$ MPa
$128 \Delta P_1 = \Delta P_2$
$128(5 \text{ MPa-}P') = P' - 0.1$ MPa
$P' = 4.96$ MPa
$\Delta P_1 = 0.04$ MPa $\Delta P_2 = 4.86$ MPa

Notice that the most of the pressure drop occurs in the second section where the flow is constricted. The volumetric flow rate is therefore:

$$Q_T = Q_1 = \frac{\pi R_1^4 \Delta P_1}{8 \eta L_1} = \frac{\pi (0.02 \text{ m})^4 (0.04 \times 10^6 \text{ Pa})}{8(1000 \text{ Pa} \cdot \text{s})(0.04 \text{ m})} = 6.3 \times 10^{-5} \text{ m}^3/\text{s}$$

The linear output velocity is found by dividing by the cross-sectional area of the die

$$v_{exit} = 0.80 \text{ m/s}$$

Consider the velocity profiles in the two sections. For a Newtonian liquid, these profiles are parabolic. We can compare the maximum velocity (at center) in each section.

$$v_x = \frac{\Delta P}{4 \eta L}(R^2 - r^2) \qquad v_{max} = v(0) = \frac{\Delta P R^2}{4 \eta L}$$

$$v_{max,1} = 0.1 \text{ m/s} \qquad v_{max,2} = 1.5 \text{ m/s}$$

The shear rate at the walls is also be higher in the second section as compared to the first. This effect leads to exit effects as the polymer emerges from the die. (The sign on the shear rate is not important.) The shear stress on the polymer is also greatest at the walls and higher in the constriction of the die.

$$\dot{\gamma} = \frac{dv}{dr} = -\frac{\Delta P r}{2 \eta L}$$

$$(\dot{\gamma})_{max} = \dot{\gamma}(R) = -\frac{\Delta P R}{2 \eta L} \qquad (\dot{\gamma})_{max,1} = 10 \text{ s}^{-1} \qquad (\dot{\gamma})_{max,2} = 608 \text{ s}^{-1}$$

---

these regions for a longer time, there may be melt degradation and poor quality of the product. Further, the stresses that develop in the melt are adversely affected by an abrupt design. The land is a necessary feature that aids in relaxing stresses developed in the melt as it is forced through the constriction of the die. Longer lands provide more time for stress relaxation. These features mitigate the two common exit effects—die swell and melt fracture.

Die swell is the increase in diameter or cross-sectional area that occurs as the polymer extrudate emerges from the die. The stresses introduced as the polymer makes its way from a large diameter to a small diameter lead to elongation and compression of the polymer coils in the melt. In the land, these stresses have some time to relax, but not enough to remove the stresses and the nonequilibrium configuration of the coils completely. Therefore, when the constraint of the die is removed on exit, the polymer extrudate swells. This phenomenon can be considered a memory effect—the polymer is attempting to restore its original configuration in the larger diameter extruder. Die swell is quantified as a ratio:

$$\text{Die swell} = \frac{D}{D_o} \tag{3.117}$$

where D is the diameter on exit and D_o is the die diameter. Hence, die swell is greater than 1.

The magnitude of die swell depends on two time scales. The first time scale is the time for polymer relaxation. Relaxation time scales with the viscosity divided by the modulus. So in general, relaxation time decreases as temperature increases due to the strong effect of temperature on viscosity. The second is the time scale of the process. The process time for this situation is simply the residence time in the land, which is the ratio of the land length divided by the exit linear velocity. The linear velocity is a direct function of the volumetric output, Q. The ratio of these two time scales is defined as the Deborah number, De:

$$De = \frac{\text{relaxation time}}{\text{process time}} \tag{3.118}$$

(This dimensionless number received its name from the Song of Deborah in the Bible, Judges 5:5, that also relates time scales.) To minimize die swell, the relaxation time should be short relative to the process time. In other words, De should be less than one.

Several process parameters influence De and hence the extent of die swell. Considering a die of constant dimensions, the die swell increases with the volumetric output, Q. With increased Q, the residence time drops and there is less time available for relaxation of stresses. At a fixed Q, the die swell drops as the land length increases, which allows more time for stress relaxation. These trends point to a trade-off between productivity and extrudate quality. Another limitation on production rate is a rough extrudate surface, as discussed more below.

Die designs must adapt to die swell. Accommodation of die swell in the forming of circular cross-sections is straightforward. The die diameter is simply decreased. But for other shapes, the features in the die opening shape must be also changed to accommodate swell. For example, a die with convex (curved in) features is needed to produce a square cross-section in an extrudate.

Melt fracture is a term that is used to describe a variety of exit instabilities that result in an extrudate with a rough surface. Some call this defect "sharkskin" or "bamboo." The defect is related to the relative stress on the polymer as compared to the strength of the melt. When the stress exceeds the strength, there is local fracture. However, there are several mechanisms of generating stress and modes of "melt fracture" so their origins and remedies are not universal. Additionally, certain polymer types are more susceptible to melt fracture, based on their melt strengths.

Melt fracture is more likely when *De* is greater than one. Under these conditions, the polymer does not have time to relax stresses in the land and hence these stresses are more likely to be above the critical level on exit and lead to fracture. There is a connection between melt fracture and the shear stress at the wall. This shear stress increases as the die diameter decreases and also as Q increases. Additionally, the taper in the die has been shown to be essential to decreasing melt fracture. The discontinuity induces a complex and severe stress state in the polymer that induces the surface defect.

3.5.7 Extruded Products and Solidification

Extrusion is a workhorse method in polymer processing and used for several purposes. First, extruders are used at the end of primary synthesis lines to convert the newly formed polymer to pellets. Pellets are produced by chopping a narrow diameter extrudate. Second, extrusion is used to produce a product or "profile." Shapes such as sheets, rods, and tubes are made by extrusion. Third, extrusion is used to prepare and pressurize a melt for other forming operations. For example, an extruder with a special die is used to make thin sheets by the blown film process, extruders create parisons for blow molding, and extruders with reciprocating screws are used in injection molding. Here, we focus on the shaping aspect of profile extrusion.

Examples of Special Dies. Tubular shapes require a special die with a mandrel for the interior hole in the tube. The polymer melt must flow around the mandrel, which is held in place by a support known as a spider, as shown in Figure 3.63. The polymer "knits" together after going around the "legs" of

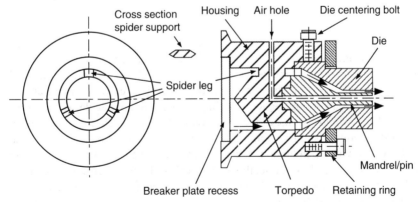

FIGURE 3.63 Die for extruding tubes or pipes. The polymer melt exiting the extruder encounters a torpedo that diverts the flow into the tubular opening that is defined by the mandrel. Streamlined spiders hold the torpedo and mandrel in place. An internal air delivery is used in some dies. *From Rauwendaal (2014), reproduced by permission of Hanser Publications.*

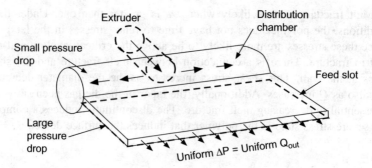

FIGURE 3.64 Diagram of a T-die geometry for extruding plastic sheet.

the spider. Tubes are also extruded with dies that are oriented vertically. In this way, the tube shape can be better stabilized through solidification.

Extrusion of sheets and films present the challenge of uniformly spreading the melt out from the extruder over the width of the die. The goal is to have a uniform output over the entire span of the die at exit. The simplest die for this purpose is a T-die, as shown in Figure 3.64. A distribution chamber that is located after the transition from the extruder. It has a relatively large diameter so that the polymer melt fills this chamber with little resistance and pressure drop. Then the melt is forced under a large ΔP between die lips to form a sheet. This approach is adequate for some applications but the finite pressure drop in the distribution chamber does lead to a non uniform output that can be improved by changing the die geometry. A better design is called the coathanger die. This die has a specially shaped chamber and varying feed slot length to achieve more uniform output. As expected from the equations in Table 3.5, maintaining a uniform gap is also essential to controlling not only the product thickness, but also the output. Adjustable die lips are used to fine tune the gap and output.

Solidification. After the polymer exits the die, it must be solidified quickly enough so that its shape is retained. As the temperature drops, the viscosity increases and the polymer undergoes a glass transition to a solid state. Some also crystallize on cooling; this transition leads to a more abrupt solidification on cooling. Several cooling methods are available: (i) air cooling, which would be used for very thin products such as blown film, (ii) water cooling, and (iii) cooling by contact with chilled metal rollers. The latter two involve conduction of heat from the polymer to the water or the chilled metal. Based on the analysis provided in the heat transfer fundamentals section, the time needed for a polymer to reach a temperature at which it can be safely handled without distortion can be found. This time is either helpful in constructing the length of a cooling bath needed for a particular product or the contact time with chilled rollers.

Controlling the profile dimensions and shape can require shaping of the extrudate after it exits from the die. Die swell, in particular, makes control of final dimensions of products, such as tubes and rods, a challenge. One approach is to exert some extensional force on the extrudate after exit. This "drawing" compensates for the swell and can lead to some molecular orientation, which can be beneficial for properties. Another method is to send the extruded product through a "sizing" die or plate, which sets the final dimensions after exit, but before complete solidification.

Thermoplastic polymer grades are made specifically with extrusion in mind. Polymers that are suitable for extrusion are designed to have high melt strength, and therefore resist melt fracture, and achieve a high viscosity on cooling relatively rapidly, a quality important to shape retention. Hence extrusion grades of polymer have higher average molecular weight and are extruded under conditions that keep the viscosity relatively high. These features in a polymer are contrasted with injection molding grades of polymer later in the chapter.

3.6 INJECTION MOLDING

3.6.1 Process Overview

Injection molding creates complex three-dimensional polymer shapes. The process begins with the preparation of a quantity of melt. Typically, the melt is created in a single screw extruder that contains a special reciprocating screw, but simpler piston extruders are also used. The melt is then forced by mechanical action into a chilled metal mold, where it solidifies to form the shape. Injection molding resembles die casting of metals. In fact, the conceptual basis for injection molding of polymers originated from die casting of metals, a process invented decades before injection molding. Likewise, there are variants of injection molding in which ceramic and metal powders are incorporated into the polymer melt. After solidification, the polymer is removed and the powders sintered to make a complex shape. These variants are addressed in a later chapter. There is also a molding process very similar to injection molding in which a metal "slush," a mixture of solid metal particles in a molten metal, is injected into a mold and solidified. Hence, the principles in this section can be applied to other processes.

Understanding the complexities of injection molding flow and solidification is a bit more challenging as compared to those events in extrusion. In injection molding, the molten polymer flows under pressure into the mold cavity, an inherently complex, 3D finite space. Additionally, concurrent heat transfer to the chilled metal mold makes the process nonisothermal with overlapping flow and solidification. Hence, some of the discussion in this section is more qualitative.

214 Materials Processing

This section begins with a description of the injection molding equipment and the molding cycle. Then the focus turns to the features of the mold and the processes that occur in the mold, including flow and solidification. A general process map is presented along with a discussion of the effects of process variables on production and product quality. Here, some issues related to part design are covered in addition to a discussion of qualities and features of polymers suited to injection molding. Lastly, reactive injection molding processes are discussed briefly. Like extruded products, injection molded polymers do not require significant post-forming operations.

3.6.2 The Injection Molding Machine and Cycle

There are three main parts to an injection molding machine: the injection unit, the mold, and the clamping unit. The *injection unit* melts and prepares the polymer for injection. The most common type resembles a single screw extruder (see Figure 3.65) and performs the same function as in the extrusion process. The starting material for the process, thermoplastic granules, is fed into the hopper of this unit. The screw design has three stages for feeding, compression, and metering. Typical L/D ratios are between 12 and 20. The rotating screw melts and conveys the material toward the end of the injection unit, but this screw has an additional feature—it is reciprocating. The screw moves forward to push polymer into the mold and backward to a position where it can rotate to prepare the next "shot" of polymer. The *mold* is two or more pieces of steel that are designed with channels to direct the melt and a mold cavity, where the part is formed. The mold also must be designed for efficient melt flow, cooling, opening, closing, and part removal. The *clamping unit* performs two functions. First, it holds the mold together and in place as the shot is pushed under pressure into the mold. Second, it controls the opening and closing of the mold. Clamping units employ toggle mechanisms.

Injection molding machines are rated based on "shot size" and clamping force. Smaller units operate with shots of about 20 g of polymer, based on a polystyrene standard, and larger units with up to 20 kg of polymer.

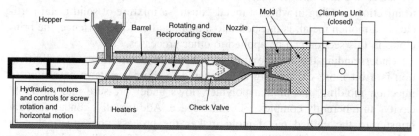

FIGURE 3.65 Schematic diagram of an injection molding machine. The reciprocating screw moves forward for injection and back for preparing the polymer for injection. The details of the three stages in the extruder screw (feed, compression and metering) are not shown.

The clamp force capacity ranges from several tons to several thousand tons. The clamping unit must be sufficient to keep the mold closed.

The molding cycle is summarized in Table 3.9. Injection molding is a cyclic operation with parts produced periodically as the process cycle is repeated. The molding cycle begins with the *mold closing*; the mold is held in place with the clamping unit, and a shot of melt is ready for injection. The next step is the rapid *injection* of the melt into the mold. During this process, the screw rotation is halted and the screw is driven forward. Melt flows through a nozzle at the end of the extruder into the mold. A "check valve" isolates the pressurized melt pool ahead of the screw from the rest of the extruder so that back flow is prevented. During the next stage, termed *hold-on*, the screw continues to push forward on the melt and the polymer in the mold solidifies. During this stage, melt continues to be pushed into the mold to compensate for solidification shrinkage there. This process is also known as "packing the mold." The gate, which separates the mold cavity from the other channels in the mold, eventually freezes and therefore stops the packing process. At this point, the next stage, *shot preparation and cooling*, begins. During this stage, the molded polymer continues to cool in the mold. The screw is retracted and it begins rotating so that the next shot can be prepared. The opening at the nozzle at the end of the extruder may be open at this point or the injection unit may be equipped with a valve that blocks output from the nozzle. The check valve is open so that melt can begin to accumulate ahead of the screw. After sufficient time, the last stage, *ejection*, takes place. The clamping unit releases the mold and opens it. Ejector pins push

TABLE 3.9 Steps in the Injection Molding Process

Step	Injection unit	Mold	Clamp unit
(1) Mold Closing	Shot is prepared, screw rotating, check valve is open.	Closed	Closed
(2) Injection	Screw, nonrotating, pushes forward, check valve is closed.	Closed, mold is filling	Closed
(3) Hold On	Screw pushes forward, check valve is closed.	Closed, packing continues, polymer is solidifying	Closed
(4) Shot Preparation and Cooling	Screw retracts and begins rotating, check valve is open.	Closed, polymer is solidified, cooling	Closed
(5) Ejection	Screw continues rotating, check valve is open.	Opened, part ejected	Opened

the part out of the mold. Then the cycle repeats. Cycle times range from less than a minute for small parts to a few minutes for large parts. In general, the most time consuming steps in the cycle involve solidification: the hold-on stage and the cooling/shot preparation step.

The features of a typical injection mold are shown in Figure 3.66. Steel is the most common mold material because it is able to withstand the high pressures associated with injection and has high thermal conductivity for fast solidification. Other mold materials, such as aluminum, are also used for prototypes and products that do not have a high target for number of parts produced. The mold design takes into account all stages in the molding cycle. During injection, the melt flows from the nozzle at the end of the extruder through the sprue, down the runners, through gates and into the mold cavities. The sprue and runners have large enough diameters so as not to slow the fluid flow to the mold cavity. Further, the taper on the sprue facilitates removal during ejection. The mold cavity is designed to create the desired geometry in the part. The gate features and placement are critical to controlling the filling of the mold cavity and the part quality, as discussed later in the section. Multiple cavities increase productivity. Molds also contain ejector pins for pushing the part out at the end of a cycle, cooling channels for temperature regulation, and guide pins for fitting of mold parts together. Figure 3.66 shows a two-part mold, but three part designs are also common. These allow for separation of the sprue and runners from the part in a single action, and therefore eliminate the need for a step to separate these extraneous pieces from the molded parts. Lastly, fine vents are present

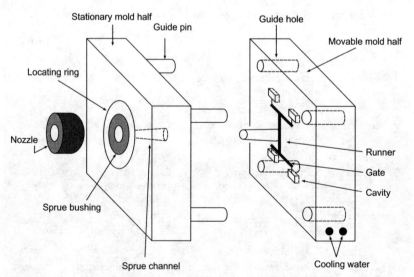

FIGURE 3.66 Schematic diagram of an injection mold. *From Strong (2000), reproduced by permission of Pearson Education.*

to allow for the air that originally occupies the channels and cavities in the closed mold to be displaced as the mold is filled.

3.6.3 Mold Flow

The events that take place as the melt fills the cavity are complex, involving 3D geometry, nonisothermal conditions, and high shear rates. In this section, we examine two critical flow issues. First, the flow that takes place in the runners is examined. Ideally, this flow is fast and results in uniform and simultaneous delivery of melt to the gates, which are the portals to the mold cavities. Second, we explore the flow into the mold cavity itself. Here, uniformity is critical; the role of the gates and general nature of the flow is described.

The flow of the molten polymer through the runners is a very rapid, pressure-driven flow. The ideal runner is large enough in cross-sectional area so that the flow is rapid, but as small as possible to minimize the shot size, decrease the scrap, and achieve rapid cycle times. While the solidified runners and sprue are recycled, the recycling process requires grinding and so minimizing scrap is important. Also important is the time that it takes to solidify the runners—if the runners are too large, their cooling and solidification can be the limiting factor in the cycle time. The melt comes in from the nozzle into a cold mold. The rapid injection leads to large shear rates and local heating of the polymer. This viscous heating lowers the viscosity for shear thinning polymer melts and increases the flow rate. The local heating helps to ensure that the polymer travels down the runner, so long as the pressure is suitable, without risk that solidification severely affects flow. In other words, we can approximate this filling process as a flow problem rather than a flow with concurrent solidification problem. The latter is the situation in the mold cavity.

With the general nature of the flow in mind, a design strategy for multicavity molds can be understood with simple principles. For example, several designs for a multicavity die are shown in Figure 3.67. The runners should

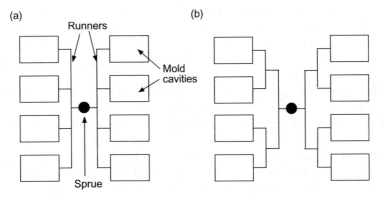

FIGURE 3.67 Schematics of (a) unbalanced runners and (b) balanced runners.

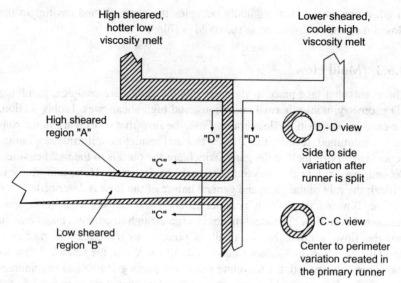

FIGURE 3.68 Diagram showing the influence of shear on viscosity and filling characteristics of a runner system. *Reproduced by permission of Carl Hanser Verlag Munich from Osswald et al. (2008).*

be designed so that the mold cavities fill simultaneously, under identical conditions, so that the final dimensions, densities, and properties of the parts in all the mold cavities are similar enough for the specifications of the part. Figure 3.67a is a simple design, but the shorter runners on the inner mold cavities cause them to fill first, while the outer ones fill later. There is a chance that the outer cavities will not fill completely or be packed with as much polymer during the hold on stage and hence the parts will have differences. Figure 3.67b is a geometrically balanced runner design. Examination of these layouts shows that the melt must travel exactly the same distance from the sprue to each mold cavity. The results from this layout are better than the simpler alternative. Interestingly, even geometrically balanced runners can still have issues because the shear condition varies as the melt changes direction. See Figure 3.68. Lastly, tuning the radius of the runners can give some improvement (Example 3.20).

Practical runner designs are worked out with the simple concepts described above and mold flow software that predicts flow patterns and solidification. Since injection molds are expensive to make, designs that allow inexpensive alterations after fabrication are preferred. Additional machining to increase the runner diameter after the main mold fabrication is one way to tune the flow. A special design option is called a "hot runner." In a hot runner, there is a separate section of the mold for the sprue and runners; this section is maintained at elevated temperatures. These parts of the

EXAMPLE 3.20 (a) Find an expression for the filling time of a cylindrical runner of length L and diameter R. Assume that the pressure is constant at the sprue and that it is atmospheric at the edge of the melt as it progresses down the runner. Also assume that the melt is Newtonian and that the temperature is constant. (b) Comment on the effects materials and geometry have on the filling time.

The volumetric flow rate in pressure-driven flow through a tube is:

$$Q = \frac{\pi R^4 \Delta P}{8\eta L}$$

Given that the pressure at the sprue inlet, P_{in}, is much greater than atmospheric pressure, then:

$$Q = \frac{\pi R^4 P_{in}}{8\eta L}$$

The filled volume of the runner is changing with time as it fills:

$$V(t) = \pi R^2 L(t)$$

Therefore,

$$\frac{dV}{dt} = \pi R^2 \frac{dL}{dt}$$

Realizing that $Q = dV/dt$,

$$\frac{\pi R^4 P_{in}}{8\eta L} = \pi R^2 \frac{dL}{dt}$$

$$\frac{R^2 P_{in}}{8\eta L} dt = L dL$$

Integrating,

$$\frac{R^2 P_{in} t}{8\eta L} = \frac{L^2}{2} \quad \text{and therefore} \quad t = \frac{4\eta L^2}{R^2 P_{in}}$$

The expression above shows the connection between the length of the runner and its fill time. Longer runners take more time to fill. Likewise there is an inverse relationship with the radius of the runner. Larger radii lead to shorter filling times. This result demonstrates the stronger effect the radius has on the flow as compared to the volume. Lastly, decreasing the viscosity decreases the fill time.

---

mold do not solidify and instead only short extensions into the chilled section of the mold and the parts themselves are cooled. This design is more intensive. For some parts, the added expense of the hot runner setup is justified by reduced cycle time, reduced waste, and improved performance.

Flow into the mold cavity is more complicated than flow through the runners. Pressure drives the melt through the gate and into the mold cavity. In the mold cavity, there is a complex, nonisothermal process in which the cavity is filled, the polymer is cooled and solidified, and more polymer is added to compensate for solidification-induced shrinkage. The filling, cooling, and packing processes overlap, which makes it difficult to model the process with simple equations. Again, mold flow software is frequently used. Nonetheless, understanding the filling and solidification in the mold cavity qualitatively is important in designing a good injection molding process.

The melt flow pattern depends on the melt properties, the gate geometry and placement, the cavity geometry, and importantly, the solidification at the mold walls. The gate geometry, especially the diameter or cross-sectional area, influences the rate of flow into the mold cavity as well as the local shear rate. Gates are typically small, which allows the part to be easily separated from the runner system. Small gates, together with rapid injection rates, lead to high shear rates, which causes local viscous heating and therefore, a drop in melt viscosity as the melt enters the mold cavity. In some circumstances, the local heating can be extreme and cause degradation. There are several geometrical patterns for the gates, including simple cylinders or pins, fans, and thin rectangles. As the melt enters the cavity, the polymer ideally spreads outward, encountering the cold walls of the mold. When the hot polymer hits the cold walls in the mold, it solidifies there. As a result, flow is faster in the center. Eventually, this centermost region turns over, as a fountain would, into a flow pattern known as fountain flow. See Figure 3.69 for the typical pattern for a gate placed at the end of a rectangular cavity. The velocity profile shows the regions of high shear rate a short distance near the interface with the solidified polymer.

The flow pattern and the rapid solidification lead to molecular orientation through the injection molded part. Because the polymer is entering the mold cavity at a high rate and it is freezing rapidly at the mold walls, there is orientation of the polymer chains. Most sources report minimal orientation in

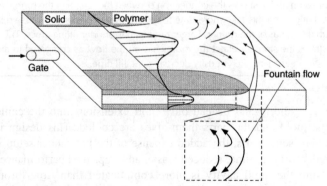

FIGURE 3.69 Schematic of polymer flowing from an end gate into a rectangular cavity and resulting velocity distribution and fountain flow. *Adapted from Middleman (1997).*

the frozen skin very near the mold wall, then an increase in orientation as a function of distance away from the wall before a peak and a drop in orientation toward the center of the piece. These gradients in structure can cause problems with stress and warping.

Some combinations of polymer properties and gate design lead to defects. If the polymer does not flow laterally on entering the cavity, then undesirable "jetting" can result. Jetting creates an extrusion-like flow through the gate, forming a stream of polymer that does not spread evenly into the cavity. Air can be entrapped in this type of flow. One solution is to use a "side gate" in which the polymer meets a cold wall head on and then turns to fill the rest of the cavity. For larger parts, there may be multiple gates in the mold cavity, and in some mold cavities, there is an obstacle that the melt must flow around. In these cases, one has to be concerned with "weld lines" that form where the flows meet.

3.6.4 Packing and Solidification

As the mold fills, the packing and solidification stage begins. Similar to metal casting, solidification leads to shrinkage, but in injection molding, the shrinkage in the mold cavity is combated by the continued forcing of the pressurized melt into the mold cavity. In this way, the process is similar to die casting. The packing and solidification is critical to the final part's dimensions and quality.

The density and size (dimensions) of the injection molded part depend on the pressure and temperature changes during the molding cycle. Figure 3.70 shows the changes in the temperature, pressure, and specific volume in an injection mold cavity. The specific volume is the inverse of density with units of volume per gram. This plot is similar to one used to describe solidification by glass transition; however, in this plot there are isobars to display the effects of pressure. Injection (0 to 2) is nearly isothermal process in which the flow of the polymer into the chamber leads to a pressure rise. The end of this stage is the start of the "hold-on" stage (2 to 3). During hold-on, the pressure is maintained at nearly a constant as the reciprocating screw continues to push forward on the melt. The temperature drops during this stage and consequently, the specific volume drops and the density increases. Eventually, the gate freezes, isolating the mold cavity. Next, the polymer, constrained to the mold cavity volume, cools and the pressure in the cavity returns to atmospheric (3 to 4). Lastly, the polymer in the mold cavity cools at atmospheric pressure without the constraint of the mold cavity. During this final cooling step (4 to 5), the part can separate from the cavity and shrink. By increasing the hold on pressure, the shrinkage can be decreased. There are limitations, however.

In practice, shrinkage factors are used to predict the mold cavity dimensions. Table 3.10 gives shrinkage factors for common polymers. These

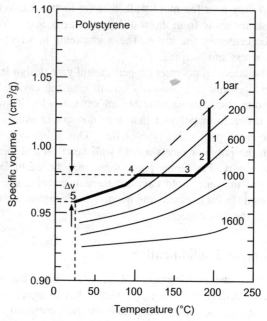

FIGURE 3.70 Effect of pressure and temperature on specific volume (1/density) of a polystyrene injection molded part. *Reproduced by permission of Carl Hanser Verlag Munich from Osswald (1998).*

TABLE 3.10 Linear Shrinkage Values for Injection Molded Polymers[a]

Polymer	Shrinkage factor (mm/mm)
Acrylonitrile butadiene styrene (ABS)	0.004–0.006
Nylon 6,6	0.015–0.020
Polycarbonate	0.005–0.007
Polyproplylene	0.010–0.020
Polyvinyl Chloride (Rigid)	0.004–0.006
Polystyrene	0.004–0.006

[a] *From data tabulated in Osswald et al. (2008).*

factors account for all the thermal contraction for typical molding conditions and hence are used to calculate the mold cavity dimensions needed for particular final part dimensions. The shrinkage factor, SF, relates the mold cavity dimension, L_m, to the part dimension, L_p by:

$$L_m = (1 + SF)L_p \qquad (3.119)$$

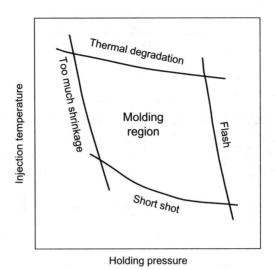

FIGURE 3.71 Operability window for injection molding. *Reproduced by permission of Carl Hanser Verlag Munich from Osswald (1998).*

The injection molding process is controlled by several processing variables, including the melt temperature, mold temperature, injection pressure, hold-on pressure, and the timing of the molding cycle. Operating diagrams, such as the one in Figure 3.71, demonstrate how conditions can be chosen for good quality parts. There is a window of temperature that provides good molding for any given polymer. At the lower end of this range, the thermoplastic viscosity is too high and filling the mold completely is difficult. This situation causes a defect, called a "short shot," in which the cavity is not completely filled. At the upper end of the range, the polymer begins to degrade. The pressure window is affected on the low end by the part shrinkage, as demonstrated in the P versus T diagram, and on the upper end, by the formation of flashing defect. Flashing occurs when the pressure on the polymer melt pushes apart the mold halves slightly and the polymer intrudes between the mold halves and solidifies into a thin sheet of polymer.

Solidification is faster in thinner parts and hence there is motivation for making part thicknesses thinner. In fact, when parts are too thick, local shrinkage defects, such as voids and sink marks, occur. Additionally, thicker parts have more issues with structural gradients and thermal stresses. Uniform thickness is best for heat transfer and solidification.

3.6.5 Reaction Injection Molding

Like thermoplastic injection molding, reaction injection molding (RIM) is a cyclic process that creates a complex, 3D polymer shape. RIM is also carried out in a sequence of steps and in a machine that has the same three basic elements (injection unit, mold, and clamping unit). But, the similarities stop

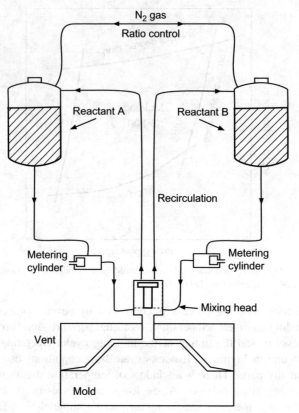

FIGURE 3.72 Reactive injection molding schematic diagram. The diagram shows the main components and circulation pattern. The mold clamping unit is not show. Often one tank, recirculation and metering system feed multiple mold via multiple mixheads. *Adapted from Johnson (1988) and Macosko (1989).*

there. The most important difference is that the liquid injected into the mold solidifies by reaction, not by cooling. And, this liquid is not a simple melt. Rather, it is a mixture of at least two low viscosity (~ 1 Pa•s) reactants.

The most common material molded by RIM is polyurethane. Polyurethane is formed by a reaction between an isocyanate and a polyol. Polyurethane structure and properties can be tuned by selection of the specific chemistries of the reactants as well as additives. Among the many variations are polyurethanes with properties tailored for automotive applications, such as bumpers, and others that are molded with a foaming agent to create rigid or flexible foams. RIM is also used to mold other polymers, including nylon (polyamide), epoxy and polyester. These other polymers are also formed by a two-part reaction in the RIM process.

The molding cycle can be described with aid of Figure 3.72, a schematic diagram of the RIM process equipment. There are two reactant tanks, the

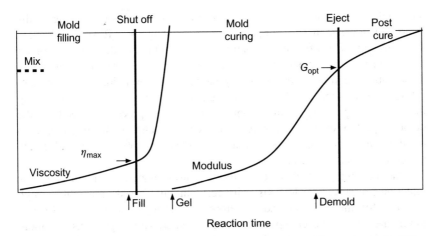

FIGURE 3.73 Development of viscosity and elasticity during the molding cycle with the filling complete (t_{fill}) while the viscosity is still low and debonding (t_{debond}) after the material has developed a sufficiently high modulus. *Reproduced by Permission of John Wiley & Sons, from Broyer and Macosko (1976), ©1976 American Institute of Chemical Engineers.*

contents of which recirculate through the system. Low pressure pumps drive the recirculation, heat exchangers keep the liquids at a controlled temperature and the mixing head is in a position that keeps the reactants isolated from one another. The molding cycle is initiated when the mold closes and the system changes from the recirculation to shot preparation. The metering cylinders are filled and moved into the correct position to be activated. These high pressure pumps control the amount of each reactant in a "shot" and provide the force needed to deliver them at high velocity to mixing head, which is open to receive the reactants at this stage of the cycle. After fast impingement mixing, the reacting mixture continues to flow, still driven by the pressure imparted by the metering cylinders, into the mold. With the mold filled, the mixhead closes and the system reverts to recirculation, while the material in the mold cures. When the molded part has developed sufficient strength, the mold is open, the part is removed and the next cycle begins. After demolding the part may be "post-cured" to continue the reaction and improve properties.

Reaction begins as soon as the reactants come into contact and therefore viscosity begins to climb. See Figure 3.73. During the very fast mixing and filling stage, the viscosity is low. In fact the viscosity may be low enough and the filling fast enough that turbulent conditions are approached. Laminar flow is desired for mold filling and high part quality. After filling is complete and the mold is closed, the continuing reaction causes the viscosity to increase dramatically and elasticity to develop. During the curing stage, the elastic modulus of the part increases. The reaction also leads to shrinkage. In thermoplastic injection molding, the "hold-on" stage leads to packing of the mold with more melt and compensation of shrinkage. For RIM, shrinkage is compensated by a different mechanism. Fine bubbles of dry nitrogen or air are mixed into one or both of the reactants. Under pressure during injection, these bubbles compress, but then they reexpand in the molded part, compensating for shrinkage. Lastly, reactions are temperature sensitive so choosing the mold temperature is important to the curing process as well as the demolding. For example, attempts to accelerate cure with an increase in mold temperature may also decrease the elasticity of the part, adversely affecting the demolding process.

One advantage of RIM over thermoplastic injection molding is low injection pressure. The low viscosity reactant mixtures do not require high pressure for mold filling and therefore the strength requirements for the molds are lower, the clamping units are smaller, and large parts are more easily achieved. Of course there are also drawbacks: materials handling and heat transfer are more complex for RIM due to the reactive nature of the process, and the materials that can be easily adapted to this process is more limited.

3.7 BLOW MOLDING

3.7.1 Process Overview

Blow molding is a process that converts a hollow parison (or preform) into a shaped container by deforming it under air pressure against a mold. Blow molding processes involve shaping while the material is in a molten state, but the melt, glass or polymer, has cooled enough that the viscosity is higher than for other melt-based processes discussed in this chapter. The process has ancient origins. Glass blowers began creating useful shapes by early versions of these methods about 3000 years ago. Today, both glass and polymer containers are made in highly automated versions of this ancient art.

Regardless of material type, blow molding has three key steps. First a controlled quantity of melt is made into parison or preform. Next, the parison is forced against a mold to increase the container volume and define the shape. Last, the molded object is cooled to retain the shape. Blow molding makes use of the steadily increasing viscosity of the glass or thermoplastic polymer melt as it cools. The viscosity of the material at the point of "blowing" must be in the appropriate range for easy deformation against the mold without gross flow under gravity, and the rate of cooling after this point must be fast enough to raise the viscosity to levels that are necessary to retain the shape. The processes themselves are highly nonisothermal with a hot melt cooling continuously throughout, but they are also automated and reproducible.

3.7.2 Blow Molding of Glass

It is interesting to begin the discussion of blow molding of glass with the artistic version of the process—glass blowing. The first step in the art of glass blowing is to gather a hot "gob" of glass melt at the end of a pipe. The pipe is immersed in a vat of glass melt in a furnace, for example, and then rotated and pulled out of the vat. At this point the gob is hot and flows under gravity and so the glass blower rotates the pipe continuously until he or she is ready to create a shape. The glass blower then sends a puff of air down the pipe. As it travels, it is heated and it expands. When it meets the gob, the gob deforms, and a pocket of air is trapped in the small blob of glass. The process is repeated to grow this pocket and push the glass outward, thinning it as it grows. Various paddles and fixtures can be used to rough out the "parison" shape. When the hollow object is large enough, the glass blower carefully places it in a mold and the final puff expands the glass shape until it meets the mold walls.

The automated version of this process begins with a machine that creates reproducible "gobs" of glass. A gob feeder is such a device. This feeder is attached to a glass melting tank, similar to the one used for production of sheet glass. Glass melt contained in tank flows through an orifice in the tank, assisted by a plunger to create a pendant drop. Two blades shear the drop to form a gob of molten glass that falls under gravity into an open mold. The gob feeder generates gobs continuously at a controlled rate.

The gob of glass drops into the first of two molds needed to make the final shape. One variant of the process in Figure 3.74; the parison is made by pressing the gob of glass against this first mold (known as the blank mold). A top on the mold prevent outflow and creates the bottom of the parison. The blank mold opens and the parison is transferred to the final mold. The parison, now right side up, is reheated before the air is blown in to deform the glass against the final mold. This mold opens and an automated lifter transfers the vessel to a conveyor belt and the as-formed glass is sent to an annealing lehr to remove thermal stresses. Coatings are also placed on the surfaces of the glass containers to help protect them from damage as they rattle down the production line.

Another variation on this process uses a puff of air to settle the gob and another puff to press the gob against the blank mold to make the parison. Then, the process continues in the same way as the "press and blow" process. Glass container manufacturing lines run as amazing speeds. A single glass tank may have several gob feeding orifices. Multiple gobs are then directed to molding lines that run in parallel, resulting in production rates of over 500 bottles per minute. Example 3.21 examines heat transfer in this process.

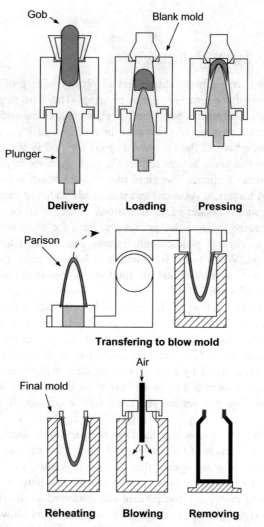

FIGURE 3.74 The press and blow process to create wide mouthed glass vessels. *From Stevens (1991), reproduced by permission of ASM International.*

EXAMPLE 3.21 Soda-lime-silica glass is blow molded using the press and blow method. (a) Estimate the average temperature of the glass after a 0.2 s contact between the plunger and mold. Assume that the initial temperature of the glass is 1100°C and that the situation can be modeled as a 2 mm thick sheet sandwiched between metal plates with initial temperature of 25°C. [Glass melt properties: thermal conductivity ~ 1.7 W/(m·K), specific heat is ~ 1200 J/(kg·K) and density is ~ 2.32 g/cm^3]. (b) Estimate the change in the viscosity of the glass from the cooling and comment on how this change impacts the process.

(a) The glass is thin and hence it is finite with respect to the heat transfer, a situation shown schematically in Figure 3.22b. We need to find the Fourier number and then use the graphical solution to the heat equation in Figure 3.24 to find the average temperature. The thermal conductivity of the glass melt is approximately. Therefore, the thermal diffusivity is:

$$\alpha = \frac{k}{\rho C_p} = \frac{1.7 \text{ W/(mK)}}{(2320 \text{ kg/m}^3)(1200 \text{ J/kgK})} = 6.1 \times 10^{-7} \text{ m}^2/\text{s}$$

The Fourier number is:

$$Fo = \frac{\alpha t}{L_c^2} = \frac{(6.1 \times 10^{-7} \text{ m}^2/\text{s})(0.2 \text{ s})}{(0.001 \text{ m})^2} = 0.12$$

Using Figure 3.24, dimensionless temperature ratio (y-axis) is ~0.4 when $Fo = 0.12$. Therefore,

$$\frac{T_{ave} - T_i}{T_o - T_i} = \frac{T_{ave} - 1100°C}{25°C - 1100°C} = 0.4$$

$$T_{ave} = 670°C$$

(b) The viscosity of soda-lime-silica glass as a function of temperature is shown in Figure 3.10. At 1100°C, the viscosity is ~180 Pa•s. On cooling to 670°C, the viscosity is ~10^7 Pa•s. This estimated viscosity is much higher, which would allow the transfer to the final mold. However, reheating is necessary to lower the viscosity back to a value that would allow the deformation.

3.7.3 Blow Molding of Polymers

Blow molding processes are also used to produce a wide variety of thermoplastic polymer containers and vessels. The general principles of the process are the same as glass blow molding, but the parison fabrication is a much different process. There are two variants based on the method used to make the parison: extrusion blow molding and injection blow molding.

In extrusion blow molding, a tubular parison is made by extrusion. The tube of polymer is extruded with a complex die that turns to flow downward and creates the tubular geometry. After the tube exits the die and descends a particular distance, a mold closes around the tube, effectively creating a parison. The closing process pinches off one end of the tube and air is blown in through the other end to inflate the parison against the mold. After cooling and solidification, the mold is opened and the part removed. One way to effectively carryout this process is on rotating platform, as shown in Figure 3.75. Multiple molds are used so that the extrusion process is continuous and productivity is increased. An alternative to the continuous process is to use a reciprocating screw, similar to that used in injection molding, and prepare a shot for each parison. This process is called intermittent extrusion blow molding.

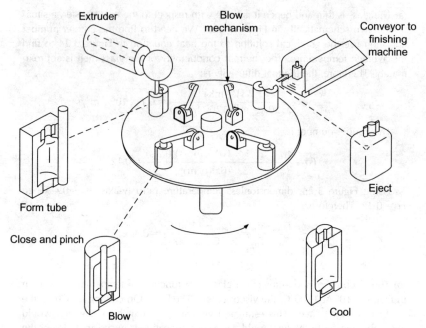

FIGURE 3.75 Extrusion blow molding on a rotating wheel. The process begins with extrusion of the parison, followed by mold closure to pinch off the bottom of the tube. Rotation brings the closed mold into position for blowing, and then to cooling and ejection. Note that this schematic shows the blowing from the top, but blow from the bottom and pinching off at the top is also practiced. *Reproduced by permission of Carl Hanser Verlag Munich from Rosato et al. (2004).*

A critical concern for extrusion blow molding is the wall thickness. There are two issues affecting the parison tube thickness: gravity induced flow or "sagging" and die swell. Gravity tends to make the bottom of the tube thicker at the onset of blowing. To prevent this problem, a special die that produces a variable wall thickness, starting thinner and ending thicker, can be used. Such programmed parison thicknesses can also be used to create uniform final thickness in complex molds. Also, the time for blowing should be as short as possible so that the time for the gravity-induced flow is minimized. The second effect, die swell, is inevitable and must be considered in the design of the process.

In injection blow molding, the parison is made by injection molding. Injection molding produces a parison with a well-controlled wall thickness and also more robust screw caps as compared extrusion blow molding. The molding of the parison may take place adjacent to the blow molding machine so that freshly molded parisons can be directly blown into final form. Or, the two processes can be decoupled and the injection molded parisons fed into a blow molding apparatus. In this case, the parison must be heated before blowing. See Figure 3.76.

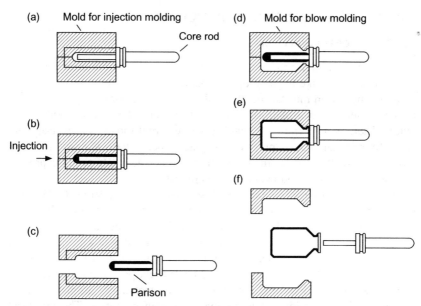

FIGURE 3.76 Injection blow molding process steps: (a) placing core rod in injection molding mold (b) injection, (c) removal of molded parison, (d) loading parison into mold for blow molding and reheating, (e) blowing through the core rod, and (f) removal of final part. *Reproduced by permission of Carl Hanser Verlag Munich from Rosato et al. (2004).*

Injection blow molding is widely used to make bottles for carbonated beverages in a variation known as stretch blow molding. The process in this case is designed to produce the best mechanical and barrier properties in the bottle. Instead if a simple expansion, the preform is deformed both radially and axially with a blowing pin that moves downward as the air pressure expands the polymer out toward the mold walls. This dual action produces molecular orientation that results in the development of superior properties, in particular decreased permeability to carbon dioxide. This process is not a true "melt-based" process because the temperature of the polymer is controlled such that it is a rubbery solid rather than a viscous liquid. In this way, its control and structure development have similarities to thermoforming, a topic in Chapter 4.

The selection of the extrusion based process or the injection molding based process depends on the requirements of the final product. Containers and vessels that do not demand outstanding precision and properties, such as milk jugs, are made by the extrusion blow molding process, the cheaper of the two processes. Injection blow molding offers much better control over wall thicknesses and screw caps, and with the addition of stretching better mechanical and barrier properties. Therefore, more demanding container applications, such as carbonated beverage bottles, are made by injection blow molding.

3.8 MELT-BASED ADDITIVE PROCESSES

3.8.1 Process Overview

In melt-based additive processes, an object is built up from a melt in a layer-by-layer fashion. Like forming operations, the first step in the process is creating the melt from the starting material. Then, unlike forming processes, the melt is printed onto a platform, one 2D layer atop the next, to build a 3D object. There is no forming or shaping with external forces, but instead there is a computer automated process that places melt where it needs to go in order to make the object. In this section, we consider the printing process that deposits the melt and creates the 3D object.

There are two basic methods. In fused deposition modeling (FDM), the melt is created from a filament supply in a printhead and extruded through a nozzle onto a platform. The printhead moves by computer control in the x-y plane to build a 2D layer and then the platform notches down and the next 2D layer is made and so on. FDM is used to make several thermoplastics, including acrylonitrile butadiene styrene (ABS) and polycarbonate. The second method is based on inkjet printing. In this technique, droplets of melt are ejected from a printhead according to computer control. The printhead rasters across a platform depositing melt droplets as it goes to create a 2D pattern, and then the platform notches down and the process repeats. The material set for this method is limited by the low viscosity requirement for generating droplets and the challenges in printing high temperature melts. Commercial processes are mainly designed for low molecular weight waxes. Consequently, the method is rarely used to make a final part, but rather it has application in making complex wax models used in metal casting.

The first step in creating a part by an additive process is to generate a computer file that contains all the details of the 3D structure of the part. This computer drawing may be constructed in computer-aided design (CAD) software or it can be produced from an actual part or model using a 3D scanner. The file is in the STL format, a type of file that was first developed for stereolithography, another additive process that is covered in a later chapter. The STL computer file is necessary to develop the printing sequence for the 2D layers or slices. With the appropriate computer file in hand, the next step is to set up the machine to print or build the part. Complex parts with undercuts require printing of both the main build material and a support material. A variety of process variables control the building process, as described more below. The building process itself is automatically controlled by the computer.

After the part is built, there are several post-processing steps. First, the support material needs to be removed either by dissolution, melting or fracture. Next the part may be treated to smooth the surface. Mechanical abrasion and chemical treatments can be used to lessen the textured appearance, which is a consequence of the layer-by-layer building process.

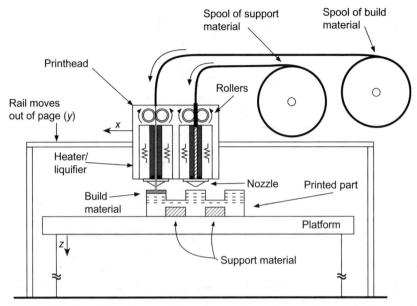

FIGURE 3.77 Fused deposition modeling machine, showing dual printhead for build material and support material. *Adapted from Stratasys, www.stratasys.com and Gibson et al. (2010).*

3.8.2 Fused Deposition Modeling (FDM)

Figure 3.77 shows a schematic of an FDM machine with a detailed rendering of a two nozzle extrusion print head. One nozzle prints the thermoplastic material of interest, sometimes called the build material, and the second prints a support material. The support material is necessary for undercuts and features that require a temporary surface for printing. Spools of both materials are fed into the extrusion head as it traverses in the x-y plane as the part is built. After a layer is complete, the platform drops in the z-direction and the next layer is constructed. A number of variables influence the process as well as the final structure and properties of the part. Important factors include the printing speed, the temperature of the melt, the nozzle diameter and feeding speed.

To understand this process further, consider the extruder head shown in Figure 3.78. The starting material, a spool of fine thermoplastic filament, is fed into a heated zone, where it melts. In this process, unlike other extrusion processes in this chapter, melting is not assisted by the mechanical action of a turning screw. Heat is transferred by conduction from the heater to the filament, which is seated in a tube that is just a slightly larger diameter than the filament. Given the fine diameter of the filament the melting process is fairly expedient. The melt is pushed out the nozzle as the feeding continues. One way to envision this process is as a piston-type extruder with the moving,

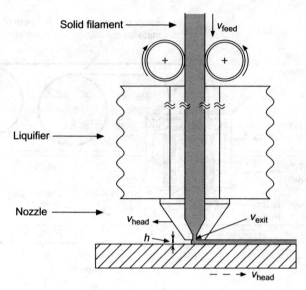

FIGURE 3.78 Simplified schematic of FDM printing showing the solid filament feeding into a thin metal tube attached to a nozzle. See text for definitions of variables.

solid filament acting as the piston. The nozzle itself is the die in the extruder, the taper serving the same purpose as in large scale extruders. On exit, die swell is expected, but given the proximity to the build surface, other factors have a greater impact on the diameter of the individual strand or "road" in the growing part. Solidification occurs as the hot melt exits the heated printhead and meets the cooler part.

The following is a simplified picture of the flow that occurs during deposition. The real situation depends on the details of the printhead design as well as operating conditions. The temperature variations are considerable as well. Figure 3.78 shows three locations in a flow path: (i) solid filament of radius R_f moving at velocity v_{feed}, which is set by the feeding rollers, (ii) melt exiting the nozzle with radius R_{exit} and average velocity v_{exit}, and (iii) solidified strand with nominal dimension of height h and width w moving away from the nozzle at velocity v_{head}.

The continuity principle provides guidance on the factors influencing process speed and strand structure. The material is fed into the extruder head with a velocity v_{feed} and exits with an average velocity v_{exit}. Ignoring the proximity of the platform for the moment, continuity requires the volumetric flow rate of the incoming filament, Q_1, to equal that of the exiting melt, Q_2:

$$Q_1 = Q_2$$
$$v_{feed}(\pi R_f^2) = v_{exit}(\pi R_{exit}^2)$$
$$v_{exit} = v_{feed}\left(\frac{R_f}{R_{exit}}\right)^2$$
(3.120)

Given the narrowing of the diameter at the nozzle exit or orifice, the material speeds up as it exits. On exit, however, the melt meets the platform or the previously deposited layer. The incremental motion of the platform downward with each subsequent layer sets the gap, h. This gap is on the order of the diameter of the orifice, so that it sets the build thickness. Now, to connect the dimensions of the strand to the operating parameters, the linear motion of the printhead is essential.

The head moves in the x-direction with a linear speed, v_{head}, but it is more convenient for this analysis to consider the equivalent situation with the head stationary and the platform moving. With this construction, the strand is moving away from the head in a plug flow mode. The strand, therefore, has volumetric flow rate of:

$$Q_3 = v_{head} hw \tag{3.121}$$

where w is the lateral width of the strand, which is assumed to nominally have a rectangular cross-section. Continuity requires that $Q_1 = Q_2 = Q_3$, so we find that the lateral dimension, d, is given by:

$$w = \frac{\pi R_f^2}{h}\left(\frac{v_{feed}}{v_{head}}\right) \tag{3.122}$$

For a given h and R_f, w is proportional to the ratio v_{feed}/v_{head}. A narrow strand can be achieved at faster printing speeds—essentially the same amount of material is spread out over a longer length and so the size scale must be smaller. There is a limit, however. The lateral dimension cannot shrink to less than the diameter of the nozzle. In this case, a discontinuity would develop. On the other end of the spectrum, a low printing speed would distribute material over a shorter length and hence make a wider strand. If printing is slow enough could overflow outside the bounds of the nozzle, which would cause some defects. Example 3.22 explores these principles.

This simple model and example give a sense for the controlling parameters for laying down a single strand of polymer. To build a part, the nozzle must follow a pattern. Since part geometries are complex, the building process involves going back and forth in the x-y plane. To keep the strand dimension nearly constant as the head slows down to turn a corner, the feeding speed must be decreased as well. Likewise, if the geometry requires the strand to end at one place and start anew elsewhere, the feeding is halted and hence the extrusion is halted. All of this complexity is programmed into the apparatus before the build begins. Once the "start" button is pressed, the machine follows a set of instructions to complete the build.

Solidification in FDM is also complex. The hot melt meets the already solid layer beneath. Heat is removed from the melt by convection into the air as well as conduction into the solid beneath. Airflow is important to heat removal and rapid solidification considering that the conduction process for

EXAMPLE 3.22 Consider the printing of polycarbonate using a nozzle with a 0.012" (0.305 mm) diameter and filament with a diameter of 0.07" (1.78 mm). The slice height or gap is set at 0.007" (0.178 mm). (a) Find the minimum feeding rate, v_{feed}, needed to print at a printhead rate of 5"/s (127 mm/s). (b) Estimate the maximum shear rate on exit from the nozzle. Assume that polycarbonate has a viscosity of 500 Pa·s and is approximately Newtonian.

(a) The minimum feed rate occurs when the lateral dimension, w, just equals the diameter of the nozzle orifice. So, $w = 0.305$ mm. We know that $h = 0.178$ mm and hence the cross-sectional area of the strand is $(0.305 \text{ mm})(0.178 \text{ mm}) = 0.0543 \text{ mm}^2$. Therefore,

$$Q_3 = v_{head} h w = (127 \text{ mm/s})(0.0543 \text{ mm}^2) = 6.90 \text{ mm}^3/\text{s}$$

$$Q_3 = Q_1 = v_{feed}(\pi R_f^2) = 6.90 \text{ mm}^3/\text{s}$$

$$v_{feed} = \frac{6.90 \text{ mm}^3/\text{s}}{\pi(1.78 \text{ mm}/2)^2} = 2.77 \text{ mm/s}$$

(b) The average velocity of the polymer at exit is:

$$v_{exit} = v_{feed}\left(\frac{R_f}{R_{exit}}\right)^2 = 2.77 \text{ mm/s}\left(\frac{1.78 \text{ mm}/2}{0.305 \text{ mm}/2}\right)^2 = 94.3 \text{ mm/s}$$

To find the maximum shear rate we would need to determine the velocity gradient. If we assume that there is cylindrical section at the end of the nozzle and that there is steady state flow in that section, then we can use the equations for pressure-driven flow through a tube. The maximum shear rate occurs at the walls of the nozzle (where $r = R$). As worked out in Example 3.19,

$$(\dot{\gamma})_{max} = \dot{\gamma}(R) = \frac{\Delta P R}{2\eta L}$$

Looking at the equation for the volumetric flow rate we can see a connection.

$$Q = \frac{\pi R^4 \Delta P}{8\eta L}$$

$$(\dot{\gamma})_{max} = \frac{4Q}{R} = \frac{4(6.90 \text{ mm}^3/\text{s})}{0.305 \text{ mm}/2} = 90.5 \text{ s}^{-1}$$

This estimate ignores the tapered section in the nozzle.

---

polymers is slow. The small dimensions are also helpful in speeding solidification. The factors influencing solidification time are the difference between the melt temperature and the glass transition temperature, airflow conditions and build thickness. While solidifying quickly is essential to overall process speed, some time at high temperature is needed for the bonding of one layer upon the next. Macroscopic dimensions play a role, considering that the small part may not be completely cooled before the next layer is deposited. Microscopic analysis of parts made by FDM show that there are relics of the strands, which results in anisotropy. Hence the rastering direction is changed layer to layer in a cross-hatch fashion.

Post-processing of FDM parts includes removal of support materials and surface finishing. Some support materials are designed to be soluble in solvents that are compatible with the build material. For these the part is immersed in the solvent until the support is dissolved. Others are friable; these supports are carefully removed by fracturing them away from the part. The surfaces of FDM parts are inherently corrugated based on the layered nature of the fabrication and therefore some surface smoothing with abrasion or solvent is sometimes carried out.

3.8.3 Inkjet Printing of Melts

Drop on demand printing of melts is essentially inkjet printing of melts with the intention of creating a 3D shape. This process is quite similar to FDM in that melt is placed onto a platform and built up 2D layer after 2D layer with the assistance of inkjet printed support material. The extrusion-based printhead for FDM is replaced with an inkjet nozzle. Droplet generation by inkjet is covered in Chapter 6, since inkjet printing finds most of its application in creating patterns on substrates using inks that are dispersions and solutions. Briefly, droplets are created from a pressure pulse in a liquid ink that is usually created by driving a piezoelectric actuator. In successful printing, a single droplet of controlled size and velocity is created and placed in a specific location on the substrate. Liquid properties, including viscosity, density and surface tension, factor into meeting this goal. Dimensionless numbers are helpful in finding the process window.

Printing of melts is more challenging than dispersions and solutions for two reasons. First the ink is hot and so thermal control becomes a challenge. For this reason, usually only polymers are considered for melt-based inkjet printing. Second, the viscosity must not be too high. While the exact limitation depends on the type of inkjet nozzle and other liquid parameters, high molecular weight, functional thermoplastics are out of range. Therefore, inkjet printing of 3D objects from melts has been limited to lower molecular weight waxes.

Complex, high-resolution wax parts can be fabricated by inkjet printing. The droplets solidify when they hit the platform or previously deposited layer. Because the droplet size is small ($\sim$70 μm) the patterns have high-resolution and smooth surfaces. Two waxes are used—one for the build and the other for support. The support either has a lower melting point and is removed by heating or is selectively soluble. Interestingly the 3D parts made by this process are rarely used as functional parts themselves. Instead they are used sacrificially in metal shape casting processes.

Investment casting, which is also known as lost wax casting, begins with a wax shape that is identical to the desired part. The wax is covered with a ceramic using a dipping or spraying operation. Then the structure is heated and the wax is removed. Lastly, molten metal is poured into the ceramic shell and solidified. 3D printed wax shapes have led to interesting possibilities and new products for investment casting. Very high-resolution metal parts can be fabricated from high-resolution wax pieces and individual, unique structures can be made. For example, dental bridges and structures created from x-rays and impressions from patients are inputs to the CAD model that drives the fabrication.

3.9 SUMMARY

Melts are the basis for a diverse set of materials processing methods. To engineer processes employing melts, the fundamentals of liquid structure and behavior must be understood. Across materials classes, liquid melt properties including structure and bonding, surface tension, and density can vary widely. Most metals form stable melts at temperatures that are accessible with standard furnace technologies. Compared with polymer and ceramic melts, metal melts are typically denser and have higher surface tension. Their structures in the liquid state are also comparatively simple. Polymer melts are easily prepared at low temperature and for thermoplastics, the melt state is of primary importance to processing. Of the three materials classes, polymer melts have the lowest density and lowest surface tension. They are also the most complex, comprised of long, typically entangled, chains. Melts prepared from crystalline ceramics are less accessible due to their high melting points, but the molten state is easily achieved and of primary importance to ceramic glass processes. Ceramic glass melts have intermediate densities and surface tensions compared with metal and polymer melts.

Rheology, the science of flow and deformation, is central to materials processing. For processing of melts, the rheological property of interest is the viscosity, which provides a measure of the resistance to flow. Viscosity decreases as temperature increases. For some melts, viscosity also depends on the shear rate. Metal melts and glass melts have no shear rate dependence; they are "Newtonian" (i.e., they follow Newton's law). Polymer melts are typically shear thinning (non-Newtonian); their viscosities decrease as the shear rate increases. This fortuitous change assists some melt processing methods. Fluid mechanics is a related topic of clear importance to melt processing. In addition to gravity, drag flow and pressure-driven flow are the key modes melt transport during processing.

Heat transfer and solidification complete a melt-based process. Melts are processed at high temperature and then solidified by cooling. The rate of cooling and temperature gradients are important to overall process times as well as to the structures that develop during solidification. The heat equation is used to predict temperature distribution during processing. For cooling, convection and conduction are two modes of heat transfer that drive solidification. Two common solidification mechanisms, crystallization and glass transition, determine the structure and therefore the properties of the final solid.

In terms of melt-based processes, shape casting and casting of flat sheets involve gravity-driven flow of low viscosity melts, such as metal and glass melts. Metal shape casting, either in sand molds or permanent metal molds, produces complex 3D objects. Controlling the rate of casting and the resulting solid structure requires application of the fundamentals of flow, heat transfer, and solidification. Sand molds offer the flexibility of making large and complex shapes but the overall production rate is limited by the slow heat extraction into the low thermal conductivity sand. Metal molds, on the other hand, offer fast solidification and better surface finish; however, the size, complexity and metal alloy selection are more limited. Casting processes are also used to fabricate flat sheet of ceramic glass by the float glass and fusion downdraw processes. The workhorse is the float glass process with its high production rates and smooth surfaces, but for thin glass, the fusion downdraw process leads the way. Control of these processes requires application of flow and heat transfer fundamentals.

Extrusion and injection molding are forming processes that involve pressure-driven flow of polymer melt. The single screw extruder consists of a rotating screw that drives melt forward in series with a constricting die that provides shape to the exiting 2D object. By application of flow fundamentals and particularly, the continuity equation, an operating diagram is built. This diagram provides the basis for understanding how process changes, such as screw rotational speed, impact the output. Equally important is the flow through the die, which leads to some interesting and potentially damaging exit effects. Injection molding uses a special type of extruder to prepare a "shot" of melt and drive it forward under pressure into a chilled metal mold. Complex polymer shapes are made by this method. Process control involves attention to the flow path in the mold and design of part dimensions. Injection molding can also be applied to reacting polymers.

Blow molding is used to produce hollow, complex shapes of ceramic glass and polymer. For glass, the process has its origins in glass blowing. In a highly automated process, gobs of glass are created and fed from a glass tank to a sequence of two molds. The first mold creates a parison, an open preform, and the second defines the shape of the hollow piece. An air jet is used to shape the parison. For polymer blow molding, the parison is made by either extrusion or injection molding. Like the glass process, an air jet deforms the polymer into the final container or vessel.

Melts are also used in additive manufacturing. In fused deposition modeling (FDM), a thin polymer strand is extruded by a moving printhead. A computer file plots the path of the printhead and lowers the platform as 2D layers are created one a top another. Complexity is built into the part by using a sacrificial support material. The same flow and heat transfer fundamentals used for other melt processes can be applied to FDM to understand how to control the structure of the printed material. Lastly, inkjet printing is used to print 3D wax melts. The resulting complex wax shapes are used in metal melt casting operations.

BIBLIOGRAPHY AND RECOMMENDED READING

Fundamentals

Adamson, A.W., Gast, A.P., 1997. Physical Chemistry of Surfaces, sixth ed. Wiley Interscience, New York, NY.

Barnes, H.A., Hutton, J.F., Walters, K., 1989. An Introduction to Rheology. Elsevier, Amsterdam, NL.

Battezzati, L., Greer, A.L., 1989. The viscosity of liquid metals and alloys. Acta Metall. 37, 1791–1802.

Callister, W.D., 1994. Materials Science and Engineering: An Introduction. John Wiley & Sons, New York, NY.

Carslaw, H.S., Jaeger, J.C., 1959. Conduction of Heat in Solids. Clarendon Press, Oxford, UK.

Cogswell, F.N., 1981. Polymer Melt Rheology. John Wiley & Sons, New York, NY.

Dealy, J.M., Wissbrun, K.F., 1990. Melt Rheology and Its Role in Plastics Processing. Van Nostrand Reinhold, New York, NY.

Gaskell, D.R., 1992. An Introduction to Transport Phenomena in Materials Engineering. Macmillan Publishing Co., Inc, New York, NY.

Geiger, G.H., Poirier, D.R., 1973. Transport Phenomena in Metallurgy. Addison-Wesley, Reading, MA.

Graessley, W.W., 1993. Viscoelasticity and flow in polymer melts and concentrated solutions. In: Mark, J.E. (Ed.), Physical Properties of Polymers, second ed. American Chemical Society, Washington, D. C., pp. 97–155.

Iida, T., Guthrie, R.I.L., 1988. Physical Properties of Liquid Metals. Clarendon Press, Oxford, UK.

Macosko, C.W., 1994. Rheology: Principles, Measurements and Applications. VCH Publishers, New York, NY.

Middleman, S., 1977. Fundamentals of Polymer Processing. McGraw-Hill, New York, NY.

Pearsen, J.R.A., 1985. Mechanics of Polymer Processing. Elsevier Applied Science Publishers, London, UK.

Pye, D., Joesph, I., Montenaro, A. (Eds.), 2005. Properties of Glass-Forming Melts. CRC Press, Boca Raton, FL.

Richardson, F.D., 1974. Physical Chemistry of Melts in Metallurgy (Vols. 1 and 2). Academic Press, London, UK.

Stokes, R.J., Evans, D.F., 1997. Fundamentals of Interfacial Engineering. Wiley-VCH, New York, NY.

White, F.M., 1999. Fluid Mechanics, fourth ed. McGraw-Hill, New York, NY.

Shape Casting of Metals

Beddoes, J., Bibby, M.J., 1999. Principles of Metal Manufacturing. John Wiley & Sons, New York, NY.

Campbell, J., 2003. Castings, second ed. Butterworth-Heinemann, Oxford, UK.

Edwards, L., Endean, M. (Eds.), 1990. Manufacturing with Materials. Butterworth Heinemann, Oxford, UK.
Flemings, M.C., 1974. Solidification Processing. McGraw-Hill, New York, NY.
Gaskell, D.R., 1992. An Introduction to Transport Phenomena in Materials Engineering. Macmillan Publishing Co., Inc, New York, NY.
Geiger, G.H., Poirier, D.R., 1973. Transport Phenomena in Metallurgy. Addison-Wesley, Reading, MA.
Kalpakjian, S., 1997. Manufacturing Processes for Engineering Materials, third ed. Addison-Wesley, Reading, MA.
Schey, J.A., 2000. Introduction to Manufacturing Processes, third ed. McGraw-Hill, New York, NY.
Stefanescu, D.M., 1988. ASM Metals Handbook: Casting (Vol. 15). ASM International, Metals Park, OH.

Casting of Flat Glass

Charnock, H., 1970. The float process. Physics Bulletin 21, 153–156.
Hynd, W.C., 1984. Flat glass manufacturing processes. In: Uhlmann, D.R., Kreidl, N.J. (Eds.), Glass: Science and Technology. Vol. 2 Processing. Academic Press, New York, NY, pp. 46–106.
Kingery, W.D., Bowen, H.K., Uhlmann, D.R., 1976. Introduction to Ceramics, second ed. Wiley, New York, NY.
Langmuir, I. (1933). Oil lenses on water and the nature of monomolecular expanded films. J. Chem. Phys., 1, 756-776.
McCauley, R.A., 1980. Float glass production: Pilkington vs. PPG. Glass Industry 61, 18–22.
McMaster, R.A., 1991. Annealed and tempered glass. In: Schneider, S.J. (Ed.), ASM Engineered Materials Handbook: Vol. 4 Ceramics and Glasses. ASM International, Materials Park, OH, pp. 453–459.
Narayanswany, O.S., 1977. A one-dimensional model of stretching float glass. J. Am. Ceram. Soc. 60, 1–5.
Seward, T.P., Varshneya, A.K., 2001. Inorganic glasses—commercial glass families, applications, and manufacturing methods. In: Harper, C.A. (Ed.), Handbook of Ceramics, Glasses and Diamonds. McGraw-Hill, New York, NY, pp. 6.1–6.140.
Stevens, H.J., 1991. Forming. In: Schneider, S.J. (Ed.), ASM Engineered Materials Handbook: Vol. 4 Ceramics and Glasses. ASM International, Materials Park, OH, pp. 394–401.
Woolley, F.E., 1991. Melting/fining. In: Schneider, S.J. (Ed.), ASM Engineered Materials Handbook: Vol. 4 Ceramics and Glasses. ASM International, Materials Park, OH, pp. 386–393.

Extrusion and Injection Molding

Agassant, J.F., Avenas, P., Sergent, J.P., Carreau, P.J., 1991. Polymer Processing: Principles and Modeling. Hanser Publishers, Munich, Germany.
Dealy, J.M., Wissbrun, K.F., 1990. Melt Rheology and Its Role in Plastics Processing. Van Nostrand Reinhold, New York, NY, Chapters 8, 14, and 15.
Macosko, C., 1989. RIM: Fundamentals of Reaction Injection Molding. Hanser Publishing, New York, NY.
Middleman, S., 1977. Fundamentals of Polymer Processing. McGraw-Hill, New York, NY.
Morton-Jones, D.H., 1989. Polymer Processing. Chapman and Hall, London, UK.
Osswald, T.A., 1998. Polymer Processing Fundamentals. Hanser Garner Publications, Munich, Germany.
Osswald, T.A., Turng, L.S., Gramann, P., 2008. Injection Molding Handbook. Hanser Garner Publications, Munich, Germany.

Pearsen, J.R.A., 1985. Mechanics of Polymer Processing. Elsevier Applied Science Publishers, London, UK.

Rauwendaal, C., 2001. Polymer Extrusion. Hanser Publishers, Munich, Germany.

Rauwendaal, C., 2014. Polymer Extrusion, fifth ed. Hanser Publishers, Munich, Germany.

Rosato, D., 1997. Plastics Processing Data Handbook. Chapman and Hall, London, UK.

Stevens, M.J., Covas, J.A., 1995. Extruder Principles and Operation, second ed. Chapman and Hall, London, UK.

Strong, A.B., 2000. Plastics Materials and Processing, second ed. Prentice-Hall, New York, NY.

Wilkinson, A.N., Ryan, A.J., 1999. Polymer processing and structure development. Dordrecht, NL. Kluwer Academic Publishers.

Blow Molding

Morton-Jones, D.H., 1989. Polymer Processing. Chapman and Hall, London, UK.

Rosato, D.V., Rosato, A.V., DiMattia, D.P., 2004. Blow Molding Handbook, second ed. Hanser Garner Publications, Munich, Germany.

Stevens, H.J., 1991. Forming. In: Schneider, S.J. (Ed.), ASM Engineered Materials Handbook, Vol. 4, Ceramics and Glasses. ASM International, Materials Park, OH, pp. 394–401.

Strong, A.B., 2000. Plastics Materials and Processing, second ed. Prentice-Hall, New York, NY.

Melt-based Additive Processes

Gibson, I., Rosen, D.W., Stucker, B., 2010. Additive Manufacturing Technologies: Rapid Prototyping to Direct Digital Manufacturing. Springer, New York, NY.

Hutchings, I.M., Martins, G.D. (Eds.), 2012. Inkjet Technology for Digital Fabrication. Wiley, New York, NY.

Kamrani, A.K., Nasr, E.A., 2010. Engineering Design and Rapid Prototyping. Springer, New York, NY.

Yardimci, A. M. (1999). *Process Analysis and Planning for Fused Deposition*, Ph.D. Thesis, University of Chicago.

CITED REFERENCES

Badger, A.E., Parmelee, C.W., Williams, A.E., 1938. Surface tension of various molten glasses. J. Am. Ceram. Soc. 20, 325–329.

Battezzati, L., Greer, A.L., 1989. The viscosity of liquid metals and alloys. Acta Metall. 37, 1791–1802.

Beddoes, J., Bibby, M.J., 1999. Principles of Metal Manufacturing. John Wiley & Sons, New York, NY.

Blair, M., Stevens, T.L., 1995. Steel Castings Handbook, sixth ed. Steel Founders' Society of America, Materials Park, OH.

Brandes, E.A. (Ed.), 1983. Smithells Metals Reference Book. sixth ed. Butterworths, Boston, MA.

Brandrup, J., Immergut, E.H., Grulke, E.A., 1999. Polymer Handbook, fourth ed. Wiley Interscience, New York, NY.

Broyer, E., Macosko, C.W., 1976. Heat transfer and curing in polymer reaction molding. AIChE J. 22 (2), 268–276.

Callister, W.D., 2007. Materials Science and Engineering: An Introduction. John Wiley & Sons, New York, NY.

Carslaw, H.S., Jaeger, J.C., 1959. Conduction of Heat in Solids. Oxford University Press, New York, NY.

Clare, A.G., Wing, D., Jones, L.E., Kucuk, A., 2003. Density and surface tension of borate containing glass melts. Glass Technol. 44, 59–62.
Dockerty, S., 1967. U.S. Patent No. 3,338,696. Washington, D. C.: U.S. Patent and Trademark Office.
Flinn, R.A., 1963. Fundamentals of Metal Casting. Addison-Wesley Pub. Co., Reading, MA.
Giles, H.F., Wagner, J.R., Mount, E.M., 2005. Extrusion: The Definitive Processing Guide and Handbook. William Andrews Publishing, Norwich, NY.
Hatch, J.E. (Ed.), 1984. Aluminum: Properties and Physical Metallurgy. ASM International, Materials Park, OH.
Hawkins, D.T., Hultgren, R., 1974. Constitution of binary alloys. In: Lyman, T. (Ed.), Metals Handbook (eighth ed.): Vol. 8 Metallography, Structures and Phase Diagrams. American Society for Metals, Metals Park, OH, pp. 251–376.
Hynd, W.C., 1984. Flat glass manufacturing processes. In: Uhlmann, D.R., Kreidl, N.J. (Eds.), Glass: Science and Technology. Vol. 2 Processing. Academic Press, New York, NY, pp. 46–106.
Johnson, C.F., 1988. In: Dostal, C.A. (Ed.), ASM Engineered Materials Handbook: Vol. 2 Engineering Plastics. ASM International, Materials Park, OH, pp. 344–351.
Kalpakjian, S., 1997. Manufacturing Processes for Engineering Materials, third ed. Addison-Wesley, Reading, MA.
Kingery, W.D., 1959. Surface tension of some liquid oxides and their temperature coefficients. J. Am. Ceram. Soc. 42, 6–10.
Li, Z., Mukai, K., Zeze, M., Mills, K.C., 2005. Determination of the surface tension of liquid stainless steel. J. Mater. Sci. 40, 2191–2195.
Macosko, C., 1989. RIM: Fundamentals of Reaction Injection Molding. Hanser Publishing, New York, NY.
Martlew, D., 2005. Viscosity of molten glasses. In: Pye, L.D., Montenero, A., Joseph, I. (Eds.), Properties of Glass Forming Melts. CRC Press, Boca Raton, FL, pp. 75–142.
Mazurin, O.V., 2005. Density of glass melts. In: Pye, L.D., Montenero, A., Joseph, I. (Eds.), Properties of Glass Forming Melts. CRC Press, Boca Raton, FL, pp. 143–192.
Narayanswany, O.S., 1977. A one-dimensional model of stretching float glass. J. Am. Ceram. Soc. 60, 1–5.
Nemoto, N., Moriwaki, M., Odani, H., Kurata, M., 1971. Shear creep studies of narrow-distribution poly(cis-isoprene). Macromolecules 4, 215–219.
Nemoto, N., Odani, H., Kurata, M., 1972. Shear creep studies of narrow-distribution poly(cis-isoprene). II. Extension to low molecular weights. Macromolecules 5, 231–235.
Osswald, T.A., 1998. Polymer Processing Fundamentals. Hanser Garner Publications, Munich, Germany.
Osswald, T.A., Turng, L.S., Gramann, P., 2008. Injection Molding Handbook. Hanser Garner Publications, Munich, Germany.
Penwell, R.C., Graessley, W.W., 1974. Temperature dependence of viscosity-shear rate behavior in undiluted polystyrene. J. Polym. Sci. Part B: Polym. Phys. 12, 1771–1783.
Piwonka, T.S., 1998. Molding methods. In: Davis, J.R. (Ed.), Metals Handbook Desk Edition. ASM International, Materials Park, OH, pp. 736–751.
Rauwendaal, C., 2001. Polymer Extrusion. Hanser Publishers, Munich, Germany.
Rauwendaal, C., 2014. Polymer Extrusion, fifth ed. Hanser Publishers, Munich, Germany.
Rosato, D.V., Rosato, A.V., DiMattia, D.P., 2004. Blow Molding Handbook, second ed. Hanser Garner Publications, Munich, Germany.
Sauer, B.D., Dee, G.T., 2002. Surface tension and melt cohesive energy density of polymer melts including high melting and high glass transition polymers. Macromolecules 35, 7024–7030.

Schey, J.A., 2000. Introduction to Manufacturing Processes, third ed. McGraw-Hill, New York, NY.

Stevens, H.J., 1991. Forming. In: Schneider, S.J. (Ed.), ASM Engineered Materials Handbook: Vol. 4 Ceramics and Glasses. ASM International, Materials Park, OH, pp. 394–401.

Strong, A.B., 2000. Plastics Materials and Processing, second ed. Prentice-Hall, New York, NY.

Theilke, J., 1998. Automatic pouring systems. In: Stefanescu, D.M. (Ed.), Metals Handbook (9th Ed): Vol. 15 Casting. ASM International, Materials Park, OH, pp. 497–501.

West, C.E., Grubach, T.E., 1998. Permanent mold casting. In: Stefanescu, D.M. (Ed.), Metals Handbook (ninth ed.): Vol. 15 Casting. ASM International, Materials Park, OH, pp. 275–285.

Wilczynski, K., White, J.L., 2003. Melting model for intermeshing counter-rotating twin-screw extruders. Polym. Mater. Sci. Eng. 43, 1715–1726.

QUESTIONS AND PROBLEMS

Questions

1. Describe the differences in the melt structures of polymers, metals, and ceramic glasses?
2. Why are polycrystalline ceramics, like alumina, not processed from the melt?
3. What are the mechanical and energetic definitions of surface tension? Why is surface tension not a term used with solids?
4. Does wetting behavior of a liquid on a solid depend on the surface tension of the liquid?
5. What is the difference between Newtonian and non-Newtonian rheological behavior?
6. Compare typical viscosities for metal melts, polymer melts, and ceramic glass melts. How are these typical values related to the melt structures?
7. Explain the temperature dependence of viscosity.
8. Explain why polymer melts tend to be shear thinning.
9. What are the three main principles governing liquid flow?
10. Give two examples from the chapter where mass conservation (continuity principle) is used.
11. Sketch the velocity gradients that arise from drag flow and pressure-driven flow for (a) a Newtonian liquid and (b) a shear thinning liquid.
12. What is the heat equation? Define thermal diffusivity.
13. What conditions lead to "Newtonian cooling"? Would you expect a ceramic glass sheet after exiting a float glass operation to cool in this way?
14. Under what conditions is the plot in Figure 3.24 used to predict cooling time?
15. List three forms of shape casting of metals with advantages and disadvantages.
16. Why is the sprue in a sand mold tapered?
17. What factors influence solidification time in sand casting?

18. Describe the origin of dendrites in alloys using a phase diagram. In the process define constitutional undercooling.
19. List the similarities between die casting and injection molding.
20. What is the sequence of events in a float glass process. How is thickness controlled?
21. What are the advantages of the fusion downdraw process?
22. Make a sketch of a single screw extruder and label all the parts.
23. What are the main operating parameters in a single screw extruder? What is the operating point?
24. What are the advantages and uses of twin screw extrusion?
25. List two defects in extrusion and their origins.
26. What is the sequence of step in injection molding? Which steps is likely the longest?
27. How are runners designed?
28. Define the process window in injection molding using a diagram of injection pressure versus holding pressure.
29. Compare blow molding of polymers and glasses.
30. In fused deposition modeling, what factors control the width of the deposited strand?
31. What are the advantages and disadvantages of FDM compared to injection molding?

Problems

1. The measurement of surface tension by the method described in Example 3.1 requires the melt density to be known. How could the experiment be carried out so that both surface tension and density could be found?
2. (a) The viscosity of borosilicate glass melt (in Poise) follows the VFT equation in the form $\log \eta = A + \frac{B}{T - T_o}$ with $A = -1.0836$, $B = 5433$, and $T_o = 179°C$ (as in Example 3.3). The parameters hold for melt temperatures between $531°C$ and $1400°C$. Estimate the activation energy for viscous flow at $1000°C$. (b) Using the parameters for viscosity as a function of temperature for polymers given in Table 3.4, estimate the activation energy for viscous flow for polystyrene at $200°C$. State assumptions.
3. Consider the pressure-driven flow of a LDPE melt at $200°C$ through a tube with diameter of 2 mm and length of 5 cm. The pressure drop is 10 MPa. (a) Plot the velocity profile (velocity vs. r). (b) Calculate the maximum shear stress on the melt. Where is this shear stress? (c) Is the flow laminar or turbulent?
4. Capillary rheometers measure the steady state viscosity of polymer melts at temperature and pressure typical in injection molding and extrusion. Shear rate ranges typically from 0.1 to $1000\,s^{-1}$. The

table below gives the experimental data of LDPE melts at 180°C measured by the capillary rheometer. This rheometer looks like a piston extruder. It is operated with different dies to gather data. In the data set, three dies are used. All have a diameter, D, of 1.4 mm and their land lengths, L, are 10, 20, and 30 mm. The barrel diameter is 12 mm. Pressure drops across the dies, ΔP, are measured.

Piston velocity v (mm/s)	Pressure drop ΔP (bar)		
	L = 10 mm	L = 20 mm	L = 30 mm
1.3889	42.903	71.754	97.801
0.7716	31.143	52.769	71.354
0.4287	22.024	38.817	51.357
0.2382	15.447	28.153	38.067
0.1323	10.763	20.081	27.425
0.0735	7.4743	14.102	20.770
0.0408	5.0826	9.7167	13.550

The true shear rate at the wall is given by:

$$\dot{\gamma}_{w,\text{true}} = \frac{8Q}{\pi D^3}\left(3 + \frac{d\log Q}{d\log \Delta P}\right) = \frac{8Q}{\pi D^3}\left(3 + \frac{1}{\frac{d\log \Delta P}{d\log Q}}\right)$$

where D is the die diameter and Q is the volumetric flow rate.
The true shear stress is given by:

$$\tau_{w,\text{true}} = \frac{\Delta P}{4(L/D + L_e/D)}$$

where L_e/D is the normalized entrance length discussed below.
The true viscosity is:

$$\eta_{\text{true}} = \frac{\tau_{w,\text{true}}}{\dot{\gamma}_{w,\text{true}}}$$

(a) Convert piston velocity to flow rate, Q. Plot log ΔP versus log Q for L = 10 mm, L = 20 mm, L = 30 mm and then get the local slope (dlogΔP/dlogQ) for each data point. (b) Plot ΔP versus L/D at different Q's (corresponding to v = 1.3889, 0.4287, 0.1323, and 0.4080 mm/s). Find the normalized entrance length, which is the common x-intercept ($-L_e/D$); average the x-intercepts at different shear rates. (b) Plot the true viscosity versus the true shear rate in log-log coordinates for each die. Find the power law relation for each die and then obtain the average power law index n and consistency index K values.

5. An iron cookie sheet with a thickness of 1 mm is in an oven at 375°F. It is removed from the oven and cooled under conditions with heat transfer coefficient of 10 W/(m² •K). How long should you wait before touching the sheet with your hands (e.g., temperature below 100°F)?

6. A cast steel slab with a thickness 10 cm in thickness is put into a furnace at 700°C. How long must the slab remain in the furnace before it reaches a temperature of 600°C? Assume heat transfer coefficient is 130 W/(m² •K).
7. The iron slab in problem 6 is removed from the mold when it is at 1000°C. Estimate the time needed to cool the slab to 40°C in an air stream at 25°C. Assume the heat transfer coefficient (h) is 50 W/(m² •K) and the average thermal conductivity of the iron is 45 W/(m •K) over this temperature range.
8. Liquid iron is undercooled until homogeneous nucleation occurs. Calculate the critical radius of the nucleus required and the number of iron atoms in the nucleus. Is homogeneous nucleation common? Why or why not?
9. An iron melt at its freezing temperature of 1536°C is cast to form a slab of thickness 0.1 m. Calculate the times required for completion of the freezing process when the mold is (a) a thick sand mold at 25°C and (b) a water-cooled copper mold at 25°C. For the iron, thermal diffusivity is 5.8×10^{-6} m²/s, density (liquid) is 7024 kg/m³, density (solid) is 7265 kg/m³, heat of fusion is 2.80×10^5 J/kg, and specific heat is 450 J/(kg•K). For the sand, thermal conductivity is 0.3 W/(m•K) and thermal diffusivity is 2.5×10^{-7} m²/s.
10. (a) Draw a schematic diagram of the microstructure formation during casting. Show the mold and the metal structure as a function of distance away from the mold; discuss the origin of the different microstructural features. (b) What is constitutional undercooling and how does it affect the microstructure? (c) How are uniform equiaxed microstructures formed in practice? (d) List three defects found in cast pieces and discuss how they can be minimized or eliminated.
11. Make a cross-sectional sketch of a sand mold that could be used to cast a slab of metal 10 cm on a side and 3 cm thick. Make your sketch to scale to show the size of all the features. Be sure to provide the following (using calculations to justify your choices): (a) the shape and dimensions of the riser and its position relative to the casting and (b) the shape and dimensions of the sprue.
12. Liquid aluminum is cast at its melting point (660°C) into a sand mold to form a slab of thickness 0.15 m. (a) Calculate the solidification time, assuming that the sand is "semi-infinite," (b) calculate the minimum thickness of the sand needed to ensure that the semi-infinite model can be used, and (c) plot the temperature gradient into the sand at a time that is half the solidification time. For sand, $\alpha = 2.5 \times 10^{-7}$ m²/s and $k = 0.3$ W/(m•K). For aluminum, $\alpha = 9.81 \times 10^{-5}$ m²/s, $C_p = 1048$ J/(kg•K), $\Delta H_{fusion} = 400,100$ J/kg, and $\rho = 2390$ kg/m³.
13. An aluminum melt at its freezing temperature of 660°C is cast to form a slab of thickness 0.15 m. Calculate the time required for completion

of the freezing process when the mold is a water-cooled copper mold at 25°C. Information on aluminum is given above.

14. A cast slab of aluminum (from problem above) is removed from the mold after it has cooled to 600°C and then it is cooled in air at 50°C. The heat transfer coefficient during air cooling is 20 W/(m²·K). Show that the conditions are consistent with Newtonian cooling and calculate the time needed for the slab to reach 100°C.

15. Describe the float glass process. Include the differences in delivery of melt, rheology of the melt, heat transfer, structural development on cooling (along with important characteristic temperatures), and control of thickness.

16. A typical float glass composition is provided below. (a) Formulate a glass batch using standard raw materials from Chapter 2 in order to produce 1 ton of glass. (b) Plot the viscosity as a function of temperature, using the empirical expression given below. (c) Above what temperature would the glass be considered a "melt"? (d) Identify a softening point, annealing point, and strain point.

Component	Wt%
SiO_2	74
Al_2O_3	0.2
CaO	8.6
MgO	3.78
Na_2O	13.4
K_2O	0.02

Viscosity as a function of temperature follows: $\log \eta = -A + \frac{B}{T-T_o}$ where the following are empirically determined for soda-lime-silicate glasses*:

$B = -6040Na_2O - 1440 K_2O - 3919CaO + 6285MgO$
$\quad + 2253Al_2O_3 + 5736$
$A = -1.4788Na_2O - 0.835 K_2O - 1.603CaO + 5.494MgO$
$\quad - 1.528Al_2O_3 + 1.455$
$T_o = 25.07Na_2O - 321.0 K_2O - 544.3CaO + 384.0MgO$
$\quad + 294.4Al_2O_3 + 198.1$

(Oxide symbols represent mole oxide ratio to mole of SiO_2.)

17. Explain how you would estimate the time to cool a sheet of Corning® Willow™ Glass that is 100 microns thick. Assume that the initial melt temperature is 1000°C and the final temperature required is 50°C. The heat transfer coefficient is 20 W/(m²·K). Consult a reference book on heat transfer.

18. (a) Plot melt viscosity as a function of shear rate (in range of 10 to 10,000 s⁻¹) for HDPE and LDPE at 200°C. Use a log scale. (b) Which polymer is more shear thinning? Why? (c) Considering the

pressure-driven flow of a Newtonian fluid through a tube, derive an equation for the shear rate at the wall as a function of the tube radius and the volumetric flow rate. (d) Calculate the volumetric flow rate and shear rate at the wall for the HDPE melt (at 200°C) down a 1 meter length of tube (diameter = 10 cm) under a pressure drop of 27 MPa. Do not assume that the melt is Newtonian.

19. (a) An engineer would like to ensure that a commercial polypropylene in an extrusion process remains Newtonian as it exits a circular die with radius 1 cm and land length of 4 cm. The engineer is using PP at 200°C. Using the data in Figure 3.12, specify the maximum pressure that can be used to drive the PP through the die. (b) The engineer mentioned in (a) tells his boss about this effort and explains that keeping the melt Newtonian will eliminate die swell. Is the engineer's argument correct? Explain. (c) The boss counters that the engineer's efforts have lowered the production rate substantially and tells the engineer to compensate for die swell in his die design. Is the boss's argument correct?

20. The goal of an extrusion process is to produce polycarbonate strips at a rate of 1 km/hr. We want to produce these strips at a rate of 1 km/hr. Assume a die with a width of 100 mm and gap of 1 mm and land length of 10 mm is used and that the rheology of polycarbonate melt at 290°C is approximately Newtonian with viscosity of 300 Pa•s. (a) What pressure must the extruder deliver to get this output? (b) How long should the cooling bath be if cooling water with an inlet temperature of 10°C is used?

21. Can the single screw extruder with the given parameters be used to produce the polymer described in the previous problem at a rate of 1 km/hr? $D = 2.54$ cm; $H = 1.65$ mm; helix angle = 17.7°; Metering zone length = 0.254 m. $N = 100$ RPM. Create an operating diagram for the extruder and die combination. Use this diagram to answer the question.

22. (a) Estimate the cooling time for injection molding of a polyethylene part with at section thickness of 0.5 mm. The melt is at a temperature of 190°C and the steel mold is at 10°C. (b) Repeat for a section thickness of 1 mm. (c) Discuss the rationale behind the design of injection molded parts with constant section thickness.

23. An engineer in an injection molding plant finds that a polyethylene part (LDPE) has a greater shrinkage than his calculations predict. The important dimension of the part is specified to be 4.500 ± 0.010 cm. The dimension of the part from the last production run is 4.480 cm. (a) What is the correct dimension for the mold cavity? (b) What could be done to reduce the shrinkage of the polyethylene part? Give all important process changes.

24. Contrast the mold design issues for sand casting of metals with injection molding of polymers. Include design of sprue and runner system, gate location, and method to accommodate shrinkage.

Chapter 4

Solid Processes

4.1 INTRODUCTION

A solid process converts a bulk solid starting material into a new shape. There are two main types of solid processes: (i) subtractive processes and (ii) deformation forming processes. Subtractive processes involve removing material from a solid block of starting material. Machining, milling, and boring are all subtractive processes that create or modify shapes. The focus of this chapter is on the second category, solid deformation processes in which a dense, solid starting material is converted into a shape by the application of mechanical forces.

Solid deformation processes, like all forming operations, involve flow, shape definition, and shape retention. How can a solid flow? Geologists describe movement of solid rock masses over time as flow. Likewise, materials scientists describe the plastic deformation of metals as *flow*. In fact, the stress needed to cause flow (plastic deformation) in metals is called the flow stress. *Shape definition* is dictated by a "tool" that applies force and sometimes by a die or mold that restrains flow. *Shape retention* is straightforward—shape is retained by the removal of the force that caused the flow. In some cases, cooling is also needed to lock in the final structure. For example, in thermoforming a rubbery polymer is deformed at elevated temperature and then is cooled rapidly to make the final object.

Solid deformation processes are commonly used to shape metals. Forging and rolling are examples of deformation processes used to create metal shapes. The starting materials are bulk metal slabs, sheets, or rods. Many metal alloys have a large capacity for plastic deformation, especially if they are heated above a characteristic temperature, the recrystallization temperature. The terms "cold forming" and "hot forming" are used for deformation processes well below and well above this temperature. In metal forming, solid deformation processes usually involve multiple steps in order to achieve the desired dimensions. In addition, post-forming operations, such as machining and joining, are typically necessary.

Solid deformation processes are not typically used to shape ceramics. Polycrystalline ceramics and glasses in the solid state have very little or no capacity for plastic deformation. On application of a load, solid ceramics fracture before they plastically deform. Plastic deformation can occur at high temperature and does play a role in high temperature hot pressing of some ceramic that originate from powder compacts. Also, mixtures of ceramic powders, particularly clays, and water develop plasticity, and these "plastic masses" are formed by deformation processes. These powder-based processes are covered in later chapters.

Polymers can also be plastically deformed in the solid state. The most important process for solid deformation of polymers is thermoforming. Thermoplastic polymer sheets are heated to a high enough temperature to enter the rubbery solid state, where they are easily deformed. The deformation therefore is elastic, rather than plastic. Then, the part is cooled to retain the shape. As with many other polymer forming operations, there are no significant post-forming operations.

The engineering of solid deformation processes revolves around the prediction of the loads needed to cause the required amount of deformation, without surpassing the failure limit of the material. Solid mechanics analysis methods provide good predictions so long as the effects of stress, strain rate, and temperature on the mechanical properties of the solid are taken into account, along with the effects of friction. The typical goal of the engineering analysis is the design of the tool and force application method needed to produce the required deformation.

In this chapter, the fundamentals of solid deformation are covered first. Then, the solid processes are described, beginning with processes that involve bulk deformation: wire drawing, extrusion, forging, and rolling. Following that is the solid deformation processes involving sheets, a topic of interest to sheet metal forming and thermoforming of polymers. Post-forming operations are mentioned throughout the discussions of the deformation processes.

4.2 FUNDAMENTALS

4.2.1 Deformation and Plastic Flow under Uniaxial Tension

Uniaxial tensile testing is commonly used to characterize the mechanical properties of engineering solids. A specimen with a standard geometry is mounted, with clamps, pins, or other fixtures, between a fixed base and a moving crosshead. The crosshead is then driven at a constant rate, such that the specimen is pulled in one direction while a load cell records the force required and an extensometer measures the length in the gauge section of the specimen. See Figure 4.1.

The raw force and length data are converted to stress and strain to remove the influence of specimen dimensions and uncover the fundamental

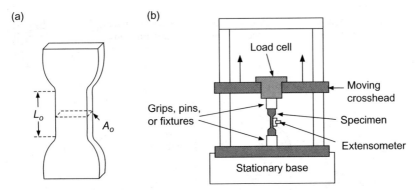

FIGURE 4.1 (a) Example of a specimen for tensile testing with a "dogbone" geometry consisting of a gauge with cross-sectional area A_o and length L_o and wider end zones for gripping. (b) Schematic of a tensile testing apparatus with extensometer for measuring gauge length and load cell for measuring force as crosshead moves at a constant speed.

mechanical properties. The engineering stress, s, and engineering strain, e, are defined as:

$$s = \frac{F}{A_o} \quad (4.1)$$

$$e = \frac{(L - L_o)}{L_o} \quad (4.2)$$

where F is the measured force or load, A_o is the initial cross-sectional area in the "gauge," L is the measured or instantaneous gauge length, and L_o is the initial gauge length. The specimen geometry is chosen so that the highest stress occurs in the gauge section where the cross-sectional area is lower than near the grips. As the specimen is pulled, the instantaneous cross-sectional area in the gauge, A, decreases; hence, a different measure of stress, the true stress, is also useful. True stress, σ, is defined as the force required for deformation divided by the instantaneous cross-sectional area:

$$\sigma = \frac{F}{A} \quad (4.3)$$

True strain, ε, is defined as the summation or integral of incremental increases in gauge length:

$$\varepsilon = \int_{L_o}^{L} \frac{dL}{L} = \ln\left(\frac{L}{L_o}\right) \quad (4.4)$$

where L is the instantaneous gauge length. Using true stress and true strain in deformation processes is necessary for accurate calculation of load requirements as well as dimensional changes. The "true" approach also facilitates analysis of multistep deformation processes.

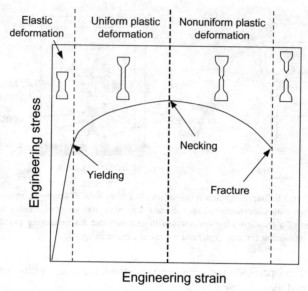

FIGURE 4.2 Schematic diagram of a tensile test of a ductile metal. The specimen sketches show shape changes that occur in the three regimes: elastic deformation, uniform plastic deformation, and nonuniform plastic deformation.

The stress–strain relationships derived from uniaxial tensile tests reveal the mechanical properties and behaviors of materials that are important for predicting condition for solid deformation processing. Below these characteristics are discussed for metals, polymers, and ceramics.

Metals. Metal alloys range from highly ductile and deformable to brittle. For deformation processing, the ductile metal is more useful and therefore we consider ductile metals here.

Figure 4.2 is a schematic diagram of the engineering stress–strain data derived from a uniaxial tensile test of a ductile metal alloy. Three regions are found. First, at low levels of strain, the deformation is elastic and recoverable if the mechanical load is removed. The elastic modulus, or Young's modulus, E, is the slope of the stress–strain curve in the elastic regime:

$$s = Ee \qquad (4.5)$$

This relationship is known as to Hooke's law. As deformation continues, the engineering stress in the specimen reaches the yield stress, where plastic (permanent) deformation and the second regime of uniform plastic deformation begin. As deformation proceeds uniformly, the cross-sectional area reduces and the length extends, such that the volume is conserved:

$$A_o L_o = AL \qquad (4.6)$$

Typically, the engineering stresses needed to continue the deformation increase in this region. That is, higher and higher loads are supported even though the cross-sectional area is shrinking. This behavior shows the material becomes stronger as it is deformed. Eventually, however, there is an instability and deformation becomes localized in a "neck." From this point on, the deformation is nonuniform, occurring only in the necked region. This third region ends when the specimen fractures at the neck.

An example of engineering stress–strain data for a ductile alloy is shown in Figure 4.3. The elastic region is barely visible on the plot that displays all the data, but is more apparent in the close up of the low strain region (Figure 4.3b). The clearly linear stress–stain relationship follows Eq. (4.3); analysis reveals the Young's modulus. The yield stress or yield strength is typically

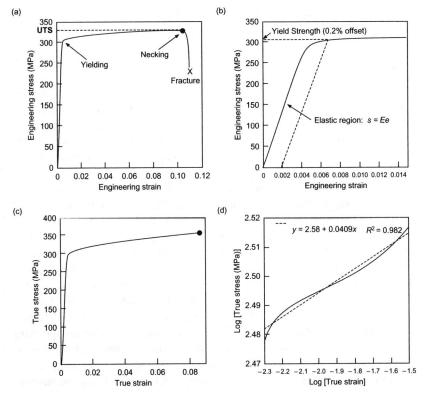

FIGURE 4.3 (a) Engineering stress–engineering strain data for an aluminum alloy, showing the three regimes of deformation. The necking point is marked with a closed circle and fracture with an "X." The definition of the ultimate tensile strength (UTS) as the stress at necking is shown. (b) A higher resolution plot of (a), showing the application of Hooke's Law (Eq. (4.5)) to the elastic region and the definition of yield strength. (c) True stress-true strain plotted from the same data as (a) up to the point of necking. (d) An analysis of the region of uniform plastic deformation plastic deformation using Eq. (4.10). The curve fit equation is $\log\sigma = \log K + n\log\varepsilon$. This alloy has a strength coefficient of 380 MPa (log K = 2.58) and a strain hardening coefficient of 0.041.

defined as the stress needed to cause 0.2% plastic deformation. Notably, the strain at yield is small, on the order of a few tenths of a percent. A few materials, such as low carbon steels, show an anomalous behavior with a distinct yield point after which the engineering stress drops, remains steady, and then rises. For these materials, the stress at the yield point is taken as the yield stress. In all ductile alloys, a broad range of uniform deformation occurs before the engineering stress reaches a maximum (i.e., the ultimate tensile strength, UTS) just when the neck forms. After necking, the deformation is nonuniform.

The ductility of the metal is defined as the percent change in elongation or percent reduction in cross-sectional area at the point of fracture:

$$\text{Ductility}(\% \text{ Elongation}) = \frac{L_f - L_o}{L_o} \times 100 \tag{4.7}$$

$$\text{Ductility}(\% \text{ Area Reduction}) = \frac{A_o - A_f}{A_o} \times 100 \tag{4.8}$$

where A_f and L_f are the cross-sectional area and the gauge length at fracture, respectively. When the value of ductility is given, the data must refer to a standard gauge length or initial cross-sectional area because the deformation after necking is nonuniform. While one might be tempted to look at ductility as a measure of a metal's ability to be processed by plastic deformation; however, this measure does not give the entire story. Only uniform plastic deformation (before necking) is useful for deformation processing. Therefore, a key property of the metal is its ability to become stronger as the material is deformed so that necking is pushed to higher strains.

Mechanical properties of typical metals and alloys determined from uniaxial tensile tests are given in Table 4.1. These data show a range in behavior but in general the elastic modulus of metals is in the range of ~50 to 400 GPa, with the upper end reserved for brittle metals, such as tungsten. The yield strengths and UTS values vary as well, depending on the metal composition and structure. The connection between the mechanical properties and structure, as described in detail in physical metallurgy texts, is related to the presence of strengthening mechanisms that are responsible for inhibiting dislocation motion. For example, pure metals such as electrolytic copper (C11000) have lower strength than alloys such as cartridge brass, which contains 70% Cu and 30% Zn, because the solid solution structure inhibits dislocation motion. Likewise, multiphase complexity in a microstructure, such as the combinations of ferrite and Fe_3C found in steel, inhibits dislocation motion as does a general reduction in the grain size. These factors come into play in determining the deformability of metals, a topic of interest for solid deformation processing as well as compaction processes used in powder metallurgy (Chapter 5).

The true stress—true strain behavior provides more information on how a material deforms plastically in the uniform plastic flow regime, and is the

TABLE 4.1 Mechanical Properties of Metals and Alloys at Room Temperature[a]

Material	Elastic modulus (GPa)	Yield strength (MPa)	Ultimate tensile strength (MPa)
Aluminum Alloys:			
1100 − O (annealed)	69	35	90
6061 − O	69	55	125
6061 − T6 (tempered)	69	275	310
Copper Alloys			
C11000, annealed	110	69	220
C26000 (cartridge brass), annealed	115	133	357
Steel Alloys:			
1020, hot rolled bar	207	205	380
1020, cold drawn bar	207	350	420
4130, annealed	207	460	560
4130, water quenched, tempered	207	979	1040
Tungsten	410	760	970

[a]Data tabulated in Davis (1998).

key to understanding and designing deformation processes for metals (Example 4.1). Figure 4.3c shows the true stress−true strain behavior corresponding to the engineering stress−engineering strain plot provided in Figure 4.3a. Since the change in the cross-sectional area during elastic deformation is very small, the true stress is approximately equal to the engineering stress in the elastic regime. Likewise, true strain is very nearly the same as the engineering strain in the elastic regime. The value of the yield stress is also the same in the true and engineering systems. Hence the elastic deformation regime is identical in the two representations. A deviation is clear in the regime of uniform plastic deformation. Here the true stress is higher than the engineering stress and climbs with more and more deformation, an indication of strengthening during deformation. In the regime of uniform plastic deformation, where the volume of the specimen is constant, the true and engineering forms of stress and strain are related by:

$$\sigma = s(1 + e) \quad \text{and} \quad \varepsilon = \ln(1 + e) \tag{4.9}$$

These conversions assume the traditional sign convention for stress and strain with positive values indicating tension and negative value indicating compression.

The necking point marks the end of the uniform deformation region and also typically the end of the true stress–true strain plot. Beyond this point continuing the calculation of true stress and true strain requires information on the geometry of the localized neck region and a correction that accounts for the complex stress state in the neck. Usually, extending the plot beyond necking adds little additional information that would be useful in metal forming. Additionally, the UTS is defined as an engineering stress and therefore not shown on the true stress–true strain plot.

Most metals become stronger as they are plastically deformed, as evidenced by the increase in stress needed to continue plastic deformation beyond the yield stress. This phenomenon is called strain hardening or work hardening. Strain hardening is an effective means of strengthening a metal and linked to structural changes.

EXAMPLE 4.1 A metal bar is plastically deformed in tension such that its gauge length increases from 10 to 15 mm. The specimen is then unloaded and loaded again in tension such that its gauge length further increases from 15 to 20 mm. Calculate the engineering strains and true strains for each deformation step and for the total deformation. Which form of strain is most useful for describing multistep plastic deformation operations?

For the first step, the initial length is 10 mm and final length is 15 mm. Therefore, the engineering and true strains can be calculated using Eqs. (4.2) and (4.4), respectively.

$$e_1 = \frac{15-10}{10} = 0.5 \qquad \varepsilon_1 = \ln\left(\frac{15}{10}\right) = 0.4$$

For the second step, the initial length is 15 mm and final length is 20 mm.

$$e_2 = \frac{20-15}{15} = 0.3 \qquad \varepsilon_2 = \ln\left(\frac{20}{15}\right) = 0.3$$

For the total deformation, the initial length is 10 mm and final length is 20 mm.

$$e_{total} = \frac{20-10}{10} = 1.0 \qquad \varepsilon_{total} = \ln\left(\frac{20}{10}\right) = 0.7$$

Notice that true strains are additive but engineering strains are not. This feature makes the true strain convention more useful for describing plastic deformation processes.

$$e_1 + e_2 = 0.8 \neq e_{total} \qquad \varepsilon_1 + \varepsilon_2 = \varepsilon_{total}$$

The structure of a metal changes considerably during plastic deformation. Plastic deformation arises from the motion of dislocations, defects in the crystal structure. As deformation proceeds, the dislocation density increases and these dislocations interfere with each other so that continued dislocation motion and plastic deformation are hindered. This interaction is responsible for strain hardening. The higher dislocation density of the deformed metal makes the metal stronger (i.e., the metal has a higher yield strength). In addition, the microstructure changes. An annealed metal microstructure typically consists of equiaxed grains. After plastic deformation, the grains become elongated because stresses are applied in a specific way relative to the microstructure. Because dislocation motion and deformation depend on crystallographic direction, not only is the microstructure distorted, but the overall crystallographic orientation of the grains is also changed. The original polycrystalline annealed metal typically contains grains with a random orientation, but a deformed metal tends to have a preferred crystallographic orientation or "texture."

The strain hardening behavior, that is, the effect of deformation (true strain) on the strength (or true stress needed to continue deformation) can be fit to a simple relationship for many metals. A standard form is:

$$\sigma = K\varepsilon^n \tag{4.10}$$

where K is the strength coefficient and n is the strain hardening exponent. Figure 4.3d shows that this relationship closely approximates the true stress–true strain behavior of the example metal. The approximation breaks down in the elastic regime as it predicts nonlinear behavior; however, for the purposes of metal deformation calculations, this difference is not significant. Likewise, the elastic deformation of the metal is very small in comparison to the plastic and elastic recovery can be ignored in predictions of the amount of deformation after a metal working operation (Example 4.2).

Figure 4.4 shows various types of strain hardening behavior. A strain hardening exponent of zero indicates a perfectly plastic material that has no strain hardening behavior. In practice a metal that displays no strain hardening ($n = 0$) has an elastic region followed by a plastic region with a constant yield stress and hence such materials are called "elastic–perfectly plastic," but here we refer to this sort of material as "perfect plastic." That is, as soon as the yield stress is reached, the elastic behavior switches to plastic with deformation taking place at the yield stress. The other extreme is a strain hardening exponent of one, a perfectly elastic material. With $n = 1$, Eq. (4.10) reverts to Hooke's law, which would indicate that dislocations are not mobile at all. (Note that that elastic modulus is not affected by strain hardening.) The behavior of real metals lies in between these extremes ($0 < n < 1$); typically, n is less than 0.5. Table 4.2 gives data for a variety of metals. For deformation under a tensile load, materials with higher n provide a larger amount of uniform plastic deformation. This conclusion is intuitive but is also proved in Example 4.3.

EXAMPLE 4.2 An annealed copper rod (original dimensions: diameter = 1 cm, length = 10 cm) is loaded in uniaxial tension until its length under load is 12 cm. Calculate the load needed to cause this deformation and the length of the rod after the load is removed. Use data for C11000 in Table 4.1 and annealed copper in Table 4.2.

First, the regime of the deformation should be established by analyzing the data provided. The length under load is provided and so the true strain can be calculated:

$$\varepsilon = \ln\left(\frac{L}{L_o}\right) = \ln\left(\frac{12}{10}\right) = 0.182$$

The stress (true or engineering) that the material would be under, if it was in the elastic regime, can be calculated with Hooke's law:

$$\sigma = E\varepsilon = (110 \text{ GPa})(0.182) = 20 \text{ GPa}$$

Clearly, there has been some plastic deformation because this calculated stress far surpasses the yield stress. Now the question is whether the deformation is uniform or nonuniform. Let's assume it is uniform and carry out some calculations. First, the volume is conserved so the area under load can be found.

$$A_o = \pi(0.005 \text{ m})^2 = 7.85 \times 10^{-5} \text{ m}^2$$
$$A_o L_o = AL$$
$$A = (7.85 \times 10^{-5} \text{ m}^2)\left(\frac{10}{12}\right) = 6.54 \times 10^{-5} \text{ m}^2$$

Then from the true stress, the load causing the deformation and engineering stress can be calculated.

$$\sigma = K\varepsilon^n = (315 \text{ MPa})(0.182)^{0.54} = 125 \text{ MPa}$$
$$F = \sigma A = (125 \times 10^6 \text{ Pa})(6.54 \times 10^{-5} \text{ m}^2) = 8175 \text{ N}$$
$$s = \frac{F}{A_o} = \frac{8175 \text{ N}}{7.85 \times 10^{-5} \text{ m}^2} = 104 \text{ MPa}$$

These calculations are consistent with uniform deformation since the engineering stress is less than the UTS. To find the length of the rod after the load has been removed, the amount of elastic strain must be found and subtracted from the strain under load, which is a combination of elastic and plastic (permanent) strain.

$$\varepsilon_{elastic} = \frac{\sigma}{E} = \frac{125 \text{ MPa}}{110 \text{ GPa}} = 0.00114$$
$$\varepsilon_{plastic} = \varepsilon_{total} - \varepsilon_{elastic} = 0.182 - 0.00114 = 0.181$$
$$\varepsilon_{plastic} = \ln\left(\frac{L}{10}\right) = 0.181 \rightarrow L = 11.98 \text{ cm}$$

This calculation shows that the amount of plastic deformation is much greater than the amount of elastic deformation. When the load is removed, the elastic part is recoverable, so the final dimensions with the load removed are slightly different than the dimensions under load.

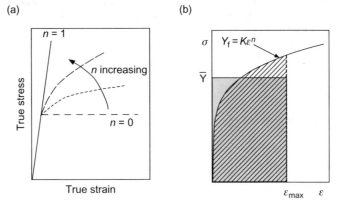

FIGURE 4.4 (a) Schematic of true stress–true strain behavior of metals. Elastic–perfectly plastic behavior ($n = 0$), strain hardening with lower n, strain hardening with higher n, and elastic behavior ($n = 1$). (b) Method to determine average flow stress for a deformation from $\varepsilon = 0$ to $\varepsilon = \varepsilon_{max}$. See Eq. (4.12). The two shaded areas are equal.

TABLE 4.2 Strength Coefficient and Strain Hardening Exponent for Metals at Room Temperature[a]

Metal alloy	Strength coefficient (MPa)	Strain hardening exponent
Aluminum: 1100–O	180	0.20
Aluminum: 6061–O	205	0.20
Aluminum: 6061–T6	410	0.05
Brass: 70Cu-30Zn, annealed	895	0.49
Copper: annealed	315	0.54
Steel: low carbon, annealed	530	0.26
Steel: 1112, annealed	760	0.19
Steel: 1112, cold-rolled	760	0.08
Steel: 4135, annealed	1015	0.17
Steel: 4135, cold-rolled	1100	0.14
Stainless Steel: 302, annealed	1300	0.30
Stainless Steel: 304, annealed	1275	0.45
Stainless Steel: 410, annealed	960	0.10

[a]*Data tabulated in Kalpakjian (1997).*

EXAMPLE 4.3 Show that the true strain at necking to the strain hardening exponent.

The necking point is the maximum in the engineering stress—engineering strain plot. The derivative of engineering stress with respect to engineering strain at the necking point is zero and therefore at the necking point: $ds = 0$. Since $s = \frac{F}{A_o}$, $dF = 0$.

From the definition of true stress, $F = \sigma A$; therefore, at necking:

$$dF = \sigma dA + A d\sigma = 0$$
$$\frac{d\sigma}{\sigma} = -\frac{dA}{A}$$

Since the volume is conserved, the product AL is constant. Therefore, $d(AL) = 0$, so $AdL + LdA = 0$ and

$$-\frac{dA}{A} = \frac{dL}{L} = d\varepsilon.$$

Hence at the necking point, $\frac{d\sigma}{\sigma} = d\varepsilon$ (where σ is the true stress at necking)

$$\frac{d\sigma}{d\varepsilon} = \sigma$$

The strain hardening behavior for metals follows: $\sigma = K\varepsilon^n$. Therefore at necking, $\sigma = K\varepsilon_x^n$ and $\frac{d\sigma}{d\varepsilon} = nK\varepsilon_x^{n-1}$, where ε_x is the true strain at necking. And based on the expression above,

$$nK\varepsilon_x^{n-1} = K\varepsilon_x^n$$
$$n = \varepsilon_x$$

This example illustrates the correlation between strain hardening and achieving uniform plastic strain in tension. The method to find the strain hardening coefficient from tensile data involves fitting the true stress—true strain data to Eq. (4.10) in the uniform plastic deformation region.

Due to strain hardening, the stress needed to cause plastic deformation, the "flow stress," increases as the deformation proceeds. The flow stress, at a particular level of deformation (*i.e.*, true strain), is therefore none other than the true stress at that level of true strain. For a perfectly plastic material, the flow stress is equal to the yield stress and is constant as the deformation proceeds. For work hardening materials, the flow stress, Y_f, usually follows:

$$Y_f = K\varepsilon^n \qquad (4.11)$$

A useful quantity for some calculations is the average flow stress, $\overline{Y}$, which is the average true stress needed to cause deformation to reach a final value of strain, ε_{max}:

$$\overline{Y} = \frac{\int_0^{\varepsilon_{max}} K\varepsilon^n d\varepsilon}{\varepsilon_{max}} = \frac{K\varepsilon^{n+1}}{\varepsilon_{max}(n+1)}\bigg|_0^{\varepsilon_{max}} = \frac{K\varepsilon_{max}^n}{n+1} \qquad (4.12)$$

TABLE 4.3 Nomenclature for Yielding and Flow in Metal Deformation Processes

Symbol	Name	Definition or usage
Y	Yield strength or yield stress	Denotes the onset of plastic deformation
Y	Flow stress (perfect plastic)	True stress needed to continue plastic deformation beyond the yield stress. Y is a constant for a perfect plastic material.
Y_f	Flow stress (strain hardening)	True stress needed to continue plastic deformation beyond the yield stress: Y_f increases with extent of deformation (strain). Flow stress at the end of a deformation is the yield strength of the deformed metal.
$\overline{Y}$	Average flow stress (strain hardening)	Average value of flow stress over the range of deformation occurring in a process, such as wire drawing and extrusion. Only needed for work hardening materials.
Y'	Plane strain flow stress (perfect plastic)	Flow stress modified for plane strain conditions; Y' is a constant for a perfect plastic material.
Y'_f	Plane strain flow stress (strain hardening)	Flow stress modified for plane strain conditions; Y'_f varies with extent of deformation (strain).
$\overline{Y'}$	Average plane strain flow stress (strain hardening)	Average value of flow stress over the range of deformation occurring in a plane strain process, such as rolling. Only needed for work hardening materials.

Figure 4.4b shows the determination of average flow stress graphically. These two new forms of "Y" are just two of several encountered in this chapter. A guide to the various notations for yielding and flow stresses is given in Table 4.3.

The work done during plastic deformation of a metal can be calculated from the true stress−true strain behavior. Work done per unit volume (J/m^3), which is equivalent to the energy expended per unit volume, is the product of true stress and true strain. [Note that J/m^3 = (N•m)/m^3 = N/m^2 = Pa]. So the work done is simply the area under the true stress−true strain curve. Therefore, the energy expended per unit volume, u, for the deformation of a metal to a final strain of ε_{max} can be found by integration:

$$u = \int_0^{\varepsilon_{max}} \sigma d\varepsilon = \int_0^{\varepsilon_{max}} K\varepsilon^n d\varepsilon = \frac{K\varepsilon^{n+1}}{(n+1)} \bigg|_0^{\varepsilon_{max}} = \frac{K\varepsilon_{max}^{n+1}}{n+1} = \overline{Y}\varepsilon_{max} \quad (4.13)$$

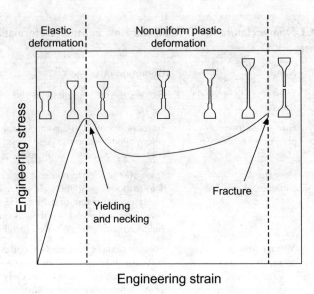

FIGURE 4.5 Schematic diagram of a tensile test of a thermoplastic polymer exhibiting cold drawing. The specimen sketches show shape changes that occur during deformation.

Note that the above calculations of average flow stress and energy expenditure or work done per unit volume calculation are based on the deformation of a metal under uniaxial tension. In a later section, the triaxial stress states encountered in plastic deformation processes are described and these equations are revisited. Additionally, we have assumed a single deformation starting from zero strain. In practice, multiple deformations are used and hence the integration limits on Eqs. (4.12) and (4.13) changed according to the deformation sequences (see Section 4.3.2).

Polymers. The mechanical response of thermoplastic polymers has some similarity to that of metals. As shown in Figure 4.5, the initial deformation is elastic and followed by a yielding event, which is typically characterized by a "yield point," followed by plastic deformation and failure. The elastic region is more significant in polymers due to their lower elastic moduli. See Table 4.4. Additionally, the elastic response of some polymers is nonlinear, requiring the use of a tangent or secant to the stress–strain data to define a modulus. Yielding involves relative motion of the polymer molecules. In semicrystalline polymers, this motion can involve dislocations. But these defects aren't necessary as amorphous polymers yield as well. The yield point is a convenient location of the onset of plastic deformation, but many polymers experience permanent deformation before that point. The yield point, in fact, marks the formation of a neck and so there is no "uniform plastic deformation" regime but rather a localized deformation in the neck dominates. After the initial instability and necking, the polymer molecules in the neck

TABLE 4.4 Mechanical Properties of Polymers and Ceramics at Room Temperature[a]

Material	Elastic modulus (GPa)	Yield stress or yield point (MPa)	Tensile strength or modulus of rupture[b] (MPa)
Polymers			
High-Density Polyethylene	1.08	26–33	22–31
Low-Density Polyethylene	0.17–0.28	9–14	8–32
Polypropylene	1.14–1.55	31–37	31–41
Polycarbonate	2.38	63	63–73
Polystyrene	2.28–3.28	–	36–52
Epoxy	2.41	–	28–90
Phenol Formaldehyde	2.76–4.83	–	34–62
Ceramics			
Aluminum oxide	380	–	282–551
Silicon carbide, hot pressed	208–483	–	230–825
Zirconia, yttria (3%) stabilized	205	–	800–1500
Soda–Lime Glass	69	–	69

[a] Data tabulated in Callister (2007); properties vary with specific grade, method of preparation, microstructure.
[b] Modulus of rupture (MOR) is a strength in flexure (values for ceramics are MOR values)

become aligned in the direction of tensile stress application, a phenomena that causes the necked region to become stronger. Therefore, polymers experience strain hardening as well, but the force needed for further deformation does not generally increase because the material adjacent to the neck yields. This process effectively extends the range of the neck as deformation continues at about the same level of stress. This phenomenon is called "cold drawing." Eventually the neck consumes the gauge length. Further deformation is elastic stretching of the neck, lastly failure occurs. As described later, the mechanical response of polymers is very sensitive to temperature and stain rate.

Figure 4.6 shows an example of engineering stress–strain behavior for a thermoplastic polymer. The elastic region persists to higher strains as

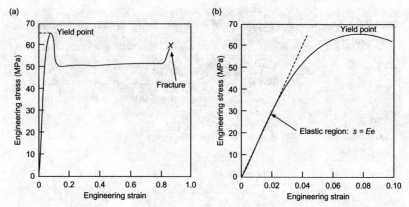

FIGURE 4.6 Engineering stress—engineering strain behavior for polycarbonate. (a) Engineering stress—engineering strain data and (b) a higher resolution plot of (a), showing the elastic regime with a dotted line for the region obeying Hooke's Law (Eq. (4.5)) and the yield point.

compared to metals, on the order of several %. The elastic behavior for this polymer follows Hooke's law initially. The yield point occurs at ~65 MPa, followed by cold drawing. The drawing continues until the necked zone encompasses the entire gauge. Then, the necked and highly deformed zone extends elastically and finally breaks. It is clear that there is considerable strain hardening during the nonuniform cold drawing stage. Unfortunately, one needs to have local information on neck geometry to produce a true stress—true strain plot, and there is no useful equation relating true stress and true strain for plastically deforming polymers.

Table 4.4 shows some basic mechanical properties of polymers. The elastic moduli are much lower than those found in metal alloys, on order 1—5 GPa. As well yielding and fracture in polymers occurs at lower values of stress compared to metals. The plastic deformation and cold drawing behavior is not the only mode of deformation experienced by polymers. Some polymers have very limited plastic deformation and are brittle. That is, they fail before any significant plastic deformation. For example, thermoplastics with high glass transition temperatures and highly cross-linked thermosets (*e.g.*, epoxy and phenolic resin) are brittle at room temperature. Other polymers have a rubbery response in which there is large, reversible deformation.

Ceramics. Ceramic materials, including polycrystalline ceramics and glasses, are not processed as bulk solids. The brittle nature of ceramics at room temperature, as shown in Table 4.4, prevents them from being formed by solid deformation processes. The strength values vary broadly due to the sensitivity of brittle fracture to microscopic flaws. However, it is possible to plastically deform polycrystalline ceramics in the solid state at elevated temperature. This high temperature deformation, however, is not used in forming of ceramics from bulk solids but rather it is connected the methodology of powder processing and so it is discussed in the next chapter.

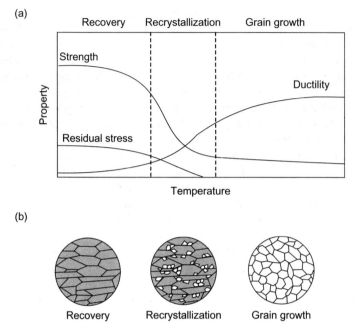

FIGURE 4.7 Schematic diagram showing (a) microstructure changes on heating of a deformed metal and (b) property. *Adapted from Sachs and Van Horn (1940).* During recrystallization new, stress-free equiaxed grains nucleate and grow in the matrix of deformed grains, eventually encompassing the entire sample and subsequently growing in size. Only one representation of these grains after some amount of growth is shown.

4.2.2 Effects of Temperature and Strain Rate on Deformation

Metals. Not surprisingly, the plastic deformation behavior of metals is sensitive to temperature. First, the mechanical properties of metal change with temperature. In general, as temperature increases the elastic modulus, yield strength and strain hardening exponent decrease and the ductility increases. The effect of temperature is linked to structural changes on heating. A key factor is whether the metal has time to recrystallize as it is deformed; therefore, both the temperature of the deformation and the strain rate are important. A good place to begin this section is with a discussion of the effect of heating an already plastically deformed metal.

A plastically deformed metal has a high dislocation density, crystallographic texture, and a microstructure composed of elongated grains. The defective state of the structure increases its Gibbs free energy; therefore, the structure transforms to a lower free energy, more stable state on heating. Figure 4.7 demonstrates the microstructure and property changes that occur as a plastically deformed metal is heated. As the metal is first heated, recovery takes place. While recovery does not change microstructure and properties much, there is a reduction in any residual stresses in the metal. (The origin of residual stress is nonuniform deformation that occurs in many forming operations.) The most significant changes occur during recrystallization. Recrystallization involves nucleation and growth of new, less defective grains in the matrix of deformed grains. This process entails diffusion, the rate of which increases with temperature. During recrystallization, the metal's dislocation density decreases and hence its strength drops and its ductility increases. The recrystallized microstructure consists of equiaxed grains. If further heating occurs, then these grains grow. The rate of recrystallization increases with temperature. To provide some guidance, a recrystallization temperature is defined as the temperature needed to recrystallize the microstructure in one hour. This temperature is roughly $0.3-0.5T_m$, where T_m is the melting temperature of the metal (in Kelvin) but is sensitive to purity, grain size, and initial state of deformation. For example, the recrystallization temperature for aluminum is much lower than that of iron. In alloys, thermal annealing conditions designed to soften an alloy by recrystallization are specified. For example, a deformed aluminum 2024 alloy can is returned to annealed state by heating at 415°C for two to three hours.

After recrystallization, the metal has low dislocation density and a uniform microstructure. The yield stress returns back to that of the original annealed state. Hence, after recrystallization, the metal is more deformable. The grain size of the metal after recrystallization depends on the original extent of plastic deformation, as well as the time and temperature. For example, if two specimens of the same metal are deformed to different extents, the metal that was originally deformed to a greater extent will recrystallize into a finer grained microstructure as compared with its less deformed counterpart. The more deformed metal is more defective and has a higher driving force for recrystallization.

Therefore, temperature relative to the recrystallization or annealing temperature has a critical influence on the metal deformation process. Deformation processes are termed "cold work" or "cold deformation" when they take place below the recrystallization temperature ($T < 0.3\ T_m$). In this case, recrystallization does not occur during deformation. At higher temperatures ($0.5T_m < T < 0.75T_m$), deformation and recrystallization occur concurrently. These processes are known as "hot working" or "hot deformation." There are advantages and disadvantages to each of these options. For alloys, specific temperature ranges for hot working are frequently specified for each alloy designation.

Under cold working conditions, there is no recrystallization during deformation and therefore work hardening increases the strength of the metal as the deformation takes place. The metal after deformation is stronger than before, a useful change for many applications. The final metal structure is textured, which may or may not be desired for a particular application. On the downside, cold working requires larger forces as compared with hot working, and less deformation is possible per deformation cycle. Controlling the dimensions of the metal, however, is easier than in the case of hot working.

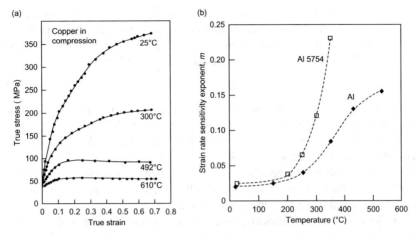

FIGURE 4.8 (a) Effect of temperature on the compressive true stress−true strain behavior of copper at various temperatures. Compressive stresses and strains are shown as positive values. The strain rate was low (∼50% height reduction in 20 minutes). (b) Effect of temperature on strain rate sensitivity exponent for aluminum and an aluminum alloy. *(a) Data from Mahrab et al. (1965). (b) Data from Fields and Backofen (1959) and Li and Ghosh (2003).*

Hot working allows more deformation with less force. Figure 4.8a shows the effect of temperature on the deformation behavior of copper in compression. In this data, the compressive stress and compressive strain are plotted as positive numbers. The yield stress drops with temperature increase and the strain hardening characteristics disappear. As demonstrated in this data, modeling metals as perfect plastics under hot working conditions is a reasonable assumption. Hot working is the method of choice when large amounts of deformation are required. For example, large ingots are reduced to billets and slabs by hot rolling and hot forging. After the hot deformation, the metal is not strengthened, but its structure and properties are isotropic, which is desirable for some applications. On the other hand, hot working requires heating, which entails additional energy and leads to complications with oxidation and controlling final dimensions.

Frequently deformation forming of a metal includes a hot working stage followed by cold working. Most of the deformation is achieved during the hot working step or steps with the cold working providing the final shape and lending strength to the product.

Plastic deformation of metals also depends on strain rate. The strain rate has units of inverse time and is defined as the speed of deformation, v, divided by the instantaneous dimension of the specimen or workpiece:

$$\dot{\varepsilon} = \frac{d\varepsilon}{dt} = \frac{d}{dt}\left[\ln\frac{L}{L_o}\right] = \frac{1}{L}\frac{dL}{dt} = \frac{v}{L} \qquad (4.14)$$

It is important to note that the strain rate depends on the size of the workpiece and therefore varies continuously during deformation. In a tensile test geometry, the speed of deformation is constant and the length increases, leading to a decrease in the strain rate during deformation. The opposite is true of uniaxial compression—strain rate increases with deformation. The change in flow stress with strain rate is given by the following:

$$Y_f = C[\dot{\varepsilon}]^m \qquad (4.15)$$

where C is a strength constant that is a function of strain, temperatures, and material, and m is the strain rate sensitivity exponent. The strain rate sensitivity exponent is low at low temperatures, but higher under hot working conditions, as shown in Figure 4.8b. The increased sensitivity with temperature is related to recrystallization. Near and above the recrystallization temperature, at faster strain rates, recrystallization cannot keep up with the deformation and hence the flow stress increases.

Polymers. The mechanical properties of polymers strongly depend on temperature and strain rate. A good starting point for this discussion is to recall the nature of the viscoelastic behavior introduced in Chapter 3. In solid deformation processes, viscoelasticity can be considered as a means to add viscous and hence time dependent effects of an elastic solid. For a linear viscoelastic material, the response of the solid is modeled by linearly combining dashpots for the viscous effects and springs for the elastic effects.

For a linear viscoelastic solid, a modified Maxwell model with a spring added in parallel is a simple approximation. See Figure 4.9. The additional spring in parallel the Maxwell series pair of spring and dashpot puts a restriction on the dashpot and allows for a closer representation to the response of a solid. Importantly, the stress in the dashpot is the product of the viscosity of the element and the strain rate; hence its contribution is strain rate sensitive. Without going through the mathematics, at higher strain rates, the dashpot does not have time to respond and relax stress and hence the overall stress experienced by the solid is higher. The role of temperature can be suggested by this model as well, because the influence of the viscous nature in the solid increases with temperature because the viscosity drops.

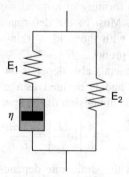

FIGURE 4.9 Spring and dashpot model for a standard linear viscoelastic solid.

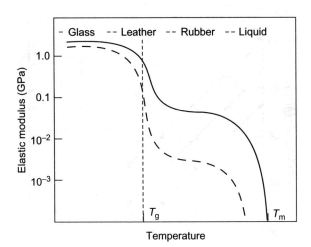

FIGURE 4.10 Schematic showing the effect of temperature on the elastic modulus of a thermoplastic, semicrystalline polymer (solid line) and a thermoplastic amorphous polymer (dashed line).

Figure 4.10 is a schematic diagram showing the change in the elastic modulus of a thermoplastic with temperature. Two states are shown for the same general chemistry of polymer: amorphous and semicrystalline. The benchmark temperature is the glass transition temperature, T_g. At temperatures below T_g, the polymer is glassy and has a high modulus that is less sensitive to temperature. As temperature increases through the glass transition, the polymer modulus drops. For semicrystalline polymers this drop is less than in the case of an amorphous polymer due to the influence of the crystallites on the mechanical response. In this transitional regime, the response is viscoelastic and very sensitive to temperature as well as strain rate. Materials are termed "leathery" in this regime as they are mechanically tough and flexible. Further temperature increase leads to a rubbery plateau in which extensive elastic deformation is possible, similar to a rubber band. Beyond this regime, the modulus falls, gradually for an amorphous polymer and more precipitously for a semicrystalline polymer when the melting point of the crystalline regions is reached. Beyond these temperatures the material is a melt.

An examination of tensile test data gathered at different temperatures also shows the effects mentioned above. Namely, at low temperature the slope on the stress–strain curve is the highest, indicating a high modulus. As temperature increases, the modulus drops and above T_g a yield point appears (e.g., refer to general plot in Figure 4.5). Extensive deformation is possible at higher temperatures. However, if strain rate were to increase, the behavior would change, as demonstrated in Figure 4.11, which shows the effect of strain rate and temperature on the yield point of polycarbonate.

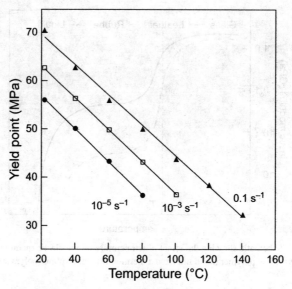

FIGURE 4.11 Effect of strain rate and temperature on yield point of polycarbonate. *Data from Bauwens-Crowet et al. (1969).*

4.2.3 Deformation and Yielding under Triaxial Stresses

Practical deformation processes involve more complex stress states than uniaxial stress, and therefore the stress state and yielding must be described in more general terms. Figure 4.12a shows a general state of triaxial stress. The state of stress is described by a stress tensor:

$$\sigma_{ij} = \begin{vmatrix} \sigma_{11} & \sigma_{12} & \sigma_{13} \\ \sigma_{21} & \sigma_{22} & \sigma_{23} \\ \sigma_{31} & \sigma_{32} & \sigma_{33} \end{vmatrix} \qquad (4.16)$$

The components of the general stress tensor include normal stresses (σ_{11}, σ_{22}, σ_{33}) and shear stresses (σ_{ij}, $i \neq j$). In the description below, shear stresses are given the notation τ rather than σ, as shown in Figure 4.12b. To keep equilibrium (i.e., prevent rotation), the shear stresses must be balanced ($\tau_{ij} = \tau_{ji}$, $i \neq j$), which lowers the number of independent stress components to six. Furthermore, for any complex stress state, there is a choice of coordinate system that results in the reduction of stress components to three principal stresses, which are normal to the faces of the cube (Figure 4.13). In this configuration, there are no shear stresses on the faces of the cube and the stress state becomes:

$$\sigma_{ij} = \begin{vmatrix} \sigma_1 & 0 & 0 \\ 0 & \sigma_2 & 0 \\ 0 & 0 & \sigma_3 \end{vmatrix} \qquad (4.17)$$

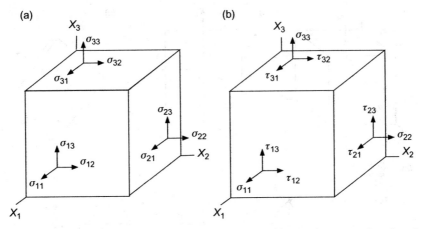

FIGURE 4.12 Representations of a triaxial stress state: (a) with general σ_{ij} convention where i represents the normal to the plane on which the stress component acts and j represents the direction in which the force that causes the stress component acts, and (b) with shear stresses represented as τ_{ij} where $i \neq j$. Note that positive stresses are shown.

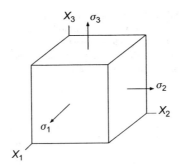

FIGURE 4.13 Stress state in which the axes are chosen such that there are no shear stresses on the planes perpendicular to the axes. The principal stresses are σ_1, σ_2, and σ_3. The principal planes are those perpendicular to the principal axes.

where σ_1, σ_2, and σ_3 are principal stresses. The planes perpendicular to the principal axes are known as principal planes. So, given the appropriate coordinate system, the state of stress imparted in any deformation process can be represented by at most three stress values.

Given knowledge of the stress state in terms of the stress tensor (or the principal stresses), the stresses on any arbitrary plane can be found by resolving forces and applying conditions for equilibrium (no motion). This task, known as the transformation of stress, is made easier by using a Mohr's circle construction. As an example, consider the case shown in Figure 4.14a (i.e., $\sigma_2 = 0$, and $\sigma_1 > \sigma_3$). This stress state is known as plane stress. Since the coordinate system is chosen with σ_1 and σ_3 as principal stresses, there

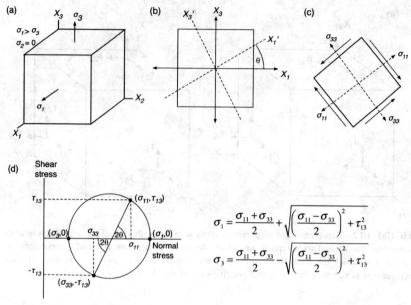

FIGURE 4.14 Use of Mohr's circle to find stresses on nonprincipal planes in a system with two nonzero principal stresses ($\sigma_2 = 0$, and $\sigma_1 > \sigma_3$). (a) Diagram of coordinate system and principal stresses, (b) rotation around X_2 produces a new coordinate system, (c) stresses on planes perpendicular to X_1' and X_3', and (d) Mohr's circle construction for the X_1X_3 plane and equations finding principal stresses from a plane stress state defined by σ_{11}, σ_{33}, and τ_{13} with all other stresses in the stress tensor zero.

are no shear stresses noted on the unit cube faces. There are, however, shear stresses on planes that are not normal to the principal axes. For example, a rotation of the coordinate axes around the X_2 axis reveals planes, which are perpendicular to the X_1' and X_3' axes, with nonzero shear stresses. See Figure 4.14b and c.

The relationship between the principal stresses and the stresses on these new planes can be determined by plotting the Mohr's circle (Figure 4.14b). The abscissa of the plot represents the normal stress with a positive value for a tensile stress and negative for compressive. The ordinate represents shear stress with shear stresses causing a clockwise rotation plotted as positive and those causing a counterclockwise rotation as negative (note that this convention is opposite to that used in Figure 4.12). The two nonzero principal stresses (σ_1 and σ_3) are plotted as points on the abscissa (i.e., there are no shear stresses on these planes). These two points lie on a circle with a radius of $(\sigma_1 - \sigma_3)/2$. To find the stresses on the plane perpendicular to X_1', one goes around the circle by twice the angle of rotation (i.e., angle between plane perpendicular to X_1 and that perpendicular to X_1'). That point defines the magnitude of the normal stress (σ_{11}) and the shear stress (τ_{13}). A similar construction gives the stresses on the plane perpendicular to X'_3. Note that as the angle increases the magnitude of the shear stress increases until 2θ is 90° or θ is 45°, where the shear stress reaches its maximum value, τ_{max}. Additionally, given information about a general plane stress state, the principal stresses can be found using Mohr's circle or the equations in Figure 4.14c.

The elastic response of an isotropic material to a triaxial stress state can be found using a generalized Hooke's law. Considering a stress state represented by principals stresses σ_1, σ_2, and σ_3, the elastic strain response can be calculated:

$$\varepsilon_1 = \frac{1}{E}[\sigma_1 - \upsilon(\sigma_2 + \sigma_3)]$$

$$\varepsilon_2 = \frac{1}{E}[\sigma_2 - \upsilon(\sigma_1 + \sigma_3)] \quad (4.18)$$

$$\varepsilon_3 = \frac{1}{E}[\sigma_3 - \upsilon(\sigma_1 + \sigma_2)]$$

where ε_1, ε_2, and ε_3 are the strains in the X_1, X_2, and X_3 directions, E is the Young's modulus, and υ is the Poisson's ratio. The plastic yielding condition under a triaxial stress state is more challenging to define.

Several assumptions are made in the following description of plastic yielding under a triaxial stress state. The material is assumed to be isotropic. In addition, plastic deformation is uniform with a plastic equivalent of the Poisson's ratio of 0.5 (i.e., no volume change on deformation). Finally, yielding is not affected by the hydrostatic or mean normal stress $[\sigma_m = 1/3(\sigma_1 + \sigma_2 + \sigma_3)]$.

To understand yielding under triaxial stresses, let's start with yielding under uniaxial tension. A state of uniaxial tension is described by the stress state with one positive principal stress ($\sigma_1 > 0$) and the other two principal stresses equal to zero ($\sigma_2 = \sigma_3 = 0$). Yielding occurs when σ_1 reaches the yield stress, Y. See Figure 4.15a. As the value of the applied tensile stress (σ_1) increases, so does the value of τ_{max}, where $\tau_{max} = \sigma_1/2$. Eventually, the material yields when the uniaxial tensile stress reaches a value of Y and the τ_{max} reaches $Y/2$. Therefore the shear yield stress, k, is defined as:

$$k = \frac{Y}{2} \quad (4.19)$$

Hence, the yielding criterion for uniaxial tension is expressed as $\sigma_1 > Y$ or $\sigma_1 > 2k$.

A somewhat more complicated situation occurs when a second normal stress is added to the uniaxial tension described above. For example, consider biaxial tension represented by $\sigma_1 > 0$, $\sigma_3 > 0$, $\sigma_1 > \sigma_3$, $\sigma_2 = 0$. As shown in Figure 4.15b, there are three Mohr's circles representing stresses viewed from three orientations. The maximum shear stress due to the combination of σ_1 and σ_3 (X_1X_3 plane) is always less than the maximum shear stress due to

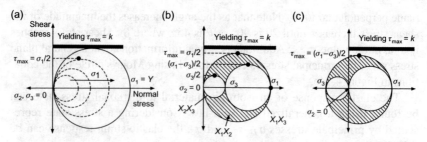

FIGURE 4.15 Use of Mohr's circle to demonstrate yield conditions: (a) uniaxial tension ($\sigma_1 > 0$, $\sigma_2 = \sigma_3 = 0$; Mohr's circle construction for the X_1X_3 or X_1X_2 plane); dotted lines show affect of increasing σ_1 on the circle, (b) biaxial tension, plane stress ($\sigma_1 > \sigma_3 > 0$, $\sigma_2 = 0$; Mohr's circle constructions for the X_1X_3, X_1X_2, and X_2X_3 planes), (c) tensile-compressive, plane stress ($\sigma_1 > 0$, $\sigma_3 < 0$, $\sigma_2 = 0$; Mohr's circle constructions for the X_1X_3, X_1X_2, and X_2X_3 planes). In (b) and (c), x-axis and y-axis labels are as in (a) and the shaded portions show possible combinations of shear and normal stresses for all orientations.

higher tensile stress alone (X_2X_3 plane). That is, the presence of the additional, tensile stress acting on the material does not impact yielding. On the other hand, if the second stress is compressive, then the maximum shear stress due to the combination is always greater than that due to the tensile stress alone (Figure 4.15c).

In a general triaxial stress state, the value of the maximum shear stress is one half the difference between the largest and smallest principal stresses:

$$\tau_{max} = \frac{1}{2}(\sigma_{largest} - \sigma_{smallest}) \qquad (4.20)$$

Hence, one way to find the condition for yielding in this more complex stress state is to assume that yielding takes place when τ_{max} reaches the shear yield strength. This is known as the Tresca criterion. The Tresca criterion states that yielding takes place when the maximum shear stress reaches a critical value—the shear yield strength. The relationship between the shear yield strength (k) and the yield strength in uniaxial tension, Y, is still valid, and hence

$$\sigma_{largest} - \sigma_{smallest} = 2k = Y \qquad (4.21)$$

By this convention, yielding takes place when the difference between the largest principal stress and the smallest principal stress is equal to the yield strength in tension.

Now, a map of yielding for all possible combinations of two nonzero principal stresses can be constructed, as shown in Figure 4.16. The solid lines are the Tresca criteria, Eq. (4.20), they define the conditions for yielding (i.e., combinations of σ_1 and σ_3). The map shows that a combination of tension and compression results in yielding at lower stress magnitudes and therefore this combination is found in many deformation processes.

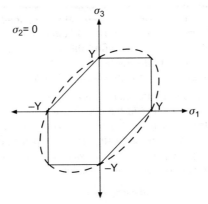

FIGURE 4.16 Map of the yield criteria for the plane stress case ($\sigma_2 = 0$). The Tresca (solid line) and von Mises (dashed line) conditions are shown. If the state of stress, given by principal stresses σ_1 and σ_3 lies inside the hexagon (Tresca) or ellipse (von Mises), then the material responds elastically. If the stress state reaches the lines, then yielding occurs. For a three-dimensional stress state, a two-dimensional yielding surface would be needed.

The Tresca criterion for yielding states that when the difference between the largest and smallest principal stresses reaches the yield stress determined under uniaxial conditions (Y), then yielding takes place. Hence, according to the Tresca criterion, the value of the intermediate principal stress does not play a role in yielding. The Tresca criterion is simple to understand and use, but tends not to match experimental data as well as an alternative criterion known as the von Mises criterion.

The von Mises criterion states that yielding occurs when the shear strain energy reaches a critical value. For a general case with three principal stresses:

$$(\sigma_1 - \sigma_2)^2 + (\sigma_2 - \sigma_3)^2 + (\sigma_3 - \sigma_1)^2 = \text{constant} \tag{4.22}$$

The constant is found by considering the uniaxial case ($\sigma_1 = \sigma_1$, $\sigma_2 = \sigma_3 = 0$) where yielding takes place when $\sigma_1 = Y$. An alternative constant can be found by considering pure torsion ($\sigma_1 = -\sigma_3$, $\sigma_2 = 0$) where yielding takes place when $\sigma_1 = k$ and $\sigma_3 = -k$.

$$(\sigma_1 - \sigma_2)^2 + (\sigma_2 - \sigma_3)^2 + (\sigma_3 - \sigma_1)^2 = 2Y^2 = 6k^2 \tag{4.23}$$

Notice that the relationship between yield strength under uniaxial tension and the shear yield strength is different with the von Mises criterion ($Y = \sqrt{3}k$) compared with the Tresca criterion ($Y = 2k$). The von Mises yield criterion also produces a yielding map as shown in Figure 4.16 for the plane stress case ($\sigma_2 = 0$). It is clear that the intermediate stress is important in establishing the yielding condition. One can also envision a map of yielding in three dimensions and a yield surface (i.e., combinations of σ_1, σ_2, and σ_3). Since the von Mises

EXAMPLE 4.4 The state of stress of a material is described by: $\sigma_{11} = 200$ MPa, $\tau_{12} = 50$ MPa, $\sigma_{22} = 100$ MPa and $\sigma_{33} = 0$. (a) Draw a Mohr's circle for the X_1X_2 plane and find the principal stresses and the maximum shear stress for that orientation. (b) The yield stress of the material under uniaxial tension is 200 MPa. Will the material yield?

a. There are shear stresses on the planes perpendicular to σ_{11} and σ_{22}, so there needs to be a transformation in the coordinate system to find the principal stresses. It is clear though that such a transformation involves a rotation around X_3 (since there are no shear stresses on the plane perpendicular to X_3, that direction is a principal direction). A Mohr's circle can be constructed by plotting the information given, as shown below.

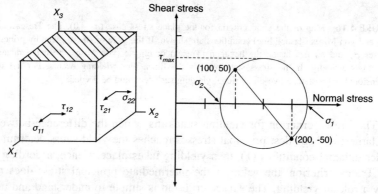

FIGURE E4.4

To find the principal stress, we can use the geometry of the circle to find the radius ($50^2 + 50^2 = R^2$) and therefore the principal stresses or the equations in Figure 4.14. The answer is $\sigma_1 = 221$ MPa and $\sigma_2 = 79$ MPa

b. To determine if the material will yield, the principal stresses are plugged into the von Mises condition

$$(\sigma_1 - \sigma_2)^2 + (\sigma_2 - \sigma_3)^2 + (\sigma_3 - \sigma_1)^2 = 2Y^2$$
$$(221 - 79)^2 + (79 - 0)^2 + (0 - 221)^2 = 75,246$$
$$2Y^2 = 2(200)^2 = 80,000$$
$$75,246 < 80,000 \quad \text{No yielding}$$

criterion matches experiments better and is a simple function of the three principal stresses, it is used exclusively from here on (Example 4.4).

The deformation of polymers under triaxial stresses is a bit different than metals. In metals, the yield stress in tension is usually the same as the yield stress in compression and therefore, the yield map, for example in Figure 4.16, is symmetric. The yield stress for polymers in compression is typically ~1.3 times that measured in tension. Hence the map of yielding is asymmetric. Likewise, cold worked metals that develop anisotropy also develop asymmetry in their yield behavior.

The complexity of triaxial stress states and deformations can be simplified by defining an effective stress, $\bar{\sigma}$, and an effective strain, $\bar{\varepsilon}$. (In these symbols, the bar denotes effective rather than average.) The effective stress is a function of the three principal stresses such that when it reaches Y, yielding takes place. Assuming the von Mises criterion, the effective stress is given by:

$$\bar{\sigma} = \frac{1}{\sqrt{2}}[(\sigma_1-\sigma_2)^2+(\sigma_2-\sigma_3)^2+(\sigma_3-\sigma_1)^2]^{1/2} \tag{4.24}$$

The effective strain, again assuming the von Mises criterion, is defined in incremental form as:

$$d\bar{\varepsilon} = \frac{\sqrt{2}}{3}[(d\varepsilon_1-d\varepsilon_2)^2+(d\varepsilon_2-d\varepsilon_3)^2+(d\varepsilon_3-d\varepsilon_1)^2]^{1/2} \tag{4.25}$$

(The differential form arises the definition of incremental work $dw = \bar{\sigma}d\bar{\varepsilon}$.) Assuming volume constancy, $d\varepsilon_1 + d\varepsilon_2 + d\varepsilon_3 = 0$, Eq. (4.25) can be reduced to:

$$d\bar{\varepsilon} = \left[\frac{2}{3}(d\varepsilon_1^2+d\varepsilon_2^2+d\varepsilon_3^2)\right]^{1/2} \tag{4.26}$$

In the special case of proportional straining in which there is a constant ratio of $d\varepsilon_1$ to $d\varepsilon_2$ to $d\varepsilon_3$, the differential strains can be replaced with true strains:

$$\bar{\varepsilon} = \left[\frac{2}{3}(\varepsilon_1^2+\varepsilon_2^2+\varepsilon_3^2)\right]^{1/2} \tag{4.27}$$

As illustrated in Example 4.5, these definitions allow effective stress–effective strain to map directly to uniaxial tension test data. They can therefore be used to predict yielding (Example 4.6). Effective stress and effective strain also enter into the description of plastic deformation or plastic flow after yielding.

Flow rules are used to describe plastic flow after yielding. The simple analog in uniaxial tensile testing is given in Eq. (4.10) and given the definitions above, the following holds:

$$\bar{\sigma} = K\bar{\varepsilon}^n \tag{4.28}$$

Essentially, this equation describes how the effective stress increases with deformation (work hardening). Since the effective stress originates from the applied principal stresses, one can envision that the yield surface must expand as the deformation proceeds. Flow rules turn this situation around

EXAMPLE 4.5 Find the effective stress for the uniaxial tensile test. Assume plastic deformation is uniform such that the volume is constant.

In a uniaxial tensile test, the load is applied in one principal direction:

$$\sigma_1 = \sigma_1 \quad \text{and} \quad \sigma_2 = \sigma_3 = 0$$

Since $d\varepsilon_2 = d\varepsilon_3$, volume consistency requires $d\varepsilon_1 + d\varepsilon_2 + d\varepsilon_3 = 0$.

$$d\varepsilon_1 = -2d\varepsilon_2 = -2d\varepsilon_3$$

Using the definitions of effective stress and effective strain:

$$\bar{\sigma} = \frac{1}{\sqrt{2}}[(\sigma_1)^2 + (-\sigma_1)^2]^{1/2} = \sigma_1$$

$$d\bar{\varepsilon} = \left[\frac{2}{3}\left(d\varepsilon_1^2 + \frac{1}{4}d\varepsilon_1^2 + \frac{1}{4}d\varepsilon_1^2\right)\right]^{1/2} = d\varepsilon_1$$

Since the straining is proportional, $\bar{\varepsilon} = \varepsilon_1$

Therefore, the true stress–true strain plot from a tensile test is a plot of effective stress–effective strain.

EXAMPLE 4.6 Revisit Example 4.4 and use the effective stress to determine if yielding occurs.

The example describes a plane stress situation with $\sigma_{11} = 200$ MPa, $\tau_{12} = 50$ MPa, $\sigma_{22} = 100$ MPa and $\sigma_{33} = 0$. The principal stresses were found to be $\sigma_1 = 221$ MPa and $\sigma_2 = 79$ MPa. The material has a yield stress of 200 MPa in uniaxial tension. So, we can find the effective stress to be:

$$\bar{\sigma} = \frac{1}{\sqrt{2}}[(\sigma_1 - \sigma_2)^2 + (\sigma_2 - \sigma_3)^2 + (\sigma_3 - \sigma_1)^2]^{1/2}$$

$$= \frac{1}{\sqrt{2}}[(221-79)^2 + (79)^2 + (-221)^2]^{1/2} = 194 \text{ MPa}$$

So, the effective stress is 194 MPa, which is less than the yield stress and hence yielding does not occur.

---

and consider the incremental changes in the principal strains that result from the application of a change in the stress state:

$$d\varepsilon_1 = \frac{d\bar{\varepsilon}}{\bar{\sigma}}\left[\sigma_1 - \frac{1}{2}(\sigma_2 + \sigma_3)\right]$$

$$d\varepsilon_2 = \frac{d\bar{\varepsilon}}{\bar{\sigma}}\left[\sigma_2 - \frac{1}{2}(\sigma_1 + \sigma_3)\right] \quad (4.29)$$

$$d\varepsilon_3 = \frac{d\bar{\varepsilon}}{\bar{\sigma}}\left[\sigma_3 - \frac{1}{2}(\sigma_1 + \sigma_2)\right]$$

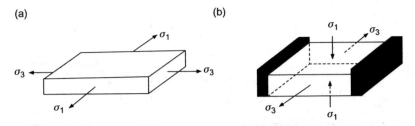

FIGURE 4.17 Schematic diagrams of complex stress states: (a) biaxial tensile stress (plane stress) and (b) plane strain.

In these equations, $d\bar{\varepsilon}/\bar{\sigma}$ is a proportionality constant much the same as the $1/E$ factor in the generalized Hooke's Law.

Two special stress states frequently encountered in deformation processing are plane stress and plane strain. In plane stress, one of the principal stresses is zero. Figure 4.17a shows a schematic of plane stress deformation by biaxial stretching. Yielding under plane stress conditions is shown in Figure 4.16; the stretching example would correspond to the first quadrant in this diagram and the von Mises criterion for yielding simplifies to $\sigma_1^2 + \sigma_3^2 - \sigma_1\sigma_3 = Y^2$. Considering the case of biaxial stretching, the flow laws show that there are extensions in the directions of the principal stresses ($d\varepsilon_1 > 0$, $d\varepsilon_3 > 0$) and contraction perpendicular to the stress-free plane ($d\varepsilon_2 < 0$). In plane strain, the dimensions of the work piece are held fixed in one dimension ($d\varepsilon_2 = 0$). Figure 4.17b shows an example of plane strain compression in which the dimension in one direction is held fixed by confinement in a channel. In plane strain, there are stresses acting on all three principal planes; using the plane strain constraint ($d\varepsilon_2 = 0$) and the flow laws (Eq. (4.29)), the stress in the plane perpendicular to the constraint can be found:

$$\sigma_2 = \frac{\sigma_1 + \sigma_3}{2} \qquad (4.30)$$

Substituting this relationship into the von Mises yield criterion (Eq. (4.23)) results in the yield condition for plane strain:

$$\sigma_1 - \sigma_3 = \frac{2}{\sqrt{3}} Y \qquad (4.31)$$

The work done per unit volume during deformation under triaxial stresses can be calculated using Eq. (4.13) with the effective stress and effective strain substituted for the uniaxial values:

$$u_{\text{ideal}} = \int_0^{\bar{\varepsilon}_{\max}} \bar{\sigma} d\bar{\varepsilon} = \int_0^{\bar{\varepsilon}_{\max}} K\bar{\varepsilon}^n d\bar{\varepsilon} = \frac{K\bar{\varepsilon}_{\max}^{n+1}}{n+1} \qquad (4.32)$$

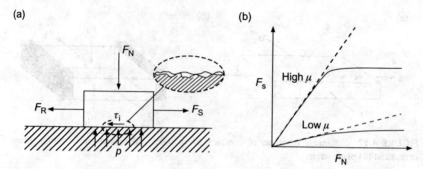

FIGURE 4.18 (a) Schematic of frictional effects on sliding a block over a surface with inset, showing the details of the contact between the materials. (b) Effect of normal force on the shear force needed to initiate motion.

The work done (or energy expended) per unit volume calculated above is considered ideal because it only accounts for the deformation that leads to shape change.

4.2.4 Friction

Friction occurs at contacting surfaces during solid deformation processes. Typically when a tool, such as a die or roll, transmits a load, the solid plastically deforms and moves relative to the tool. This motion is affected by friction at the interface; notably, the following factors influence the effect of friction on the process: the chemistry and microscopic features of the surfaces in contact, the presence or absence of a lubricant, and the load or pressure that exists between the surfaces.

To quantify friction, consider a block of one material sitting atop a rigid surface of another material, as in Figure 4.18a. To set the block in motion, an applied shear force at the interface, F_S, must surpass the resistive force presented by friction, F_R. If, by some means, there were no friction at the interface then the block could be moved effortlessly regardless of its size, i.e., regardless of the normal force on the contact between the block and the material beneath, F_N. With friction at the interface, it is intuitive that as F_N increases so does the shear force F_S needed to move the block. The coefficient of friction, μ, is defined to capture this trend.

$$\mu = \frac{F_S}{F_N} \qquad (4.33)$$

See Figure 4.18b. The friction coefficient varies between near zero to near one. Table 4.5 gives some example values. The friction coefficient can vary with the microstructure of the surface and with the test method.

The coefficient of friction is dimensionless, the ratio of two forces. Since both forces act on the area of contact, the coefficient of friction can also be considered as a ratio of an interfacial shear stress ($\tau_i = F_S A$) to a normal

TABLE 4.5 Examples of Friction Coefficient[a]

Fixed material	Moving material	Static friction coefficient	Kinetic friction coefficient
Al Alloy 6061-T6	Al Alloy 6061-T6	0.42	0.34
Al Alloy 6061-T6	Steel 1032	0.35	0.25
Steel 1032	Al Alloy 6061-T6	0.47	0.38
Steel 1032	Steel 1032	0.31	0.23
Stainless Steel 304	Copper	0.23	0.21
Alumina	Alumina	na	0.5
Iron	Alumina	na	0.45
Tempered Glass	Al Alloy 6061-T6	0.17	0.14
PMMA	PMMA	0.8	na
Nylon 6/6	Nylon 6/6	0.06	0.07
PS	PS	0.5	na
PE	PE	0.2	na
Teflon	Teflon	0.08	0.07
Mild Steel	PE	0.09	0.13
Carbon Steel	PS	0.43	0.37
Al Alloy 6061-T6	Teflon	0.19	0.18
Molded Graphite	Al Alloy 2024	0.16	na
Molded Graphite	Teflon	0.18	na

[a] Data tabulated in Blau (1992).

stress, which is best considered an interfacial pressure ($p = F_N A$). In solid deformation processes, this interfacial pressure is imparted purposefully by tools and dies to cause the deformation and hence it can be very large.

The interface between two solids is not simple, as depicted in the inset of Figure 4.18a. Depending the surface roughness, there may be microscopic or not so microscopic hills and valleys on the surfaces such that the contact, especially at low interfacial pressure, occurs selectively on the raised features. Further, lubricants may be added at the interface, creating an even more complex three-body problem.

The proportionality that defines the coefficient of friction eventually breaks down. For example, as higher and higher normal forces are applied at the interface, the shear stress at the interface needed for motion eventually reaches the shear yield strength of the material. At this point, the material at the interface yields and flows by shear. This is called *sticking*. According to the von Mises criterion, this condition occurs when:

$$\tau_i = k = \frac{Y}{\sqrt{3}} \tag{4.34}$$

where Y represents the yield strength of the sliding material under uniaxial tension. For the sliding block, one would find the interface to be stable, not sliding, but the block itself deforming by shear. Therefore, the term "sticking" refers to immobilization and not necessarily a physical or chemical adherence.

There is more than one type of friction coefficient. In the description above, the coefficient of friction would be called the *static* coefficient of friction as the situation involves the initiation of motion from an initially static state. Once motion begins, a different and frequently lower coefficient of *sliding or kinetic* friction characterizes the motion. Further, there is also a coefficient of *rolling* friction for situations involving that type of motion.

Friction plays an important role in most solid deformation processes. On one hand, it interferes with shape formation because it presents an extra force that must be overcome for the deformation to occur. Additionally, the rate of wear of the tools and dies increases with friction and the energy dissipation associated with friction causes heating. For these reasons, lubricants are used in many deformation processes to reduce friction coefficient. Frequently, lubricants also have the effect of dissipating some of the heat generated during deformation. On the other hand, friction is essential to starting and maintaining contact in processes. For example, friction is essential to the gripping action of the two counterrotating rolls in the rolling process.

4.2.5 Efficiency and Temperature Rise

Deformation processes vary in their efficiency. In a theoretical deformation process with 100% efficiency, all of the work done during the process would go into the deformation or shape change. In reality, there is additional work done to overcome friction, and due to die geometry, there can be redundant work. Redundant work is internal plastic deformation that does not result in shape change. So the total work done (or energy expended) per unit volume, u_{total}, is greater than u_{ideal}. The efficiency of a deformation process is therefore given as:

$$\text{Efficiency} = \frac{u_{\text{ideal}}}{u_{\text{total}}} \tag{4.35}$$

The efficiency of deformation processes expressed as a percentage varies from around 30–60% for extrusion from 75–95% for rolling.

Plastic deformation can frequently lead to an increase in the temperature of the workpiece. Only a small amount of the energy expended to create a new shape is stored as elastic energy or in dislocations and other defects. Similar to the origin of viscous heating in flowing melts, the origin of thermal effects in plastic deformation can be traced to the first law of thermodynamics, which defines the change in the internal energy (per volume) of a system, ΔU, as:

$$\Delta U = q - w \qquad (4.36)$$

where positive q refers to heat added to the system from the surroundings and positive w refers to work done *by the system* on the surroundings. When work is done *on the system*, then that work done per unit volume either raises the internal energy and hence the temperature of the system or it causes heat to flow from the system or both. If the material is deformed rapidly such that there is little time for heat flow from the deformed material to the surroundings (i.e., an adiabatic condition, $q = 0$), then all of the work done on the material causes an increase in internal energy:

$$\Delta U = -w = \frac{\text{Work done on system}}{\text{Volume}} = u_{\text{total}} \qquad (4.37)$$

The increase in internal energy leads to an increase in the temperature of the work piece. The extent of the temperature increase is determined by the heat capacity. By definition,

$$c_v = \left(\frac{dU}{dT}\right)_V \qquad (4.38)$$

The heat capacity at constant volume, c_v, is approximately the heat capacity at constant pressure. If the specific heat capacity, C_p, given in units of J/(kg•K), is used, then integration gives:

$$\Delta U = \rho C_p \Delta T \qquad (4.39)$$

Therefore, the increase in temperature of the workpiece can be found:

$$\Delta T = \frac{u_{\text{total}}}{\rho C_p} \qquad (4.40)$$

where ρ is the density of the workpiece metal. Temperature rises in the tens of degrees or more are common. Equation (4.40) does not include the conduction of heat away from the workpiece. Hence, processes most susceptible to large temperature rises are those that involve large amounts of deformation that occur quickly so that the heat does not have time to be conducted away by the tool.

4.3 SOLID PROCESSES

4.3.1 Process Overview

Solid deformation processes are categorized as either bulk deformation or sheet deformation. Bulk deformation processes convert solid starting materials into new shapes by the application of mechanical forces. What differentiates bulk processes from sheet processes is the shape of the starting material and the mode of the deformation. The starting material for bulk deformation is a billet, block, slab or other simple shape that has a low surface area to volume ratio. Bulk deformation involves a large reduction in thickness or diameter under a compressive or combination of compressive and tensile stresses or gross changes that result in a new 3D shape. The changes can cause a slab to become thinner, a rod or wire to become narrower or a crude shape to take on details.

The final goal of a bulk deformation process may be a complex shape, such as a wrench or railroad wheel, or a simple one, such as a wire. Often, the shape develops over a series of deformation steps. For example, a thin sheet of aluminum foil is created by passing a slab of aluminum starting material through a series of rolls, with the gap between each subsequent set of rolls decreasing. By contrast, a fairly complex extruded profile is formed with a single extrusion operation. The importance of limits of deformability and the stresses placed on the part as it is being formed are key. Additionally, the role of temperature is central. Large deformations are possible under hot working conditions.

Figure 4.19 shows simple schematics of the bulk deformation processes that are covered in this section. In each there is a "workpiece," the material being shaped or deformed, and a "tool," the apparatus or part that is transmitting the mechanical forces that cause the deformation. Wire drawing (Figure 4.19a) is a relatively simple process that involves pulling a wire through a die. Compression placed on the wire by the die assists the deformation. This process is covered in some detail to show the utility of simple force balances in predicting the conditions of deformation and also to show the influence of friction. Extrusion (Figure 4.19b) is roughly the reverse of drawing with the starting material pushed through a die. Here, however, fairly complicated cross-sectional geometries are possible. The simple force balance method falls short of predicting the conditions for deformation because large redundant work is typical. Forging (Figure 4.19c) is the deformation under compressive loads. Hot forging with open dies is a widely used process for reduction of cast ingots to smaller shapes and also a first step in many deformation processes. Forging with shaped dies produces three dimensional parts with shape complexity. Rolling (Figure 4.19d) is a method for thickness reduction of sheets as well as for creating simple cross-sections with shaped rolls. It has analogies to forging due to the compression between the rolls.

Sheet deformation processes involve starting material that has already been formed into a sheet with a uniform thickness. See Figure 4.19e. For polymer processes, the sheet may have been formed by melt extrusion, for example, and for metal sheet process, rolling is frequently used. The sheet itself then is formed into a shape. There are various means to do sheet deformation processing. In this chapter, bending (Figure 4.19e), thermoforming

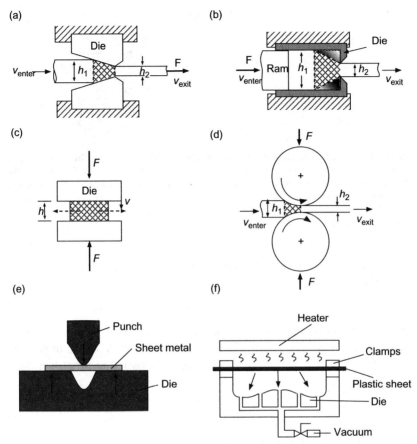

FIGURE 4.19 Schematic diagrams of solid processes: (a) wire drawing, (b) extrusion, (c) open die forging, (d) rolling, (e) bending, and (f) thermoforming. Processes (a)–(d) are bulk deformation processes. These diagram emphasizes how a critical dimension, h, is reduced by each process, and the deformation zones are noted in cross-hatch. Processes (e) and (f) are sheet deformation processes.

(Figure 4.19f), and superplastic forming of metals, a process that is similar to thermoforming, are discussed.

4.3.2 Wire Drawing

Wire drawing reduces the diameter of a metal wire or rod by pulling the wire or rod through a die. It is a fairly straightforward operation that serves as a good starting point for an exploration of deformation processing. This section is based on a thorough analysis given by Hoffman and Sachs (1953).

Consider a wire with an initial diameter of D_1 drawn through a die with a conical channel to create a wire with a reduced diameter of D_2. See

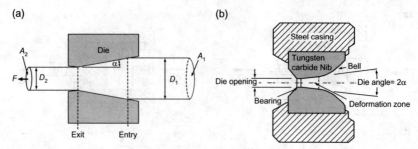

FIGURE 4.20 (a) Simplified schematic of wire drawing process, showing the change in wire diameter and critical parameters. The deformation zone is from entry to exit. In this zone, the strain as well as flow stress (in the case of a work hardening material) increases. The coordinate system discussed in the text has x as the axial direction with y and z mutually perpendicular axes in the plane perpendicular to x. (b) Example of a wire drawing die. *Adapted from Fort Wayne Wire Dies product literature, www.fortwaynewiredie.com.*

Figure 4.20. Volume conservation requires the wire to lengthen as its diameter is reduced:

$$A_1 L_1 = A_2 L_2 \quad \text{or} \quad D_1^2 L_1 = D_2^2 L_2 \quad (4.41)$$

where A_1 and A_2 are the cross-sectional areas of the wire before and after the draw, respectively. The action of drawing is both tensile and compressive. The amount of deformation in wire drawing is referred to as the reduction, r, as defined by the ratio:

$$r = \frac{A_1 - A_2}{A_1} \quad (4.42)$$

The percent reduction is simply $100r$.

A stress is applied to pull the wire through the die, but the action of the die walls is compressive. The draw stress is defined as the force required for drawing, F (i.e., the draw force), divided by the cross-sectional area on exit:

$$\sigma_d = \frac{F}{A_2} \quad (4.43)$$

The amount of deformation for a single pass through a die is quantified as a positive axial strain.

$$\varepsilon_x = \ln\left(\frac{L_2}{L_1}\right) = \ln\left(\frac{A_1}{A_2}\right) = \ln\left(\frac{1}{1-r}\right) \quad (4.44)$$

Concurrently the diameter reduces. This deformation is axisymmetric. Referring to the Cartesian coordinates mentioned in the caption of Figure 4.20, the strains in the y and z direction are equal ($d\varepsilon_y = d\varepsilon_z$) and based on volume conservation, $d\varepsilon_x = -2d\varepsilon_y = -2d\varepsilon_z$. According to Example 4.5, the axial strain, ε_x, is therefore equal to the effective strain

($\bar{\varepsilon} = \varepsilon_x$), a condition that allows the use of axial strain directly in the analysis of the deformation of strain hardening materials.

The goal of a wire drawing analysis is to find the force needed to pull the wire through a given die and also to design the process that produces the required diameter reduction most efficiently (i.e., with low loads, low power consumption, and fewest number of dies). A basic criterion for the process is that the uniaxial tensile stress created in the wire after exit cannot exceed the yield strength of the wire. (The stress inside the die is not uniaxial.) This criterion limits to amount of diameter reduction that can take place from any given die. Below, the wire drawing process is analyzed, progressing from a simple analysis of an ideal process without consideration of friction or nonuniform deformation, also known as redundant work, to one that includes friction. Lastly, both friction and redundant work are considered.

Ideal Deformation. The simplest way to analyze an ideal deformation process is through an energetic analysis. For ideal deformation, the energy dissipated in plastic deformation per unit volume, u, is the area under the true stress−true strain curve, regardless of how the shape change took place. Hence, if we consider a perfectly plastic wire that changes from a diameter of D_1 and length of L_1 to diameter of D_2 and length of L_2,

$$u = Y\varepsilon_x = Y \ln\left(\frac{L_2}{L_1}\right) = Y \ln\left(\frac{A_1}{A_2}\right) \tag{4.45}$$

where Y is the flow stress, which remains constant during deformation, and ε_x is the axial strain. The work done on the wire is the product of u and the volume of the wire. This work is equal to the work done by the drawing apparatus, which applies a force, F, over the distance traveled, L_2.

$$\text{Work} = u(A_2 L_2) = FL_2 \tag{4.46}$$

$$u = \frac{F}{A_2} = \sigma_d$$

$$\sigma_d = Y \ln\left(\frac{A_1}{A_2}\right) = Y \ln\left(\frac{D_1^2}{D_2^2}\right) = Y\varepsilon_x \tag{4.47}$$

From these expressions, it is clear that the force needed for wire drawing ($F = \sigma_d A_2$) increases with the flow stress of the material and the amount of strain during the deformation.

For a work hardening material, the flow stress increases with deformation and the extent of deformation varies as the wire goes through the die. To handle this problem, Y is replaced with the average flow stress, $\bar{Y}$, in the analysis above. See Example 4.7. The draw stress calculated without the influence of friction or redundant work sets the lower bound on the force required.

EXAMPLE 4.7 Assume ideal deformation when answering the following. (a) Find the draw force needed for a 30% reduction of a 2 mm diameter annealed aluminum wire with K = 180 MPa and n = 0.2. (b) Find the draw force needed to reduce the wire formed in part (a) by a reduction of 30%.

a. A 30% reduction requires:

$$r = 0.30 = \frac{A_1 - A_2}{A_1} = 1 - \frac{A_2}{A_1} \quad \text{Therefore,} \quad \frac{A_2}{A_1} = 0.70$$

$$A_2 = 0.70 A_1 = 0.70 \left(\frac{\pi (2 \text{ mm})^2}{4} \right) = 2.12 \text{ mm}^2 \quad [D_2 = 1.64 \text{ mm}]$$

$$\varepsilon_x = \ln\left(\frac{L_2}{L_1}\right) = \ln\left(\frac{A_1}{A_2}\right) = \ln\left(\frac{1}{0.7}\right) = 0.357$$

The draw takes place from an initial strain of 0 to a final strain of 0.357. So, recalling that $\varepsilon_x = \bar{\varepsilon}$ and letting $\varepsilon = \varepsilon_x = \bar{\varepsilon}$ the average flow stress can be found using the strain hardening law:

$$\bar{Y} = \frac{\int_0^{\varepsilon_{max}} K\varepsilon^n d\varepsilon}{\varepsilon_{max}} = \frac{K\varepsilon_{max}^n}{n+1} = \frac{(180 \text{ MPa})(0.357)^{0.2}}{1.2} = 122 \text{ MPa}$$

The draw stress and draw force are:

$$\sigma_d = \bar{Y} \ln\left(\frac{A_1}{A_2}\right) = (122 \text{ MPa})(0.357) = 43.6 \text{ MPa}$$

$$F = \sigma_d A_2 = (43.6 \text{ MPa})(2.12 \text{ mm}^2) = 92.43 \text{ N}$$

b. After drawing through the first die, the diameter is reduced from 2 mm to 1.64 mm, and the cross-sectional area is reduced from 3.14 mm² to 2.12 mm². A second draw causes another 30% reduction. Therefore,

$$A_3 = 0.70 A_2 = 0.70(2.12 \text{ mm}^2) = 1.484 \text{ mm}^2 \quad [D_3 = 1.375 \text{ mm}]$$

For this draw, the strain is the same as for the first draw because the amount of reduction is the same for both. The total strain after the second draw is:

$$\varepsilon_{total} = 0.357 + 0.357 = 0.714$$

Therefore the average flow stress can be found, realizing that the strain after the first draw was 0.357 and therefore the average is taken from this strain to the final strain.

$$\bar{Y} = \frac{\int_{\varepsilon_{initial}}^{\varepsilon_{total}} K\varepsilon^n d\varepsilon}{\varepsilon_{total} - \varepsilon_{initial}} = \frac{K\varepsilon_{total}^{n+1} - K\varepsilon_{initial}^{n+1}}{(\varepsilon_{total} - \varepsilon_{initial})(n+1)}$$

$$= \frac{(180 \text{ MPa})[(0.714)^{1.2} - (0.357)^{1.2}]}{(0.714 - 0.357)(1.2)} = 158.4 \text{ MPa}$$

The draw stress and draw force are:

$$\sigma_d = (158.4 \text{ MPa})(0.357) = 56.5 \text{ MPa}$$
$$F = \sigma_d A_3 = (56.5 \text{ MPa})(1.484 \text{ mm}^2) = 83.85 \text{ N}$$

Notice that the draw stress increases for the second draw due to work hardening, but the draw force is lower because the area on exit has dropped.

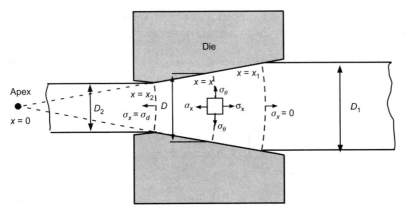

FIGURE 4.21 Schematic of wire drawing process with a spherical shell superimposed for analysis of the change in stress state through the deformation zone. The volume element shown is at an axial position of x. The diameter of the die at that position is D. Therefore, x and D vary continuously through the die. *Adapted from Hoffman and Sachs (1953)*.

A more complete analysis of ideal deformation that considers the stress state in the wire is a useful prelude to the more complicated analysis with friction. Figure 4.21 shows the coordinate system and a volume element in the wire. The wire is being pulled to the left. The wire deforms continuously from a diameter D_1 on entrance to a diameter D_2 on exit. Assuming that the deformation can be represented in terms of a plastic flow radially toward the apex ($x = 0$). A spherical coordinate system developed for thick wall pressure vessels can be used to analyze the deformation. See the Appendix for a description of the pressure vessel. A section through the three-dimensional volume element is shown in Figure 4.21. The radial direction, one of the three principal directions, is given by the symbol x here to represent the axial direction. The other two principal directions are defined as two mutually perpendicular tangential directions. The stress in one of the tangential directions σ_θ is shown in the figure and the other, equivalent in magnitude, is acting out of the page. This tangential stress represents the compressive action of the die on the wire.

The equation governing the state of stress is the same as that given in the Appendix, but rewritten here with x, the axial direction, replacing r.

$$\frac{d\sigma_x}{dx} + \frac{2}{x}(\sigma_x - \sigma_\theta) = 0 \tag{4.48}$$

For the case of radial flow toward the apex, the axial stress (tension) is greater than the tangential stress (compression). Since the two tangential stresses are equal, the condition for yielding, by the Tresca or von Mises criterion, is:

$$\sigma_x - \sigma_\theta = Y \tag{4.49}$$

where Y is the flow stress for a perfectly plastic solid. For a strain hardening material, Y is replaced by the average flow stress, $\overline{Y}$. Therefore,

$$\frac{d\sigma_x}{dx} + \frac{2Y}{x} = 0 \tag{4.50}$$

A boundary condition is needed to solve this equation. The stress σ_x at the entry ($x = x_1$) is equal to zero and therefore,

$$\int_0^{\sigma_x} d\sigma_x = -2Y \int_{x_1}^{x} \frac{dx}{x}$$

$$\sigma_x = Y \ln\left(\frac{x_1^2}{x^2}\right) \tag{4.51}$$

$$\sigma_\theta = \sigma_x - Y = Y\left[\ln\left(\frac{x_1^2}{x^2}\right) - 1\right] \tag{4.52}$$

Noting that geometry requires $x_1/x = D_1/D$

$$\sigma_x = Y \ln\left(\frac{D_1^2}{D^2}\right) \tag{4.53}$$

$$\sigma_\theta = Y\left[\ln\left(\frac{D_1^2}{D^2}\right) - 1\right] \tag{4.54}$$

These equations show a smooth increase in the tensile axial stress from zero to a maximum as x varies from entry (x_1, D_1) to exit (x_2, D_2) and a smooth decrease in the magnitude of the compressive radial stress over the same range. See Example 4.8. The draw stress is the axial stress at exit (Eq. (4.53)) and can be found be identical to Eq. (4.47) using the expressions above.

Deformation with Friction. Friction plays an important role in wire drawing. The wire slides across the die surface during drawing; therefore, the draw stress must increase over that required for ideal, frictionless deformation to overcome the frictional resistance. The coordinate system set up for analysis of the ideal case, Figure 4.21, is not convenient for analyzing wire drawing with friction. However, using that stress state as a starting point, we can simplify it to one that is easier to analyze. The wire is assumed to be in a uniform state of stress perpendicular to the draw axis. That is, the previously curved surfaces of the volume element from spherical symmetry are now straight. As shown in Figure 4.22, there is an axial tensile stress, σ_x, and in the plane perpendicular, there is a compressive stress, σ_θ, that is the same through the entire rotation in the perpendicular plane (dotted circle). The figure shows this stress in two arbitrarily chosen, mutually perpendicular directions, which serve as example principal directions.

The stress between the die and the wire, perpendicular to the axis, is compressive. If we assign that a positive pressure p to this stress, then the

EXAMPLE 4.8 Consider the drawing of a wire of diameter D_1 to a diameter D_2 according to the assumption of ideal deformation with no friction. Plot the axial and tangential stresses in the deformation zone using the geometry defined in Figure 4.21. The wire is a perfect plastic with a flow stress of Y.

Notice that the apex is at $x = 0$ and therefore the deformation in the die occurs over an axial length from $x = x_2$ to $x = x_1$ or from a diameter of $D = D_2$ to $D = D_1$. For plotting purposes, we can choose $x_2 = 1$ and $x_1 = 1.5$, which corresponds to $D_1 = 1.5$ and $D_2 = 1.0$ ($r = 0.55$). The ratios of the axial and tangential stress to the flow stress (normalized stresses) are:

$$\frac{\sigma_x}{Y} = \ln\left(\frac{x_1^2}{x^2}\right) = \ln\left(\frac{2.25}{x^2}\right) \quad \text{and} \quad \frac{\sigma_\theta}{Y} = \frac{\sigma_x}{Y} - 1$$

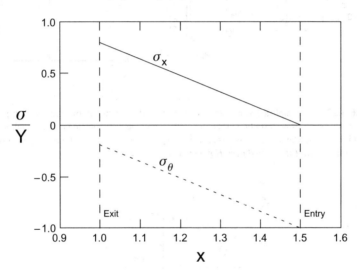

FIGURE E4.8

This plot is an example. It shows that the normalized axial stress is zero at entry and rises to its highest value, which corresponds to the σ_d/Y, at exit. Notice that the draw stress is less than the yield stress of the wire, a requirement for a successful draw. It should be recognized that the combination of axial tension and radial compression leads to deformation at an applied tensile stress that is less than that needed for deformation in uniaxial tension alone (i.e., see Figure 4.15).

stress σ_θ is equal to $-p$. Keeping track of compression as a negative stress and tension (along the axis) as a positive stress is the convention in drawing and useful in specifying the condition for yielding and flow.

To find how the stresses in the wire vary as the wire travels through the die and ultimately, to find the draw stress, a force balance is carried out on a volume element. This procedure, which is sometimes called "slab analysis,"

294 Materials Processing

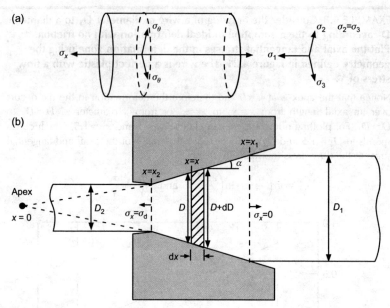

FIGURE 4.22 (a) Simplified stress state for the wire. (b) Schematic of wire drawing process for analysis of stresses with friction at the die wall. A volume element shown at an axial position of x is highlighted. The angle, α, is larger than is typical for clarity. The force is applied on the wire toward the left. *(b) Adapted from Hoffman and Sachs (1953).*

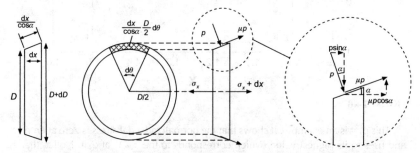

FIGURE 4.23 Schematic diagrams showing dimensions and forces on volume element in Figure 4.22. *Adapted from Meyers and Chalwa (1984).*

is the simplest method for determining metal deformation stresses. Figure 4.22b shows a slice through the volume element and the general geometry of the deformation. Figure 4.23 shows more details about the stresses and geometry of volume element. For this problem, the force balance is carried out in the axial (x) direction. There are the three contributions to sum: stresses from the draw force, the axial component of the pressure, and the axial component of the friction. Friction is assumed to be sliding friction such that an interfacial shear stress that resists motion is given by the product of the friction coefficient, μ, and the pressure, p. The friction and pressure forces act on a ring-like area of contact between the volume element and the die. This area is found by integration:

$$A = \int_0^{2\pi} \frac{dx}{\cos \alpha} \frac{D}{2} d\theta = \frac{\pi D}{\cos \alpha} dx \qquad (4.55)$$

The forces acting in the x direction are found by multiplying the stress on the element face by the area of the face, which in the following is given in square brackets.

$$(\sigma_x + d\sigma_x)\left[\pi\left(\frac{D+dD}{2}\right)^2\right] - \sigma_x\left[\pi\left(\frac{D}{2}\right)^2\right] + \\ p \sin \alpha \left[\frac{\pi D}{\cos \alpha} dx\right] + \mu p \cos \alpha \left[\frac{\pi D}{\cos \alpha} dx\right] = 0 \qquad (4.56)$$

The first two terms arise from the draw force. Neglecting higher order terms, which are very small (e.g., dD^2, $d\sigma_x$, dD), the following can be found for these first two terms:

$$(\sigma_x + d\sigma_x)\left[\pi\left(\frac{D+dD}{2}\right)^2\right] - \sigma_x\left[\pi\left(\frac{D}{2}\right)^2\right] = \frac{\pi D}{4}(Dd\sigma_x + 2\sigma_x dD) \qquad (4.57)$$

The geometry requires a connection between the axial dimension, x, and the diameter, D: $dx = dD/(2 \tan \alpha)$. With that transformation and Eq. (4.57), Eq. (4.56) can be simplified to:

$$Dd\sigma_x + 2\sigma_x dD + 2pdD\left(1 + \frac{\mu}{\tan \alpha}\right) = 0$$

A constant, B, is defined as:

$$B = \frac{\mu}{\tan \alpha} = \mu \cot \alpha \qquad (4.58)$$

Therefore:

$$Dd\sigma_x + 2\sigma_x dD + 2p(1+B)dD = 0 \qquad (4.59)$$

To solve this differential equation and discover how σ_x and p vary with position in the die (as represented by the local diameter, D), the yielding condition must be used to eliminate one of the unknowns.

The choice of yielding condition requires a simplification. A principal stress state requires that there are no shear stresses on the faces; however, assuming that the shear stresses due to friction are small and that the angle, α, is also small, then we can select the principal stresses (σ_1, σ_2, σ_3) simply based on the volume element. The axial stress is the first principal stress ($\sigma_1 = \sigma_x$) and the other two principal stresses are equal to each other $\sigma_2 = \sigma_3 = \sigma_\theta = -p$. Since $\sigma_2 = \sigma_3$ and $\sigma_1 > \sigma_2$, the condition for yielding is the same for both the Tresca and von Mises criteria: $\sigma_1 - \sigma_2 = Y$. Therefore, the yield condition is:

$$\sigma_x + p = Y \tag{4.60}$$

Substituting into Eq. (4.59),

$$\frac{d\sigma_x}{\sigma_x B - Y(1+B)} = 2\frac{dD}{D} \tag{4.61}$$

This equation can now be solved by integration, noting that the axial stress at the entry ($D = D_1$) is zero:

$$\sigma_x = Y\left(\frac{1+B}{B}\right)\left[1 - \left(\frac{D^2}{D_1^2}\right)^B\right] \tag{4.62}$$

$$p = Y - \sigma_x$$

The draw stress is the stress at a diameter position of D_2; therefore:

$$\sigma_d = Y\left(\frac{1+B}{B}\right)\left[1 - \left(\frac{D_2^2}{D_1^2}\right)^B\right] \quad \text{(perfect plastic)} \tag{4.63a}$$

$$\sigma_d = \overline{Y}\left(\frac{1+B}{B}\right)\left[1 - \left(\frac{D_2^2}{D_1^2}\right)^B\right] \quad \text{(strain hardening)} \tag{4.63b}$$

It is important to note that Eqs. (4.62) and (4.63a, 4.63b) are not valid when B is zero. The ideal equation, Eq. (4.47), is used in this case.

The effect of friction and reduction on the drawing process is shown in Figure 4.24. As anticipated, the draw stress increases with reduction, reaching a limit where the draw stress reaches the flow stress (yield stress) and plastic deformation of the wire on exit occurs. The draw stress also increases with an increase in the coefficient of friction for a given reduction. As such, less total deformation is possible before reaching the limit where $\sigma_d = Y$ (Example 4.9).

Other factors. A final factor to consider in the prediction of draw stress is redundant work or nonuniform deformation. The previous derivations assumed uniform plastic deformation. That is, the extent of deformation did not vary from the surface to the middle of the wire. Figure 4.25 compares uniform and nonuniform deformation in a drawn wire. The term "redundant work" or "redundant deformation" refers to the fact that work is put into the metal to make these internal distortions and hence we expect the draw stress to increase in a wire that experiences nonuniformity compared to one that is uniform.

Unfortunately, there are no simple models to predict the effect of nonuniform deformation on the draw stress. However, empirical evidence and more detailed analytical analyses have led to some conclusions. Importantly, the effect of nonuniform deformation on the draw stress increases with the angle α.

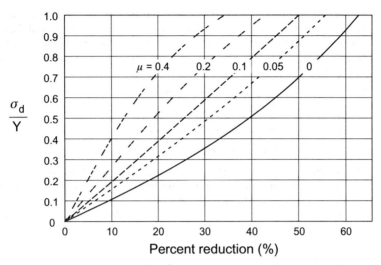

FIGURE 4.24 Effect of friction coefficient and amount of reduction on draw stress for a perfectly plastic metal with a flow stress of Y. The extreme of $\sigma_d = Y$ represent the limitation of plastic deformation of the wire after exits. The data shown was calculated for a die with $\alpha = 6°$ and using Eq. (4.63a) for the case with friction and Eq. (4.47) for the frictionless (ideal) case.

EXAMPLE 4.9 Assume deformation with a friction coefficient of 0.1 and a die with $\alpha = 5°$ when answering the following. (a) Find the draw force needed for a 30% reduction of a 2 mm diameter annealed aluminum wire with K = 180 MPa and n = 0.2. (b) If the wire is travelling at 2 m/s through the die, find the power required for the process described in (a). (c) Find the smallest diameter die that can be successfully used to draw the 2 mm diameter annealed wire described in part (a).

a. As in Example 4.7, a 30% reduction (r = 0.30) requires a die with diameter (D_2) of 1.64 mm and a strain of 0.357. The average flow stress during the deformation is 122 MPa. Given that $\mu = 0.1$ and $\alpha = 5°$, then

$$B = \frac{\mu}{\tan \alpha} = \frac{0.1}{\tan 5°} = 1.143$$

$$\sigma_d = \overline{Y}\left(\frac{1+B}{B}\right)\left[1 - \left(\frac{D_2^2}{D_1^2}\right)^B\right]$$

$$= (122 \text{ MPa})\left(\frac{1+1.143}{1.143}\right)\left[1 - \left(\frac{1.64^2}{2.0^2}\right)^{1.143}\right] = 83.4 \text{ MPa}$$

$$F = \sigma_d A_2 = (83.4 \text{ MPa})(2.12 \text{ mm}^2) = 176.9 \text{ N}$$

The draw stress and draw force are nearly double relative to the values from Example 4.7 for the frictionless ideal case. One might wonder if the wire will yield after exit, so we can check to make sure we have not violated that criterion. We can find this value by recalling that $\varepsilon_x = \bar{\varepsilon} = \varepsilon$ and therefore, we can plug in the maximum axial strain into strain hardening law and find the flow stress exit, which is equivalent to the yield strength of the freshly drawn wire,

$$Y_f = K\varepsilon_{max}^n = 180(0.357)^{0.2} = 146.5 \text{ MPa}$$
$$Y_f > \sigma_d$$

b. The power is the force multiplied by the velocity.

$$\text{Power} = Fv = (176.9 \text{ N})(2 \text{ m/s}) = 353.8 \text{ W}$$

c. The maximum extent of reduction is related to yielding of the wire after exiting. For a work hardening metal, we should compare the yield strength of the metal after exit with the draw stress.

The yield strength of the wire on exit is the flow stress at maximum deformation:

$$Y_f = K\varepsilon_{max}^n \quad \text{where} \quad \varepsilon_{max} = \ln\left(\frac{D_1^2}{D_2^2}\right) \quad \text{and} \quad D_1 = 2 \text{ mm}$$

The draw stress is:

$$\sigma_d = \bar{Y}\left(\frac{1+B}{B}\right)\left[1 - \left(\frac{D_2^2}{D_1^2}\right)^B\right]$$

where $\bar{Y} = \dfrac{K\varepsilon_{max}^n}{n+1}$ (initial draw from annealled state)

Equating the two:

$$K\varepsilon_{max}^n = \left(\frac{K\varepsilon_{max}^n}{n+1}\right)\left(\frac{1+B}{B}\right)\left[1 - \left(\frac{D_2^2}{D_1^2}\right)^B\right]$$

Substituting in for n, B, and D_1 and solving gives $D_2 = 1.24$ mm, resulting in a reduction of 59%. Note that this relationship does not include a safety factor, which would be advisable in a practical application.

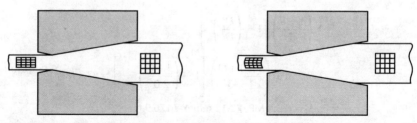

FIGURE 4.25 Comparison of uniform (left) and nonuniform (right) deformation during wire drawing. The final amount of reduction is the same, but in the case of nonuniform deformation, there is internal distortion in the metal. *Adapted from Hosford and Caddell (1993).*

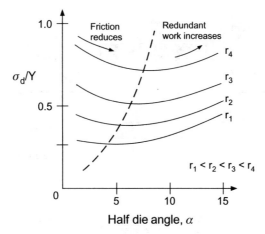

FIGURE 4.26 Effect of die angle on draw stress requirements for various reductions, r. *Adapted from Wistreich (1955)*.

This trend is in contrast with the effect of friction. As die angle increases, the contact length between the die and the wire decreases and hence frictional resistance decreases. So, there is an "optimum" die angle based on this compromise where the combined effects friction and redundant work are minimized. See Figure 4.26. The optimum value of α is typically around 6° to 10°.

In the practical application of wire drawing, the wire is reduced gradually through a series of dies that are arranged in a process line. Since the drawing conditions vary from die to die, intermediate capstands are used to collect the wire on a spool and feed it to the next process step. Lubricants, including liquid baths, are used to reduce friction and sometimes dissipate heat.

4.3.3 Extrusion

Extrusion deforms a solid billet by pushing it through a die to create a constant cross-section shape or profile. The process has some similarities to the extrusion of thermoplastic polymer melts, but extrusion of solids is a batch process rather than a continuous one and involves solid deformation rather than melt flow and solidification. Like the other processes discussed in this section, solid extrusion is used for processing metal almost exclusively.

Metal extrusion can be carried out under cold or hot working conditions. Hot working conditions are needed for gross deformation of a billet into a profile with a highly reduced cross-sectional area relative to the original billet. Therefore, the technique has widespread application particularly for nonferrous metals, such as aluminum, due to their lower hot working temperature requirements compared to steel and other ferrous alloys. Cold working, however, is preferred in extrusions requiring high precision or the strength advantage gained from working hardening during the deformation.

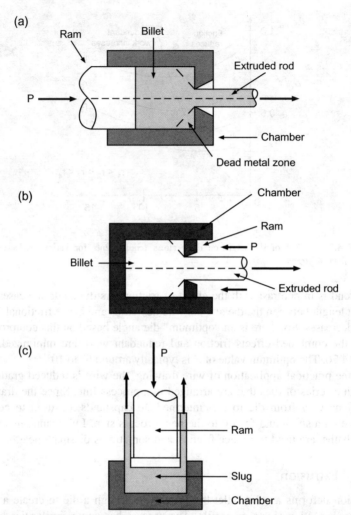

FIGURE 4.27 Three variations on solid extrusion (a) forward or direct extrusion, (b) backward or indirect extrusion and (c) impact extrusion. *From Beddoes and Biddy (1999), reproduced by permission from Elsevier.*

Frequently, a hot working process precedes a final cold working process, sometimes called finishing.

Three types of extrusion processes are shown in Figure 4.27. Forward or direct extrusion is the simplest. A metal billet is pushed through a die. The constraint of the chamber and die produces compressive stress and yielding. While dies can have engineered converging cross-sections, similar to wire drawing dies, simple 90° dies can be used and a natural "dead zone" develops during deformation, resulting in the channeling of the metal to the die exit. Backward or indirect extrusion involves pushing a shaped ram into the billet. In this process, the chamber holds the billet and a dead zone also develops. In a similar process, impact extrusion, a ram is forced into a billet and the extruded metal flows on the outside of the ram to produce a hollow shape. Another process, hydrostatic extrusion (not shown), employs a fluid between the work piece and the die, and the motion is analogous to forward extrusion.

Friction is a key concern in extrusion processes. Examination of the schematic diagrams in Figure 4.27 reveals considerable contact area between the billet and the chamber and hence large frictional effects in forward extrusion. Backward extrusion limits the friction to the moving ram and the billet. In both forward and backward extrusion, the die may be angled or a dead zone developed to limit contact. Hydrostatic extrusion with its intervening liquid limits friction almost entirely. This process is used for cold extrusion processes.

The mechanics of extrusion have some similarity to those of wire drawing. Starting with an energetic analysis, consider a rod with diameter D_1, cross-sectional area A_1, and length L_1 that is deformed by extrusion to a rod with diameter D_2, cross-sectional area A_2 and length L_2. For this deformation, the extrusion ratio, R, is defined as

$$R = \frac{A_1}{A_2} = \frac{D_1^2}{D_2^2} \tag{4.64}$$

and the axial strain associated with the length increase due to the deformation is:

$$\varepsilon = \ln\left(\frac{A_1}{A_2}\right) = \ln R \tag{4.65}$$

The extrusion ratio is the preferred metric for this process as compared to the use of the reduction, r, which is used for wire drawing. This preference allows large area reductions that are common in extrusion processes to be easily discerned. For example, a 20-to-1 and a 40-to-1 area reduction have R values of 20 and 40, respectively, while they have r values of 0.95 and 0.975, respectively.

The compressive stress required for deformation is generated by a force pushing on the billet (area A_1). By convention, this compressive stress is expressed as a positive pressure, p. Following the same analysis as wire drawing (noting that this process is also axisymmetric with the axial strain equal to the effective strain, $\bar{\varepsilon} = \varepsilon$), the required pressure for deformation of a perfect plastic metal with flow stress, Y, is

$$p = \frac{F}{A_1} = Y \ln R \tag{4.66}$$

or for a work hardening metal

$$p = \bar{Y} \ln R \tag{4.67}$$

> **EXAMPLE 4.10** Find the extrusion ratio corresponding to uniaxial compressive plastic flow of the billet. Assume no friction.
>
> Because there is no friction in this case, we can use the energy analysis result
>
> $$p = Y \ln R$$
>
> Yielding of the billet occurs at $p = Y$ for uniaxial compression and so
>
> $$\ln R = \ln\left(\frac{A_1}{A_2}\right) = \ln\left(\frac{D_1^2}{D_2^2}\right) = 1$$
>
> Therefore $R = 2.7$ and the extruded rod area A_2 is $0.37 A_1$ and the extruded diameter D_2 is $0.61 D_1$.

Therefore, in addition to the size of the billet, a central factor influencing the force needed for extrusion is the extent of area reduction in the process, R. Example 4.10 shows a special case.

The practical limitation in an extrusion process, the maximum area reduction, is different than that for wire drawing. Recall in wire drawing, plastic deformation of the exiting wire under tension sets the limit for the maximum reduction per die. In extrusion, the analogous limitation would be the plastic deformation of the entering rod or billet under compression; however, this deformation is not an issue in the process but in fact a common occurrence. The chamber walls prevent the expansion of the billet as it plastically deforms under compression. Instead, the strength of the chamber to withstand those radial forces is one limitation; another is the maximum force that can be generated by the ram. The size of the billet is another concern. Billets that are long and narrow may buckle when they are pushed against the die.

When friction is included, a more detailed description of the mechanics is required. In the simplest example, forward extrusion through a conical die can be analyzed as the reverse of wire drawing with a force applied on the material entering the die rather than on the exiting material. Considering this scenario and referring to Figure 4.22, the slab analysis with a die having an angle, α, and $B = \mu/\tan \alpha$ gives the following expression for extrusion pressure:

$$p = Y\left(\frac{1+B}{B}\right)\left[1 - \left(\frac{D_1^2}{D_2^2}\right)^B\right] \tag{4.68}$$

For a work hardening material, $\overline{Y}$ would be used in place of Y. This analysis, however, does not take into account the finite size of the billet, the influence of the chamber, particularly friction and the possibility of plastic deformation before the die, or the extensive redundant work that is typical for extrusion.

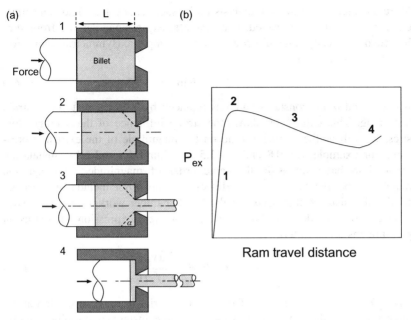

FIGURE 4.28 Change in extrusion pressure with ram travel: (a) sequence of ram positions and (b) extrusion pressure as a function of ram travel.

The frictional and redundant work effects can be understood qualitatively using Figure 4.28. At the beginning of the process, the extrusion pressure climbs as the billet is compressed in the chamber, reaching a peak pressure just at the point where the extrudate emerges from the die. The peak pressure required depends on a variety of factors, including the mechanical properties of the billet at the temperature of extrusion, and the extrusion ratio, but also friction and inhomogeneous deformation or redundant work. The friction between the billet and the chamber can be lowered with a lubricant such that sliding conditions prevail. However, the internal friction at the "dead zone" is a sticking friction process in which there is internal shear yielding between the moving billet and immobilized dead zone metal. Clearly, there is considerable nonuniformity in the deformation, leading to redundant work. The extrusion pressure drops as the extrusion continues because the interfacial area that causes the friction drops as the ram travels. This trend continues until the very end of the process when the last section of the billet is compressed into the die. Here the flow is disrupted and the metal must squeeze between the ram and the end of the chamber to the exit. In practice the process is halted before all the metal exits.

Developing more in-depth predictions of the pressure and hence force requirements for extrusion processes is challenging. The slab method has only limited application because it does not account for redundant work.

More advanced deformation analysis tools, such as slip-line fields and finite element modeling, are needed. For example, expressions derived from slip-line fields of steady state, direct extrusion of simple axisymmetric shapes follow this general form:

$$p = Y(a + b \ln R) \tag{4.69}$$

where a and b are constants and Y is replaced by $\overline{Y}$ is the material is strain hardening. This expression shows the strong influence of the average flow stress as well as the extrusion ratio on the magnitude of the extrusion pressure. For example, $a = 0.8$ and $b = 1.5$ for aluminum and other metals are reported to have values in the same order of magnitude. This equation matches experimental data well. However, it does not include the variable frictional resistance that occurs at the billet/chamber workpiece but an extra term can be included for that effect. A complete expression for extrusion pressure has an extra term:

$$p = Y(a + b \ln R) + \frac{4Y_t L}{\sqrt{3}D} \tag{4.70}$$

where Y_t is the yield strength of the billet material in uniaxial tension at the temperature of the extrusion process, L, is the instantaneous billet length, and Y is replaced by $\overline{Y}$ if the material is strain hardening. See Example 4.11 for the derivation. Notice that the second term in the equation diminishes as the extrusion proceeds and L decreases.

The extrusion of more complex cross-sections, such as those shown in Figure 4.29, requires additional consideration. First, experimental evidence suggests that the extrusion pressure increases with the extrusion ratio, which is calculated using the ratio of the billet cross-sectional area to the total cross-sectional area in the die opening. However, complexity typically has extra penalties. One way to estimate the addition work required to form a complex shape is to modify the extrusion ratio by a factor related to the geometry.

$$R' = \left(\frac{L_{\text{total, complex}}}{\pi D_{\text{round}}}\right) R \tag{4.71}$$

where R is the extrusion ratio based on the ratio of the billet to extrusion cross-section areas, R' is the modified extrusion ratio, D_{round} is the diameter of a round bar with the same R as the complex extrusion and $L_{total,\ complex}$ is the total length of the exterior and interior perimeters of the complex shape (Example 4.12).

Hot extrusion allows for more deformation and higher extrusion ratios. Analysis is complicated by the transient nature of the temperature; not only is the hot metal subject to cooling due to heat transfer to the surroundings, but also to temperature increases during the deformation process. Qualitatively, the average flow stress drops with temperature but the strain rate sensitivity exponent increases. For a given temperature and extrusion ratio, the extrusion pressure

EXAMPLE 4.11 Use the slab method to find an expression for the maximum contribution of friction at the billet/chamber interface to the extrusion pressure as a function of the extent of ram travel or instantaneous billet length, L. The geometry is presented below.

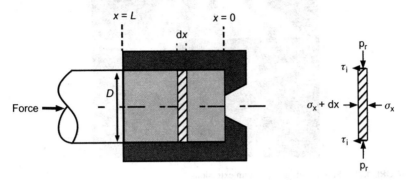

FIGURE E4.11

The force balance for the frictional effect as the billet moves in the chamber is:

$$(\sigma_x + d\sigma_x)\pi\left(\frac{D}{2}\right)^2 - \sigma_x\pi\left(\frac{D}{2}\right)^2 - \pi D \tau_i dx = 0$$

$$Dd\sigma_x = 4\tau_i dx$$

With sliding friction, $\tau_i = \mu p_r$, where p_r is the pressure exerted by the chamber around the radius. The maximum frictional effect corresponds to the "sticking" condition when there is deformation at the interface. This condition occurs when $\mu p_r > k$, where k is the shear yield stress. For the von Mises yield criterion, the shear yield strength is equal to the yield strength in uniaxial tension divided by the $\sqrt{3}$. In the expression below, the yield strength in uniaxial tension is noted by Y_t so that it may be distinguished from the flow stress involved the extrusion process.

$$\tau_i = k = \frac{Y_t}{\sqrt{3}}$$

$$Dd\sigma_x = \frac{4Y_t}{\sqrt{3}} dx$$

To solve this equation, we need to keep in mind that the goal here is to find the contribution of the billet/chamber interfacial friction to the extrusion pressure, not the total extrusion pressure. Hence, this contribution appears in the stress in the axial or x direction. The integration is from $x = 0$ and $\sigma_x = 0$ to $x = L$ and $\sigma_x = p_{ex,f}$, which is the frictional term in Eq. 4.70:

$$\int_0^{p_{ex,f}} Dd\sigma_x = \int_0^L \frac{4Y_t}{\sqrt{3}} dx$$

$$p_{ex,f} = \frac{4Y_t L}{\sqrt{3} D}$$

FIGURE 4.29 Photograph of aluminum extrusions.

EXAMPLE 4.12 (a) A billet of an Al alloy with a diameter of 10 cm is cold extruded to a rod with a diameter of 8 cm. Estimate the extrusion pressure and required extrusion force using Eq. (4.67) and using Eq. (4.69) with $a = 0.88$ and $b = 1.5$. (b) Compare the extrusion requirements for extruding from the same size billet through a die to create a square cross-section 5×5 cm with a central hole with a diameter of 2.5 cm to those of a round bar of the same cross-section area. The Al alloy has a $K = 150$ MPa and $n = 0.24$.

a. The first step is to find the extrusion ratio and the strain.

$$R = \frac{A_1}{A_2} = \frac{D_1^2}{D_2^2} = \frac{100}{64} = 1.56$$

(Note: $A_1 = \frac{\pi \cdot 10^2}{4} = 78.5$ cm^2 and $A_2 = \frac{\pi \cdot 8^2}{4} = 50.2$ cm^2)

$$\varepsilon = \ln R = \ln(1.56) = 0.44$$

This axial strain at the end of the extrusion process is 0.44; therefore, the average flow stress can be found using the mechanical properties of the alloy.

$$\overline{Y} = \frac{1}{\varepsilon_2 - \varepsilon_1} \int_{\varepsilon_1}^{\varepsilon_2} \sigma d\varepsilon = \frac{1}{0.44 - 0} \int_{0}^{0.44} K\varepsilon^n d\varepsilon = \frac{1}{0.44}\left[\frac{K\varepsilon^{n+1}}{n+1}\right]_0^{0.44}$$

$$= \frac{1}{0.44}\left[\frac{(150 \text{ MPa})(0.44)^{1.24}}{1.24}\right]$$

$$= 99.3 \text{ MPa}$$

Then, the extrusion pressure (lower bound, ideal) is given by Eq. (4.67):

$$p = \overline{Y} \ln R = (99.3 \text{ MPa})(0.44) = 43.7 \text{ MPa}$$

$$F = pA = (43.7 \times 10^6 \text{ Pa})\left(\frac{\pi(0.1 \text{ m})^2}{4}\right) = 3.43 \times 10^5 \text{ N} = 0.343 \text{ MN}$$

Using Eq. (4.69), which accounts for friction

$$p = \overline{Y}(a + b \ln R) = 99.3 \text{ MPa}[0.88 + 1.5(0.44)] = 153 \text{ MPa}$$

$$F = pA = (153 \times 10^6 \text{ Pa})\left(\frac{\pi(0.1 \text{ m})^2}{4}\right) = 1.20 \times 10^6 \text{ N} = 1.2 \text{ MN}$$

b. This extrusion has a complex cross-section – a 5×5 cm square with a 2.5 cm diameter hole.

$$A_1 = 78.5 \text{ cm}^2 \quad \text{and} \quad A_2 = 5^2 - \frac{\pi \cdot 2.5^2}{4} = 20.1 \text{ cm}^2$$

$$R = \frac{A_1}{A_2} = \frac{78.5}{20.1} = 3.9$$

The diameter of a round rod with the same R would be close to the situation in part (a):

$$A_2 = \frac{\pi \cdot D^2}{4} = 20.1 \text{ cm}^2 \rightarrow D = 5.06 \text{ cm}$$

The perimeter of the complex die is $4 \times 5 + \pi(2.5) = 27.85$ cm. So the new extrusion ratio can be found:

$$R' = \left(\frac{L_{\text{total, complex}}}{\pi D_{\text{round}}}\right) R = \left(\frac{27.85}{\pi \cdot 5.06}\right) 3.9 = 6.83$$

For the comparison consider the ratio of the extrusion pressure to the average flow stress. For the complex cross-section:

$$p/\overline{Y} = \ln R' = 1.92$$

For a round bar with an equivalent cross-sectional area:

$$p/\overline{Y} = \ln R = 1.36$$

Hence the complexity boosts the pressure requirement as the two examples would have the same axial strain and hence the same average flow stress.

---

increases with the extrusion speed. There is a maximum velocity beyond which the press can no longer supply the needed force. Or, we could consider a fixed velocity and the press would place an upper limit on the extrusion ratio. Another limitation is surface quality of the extrudate. For a given temperature, formation of surface cracks also limits the velocity at constant extrusion ratio (or the extrusion ratio at constant velocity).

The limitations in conditions of hot extrusion can be displayed on an extrusion limit diagram, as shown in Figure 4.30. At fixed extrusion ratio, the capacity of the extruder limits the speed at low temperature and surface quality limits the speed at high temperature. From Figure 4.30a, the reduction in the flow stress with temperature is a more significant factor than the

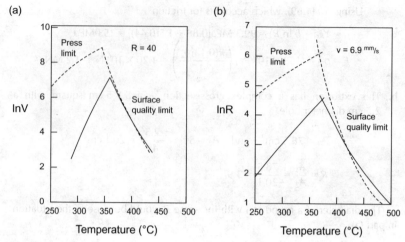

FIGURE 4.30 Extrusion limit diagrams for aluminum alloy 2024 showing the effect of temperature on (a) highest speed to achieve $R = 40$ and (b) largest R possible for an extrusion velocity of 6.8 mm/s. The dotted lines represent indirect extrusion and the solid lines are direct extrusion. *Adapted from Sheppard (1993).*

strain rate sensitivity increase, at least in this case, as the velocity limit due to press capacity increases with temperature. However, the surface cracking issues take over at higher temperature, leading to an optimum temperature for high velocity extrusion. Similarly, conditions to maximize the extrusion ratio at a fixed velocity can be found (Figure 4.30b).

4.3.4 Forging

Forging processes involve the deformation of a workpiece by compressive forces to form a new shape. There are two general types of forging processes: open die and closed die. In open die forging, the workpiece is compressed between parallel plates or two platens with no restrictions in the deformation. Material is free to "flow" out between the open edges. See Figure 4.31. This form of forging is used to reduce specimen height. In hot deformation mode, open die forging is also known as "upsetting."

Closed die forging usually involves a series of shaped dies. Multiple deformations progressively shape the metal from a crude form to a more detailed shape. In one variant of closed die forging, impression die forging, excess metal is used so that features are filled in the shaped die and as a result of the excess, "flash," is pushed out between the die halves. Post-forming machining is used to remove this excess. Another variant, precision closed die forging, uses an exact amount of metal such that no flash is formed. Like open die forging, closed die processes are usually carried out under hot working conditions.

FIGURE 4.31 Photos of open die hot forging. *Courtesy of Scot Forge, www.scotforge.com.*

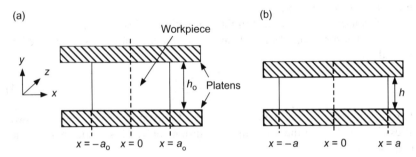

FIGURE 4.32 Schematic of plane strain open die forging (a) initial configuration and (b) after some deformation. Deformation is restrained in the z-direction.

Forging forces are applied with a hammer, which is raised and dropped on the workpiece, or a mechanized press (e.g., hydraulic press). Much like a blacksmith operation, multiple deformations or drops of the hammer can be used. As the workpieces themselves may be large, the forging equipment can be massive (see Figure 4.31), applying up to several tens of thousands of tons in a single step.

In this section, forging is explored starting with the simple situation, open die forging with no friction. From the simple case, we learn the parameters that control forging, in particular the factors that influence the forces required for deformation and the temperature rise. Then the role of friction is explored. Lastly, closed die forging is treated in a more qualitative fashion.

Open Die Forging in Plane Strain with No Friction. Figure 4.32 shows the open die deformation of a rectangular workpiece in which one of the lateral dimensions is kept constant (plane strain). When the forging force is applied, there is a reduction in thickness (y-direction). By convention, the definition of true strain is inverted so that the reduction is noted as positive strain.

$$\varepsilon_y = \ln\left(\frac{h_o}{h}\right) \tag{4.72}$$

There is also lateral expansion in the x-direction and no strain in the z-direction ($\varepsilon_z = 0$). Because of volume conservation, the lateral expansion, ε_x, is equal and opposite to that of the thickness reduction strain, ε_y, so $\varepsilon_x = -\varepsilon_y$. Therefore, the effective strain for this process is found using Eq. (4.26) to be $\bar{\varepsilon} = 2/\sqrt{3}\varepsilon_y$.

Force is placed on the platen to accomplish the forging. Without friction, the stress is uniform under the platen. This stress is called a pressure ($\sigma_y = p$). Note that in this case, we define compressive stress as a positive value.

$$\sigma_y = p = \frac{\text{force}}{\text{instantaneous area}} \tag{4.73}$$

There are no forces acting in the x-direction because there is no friction. Therefore, $\sigma_x = 0$. Note now that σ_x, σ_y, and σ_z are principal stresses because there are no shear stresses.

Deformation is constrained in the z-direction and so there must be a force acting in that direction. As noted earlier, for plane strain, σ_z is determined by σ_x and σ_y:

$$\sigma_z = \frac{\sigma_y + \sigma_x}{2} \tag{4.74}$$

For the frictionless case, therefore, $\sigma_z = \sigma_y/2$. Given this stress state, the flow stress is found using the von Mises condition for yielding under plane strain (Eq. (4.31)):

$$\sigma_y - \sigma_x = \frac{2}{\sqrt{3}} Y = Y' \tag{4.75}$$

Y' is known as the plane strain flow stress. In the case of frictionless plane strain forging, $\sigma_x = 0$, and therefore yielding takes place when $\sigma_y = p = Y'$. This same condition can be derived using the effective stress. From Eq. (4.24) the effective stress is found to be $\bar{\sigma} = \sqrt{3}/2\sigma_y$. So yielding occurs when the effective stress equals Y or in other words when $\sigma_y = 2/\sqrt{3}Y = Y'$.

For a perfectly plastic material (without strain hardening), deformation begins when the pressure reaches Y' and then continues at constant pressure (or stress). However, the force required for continued plastic deformation increases because the area increases with deformation. The forging force needed to achieve a final deformed area of A_{final} and a final thickness of h_{final} is:

$$F = pA_{final} = Y'A_{final} \quad \text{(perfect plastic)} \tag{4.76a}$$

$$F = pA_{final} = Y'_f A_{final} \quad \text{(strain hardening)} \tag{4.76b}$$

For a strain hardening material, deformation begins when the pressure reaches Y' and then increases (Y'_f) as deformation continues and therefore the force needed for forging rises even more than in the perfect plastic case. Example 4.13 examines the forging force and temperature rise during frictionless forging.

Open Die Forging in Plane Strain with Friction. Friction between the workpiece and the platens raises the force needed for deformation and complicates the analysis. Consider a plane strain deformation of a rectangular workpiece, but this time with friction. See Figure 4.33. As in the frictionless case,

---

EXAMPLE 4.13 Consider frictionless, plane strain forging of a slab with initial dimensions of 10 cm × 5 cm × 2 cm such that the 2 cm dimension is kept constant and the thickness drops from 5 cm to 4 cm. (a) Calculate the force needed to hot forge an alloy that displays perfect plastic behavior with a yield stress (flow stress) of 100 MPa. (b) Calculate the force needed to forge an Al aluminum alloy (1100-O, K = 180 MPa, n = 0.2, density = 2700 kg/m³, specific heat = 904 J/(kg · K). Also, estimate the temperature rise.

a. Using volume conservation, the final dimensions of the piece will be 12.5 × 4 cm × 2 cm. So the final contact area is 12.5 cm × 2 cm. To find the forging force, the force needed to reach the final state is found. The flow stress is 100 MPa and it does not vary as deformation proceeds and so,

$$F = Y'A = \frac{2}{\sqrt{3}}(100 \times 10^6 \text{ Pa})(0.125 \text{ m} \times 0.02 \text{ m}) = 2.88 \times 10^5 \text{ N}$$

b. For the aluminum alloy that displays strain hardening behavior, the flow stress is not constant during deformation and hence we need to find the flow stress Y'_f at the end of the forging process in order to calculation the force requirement. The final strain in the thickness direction is:

$$\varepsilon_y = \ln\left(\frac{5}{4}\right) = 0.223$$

To calculate the Y'_f, we need to use the strain hardening law and so the effective strain is needed:

$$\bar{\varepsilon} = \frac{2}{\sqrt{3}}\varepsilon_y = 0.257$$

The value of $\sigma_y = p = Y'_f$ to cause the deformation to that level of strain is:

$$Y'_f = \frac{2}{\sqrt{3}}K\bar{\varepsilon}^n = \frac{2}{\sqrt{3}}\left[180(0.257)^{0.2}\right] = 158.4 \text{ MPa}$$

[Another way to arrive at this answer is to find the effective stress and convert it back to σ_y].

And the force can be found using the final area:

$$F = Y'_f A_{\text{final}} = (158.4 \text{ MPa})(0.02 \text{ m} \times 0.125 \text{ m}) = 3.96 \times 10^5 \text{ N}$$

The work of deformation per unit volume (or energy of deformation) is:

$$u = \int_0^{\bar{\varepsilon}_{max}} \bar{\sigma} d\bar{\varepsilon} = \int_0^{\bar{\varepsilon}_{max}} K\bar{\varepsilon}^n d\bar{\varepsilon} = \frac{K\bar{\varepsilon}_{max}^{n+1}}{n+1} = \frac{(180)(0.257)^{1.2}}{1.2} = 29.4 \text{ MPa} = 29.4 \text{ J/m}^3$$

For the frictionless case, this value is equivalent to u_{total} and we can find the temperature rise.

$$\Delta T = \frac{u_{total}}{\rho C_p} = \frac{29.4 \times 10^6 \text{ J/m}^3}{(2700 \text{ kg/m}^3)(904 \text{ J/(kg·K)})} = 12 \text{ K}$$

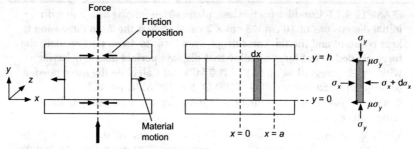

FIGURE 4.33 Set up for slab analysis of plane strain open die forging with friction. *Adapted from Kalpakian (1997)*.

workpiece is constrained in the z-direction ($d\varepsilon_z = 0$). Now, with friction, the stress in the y-direction, σ_y, the pressure, is not constant but varies with lateral position x. To find σ_y, σ_x, and σ_z, a slab analysis is used once again.

The slab analysis begins with examining a slice in the workpiece at width dx, as shown in Figure 4.33 (center). The slice is displaced from the center and therefore friction acts at the interfaces between the slice and the platens. Friction opposes the material flow, resulting in an interfacial shear stress, $\mu\sigma_y$, where μ is the friction coefficient and σ_x is a normal stress in the x-direction. A force balance of the slice in the x-direction gives:

$$(\sigma_x + d\sigma_x)h + 2\mu\sigma_y dx - \sigma_x h = 0 \quad (4.77)$$

$$d\sigma_x + \frac{2\mu\sigma_y}{h} dx = 0 \quad (4.78)$$

In this expression, the width of the slice in the two directions is present in all terms and is therefore eliminated.

Equation (4.78) has two unknowns: σ_x and σ_y. Therefore, another relationship is needed to solve the problem. Examining the coordinate system, we define the principal stresses and then find a yield criteria for the second equation. One approximation is to ignore the shear stress and define σ_x, σ_y, and σ_z as shown as the principal stresses. As in the frictionless case, the deformation is plane strain with the yield criterion given by the von Mises criterion.

$$\sigma_y - \sigma_x = \frac{2}{\sqrt{3}} Y = Y' \tag{4.79}$$

This equation requires $d\sigma_y = d\sigma_x$. Therefore,

$$d\sigma_y + \frac{2\mu\sigma_y}{h} dx = 0 \tag{4.80}$$

$$\frac{d\sigma_y}{\sigma_y} = \frac{-2\mu}{h} dx \tag{4.81}$$

This equation is solved by integration with the boundary condition: at $x = a$ (edge), $\sigma_x = 0$, so $\sigma_y = Y'$ (yield condition). The solution gives the pressure, σ_y, also known as p, as a function of position, x. For a perfect plastic,

$$\sigma_y = p = Y' \exp\left[\frac{2\mu(a-x)}{h}\right] \quad \text{(perfect plastic)} \tag{4.82}$$

The axial stress is given by:

$$\sigma_x = \sigma_y - Y' = Y'\left(\exp\left[\frac{2\mu(a-x)}{h}\right] - 1\right) \quad \text{(perfect plastic)} \tag{4.83}$$

Figure 4.34 shows the local pressure under the platen as a function of lateral position away from the center at the workpiece for various levels of friction. The pressure is highest directly under the center ($x = 0$) and lowest at the edges ($x = \pm a$). One way to understand this distribution is to consider that the deformation of material directly in the center requires pushing the workpiece from $x = 0$ to $x = a$, which involves overcoming the friction along the whole distance. By contrast, deforming the material nearer to the edge requires lower stress because there is less frictional opposition. In fact, the dependence of p on lateral position is called a "friction hill."

The uneven pressure beneath the platen presents a challenge for engineering the forging process. Only a single forging force is applied. Therefore, an average pressure is needed to find the forging force necessary for the deformation. See Example 4.14. An approximation is typically used:

$$p_{av} \cong Y'\left(1 + \frac{\mu a}{h}\right) \quad \text{(perfect plastic)} \tag{4.84}$$

The forging force is then the average pressure multiplied by the area of the workpiece that is in contact with the platen at end of the forging process:

$$F = p_{av}(2a)(\text{width}) \tag{4.85}$$

For forging, the force required increases with the amount of deformation, or in other words, the height reduction. For a perfect plastic, this increase is due to the increase in the contact area and the increasing role of friction as the specimen deforms.

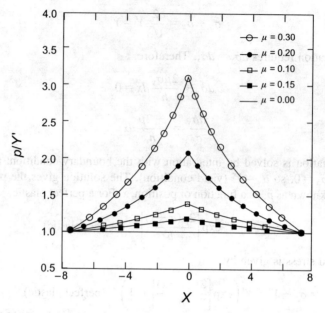

FIGURE 4.34 Example calculations showing effect of friction coefficient on distribution of local pressure beneath an open die forging platen ($a = 7.5$, $h = 4$).

EXAMPLE 4.14 Find the average pressure under a platen during plane strain forging and show how the approximation is derived.

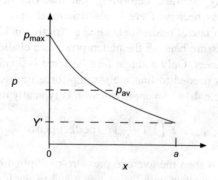

FIGURE E4.14

At $x = 0$, $p = p_{max}$ and at $x = a$, $p = Y'$. From Eq. (4.82),

$$p = Y' \exp\left(\frac{2\mu(a-x)}{h}\right)$$

Noting that the pressure is uniform in the width (z) direction, p_{av} is found by integration:

$$p_{av} = \frac{1}{a}\int_{x=0}^{x=a} Y' \exp\left(\frac{2\mu(a-x)}{h}\right) dx$$

The exact solution to this integral is: $p_{av} = \frac{hY'}{2\mu a}[\exp(2\mu a/h) - 1]$.

Using a series expansion of the exponential function, the solution can be re-written as:

$$p_{av} = \frac{hY'}{2\mu a}\left[1 + \frac{2\mu a}{h} + \frac{1}{2!}\left(\frac{2\mu a}{h}\right)^2 + \frac{1}{3!}\left(\frac{2\mu a}{h}\right)^3 + \cdots + \frac{1}{n!}\left(\frac{2\mu a}{h}\right)^n - 1\right]$$

For simplification, all terms involving a power greater than two are ignored (for small $2\mu a/h$) and the above expression is simplified to:

$$p_{av} \cong Y'\left(1 + \frac{\mu a}{h}\right)$$

For a strain hardening metal, the flow stress increases as deformation continues. To account for the variable flow stress, Y'_f is used in place of Y'. That is,

$$p = Y'_f \exp\left(\frac{2\mu(a-x)}{h}\right) \quad \text{(strain hardening)} \tag{4.86}$$

$$p_{av} \cong Y'_f\left(1 + \frac{\mu a}{h}\right) \quad \text{(strain hardening)} \tag{4.87}$$

$$F = p_{av}(2a)(\text{width}) \tag{4.88}$$

The final dimensions, a and h, are used to find p, p_{av}, and F at the end of the forging operation. Y'_f is found using the strain at the end of forging (Example 4.15).

In the analysis above, the friction is "sliding friction." That is, the workpiece moves relative to the platen. As deformation continues, the sliding stops and sticking begins when shear stress at the interface (μp) is greater than the local shear yield strength, k, of the material. In plane strain, the shear yield strength is $Y'/2$. The critical condition is reached, at constant μ, as the pressure increases, a natural consequence of increasing the extent of deformation. Additionally, the pressure increases from the edge of the specimen toward the center and so there may be partial sticking.

EXAMPLE 4.15 Repeat Example 4.13 taking into account friction with a friction coefficient of 0.1.

a. As in the previous example, the final dimensions of the piece are 4 cm × 2 cm × 12.5 cm and the thickness strain and the effective strain are:

$$\varepsilon_y = \ln\left(\frac{5}{4}\right) = 0.223 \quad \bar{\varepsilon} = \frac{2}{\sqrt{3}}\varepsilon_y = 0.257$$

The plane stain flow stress is:

$$Y' = \frac{2}{\sqrt{3}} 100 \text{ MPa} = 115 \text{ MPa}$$

Because of friction, there is a distribution of pressure under the platen. The functional form of the distribution from the center to the far edge ($x = a$) is given by:

$$p = Y'\exp\left(\frac{2\mu(a-x)}{h}\right)$$

Let us consider the extremes to understand the distribution better:

$$p = p_{max} \text{ at } x = 0 \text{ (center)}$$

$$p_{max} = Y'\exp\left(\frac{2\mu a}{h}\right) = (115 \text{ MPa})\exp\left(\frac{2(0.1)(6.25 \text{ cm})}{4 \text{ cm}}\right)$$
$$= (115 \text{ MPa})(1.367) = 157 \text{ MPa}$$

$p = p_{min}$ at $x = a$ (edge)
$p_{min} = Y'\exp(0) = 115$ MPa

The force is based on the average pressure because we can only apply one force, not a force distribution.

$$p_{av} = Y'\left(1 + \frac{\mu a}{h}\right) = (115 \text{ MPa})\left(1 + \frac{(0.1)(6.25 \text{ cm})}{4 \text{ cm}}\right) = 133 \text{ MPa}$$
$$F = p_{av}A_{final} = 133 \text{ MPa}(0.02 \text{ m} \times 0.125 \text{ m}) = 0.33 \text{ MN}$$

b. Repeating for the work hardening material, we need to find the thickness strain, effective strain and plane strain flow stress needed to achieve the desired final deformation. These conditions are the same as in the previous example.

$$\varepsilon_y = \ln\left(\frac{5}{4}\right) = 0.223 \quad \bar{\varepsilon} = \frac{2}{\sqrt{3}}\varepsilon_y = 0.257 \quad Y'_f = 158.4 \text{ MPa}$$

$$p_{av} = Y'_f\left(1 + \frac{\mu a}{h}\right) = (158.4 \text{ MPa})\left(1 + \frac{(0.1)(6.25 \text{ cm})}{4 \text{ cm}}\right) = 183 \text{ MPa}$$
$$F = p_{av}A_{final} = 183 \text{ MPa}(0.02 \text{ m} \times 0.125 \text{ m}) = 0.46 \text{ MN}$$

The force required is larger, as expected, accounting for friction. Now, to find the temperature rise with friction, we notice that the force has increase by 0.46 MN/0.396 MN = 1.16, so we can assume that the work done goes up by the same amount. Hence the total work of deformation is 1.16 (29.4 MJ/m³) = 34.1 MJ/m³ and the temperature rise is 14K.

"Sticking" means the material near the platen is stationary, shearing rather than sliding. It does not mean a physical adherence. Under sticking friction, the interfacial shear stress is equal to k ($Y'/2$) rather than μp. Using slab analysis the pressure variation for sticking friction is found to be:

$$p = Y'\left(1 + \frac{a-x}{h}\right) \quad \text{(perfect plastic)} \tag{4.89}$$

$$p = Y'_f\left(1 + \frac{a-x}{h}\right) \quad \text{(strain hardening)} \tag{4.90}$$

The average pressure is found easily for this linear function.

Axisymmetric Open Die Forging with Friction. When a cylindrical workpiece is compressed between two platens in an open die configuration, the deformation is not plane strain but rather an axisymmetric. The thickness is reduced and the radius expands. Like wire drawing, another example of axisymmetric deformation, the effective strain is equal to the strain in thickness direction: $\bar{\varepsilon} = \varepsilon_z = \ln(h_o/h)$. The slab analysis with cylindrical coordinates is used to find the pressure under the platen:

$$p = Y \exp\left(\frac{2\mu(R-r)}{h}\right) \quad \text{(perfect plastic)} \tag{4.91}$$

$$p = Y_f \exp\left(\frac{2\mu(R-r)}{h}\right) \quad \text{(strain hardening)} \tag{4.92}$$

where R is the outer radius of the cylinder and r is the radial position with $r = 0$ as the center. Therefore the 'friction hill' has a central peak and decays radially. The average pressure is approximated by:

$$p_{av} \cong Y\left(1 + \frac{2\mu R}{3h}\right) \quad \text{(perfect plastic)} \tag{4.93}$$

$$p_{av} \cong Y_f\left(1 + \frac{2\mu R}{3h}\right) \quad \text{(strain hardening)} \tag{4.94}$$

As in other forging operations, the forging force is given by the average pressure multiplied by the area of contact.

Closed Die Forging. Closed die forging is used to create shapes in the forging process. These processes are almost exclusively done with hot forging due to the high degree of deformation needed. There are two variations: impression die forging and precision closed die forging. See Figure 4.35. In both processes, the workpiece is pressed between two platens with shaped die cavities. The forging stroke deforms the metal into those cavities to create a shape. The difference between the two variations is the formation of flash. In impression die forging, flash forms between the two die halves. As such the forging force rises steeply in the last part of the deformation as the metal is squeezed in the narrow space. The flash is removed after the forging

318 Materials Processing

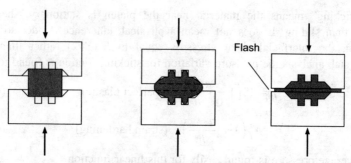

FIGURE 4.35 Schematic diagram of impression die forging, progressing from left to right. Precision closed die forging is similar but the process is engineered so that the workpiece volume is the same as the final part volume such that there is no flash. *Adapted from Kalpakjian (1997).*

is complete. The utility of the flash is that the volume of the workpiece does not need to be the same volume as the final part and that metal deformation into complex and small features in the die cavity is accomplished at the same time as the flash formation. In precision closed die forging, the process is controlled so that it is near net shape and no flash forms. Highly deformable alloys are needed for this process and hence it is usually carried out with nonferrous alloys.

Predicting the forging force for closed die forging is largely empirical. The state of stress in the die is complicated. Qualitatively the force need is the product of three terms: (i) the projected area of the part, (ii) the flow stress of the metal at the temperature, strain rate and approximate level of strain needed for the deformation, and (iii) a multiplicative factor ranging from ~3 to 12 depending on the presence of flash and the complexity of the part.

4.3.5 Rolling

Rolling is a deformation process that involves thickness reduction by the application of compressive forces applied through a set of rolls. The general idea of the process is illustrated in Figure 4.36. A slab or sheet of material with thickness h_1 and width w_1 is fed into the "nip" of a set of two counter-rotating rolls. The gap between the rolls is set to less than the thickness of the incoming workpiece, and typically much smaller than the width or the length of the workpiece. The workpiece is pulled into the nip and deformed by the converging section. The material then exits with a smaller thickness, h_2 and longer length, but with the same width ($w_2 = w_1$). The constant width is a feature of plane strain rolling processes. The width is held constant by the undeformed material adjacent to it (outside of the nip). The plane strain condition generally holds when the width is at least five times the thickness of the workpiece. The length increase requires the speed of the exiting

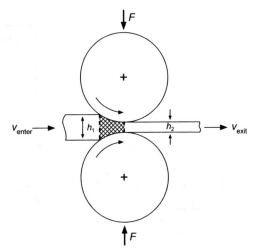

FIGURE 4.36 Schematic diagram of a rolling process.

material to be higher than the entering. The continuity equation under plane strain conditions can be applied to find the velocity increase:

$$Q_1 = Q_2 \tag{4.95}$$

$$A_1 v_1 = A_2 v_2 \tag{4.96}$$

$$v_2 = v_1 \frac{A_1}{A_2} = v_1 \frac{h_1}{h_2} \tag{4.97}$$

The force in rolling is the roll separating force. The rolls themselves are mounted in a frame that keeps them from coming apart, or separating, as the workpiece enters into the nip. Therefore, one limitation in a rolling process is the "roll separating force," which must be within the limits of the equipment. Finding this force and how it varies with the gap and mechanical properties of the workpiece is one goal of the mechanical analysis of the rolling process.

Like forging, many rolling operations are carried out under hot working conditions due to the larger amount of deformation that can be achieved, lower load requirements and isotropic nature of the metal at the end of deformation. Hot rolling can be followed with a final cold rolling step for strengthening of the final workpiece and establishing its dimensions to higher tolerance. Like wire drawing, rolling frequently occurs in a stepwise fashion with the workpiece fed sequentially through a series of rolling stands, each subsequent set with decreasing gap. Shaping is possible in a rolling process but using rolls that have features. For example, I-beams are created by rolling.

Features in the deformation zone are further explored in Figure 4.37. The workpiece enters at velocity v_1 and exits at a higher velocity v_2 and hence

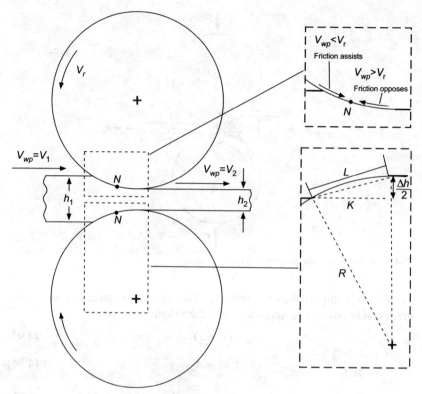

FIGURE 4.37 Schematic diagram of a rolling process, showing geometrical factors and frictional effect. The workpiece velocity, v_{wp}, increases as the sheet is deformed in the nip. *Adapted from Hosford and Caddell (1993).*

the workpiece accelerates as it traverses the nip. The rolls, however, rotate at a single velocity. The rolls must be moving faster than v_1 in order for the workpiece to be pulled into the nip. So at the entry point the roll velocity is higher than the workpiece velocity. Hence, friction between the two is in the direction of forward motion. That is, friction assists the process! As the workpiece speeds up, there is a point at which the velocities of the workpiece and the roll are the same, the neutral point. Beyond the neutral point, in the exit zone, the workpiece is slower than the roll and hence forward motion must work against friction at the interface.

Another important feature is the contact area between the workpiece and the rolls. The contact length, L, is found by examining the geometry, as shown in the inset. There are two right triangles defined in the sketch. Taking L as the hypotenuse of the smaller triangle and applying the Pythagorean theorem to both and solving for L gives:

$$L = \sqrt{R\Delta h} \qquad (4.98)$$

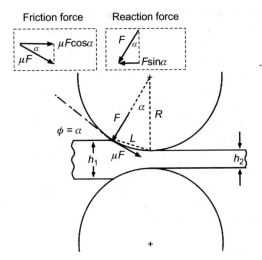

FIGURE 4.38 Schematic diagram showing the balance between friction force and the reaction force. *Adapted from Hosford and Caddell (1993)*.

The contact area then is the contact length, L, multiplied by the width, w. The thickness change or gap, $\Delta h = h_2 - h_1$, is a significant feature of a rolling process. This thickness change is also known as the draft or reduction.

The maximum possible reduction depends on the friction conditions. For unaided entry (i.e., friction supplies the force forward), the friction force must overcome the reaction force provided by the roll/workpiece interaction. According to the force diagram in Figure 4.38, the forward pull from friction is the friction force is $\mu F \cos \alpha$ and the backward reaction force is $F \sin \alpha$. So forward motion requires:

$$\mu F \cos \alpha \geq F \sin \alpha$$
$$\tan \alpha \leq \mu \quad \text{or} \quad (\tan \alpha)_{max} = \mu \qquad (4.99)$$

Using geometry, $\tan \alpha \approx L/R$ and given the relationship between L and Δh,

$$\Delta h_{max} = \mu^2 R \qquad (4.100)$$

Therefore, the max reduction possible is connected to the roll radius and the friction coefficient. While using larger rolls and higher friction conditions allows greater reductions, this benefit must be weighed against the roll cost as well as the need for higher roll separating forces, as discussed below.

Finding the roll separating force requires an understanding of the pressure between the roll and the workpiece. Various approaches can be used to find this pressure. There are significant similarities to plane strain forging that allow the useful application of forging expressions with appropriate substitutions. Also, the complexity of rolling might be approached from a starting point of plane strain extrusion through curved dies that are analogous to stationary rolls. Another approach is to carry out a slab analysis on the entry and exit regions, separately, using approximate geometry that avoids some of the complications of curved surfaces. In the entry region, frictional forces are in the forward (x) direction, pulling the workpiece in as it is compressed in the converging section supplied by the rolls. In the exit region, the frictional forces are reversed, resisting the forward motion. The full slab analysis follows the principles developed in this chapter, but is complex in its geometry. Here, we examine the solutions for the pressure variations in each zone to examine the variables that affect pressure. A full derivation is in Kalpakjian (1997).

The extent of deformation changes from zero at the start of the entry zone to some maximum value at the exit. According to Figure 4.38, the angle between the workpiece and the roll, ϕ, starts at α, and ends at the exit at 0. In the expressions that follow, ϕ is used as a variable marking the position in the gap. Additionally, the distance between the rolls, h, in the deformation zone begins at h_1, the incoming sheet thickness, and ends at h_2, the outgoing sheet thickness. By slab analysis, the pressure exerted by the rolls is given by:

$$p = Y'\left(\frac{h}{h_1}\right) e^{2\mu M (\arctan(\alpha M) - \arctan(\phi M))} \quad \text{(Entry)} \quad (4.101)$$

$$p = Y'\left(\frac{h}{h_2}\right) e^{2\mu M \arctan(\phi M)} \quad \text{(Exit)} \quad (4.102)$$

In these expressions, M is a constant defined as $M = \sqrt{R/h_2}$ and Y' is the plane strain yield stress assuming a perfect plastic material.

To incorporate work hardening, the value of the plane strain flow stress at that particular position in the nip must be used (i.e., Y'_f). Since the operation is plane strain and volume is conserved, like plane strain forging, the effective strain is given by $\bar{\varepsilon} = 2/\sqrt{3}\varepsilon_y$, where ε_y is the strain in the thickness direction. In rolling this strain is a function of lateral position, while in forging it is a function of time during the compression and usually just considered at the end point of the process.

These two expressions are plotted in Figure 4.39 schematically. The neutral point is intersection of the two pressure distributions. There is a friction hill evident in the plots. The origin of the friction hill can be understood as a combination of two effects. First, consider the pressure distribution. Similar to plane strain forging with platens converging, the neutral point would shift and the extent of deformation on either side of the neutral point would be different. Second, the changing effect of friction is also a factor with friction assisting the deformation on the entry and resisting it on the exit.

Although the pressure between the roll and the workpiece varies in the nip, achieving the proper conditions for rolling requires knowledge of a

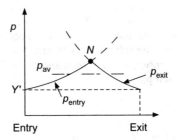

FIGURE 4.39 Schematic representation of the pressure distribution in the nip during rolling.

single roll separating force. The roll separating force is determined from the average pressure as well as the contact length and the width. Again, there are various approaches to estimating this pressure. Commonly, the same expression for plane strain compression is used with appropriate substitution of variables (i.e., a (forging) = $L/2$ (rolling) and h (forging) = h_{av}(rolling), where h_{av} is $(h_1 + h_2)/2$). So, with this analogy the average pressure is given by:

$$p_{av} = Y'\left(1 + \frac{\mu L}{2h_{av}}\right) \quad \text{(perfect plastic)} \quad (4.103)$$

$$p_{av} = \overline{Y'}\left(1 + \frac{\mu L}{2h_{av}}\right) \quad \text{(strain hardening)} \quad (4.104)$$

For a strain hardening material, the average plane strain flow stress, $\overline{Y'}$ is used because the extent of deformation, and hence the flow stress, varies from entry to exit. (This situation is different from plane strain forging, which necessitates achieving a final extent of deformation throughout and hence Y'_f is used.) The roll separating force is given by (Example 4.16):

$$F = Lwp_{av} \quad (4.105)$$

Since the size of the rolling apparatus varies with the value of the roll separating force, it is instructive to consider how to lower this force effectively. Any factor that lowers either the average pressure between the rolls and the workpiece or the contact area has the effect of lowering the force. Since the contact length L scales with the roll radius, smaller rolls lead to reduced roll separating forces; however, smaller rolls are also more susceptible to deflection. The central spindle of the rolls is held at either end and so with large forces and small rolls, the rolls (and spindles) may flex or bow to make the gap in the center larger than near the fixed ends. Under these conditions the exiting sheet is likely to nonuniform or defective. There are strategies involving small "work" rolls backed up with larger supporting rolls to gain the benefits of small rolls without the deflection and defects. Of course, lowering the flow stress of the metal effectively lowers the pressure needed for deformation. The advantage of hot rolling in this regard is clear.

EXAMPLE 4.16 Copper is rolled from 1 mm thickness to 0.8 mm thickness in a first pass and then to 0.6 mm thickness in a second pass. The width of the copper is 1.9 cm and the roll radius is 10 cm. Assume the friction coefficient is 0.4. For copper, K = 315 MPa and n = 0.54. Find the roll separating force for each pass.

$$F = Lw\overline{Y}\left(1 + \frac{\mu L}{2h_{av}}\right)$$

$L = \sqrt{R\Delta h}$, $R = 0.1$ m, $w = 0.019$ m and $\mu = 0.4$
$\overline{Y}$, Δh and h_{av} must be found for each pass

Pass 1:

We need to know the average flow stress during this pass. Copper is a work hardening metal and so we need to use the data given for K and n.

True strain in the thickness direction: $\varepsilon_y = \ln\left(\frac{1.0}{0.8}\right) = 0.223$

Effective strain: $\bar{\varepsilon} = \frac{2}{\sqrt{3}}\varepsilon_y = 0.257$

We need the average flow stress from the annealed condition to the final strain of 0.257.

$$\overline{Y}' = \left(\frac{2}{\sqrt{3}}\right)\frac{K\varepsilon^n}{n+1} = \left(\frac{2}{\sqrt{3}}\right)\frac{(315)(0.257)^{0.54}}{1.54} = 113 \text{ MPa}$$

$\Delta h = 1 - 0.8 = 0.2$ mm $= 2 \times 10^{-4}$ m

$h_{av} = \frac{1 + 0.8}{2} = 0.9$ mm $= 9 \times 10^{-4}$ m

$L = \sqrt{R\Delta h} = \sqrt{(0.10 \text{ m})(2 \times 10^{-4} \text{ m})} = 0.0045$ m.

Now the roll separating force can be found.

$$F = Lw\overline{Y}'\left(1 + \frac{\mu L}{2h_{av}}\right)$$

$$= (0.0045 \text{ m})(0.019 \text{ m})(113 \times 10^6 \text{ Pa})\left(1 + \frac{(0.4)(0.0045 \text{ m})}{2(9 \times 10^{-4} \text{ m})}\right)$$

$$= 1.94 \times 10^4 \text{ N}$$

Pass 2:

We need to know the average flow stress for this pass.

True strain for this pass: $\varepsilon_y = \ln\left(\frac{0.8}{0.6}\right) = 0.288$

Effective strain: $\bar{\varepsilon} = \frac{2}{\sqrt{3}}\varepsilon_y = 0.332$

Total effective strain: $\bar{\varepsilon} = 0.257 + 0.332 = 0.582$

We need the average flow stress from work hardened condition after the first pass to the state at the end of the second pass.

$$\overline{Y}' = \left[\frac{2}{\sqrt{3}}\right]\frac{K}{\bar{\varepsilon}_2 - \bar{\varepsilon}_1}\left[\frac{\bar{\varepsilon}_2^{n+1} - \bar{\varepsilon}_1^{n+1}}{n+1}\right]$$

$$= \left[\frac{2}{\sqrt{3}}\right]\frac{(315 \text{ MPa})}{0.582 - 0.257}\left[\frac{(0.582)^{1.54} - (0.257)^{1.54}}{1.54}\right]$$

$$= 226 \text{ MPa}$$

$$\Delta h = 0.8 - 0.6 = 0.2 \text{ mm} = 2 \times 10^{-4} \text{ m}$$
$$h_{av} = 0.7 \text{ mm} = 7 \times 10^{-4} \text{ m}$$
$$L = \sqrt{R\Delta h} = \sqrt{(0.10 \text{ m})(2 \times 10^{-4} \text{ m})} = 0.0045 \text{ m}.$$

$$F = Lw\overline{Y'}\left(1 + \frac{\mu L}{2h_{av}}\right)$$
$$= (0.0045 \text{ m})(0.019 \text{ m})(226 \times 10^6 \text{ Pa})\left(1 + \frac{(0.4)(0.0045 \text{ m})}{2(7 \times 10^{-4} \text{ m})}\right)$$
$$= 4.41 \times 10^4 \text{ N}$$

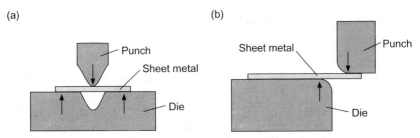

FIGURE 4.40 Two modes of bending (a) V-die and (b) wiping die. *Adapted from Beddoes and Bibby (1999).*

Another important point is the influence of an additional tensile stress placed on the sheet either at the exit or the entrance. In either case, the sheet experiences tension and compression, which lowers the pressure required between the workpiece and the rolls and hence the roll separating force.

4.3.6 Bending

Bending is a simple way to convert a sheet of metal or thermoplastic polymer into a shape. It is most frequently applied to sheet metal processing and as such is carried out at room temperature. Polymer sheet bending is less common, but would involve heating the polymer, as described more in the next section, then bending the rubbery polymer and, lastly, cooling to retain the bent shape. In this section, the basic mechanics of bending are discussed along with an example from sheet metal deformation.

Figure 4.40 shows two ways in which a sheet of metal can bent into a shape. The combination of a punch and a die determines the final shape of the bent piece. There are at least three critical questions to consider concerning bending. What is the smallest bend radius that is possible without fracture or defects? What force is required to bend the metal to a given state?

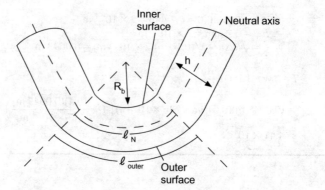

FIGURE 4.41 Geometry of a simple bending operation. *Adapted from Beddoes and Bibby (1999).*

What is the impact of the elasticity on the final shape? To answer these questions, we explore a simple analysis of the mechanics of bending.

Figure 4.41 shows a cross-section of a bent sheet. During bending, the inner surface is under compression, the outer is in tension and there is a stress-free neutral axis. Using similar triangles, the following relationship is found:

$$\frac{l_o}{l_n} = \frac{R_b + h}{R_b + h/2} \qquad (4.106)$$

where l_o and l_n are the lengths at the outer surface and at the neutral axis, respectively, R_b is the bending radius and h is the sheet thickness.

Since the outer surface is in tension, failure is initiated there. The strain at the outer surface is:

$$\varepsilon_o = \ln\left(\frac{l_o}{l_n}\right) \qquad (4.107)$$

For a benchmark, information from a uniaxial tensile test is applied to finding limitations in the bending. In a tensile test, the strain at failure, ε_f, is given by:

$$\varepsilon_f = \ln\left(\frac{A_o}{A_f}\right) = \ln\left(\frac{1}{1-RA}\right) \qquad (4.108)$$

where A_o and A_f are the initial gauge area and the final gauge area, respectively, and RA is the reduction of area at fracture, which is defined as:

$$RA = \frac{A_o - A_f}{A_o} \qquad (4.109)$$

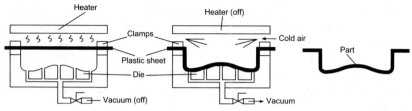

FIGURE 4.42 Schematic of vacuum thermoforming process, from left to right. *Adapted from Strong (2000).*

Setting the tensile strain in the outer surface equal to the failure strain and invoking the condition in Eq. (4.106), the minimum bending radius is found:

$$R_{b,\min} = h\left(\frac{1}{2RA} - 1\right) \qquad (4.110)$$

This expression holds when the neutral axis is at $h/2$ and for RA is less than 0.2. If RA is greater then,

$$R_{b,\min} = h\left[\frac{(1-RA)^2}{2RA - RA^2}\right] \qquad (4.111)$$

Examination of these equations reveals that the tighter bends, smaller minimum bending radii, are possible with materials that are more ductile (i.e., larger RA values).

The force required for the deformation can be estimated for simple bending according to Figure 4.42. The bending force increases as bending proceeds and reaches a maximum value that is most conveniently related to the ultimate tensile strength of the metal, *UTS*, from a tensile test:

$$F_{b,\max} = k\frac{UTS \cdot Lh^2}{W} \qquad (4.112)$$

where k is a factor related to the mode of bending, L is the length of the metal undergoing bending, and W is the bent width or die opening. For a V die, k is in the range of ~ 1.33, while wiping dies have a k of ~ 0.3.

When the forces required to form the bend are removed then the elastic deformation is recovered. This "springback" changes the final shape and therefore should be accounted for in the tool design. Elastic recovery in a simple tensile test is connected to the value of the Young's modulus, E, as well as the yield strength in tension, Y_t. Hence these two parameters appear in the expression for the final radius after release of the load, R_f.

$$\frac{R_b}{R_f} = 4\left(\frac{R_b Y_t}{hE}\right)^3 - 3\left(\frac{R_b Y_t}{hE}\right) + 1 \qquad (4.113)$$

The final radius is typically larger than the initial radius (positive springback), leading to a metal sheet that is less bent than when it was under load. However, negative springback is also possible in the V-type configuration.

Creating a shape from bending can involve several deformations to create a complex shape. For example, bending can be used to transform a sheet into a part with an open rectangular cross-section via a series of dies and punches.

4.3.7 Thermoforming

Thermoforming converts sheets of thermoplastic polymer into open form shapes, such as cups. Figure 4.42 shows an example of a thermoforming process. A thermoplastic sheet is heated and then pulled into contact with a mold via a vacuum. The mold is chilled and cold air can be added at this point so that the plastic solidifies in the new shape. The part is then removed. This variant of thermoforming is called vacuum thermoforming.

The process makes use of the thermomechanical characteristics of thermoplastics and is unlike the earlier processes in this section in that the deformation, the shape change, is not due to a *plastic* deformation, but rather to an *elastic* one. This elastic deformation occurs at high temperature and then the deformed state is retained by cooling. As Figure 4.10 shows thermoplastic polymers enter into rubbery state above T_g, where large amounts of deformation are possible. If the material were to be released from the tool at the high temperature, however, all of the deformation would be recovered and the polymer would go back to its original, featureless state. Instead, the deformed state is locked in by cooling (below T_g). The final part has "memory" of its previous undeformed state. If it is heated to near the forming temperature, it reverts to its original flat sheet.

Table 4.6 shows some characteristics of polymers that are commonly thermoformed. Thermoforming takes place above the glass transition temperature but below the point where the polymer flows as a liquid readily under gravity. The temperature range of thermoforming for amorphous materials is usually broader then for semicrystalline polymers. The crystallinity in the semicrystalline polymer restricts the deformability and so thermoforming is carried out at temperatures above the melting point. Typically, the elastic modulus falls rapidly with temperature and so the temperature window with an appropriate modulus is small.

There are six basic steps in a thermoforming process. First, the polymer sheet is loaded or fed into the apparatus. As mentioned, the polymer sheet is preformed by extrusion typically and hence it is more expensive than the starting material for the polymer forming processes that are melt-based. The sheets are classified as heavy gauge for thicknesses between 0.06 and 0.5" thick (1.52–12.7 mm) or thin gauge for 0.01–0.06" (0.25–1.52 mm). The thick gauge sheet is cut to size and clamped in place while the thin gauge is

TABLE 4.6 Properties of Thermoforming Polymers

Polymer	Glass transition temperature[a](°C)	Thermoforming temperature range[b](°C)	Shrinkage range[b](%)
PC	150	168–204	0.5–0.7
HDPE	−90	127–182	2.0–4.5
PMMA	105	149–193	0.2–0.8
PP	−18	132–166	1.0–2.5
PS	100	127–182[c]	0.5–0.8
Rigid PVC	87	104–154	0.1–0.5

[a]From Callister (2007)
[b]From Throne (2008)
[c]Range for PS containing 40% glass fibers.

typically on a roll that is fed into the machine. Second, the sheet is heated. Infrared (radiative) heating is common. The temperature is chosen to hit the correct temperature for the polymer and sheet thickness. Importantly, the temperature is not so high that the sheet droops under its own weight. The third step is the deformation itself. This step involves the most diversity in approaches. The general idea is to deform the sheet so that it conforms to the mold dimensions. Two of the many approaches are shown in Figures 4.42 and 4.43. Fourth, the polymer is cooled. The fifth and sixth step are releasing from the mold and trimming, respectively.

The thermoforming process window is important to define for a given polymer and process. Figure 4.44 shows one way to define a process window. The temperature-dependent stress-strain data are shown with overlays that represent limitations or conditions. At low temperatures, the stress-strain data shows glassy behavior, but as temperature increases yielding is observed and eventually a rubbery behavior and a lower modulus. If the temperature becomes too high then excessive flow, which is realized as gravity-induced sagging, limits the ability to thermoform. So, one overlay is the temperature window. A second consideration is the forming conditions, which lead to a maximum possible stress that can be achieved—a horizontal that cuts through the temperature window. Finally a vertical line shows the maximum design strain, which is determined by the particular mold features. The forming window is defined. Given that the required strain is fixed, the forming temperatures range from the temperature that causes the stress-strain behavior that intersects the upper right corner of the window to the temperature that leads to the stress-strain curve the intersects the lower right corner. Clearly, increasing the level of stress achievable in the apparatus (e.g., adding a pressure assist) extends the range of temperatures available for forming.

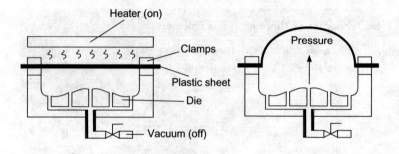

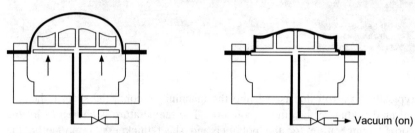

FIGURE 4.43 Schematic diagram of plug assist thermoforming process, from top left to bottom right. *Adapted from Osswald (1998)*.

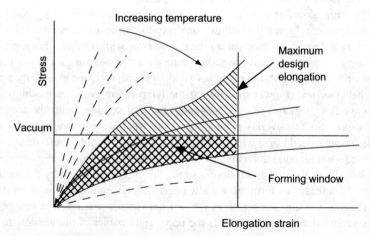

FIGURE 4.44 Processing window for vacuum thermoforming. *From Throne (2008). Understanding Thermoforming (2nd Ed.). Munich, Germany: Carl Hanser Verlag. Reproduced with permission.*

The deformation in thermoforming is usually one of biaxial tension, at least initially. The sheet is stretched. In the case of reverse draw thermoforming, there is an initial purposeful biaxial stretching followed by the deformation to the mold. Modeling that deformation using a suitable equation that predicts the stress—strain behavior of a rubber is needed. For large deformations, two options are the Mooney-Rivlin and Ogden model. These models require two to four materials constants that are found by fitting data. Once the materials constants are known, the effect of pressure, temperature, and sheet thickness on deformation can be predicted. This prediction, however, is only part of the information that is needed to design a process.

As in other polymer molding processes, such as blow molding and injection molding, thermoforming involved complex shapes and is nonisothermal. This complexity makes simple predictions impossible and the true problem solving and process development takes place with finite element models that include the geometry, mode of force application and thermal conditions. Further empirical methods based on experience and simple experiments help to establish process conditions.

Thermoforming produces a wide variety of shapes and so it is worthwhile to compare it to an other versatile polymer molding method, injection molding. Thermoforming has some advantages over injection molding. First the pressures required are much lower as are the thermal requirements. Therefore the equipment is not as massive and the molds do not require the same level of mechanical strength as the high strength steel used in injection molding. Aluminum molds are common. Also because of the lower pressures and lighter weight molds, larger parts can be prepared with greater ease. However, there is a price to pay for these advantages, namely the cost of the sheets themselves. They are much higher in price compared to the pellets that are fed into an injection molding process. Further, the shapes and polymers that can be molded by thermoforming are more limited and the precision is less. Importantly, control over the wall thickness is lacking in thermoforming. The trouble spots tend to be at corners and well as vertical thicknesses relative to horizontal ones. Lastly, the thermal stability of a thermoformed shape is less than that of an injection molded shape. Here, the problem is with the recovery of the massive elastic stresses if the object is heated. The use temperature for a thermoformed plastic part must be significantly less than its forming temperature.

4.3.8 Superplastic Forming

Some metals exhibit amazingly large degrees of plasticity. Strains as high as 2000% or more have been recorded for alloys with fine grain sizes (e.g., ~10 μm or finer) and at high temperatures and controlled strain rates. More typical strains to failure in the several hundreds of % range. The exceptionally large deformation behavior in fine grained alloys is known as superplasticity. Its origin is related to the microstructure, most importantly the disorder that exists on the grain boundaries. There are several forming operations in metal working that make use of superplasticity. Here, the most common operation based on using the phenomena to form a part from a metal sheet is discussed.

The high extent of deformation in superplasticity is very similar to the degree of deformation that is achieved in polymers, again in certain ranges of temperature. Thus, processes mirroring thermoforming are used to create metal shapes from sheets of superplastic metal. The forming equipment is similar to those shown in Figure 4.42; however, much higher temperatures and higher levels of force are needed for superplastic forming of metals and hence specially designed equipment is necessary. The process steps are similar but also have differences. In superplastic forming, the metal sheet is clamped in place and then heated along with the die. Superplastic forming is carried out at a particular temperature; the shape is formed at the temperature by plastic deformation and does not require cooling in order for the shape to be retained. Pressure from an inert gas or vacuum are used to deform the sheet, and plug assist can be used as well. After the part is formed, then the material is removed and trimmed.

In the forming operation, it is important to find the temperature and strain rate that alloy the highest deformation. As an example, Figure 4.45 shows the effect of strain rate and temperature on the strain of a zinc-aluminum alloy. From the data, there is a compromise between temperature and speed.

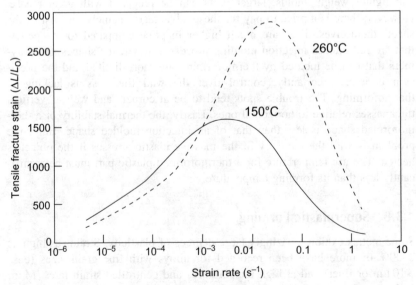

FIGURE 4.45 Effect of temperature and strain rate on tensile deformation of a Zn-22Al eutectoid alloy, demonstrating the superplastic effect. *Adapted from Mohamed et al. (1977).*

By increasing the temperature, the deformation can occur more quickly. The flow stress of superplastic materials is also sensitive to the temperature and strain rate.

Superplastic forming is used in a variety of industries, but perhaps most commonly in automotive parts that are fairly large. The most direct competitors to superplastic forming would be forging, casting or assembly of a number of smaller parts to make the desired shape. The complexity can be developed in forging but multiple dies would be needed and large forces for deforming larger shapes. While melt casting is possible for these shapes, there are challenges with thin sections and considerable surface finishing would be needed for sand castings. Smaller objects could be cast with permanent molds or with die casting more easily than superplastic forming. Of course superplastic forming has limitations. Like thermoforming, the starting material is expensive and it is not possible to achieve superplastic effects in all alloys.

4.4 SUMMARY

Solid processes involve mechanical deformation of solid engineering materials into new forms and shapes. Appreciating the fundamental mechanical properties of metals, ceramics and polymers, as accessed by uniaxial tension experiments, is a necessary starting point. Most metals experience extensive plastic deformation during a tensile test; metrics from engineering stress—engineering strain curves, such as the ultimate tensile strength, yield strength, and ductility, differentiate metals in this regard. However, to harness plastic deformation as a tool for shaping a metal, the true stress—true strain behavior is more relevant. Once a yield stress is reached, continued deformation typically requires higher and higher true stress values. The increase in true stress with continued strain is due to work hardening and follows an empirical power law. The power law allows the prediction of the stress needed to reach a certain extent of deformation—this stress is called the flow stress. By contrast, ceramics have little capacity for plastic deformation and are therefore absent from this chapter. Plastic deformation of polymers is routine, but cold drawing of a polymer is not an easily used to shape the plastic. Instead the heat and mechanical forces are used to deform in the rubbery solid state, followed by cooling to lock in the new structure.

Solid deformation processes are more complex than a uniaxial tension, involving triaxial stresses. To understand how a complex stress state might cause plastic deformation in a material, the stress state is represented in terms of three principal stresses. With this simpler representation, the critical condition for yielding can be stated. The von Mises yield criterion matches experimental data for metals. This criterion reveals a fundamental concept—yielding can occur at lower applied stresses if a compressive stress in one principal direction and a tensile stress in another principal direction are combined. While real deformation processes are not as simple as pushing in one direction while pulling in another, this framework permeates bulk deformation processes.

Wire drawing is both simple in concept—pull a wire through a die that has a smaller diameter than the wire—and complex in its full analysis. A simple energy balance, without regard to complications like friction, shows that the stress needed for deformation, the draw stress, scales with the flow stress of the metal and the strain (i.e., diameter reduction). The flow stress of interest is the average flow stress over the range of strains experienced by the metal in the draw. When friction is added, a slab analysis, which is basically a force balance carried out on a volume element, shows that friction leads to higher draw stress requirements. A final factor, redundant work, enters into the picture as well, notably factoring into the design of dies. These factors—the mechanical properties of the metal, the extent of deformation, friction, and redundant work—are also important in other deformation processes.

Extrusion, forging, and rolling are three other deformation processes used to shape solid metals. Extrusion is similar to wire drawing in reverse; a metal billet is pushed through a die. Extrusion is frequently carried out under "hot" conditions, which means that recrystallization limits strain hardening and the flow stresses are low. Accounting for some factors, such as friction and redundant work, is done empirically. Like wiredrawing, the average flow stress is the key materials property for predicting the forces needed for extrusion. In forging, the metal is compressed between flat platens (open die forging) or shaped platens. Open die forging lends itself to slab analysis. The force needed to forge a metal increases as deformation continues; in particular, the contact area with the platen increases and work hardening factors in. Friction at the workpiece/platen interface leads to nonuniform pressure, the so-called pressure hill. An average pressure is used in the design of a forging load or force, but the flow stress at the end of the forging cycle is used for the design (not the average flow stress). Rolling has similarities to forging in its analysis but the frictional effects are different due to the speeding up of the sheet as it passes though the nip. For this process, the roll separating force is calculated and like wire drawing, the average flow stress is used in calculations for rolling of work hardening materials.

Solid processes can also deform sheets in a manner that creates shapes without appreciably changing the sheet thickness. Bending is one operation that is used to change sheets or strips of metals into 3D objects. The tensile stress on the outer side of a bent material limits the extent of deformation possible before failure. Thermoforming is a process that shapes thermoplastic polymer sheets. Here, the sheet is heated until it is rubbery but not fluid, and then it is deformed against a mold surface using relatively low pressures (compared to injection molding, for example), and then the shape is locked in by cooling. A related process, superplastic forming, creates shapes by deforming metal sheets that have fine grain sizes and the unusual ability to be deformed to large extents (e.g., 500–1000%).

BIBLIOGRAPHY AND RECOMMENDED READING

Fundamentals

Callister, W.D., 2007. Materials Science and Engineering: An Introduction, seventh ed. Wiley, New York, NY.

Davis, J.R., 1998. Metals Handbook: Desk Edition, second ed. ASM International, Materials Park, OH.

Dieter, G.E., 1986. Mechanical Metallurgy, third ed. McGraw Hill, New York, NY.

Dowling, N.E., 2013. Mechanical Behavior of Materials: Engineering Methods for Deformation, Fracture, and Fatigue, fourth ed. Pearson, New York, NY.

Hoffman, O., Sachs, G., 1953. Introduction to Plasticity for Engineers. McGraw Hill, New York, NY.

Hosford, W.F., Caddell, R.M., 1993. Metal Forming: Mechanics and Metallurgy, second ed. Prentice Hall, Inc., Englewood Cliffs, NJ.

Kalpakjian, S., 1997. Manufacturing Processes for Engineering Materials, third ed. Addison-Wesley, Reading, MA.

Meyers, M.A., Chawla, K.K., 1984. Mechanical Metallurgy. Prentice-Hall, Inc., Englewood Cliffs, NJ.

Schey, J.A., 2000. Introduction to Manufacturing Processing, third ed. McGraw Hill, New York, NY.

Young, R.J., 1983. Introduction to Polymers. Chapman and Hall, New York, NY.

Metal Deformation Processing

Backofen, W.A., 1972. Deformation Processing. Addison-Wesley, Reading, MA.

Beddoes, J., Biddy, M.J., 1999. Principles of Metal Manufacturing Processes. John Wiley & Sons, New York, NY.

DeGarmo, E.P., Black, J.T., Kohser, R.A., 2003. Materials and Processes in Manufacturing, ninth ed. Wiley, New York, NY.

Dieter, G.E., 1986. Mechanical Metallurgy, third ed. McGraw Hill, New York, NY.

Groover, M.P., 1996. Fundamentals of Modern Manufacturing: Materials, Processes and Systems. Prentice-Hall, Upper Saddle River, NJ.

Hoffman, O., Sachs, G., 1953. Introduction to Plasticity for Engineers. McGraw Hill, New York, NY.

Hosford, W.F., Caddell, R.N., 1993. Metal Forming: Mechanics and Metallurgy, second ed. Prentice Hall, Inc., Englewood Cliffs, NJ.

Kalpakjian, S., 1997. Manufacturing Processes for Engineering Materials, third ed. Addison Wesley, Menlo Park, CA.

Rowe, G.W., 1965. An Introduction to the Principles of Metal Working. St. Martin's Press, New York.

Schey, J.A., 1970. Metal Deformation Processes: Friction and Lubrication. Marcel Dekker Inc., New York, NY.

Schey, J.A., 2000. Introduction to Manufacturing Processing, third ed. McGraw Hill, New York, NY.

Polymer Deformation Processing

Mascia, L., 1989. Thermoplastics: Material Engineering, second ed. Elsevier Science Publishers Ltd, New York, NY.

Morton-Jones, D.H., 1989. Polymer Processing. Chapman and Hall, New York, NY.

Osswald, T.A., 1998. Polymer Processing Fundamentals. Hanser/Gardner Publishing, Cincinnati, OH.

Strong, A.B., 2000. Plastics Materials and Processing, second ed. Prentice-Hall, Upper Saddle River, NJ.

Tadmor, Z., Gogos, C.G., 1979. Principles of Polymer Processing. Wiley Interscience, New York, NY.

Throne, J.L., 2008. Understanding Thermoforming, second ed. Carl Hanser Verlag, Munich, Germany.

Ward, I.M., 1971. Mechanical Properties of Solid Polymers. John Wiley & Sons Ltd., Bristol, England.

CITED REFERENCES

Bauwens-Crowet, C., Bauwens, J.C., Homès, G., 1969. Tensile yield-stress behavior of glassy polymers. J. Polym. Sci. [A2](7), , 735–742.

Beddoes, J., Biddy, M.J., 1999. Principles of Metal Manufacturing. John Wiley & Sons, New York, NY.

Blau, P.J., 1992. Appendix: static and kinetic friction coefficients for selected materials. In: Blau, P.J. (Ed.), Friction, Lubrication, and Wear Technology, Vol. 18 ASM Handbook. ASM International, Materials Park, OH, pp. 70–75.

Davis, J.R. (Ed.), 1998. ASM Metals Handbook: Desk Edition, second ed. ASM International, Materials Park, OH.

Li, D., Ghosh, A., 2003. Tensile deformation behavior of aluminum alloys at warm forming temperatures. Mater. Sci. Eng. A A352 (1–2), 279–286.

Fields Jr., D.S., Backofen, W.A., 1959. Temperature and rate dependence of strain hardening in aluminum alloy 2024-0. Trans. Am. Soc. Met. 51, 946–960.

Hoffman, O., Sachs, G., 1953. Introduction to Plasticity for Engineers. McGraw Hill, New York, NY.

Hosford, W.F., Caddell, R.N., 1993. Metal Forming: Mechanics and Metallurgy, second ed. Prentice Hall, Inc., Englewood Cliffs, NJ.

Kalpakjian, S., 1997. Manufacturing Processes for Engineering Materials, third ed. Addison Wesley, Menlo Park, CA.

Mahrab, F.U., Johnson, W., Slater, R.A., 1965. Dynamic indentation of copper and an aluminum alloy with conical projectile at elevated temperatures. Proc. Inst. Mech. Eng. 180 (11), 285–294.

Mohamed, F.A., Ahmed, M.M., Langdon, T.G., 1977. Factors influencing ductility in superplastic Zn-22Al eutectoid. Metall. Trans. A 8 (6), 933–938.

Osswald, T.A., 1998. Polymer Processing Fundamentals. Hanser/Gardner Publishing, Cincinnati, OH.

Sachs, G., Van Horn, K.R., 1940. Practical Metallurgy. American Society for Metals, Cleveland, OH.

Sheppard, T., 1993. Extrusion of AA 2024 alloy. Mater. Sci. Technol. 9, 430–440.

Strong, A.B., 2000. Plastics Materials and Processing, second ed. Prentice-Hall, Upper Saddle River, NJ.

Throne, J.L., 2008. Understanding Thermoforming, second ed. Carl Hanser Verlag, Munich, Germany.

Wistreich, J.G., 1955. Investigation of the mechanics of wire drawing. Proc. Inst. Mech. Eng. 169, 654–678.

QUESTIONS AND PROBLEMS

Questions

1. Why are true stress and true strain conventions used in metal working instead of engineering stress and engineering strain?
2. What does the strain hardening exponent represent? What are typical values? Does the strain hardening exponent change with temperature? If so why?
3. Why is elastic deformation frequently ignored in the design of metal working processes?
4. Is strain hardening behavior a help or a hindrance to metal deformation processes?
5. Describe the differences between cold work and hot work? List some advantages and disadvantages of each.
6. Why do deformation processes involve complex, biaxial or triaxial stress states?
7. Is friction always a disadvantage in metal working? Explain.
8. What is the utility of defining the principal stresses for a metal working operation?
9. By unit analysis, show that work done per unit volume in plastic deformation is equivalent to the energy expended per unit volume.
10. What is redundant work? How does redundant work influence the efficiency of a deformation process?
11. How does the draw stress in a wire drawing operation compare to the flow stress (yield stress) if the wire in uniaxial tension?
12. What factors determine the optimum die angle in wire drawing?
13. Is sticking friction a concern in wire drawing? Explain why or why not using an analysis of the key relationships.
14. Would wire drawing of a perfect plastic be "easier" than drawing of a work hardening metal?
15. What factors influence the required pressure for (a) direct extrusion and (b) indirect extrusion.
16. Compare extrusion and wire drawing in terms of their applications and limitations.
17. Hot extruded shapes are sometimes distorted and need to be straightened after they exit the die. Why?
18. Why does the workpiece temperature rise during metal working? Would temperature rise if there were no friction?
19. What is the difference between open die forging and closed die forging?
20. Why are hot working conditions preferred in forging operations?
21. The analysis of open die forging leads to an expression for forging force that is based on the flow stress at the final state of strain, not the

average flow stress. Why? Compare to extrusion and wire drawing, both of which have a dependence on average flow stress.
22. With respect to forging, what is meant by a "friction hill"?
23. Why are multiple shaped die sets used to make shapes in a forging operation?
24. What is difference between impression die forging and precision die forging?
25. What role does flash play in making shapes in a forging operation? What role does flash play in the force needed for closed die forging?
26. How is friction beneficial in rolling?
27. What is the origin of the plane strain condition in rolling?
28. Would large rolls or small rolls be best for attaining large reductions in thickness in a single pass? Why?
29. Would large rolls or small rolls be best for lowering the roll separating force for a given extent of reduction? Why?
30. What is the origin of the "friction hill" in rolling?
31. Why are clusters of rolls sometimes used in rolling mills?
32. What are the main steps in a thermoforming process?
33. Why are amorphous polymers more commonly thermoformed as compared with semicrystalline polymers?
34. Explain the role of heating the polymer initially and also the role of cooling the polymer in the last step.
35. Why is thermoforming called a "secondary process"?
36. What is meant by superplastic forming?

Problems

1. The following stress condition causes yielding ($\sigma_{11} = 690$ MPa, $\tau_{12} = 138$ MPa, $\sigma_{22} = 138$ MPa). (a) What are the directions and magnitudes of the maximum shear stresses? (b) What is the uniaxial yield stress (Y) of the material according to (i) the Tresca condition and (ii) the von Mises condition?
2. A material is placed under the following state of stress ($\sigma_{11} = -200$ MPa, $\tau_{12} = 30$ MPa, $\sigma_{22} = -50$ MPa). (a) Draw the Mohr's circle for the X_1X_2 plane and find the principal stresses and maximum shear stress. What are the magnitudes of the principal stresses and what is the orientation of the principal stress coordinate system relative to the original? (b). The yield strength of the material under uniaxial tension is 50 MPa. Will the material plastically deform if loaded to the given stress state?
3. Derive: $\sigma_2 = (\sigma_1 + \sigma_3)/2$ for plain strain deformation (Eq. 4.30)
4. Find the force needed to draw a steel wire with a diameter of 3 mm to a reduction (r) of 30%. Given: the steel has a strength coefficient of

530 MPa, a strain hardening coefficient of 0.26, elastic modulus 200 GPa. The drawing process has an efficiency of 50%.
5. Find the force needed to draw a steel wire with a diameter of 3 mm to a reduction (r) of 30%. Given: the steel has a strength coefficient of 530 MPa, a strain hardening coefficient of 0.26, elastic modulus 200 GPa. The drawing process has a friction coefficient of 0.15, a die angle ($2\alpha = 12°$), and no redundant work.
6. Answer the following for conditions cited in Problem 5. (a) What is the maximum reduction that can be achieved in a single pass? (b) Find the force needed to draw the wire through a second die that also provides 30% reduction (also with friction coefficient of 0.15, $2\alpha = 12°$), and no redundant work.
7. Is sticking friction a concern in wire drawing? Explain why or why not using an analysis of the key relationships.
8. A factor for redundant work in wire drawing (cylindrical shape) is $\Phi = 0.88 + 0.12 \frac{D_{avg}}{L}$, where D_{avg} is the average diameter and L is the contact length between the die and the wire. This factor is multiplicative in expression for draw stress. For a reduction of 30%, plot the effect of half die angle α on the redundant work factor and comment on the effect of redundancy and friction on choosing die angle.
9. A strip of width w_1 and thickness h_1 is drawn through a wedge shaped die to a thickness of h_2, keeping the width constant ($w_2 = w_1$). The die has a die angle of 2α (analogous to a wire drawing die) and there is friction at the interface between the strip and the die, but no redundant work. Use slab analysis to find this relationship for the draw stress:

$$\sigma_d = Y' \left(\frac{1+B}{B}\right) \left[1 - \left(\frac{h_2}{h_1}\right)^B\right]$$

where Y' is the plane strain flow stress.
10. Starting from the differential equation governing wire drawing, derive the following expression for extrusion through a conical die. Pay attention to the boundary conditions in solving the differential equation and comment on the validity of the assumptions made in the derivation:

$$p = Y \left(\frac{1+B}{B}\right) \left[1 - \left(\frac{D_1^2}{D_2^2}\right)^B\right]$$

11. The following are questions about metal deformation of annealed copper. Data: strength coefficient (K) of 320 MPa, a strain hardening coefficient of 0.54, elastic modulus 115 GPa. (a) Determine the ultimate tensile strength, the true strain and the engineering strain at maximum load. (b) A 30 mm diameter copper bar is cold extruded to 27 mm diameter in a first step and then cold extruded to 22 mm diameter in a

second step. Determine the amount of cold work (expressed in % area reduction) for each step, for the two-step process in sequence and the hypothetical case of a one-step extrusion from 30 mm to 22 mm diameter. (c) Calculate the true strains associated with the steps and total process described in (b). (d) Calculate the yield strength of the copper bar after the two-step deformation. Is this the same as the hypothetical case of a one-step extrusion?

12. Calculate the force required to extrude an aluminum alloy billet from a diameter of 6 inches to a diameter of 2 inches by direct extrusion. Assume the redundant work is 40% of the ideal work of deformation and the friction work is 2.5% of the total work of deformation. (a) Carry out the calculation for cold extrusion. For the Al alloy, $K = 180$ MPa and $n = 0.2$. (b) Carry out the calculation for hot extrusion at which the Al alloy can be consider a perfect plastic with a flow stress of 50 MPa.

13. Experiments reveal that the following empirical relationship for extrusion pressure:

$$p = \overline{Y}(0.8 + 1.5 \ln R)$$

Estimate the largest extrusion ratio possible for cold extrusion of low carbon steel ($K = 530$ MPa, $n = 0.26$) using a 2000 ton press, starting with a rod with diameter of 5 cm.

14. (a) Plot force vs. reduction in height for plane strain forging of an annealed copper specimen that is 1 cm high and 5 cm in length (width is constant at 5 cm), up to a reduction of 75% for the case of (i) no friction and (ii) $\mu = 0.2$. Use average pressure formulas. Efficiency with friction is 65%. (b) Calculate the work done for both conditions described in part a. (c) Calculate the temperature rise for both conditions described in part a. (Given for annealed copper: $E = 110$ GPa, $K = 315$ MPa, $n = 0.54$, $UTS = 220$ MPa, tensile yield strength, 0.2% offset = 69 MPa, $C_p = 386$ J/(kg K).)

15. A metal slab is hot forged from initial dimensions of $1 \times 1 \times 10$ cm to final dimensions of $1/2 \times 2 \times 10$ cm. The forging is carried out with a drop hammer. For the conditions involved, the flow stress is 20 MPa and remains constant during the deformation. Assume sticking friction at the interface and that yielding occurs by a von Mises criterion. (a) How much force is required to produce the final thickness? (b) How much work is required to produce the final thickness? (c) From what height would a 100 kg drop hammer have to fall to complete this operation?

16. An annealed metal has the following strain hardening behavior: $\bar{\sigma} = 670\, \bar{\varepsilon}^{0.4}$ (in MPa). (a) If the bar is initially cold worked 20% followed by additional cold work of 30%, determine the yield stress of the

material after the second step. (b) Estimate the heat rise in the plastic deformation, given that the density of the metal is 7870 kg/m^3 and the heat capacity is 460 J/kgK. (c) Repeat (b) assuming that the deformation has an efficiency of 70%.

17. A piece of an alloy with original dimensions of 10 cm × 5 cm × 1 cm is hot rolled to a thickness of 0.5 cm. The rolling operation has a friction coefficient of 0.4 and rolls with radius of 5 cm. The piece is fed into the rollers without front or back tension and with the 5 cm dimension as the width. If the metal can be considered as a perfect plastic with a flow stress of 100 MPa, find the final dimensions of the rolled piece, and the roll separating force. (b) Compare the roll separating force for the rolling operation to force needed for a plain strain forging to the same final dimensions. The forging process is also carried out hot, but the friction coefficient is 0.25.

18. A steel sheet is hot rolled from a thickness of 0.16 cm to a thickness of 0.14 cm using 20 cm diameter rolls. The friction coefficient is 0.05. The steel has a flow stress of 206 MPa. (a) Find the average pressure that the rolls exert on the metal. (b) Qualitatively, where is the pressure the highest in the nip? Make a sketch.

19. (a) Find the minimum bending radius for a sheet of annealed copper with a thickness of 2 mm [$E = 110$ GPa, $K = 315$ MPa, $n = 0.54$, $UTS = 220$ MPa, tensile yield strength = 69 MPa, Specific heat = 386 J/(kg K)]. (b). Select a bending radius to produce a final radius of curvature of 10 mm.

APPENDIX: STRESS IN A SPHERICAL PRESSURE VESSEL

Understanding the state of stress in a spherical shell is useful for the analysis of some metal working processes. Consider a sphere that encloses a gas or liquid that exerts a pressure, p, on the inner surface of the shell. See Figure A4.1. Due to symmetry, the state of stress at any distance, r, from the center of the sphere is the same. The radial direction is a principal direction and σ_r is a principal stress. All directions tangential to the radial direction are equivalent. Therefore, we can choose two of these directions that are perpendicular to each other as the other two principal directions and the principal stresses in those directions are the same, σ_θ. In this spherical "pressure vessel," the radial stress is compressive and the tangential stress is tensile; both stresses vary with position through the shell. To find these variations, consider a volume element shown in Figure A4.1. It is clear that the forces on the element are balanced in the tangential directions. A force balance in the radial direction followed by elimination of very small terms leads to a differential equation that governs the stress in the shell.

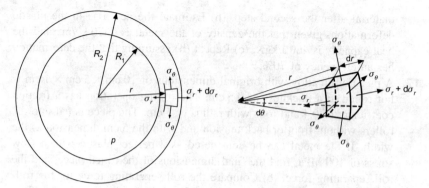

FIGURE A4.1 (Left) Spherical thick walled pressure vessel with internal pressure p. (Right) Stresses acting on a volume element.

$$\frac{d\sigma_r}{dr} + \frac{2}{r}(\sigma_r - \sigma_\theta) = 0 \qquad (4.\text{A}1)$$

This equation can be integrated to find σ_r and σ_θ as a function of r using the material properties of the shell and the boundary conditions. For an incompressible elastic solid with an internal pressure p:

$$\sigma_r = -p\left(\frac{R_1^3}{R_2^3 - R_1^3}\right)\left(\frac{R_2^3}{r^3} - 1\right) \quad \sigma_\theta = p\left(\frac{R_1^3}{R_2^3 - R_1^3}\right)\left(\frac{R_2^3}{2r^3} + 1\right) \qquad (4.\text{A}2)$$

where R_1 and R_2 are the inner and outer radii of the sphere, respectively.

Chapter 5

Powder Processes

5.1 INTRODUCTION

Powder processes convert engineering materials in a dry powdery state to solid shapes. A "powder" is a collection of solid particles with individual sizes in the range of nanometers to microns. Each individual particle has physical properties, such as yield strength and melting point; these properties depend on the chemical composition as well as the nano- and microstructure of the particle. Typically, the physical properties of a particle are similar to those of a bulk solid of the same composition. For example, ceramic particles are hard and brittle, similar to the characteristics of a final ceramic solid. Particles of metal alloys that are known for their low yield strengths also have low yield strengths and so on. A powder, however, as a collection of these individual particles, behaves much differently than the bulk solid. A powder "flows" under gravity, for example, provided that the individual particles have the right characteristics to move past each other easily.

Powders can be shaped. In this chapter, the powder processes involve powders in the dry state with little or no liquid added. These powders can be engineered so that they can be poured into molds or dies and then compacted under pressure to form shapes. Figures 5.1a and b show schematic diagrams for some of the powder forming processes that are covered in this chapter. The compacted shapes can be simple or fairly complex. Frequently, the compacted part is porous and a post-forming heat treatment step called *sintering* is needed to create a dense final part. Polymer powders can also be processed into shapes. These powders are converted to complex, hollow shapes through rotational molding. See Figure 5.1c. Lastly, several additive processes make use of powders. Figure 5.1d shows one of these processes—selective laser sintering.

With powder forming operations, we again encounter the same sequence of steps to form the engineering material into the desired geometry. After selecting and preparing the starting material, the powder, the first step is *flow*. The powder must be engineered to flow easily and reproducibly, not unlike the necessity to engineer a liquid melt for its flow properties or a solid for its plastic flow properties. The powder flows into a mold or die, which is also responsible for *shape definition*, and mechanical compaction, sometimes

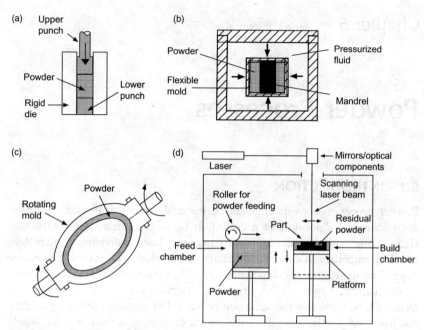

FIGURE 5.1 Schematics for some powder processes: (a) uniaxial compaction—powder is compacted, shaped into a green part with a punch and rigid die; (b) isostatic compaction—powder is compacted, shaped into a green part, using a pressurized fluid, flexible mold, and rigid mandrel; (c) rotational molding—powder is distributed on the inside of a rotating mold, heated to create a melt layer and then cooled to make a shape; and (d) selective laser sintering—powder is distributed onto build platform where it is converted to solid, layer-by-layer, using a scanning laser beam. *(d) Adapted from Gibson et al. (2010).*

accompanied by heat, takes place. *Shape retention* occurs naturally in a well-designed process. The compacted powder part retains the shape established by the mold or die, and does not revert back to its original loose state. Bonds form between the particles during compaction. For some powders, polymer binders are also used to assist in the interparticle bonding in the "green" or unfired state. Lastly, post-forming thermal treatment is critical in most cases to densify the final part.

There are several powder-based additive methods that also involve flow, shape definition, and shape retention. The starting point for these additive processes is a layer of powder on a build platform. Powder *flow* is very important to creating a uniform layer in this first step. Portions of this layer are selectively converted to solid by either adding droplets of binder solution via an ink-jet printing head, which moves over the surface of the powder bed, or by local heating with a laser, which is scanned in a 2D pattern over the powder surface. Then, the build plate is lowered and a fresh layer of powder is added and the process repeats. Hence, *shape definition* is accomplished through the CAD model that ultimately drives the creation of the part. The binder process is known as "3D Printing," but this term has grown to be synonymous with additive processing overall. The binder solution links particles together into a green state, providing a means of *shape retention*. The laser-based processes either locally melt the powder (selective laser melting, SLM) or sinter it (selective laser sintering, SLS). Both result in structural changes that retain the shape of the part.

Powder processes are vital to the fabrication of polycrystalline ceramics. The hard and brittle nature of most ceramics makes solid deformation impossible; the high melting points and demanding microstructure requirements of ceramics limit the use of melt-based forming operations. Therefore, powders are frequently the only choice. Powder processing of ceramics draws from deep historical roots. Naturally occurring ceramics, clays for example, are removed from the earth, crushed into powders, shaped into objects and then "fired" to make useful vessels and dishes, for example. Similar processes are used to make technical ceramics from more refined powders. A principal process sequence is compaction and sintering. This method can be applied to nearly any ceramic, but it is almost exclusively used for the processing of polycrystalline ceramics, and rarely to ceramic glasses, which have much more flexible melt-based processes open to them. Moreover, ceramics are prepared by powder-based additive processes.

Powder-based processing routes, collectively known as "powder metallurgy," are important for metal part fabrication. Similar to ceramics, many metal powder processes involve compaction and sintering. These techniques have grown from a niche method to prepare difficult to process metals, such as refractory metals (e.g., tungsten), to a routine route for increasingly complex shapes. With melt processes and solid deformation processes viable routes for many metals, why would powder metallurgy be gaining in importance and application? The answer is found in the concept of "near net shape" processes. From previous chapters, we know that after forming metal parts from the melt or from the solid, machining and surface finishing are commonplace. The part made by powder metallurgy has little or no post-forming machining; it is "near net shape." Material waste is reduced, and sometimes overall process time is also reduced and economic advantages are realized. The automotive industry is the driver in many developments in powder metallurgy. Powder-based additive processes, principally SLS and SLM, are also becoming more important for complex shapes.

Powder processes has some application and importance in polymer processing. One challenge is the formation of solid polymer particles. While polymer pellets on the millimeter size scale are used as starting materials for thermoplastic melt processes, such as extrusion and injection molding, powders are made up of much smaller particles. There are several processes that start with a dry polymer powder. In this chapter, we explore rotational molding and powder-based additive processes. In rotational molding, a flowable polymer powder is loaded into a hollow metal mold and then distributed by rotation onto the interior surfaces of the mold as it is heated, and converted to a solid, dense hollow object by melting and solidification. Additionally, polymer-based additive processes are used to make complex polymer shapes.

The engineering of powder processes requires attention to all process steps. First, powders must be selected and prepared for the multiple requirements. For example, in powder forming operations, the powder is engineered for flow, compaction, and sintering. Specific engineering solutions are needed to handle the inherent trade-offs. For example, fine particle sizes are best for fast sintering but fine powders do not flow as well as larger-sized particles. This trade-off is also encountered in powder-based additive manufacturing. Other process steps, such as compaction, require dies and conditions that are designed to produce uniform compacts. Lastly, the conditions for sintering, or melting in the case of the SLM additive process, must be tuned to produce high density in a short time. Dimensional control further requires attention to all the steps.

This chapter begins with fundamentals concepts and then explores several important powder processes. The fundamentals include an overview of powder characteristics and flow, a section on shrinkage and density changes, and lastly the fundamentals of sintering and microstructure control. Four types of powder processes are then covered. The first is powder compaction at room temperature under uniaxial and isostatic pressing conditions. Another section explores the "hot" counterparts to these two processes in which heat and pressure are used simultaneously. Rotational molding of polymers is then presented as unique route to large, hollow polymer shapes. Lastly, powder-based additive processes are covered.

5.2 FUNDAMENTALS

5.2.1 Powder Characteristics and Flow

For powder processes, obviously, the powders are considered starting materials. These materials, manufactured by vendors, are purchased and then sometimes further prepared for particular forming operations. Here, we consider the powder state in a general way and describe its characteristics, including particle packing and flow. Later, the important differences between metal, ceramic, and polymer powders are presented as the different powder processes are discussed.

Powder Characteristics. Each individual particle in a powder has a set of characteristics. See Table 5.1. The solid particle may be crystalline, semicrystalline, or amorphous. A crystalline particle can be a single crystal or polycrystalline. The particle can contain a single phase or several phases. It may be completely dense or contain pores. Its shape may be symmetric (e.g., spherical) or asymmetric (e.g., needle-like or plate-like). For a symmetric shape, a single particle size, such as the particle diameter, characterizes the size of the particle, but for an asymmetric particle, two or three dimensions

TABLE 5.1 Particle and Powder Characteristics[a]

Characteristics of a Single Particle	Powder Characteristics
Size	Particle Size Distribution
	Average Particle Size
	Standard Deviation in Particle Size
Shape	Characteristic Particle Shape
Composition	Average Composition
Crystallinity and Phase Content	Average Crystallinity and Phase Content
Density	Average Particle Density
	Bulk Density
	Tap Density
Geometric Surface Area	Specific Surface Area (surface area per quantity of powder)
Yield Strength	Compressibility or Compaction Behavior
----	Flow Rate
Surface Energy	----

[a]Adapted from Svarovsky (1990).

might be needed. Other less easily characterized features, like the yield strength and surface energy, are also important to powder processes. Based on the list in Table 5.1, to completely characterize a single particle is a challenge. Given that a process may call for a quantity of powder that encompasses 10^{20} or 10^{30} particles, the characterization challenge appears to be even greater for the powder.

Powder characteristics and behaviors are determined by the collective properties of many, many constituent particles. For a powder, distributions and ranges matter. The particle size, for example, is a distribution that is quantified by an average and a standard deviation. Likewise, particle shape may vary in the powder over some range but have a certain qualitative or semiquantitative description. Density is a collective property that depends on how the powder packs, so it is also related to the size distribution as noted later in this section. Other analogies between individual particles and the powder are given in Table 5.1. There are some characteristics that have no analogies. For example, the flow rate of a powder has no single particle analogy, and the surface energy of a particle has no powder analogy.

Several analytical tools are available to characterize a powder. As described in Chapter 2, the particle size distribution is determined using techniques that operate over various ranges of size. Scanning electron

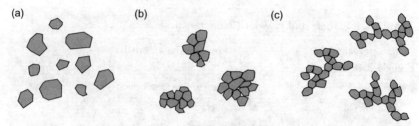

FIGURE 5.2 Schematics of (a) individual particles, and (b) and (c) different geometries of aggregates and agglomerates. The particles in some of these clusters are held together by strong chemical bonds while in others the bonds between particles are weak.

microscopy, transmission electron microscopy, and optical microscopy have an important role to play in characterizing overall morphology and shape, in addition to being one method of determining particle size distribution. Crystallinity and phase content are determined by x-ray diffraction, and the specific surface area is found by nitrogen gas adsorption. The flow and compressibility of powders are discussed later.

Since the powder is a collection of particles, individual particles in the powder interact with each other and frequently these interactions lead to the formation of clusters called agglomerates or aggregates. See Figure 5.2. Clusters held together weakly, so-called weak agglomerates, are broken apart by input of moderate agitation. These clusters frequently form spontaneously in dry powder due to van der Waals attraction or if some moisture is present in the powder, then by capillarity. By contrast, some clusters are composed of strongly bonded particles that are difficult to break down into individual particles. These clusters are typically called hard agglomerates or aggregates. Frequently, these clusters originate in the synthesis of the powder, involving bonds formed at high temperature, for example.

Van der Waals attractive force is a ubiquitous force, arising from dipole-dipole interactions between atoms on the surfaces of neighboring particles. The collective interaction of multiple atoms on the surfaces makes this force significant. The magnitude of the attraction increases as particles approach each other and with an increase in the Hamaker constant, which depends on the properties of the materials as well as the substance between them (i.e., air in the case of dry powders). This topic is discussed in detail in Chapter 6. Because the particles in a dry powder are in contact and also can be jostled about in transport, the formation of agglomerates is expected; therefore, in preparation of the powder for compaction or other powder processes, steps are usually taken to break the agglomerates up.

Particle Packing. One goal of most powder processes is to establish a high density of particles. That is, packing particles to fill space efficiently is desired since the eventual goal is a dense part. Initial packing density, even before any application of force or pressure, depends on the ability of the particles to rearrange and also on the particle size distribution.

An idealized powder composed of spherical particles, all of the same size (monodisperse), is the simplest case. In this sort of powder, the spheres pack to fill about 52% to 74% of the space, depending on how the particles are arranged. That is, the packing fraction, or the sphere volume divided by the total volume, ranges from 0.52 to 0.74. The highest packing fraction is achieved when the particles arrange themselves into hexagonal close packing. It is important to realize that the packing fraction is based not on the size of the individual particles, but rather how the particles are arranged. That is, basketballs arranged in hexagonal close packing fill 74% of the space, just the same as spheres with 1 µm diameter organized in the same way. The size of the spaces between particles, however, does change with the particle size. These spaces, which we can think of a "pores," are proportional to the size of the particles.

The most realistic arrangement is random packing in which the spheres fill about 64% of the available space. In other words, the packing fraction is 0.64. If the spheres themselves are completely dense (i.e., they contain no internal pores), then the pore fraction, or the open space volume divided by the total volume, is 0.36. The porosity, P, is usually given as a percent. The porosity of a random packing of dense particles is 36%. The density of the packing is obviously much less than the density of the individual particles. In fact, this packing would have a density that is 64% of the density of the particles. We would say that it is 64% dense or that it has density that is 64% of the theoretical density.

Most powders contain particles of various sizes and sometimes various shapes. Therefore, the packing behavior is not so uniform or predicable. As a first step, consider a mixture of relatively large particles with relatively small ones. In such a mixture, the small particles fit in the spaces between the large particles and therefore increase the packing density. The data in Figure 5.3 illustrate this concept. For the goal of high packing density, there is an optimum in not only the amount of fine particles but also the size of the fine particles relative to the coarse particles. The addition of a third particle size can also be considered as can a fourth, etc. In practice, the particle size distribution can be tailored to some extent, but given the expense of monodisperse particles it is usually not possible to engineer particle size distributions to such a degree.

Powder Flow and Density. Powder processes require the powder to be "free flowing." That is, when it is put in a funnel or other feeding device, the powder should flow readily and reproducibly under gravity. Several factors influence flow. For example, dry sand, which is a very coarse sort of powder, flows easily, but moist sand does not flow readily. Why? Water bridging between the "grains" of sand creates capillary forces that hold the grains together and prevent them from moving past each other. From this example, we learn the key to powder flow—individual particles must be free to move relative to each other. Capillary forces are eliminated by drying a

350 Materials Processing

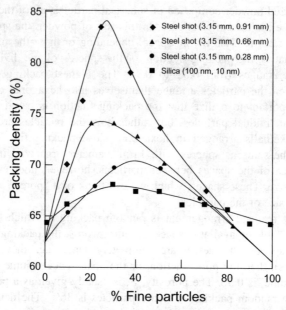

FIGURE 5.3 Packing densities in mixtures of monodisperse spheres of two sizes shown in the legend. Results are shown for mixtures of two sizes of steel shot [*data from McGeary (1961)*] and two sizes of colloidal silica [*data from Gauthier and Danforth (1991)*].

powder before it is used, but the relative motion of particles is also affected by other factors, including friction, steric effects (i.e., geometric limits in the mobility of particles), and particle-particle interactions. The latter includes van der Waals attraction. These other factors are influenced by particle characteristics, such as size and shape, as well as chemical composition.

Friction limits powder flow. Individual particles in a powder loaded into a funnel or hopper are in contact with other particles under a complex set of loading conditions based on how gravitational forces are distributed. See Figure 5.4. Each particle experiences a complex load with shear and normal forces exerted by neighboring particles. The container also enters into the picture due to friction between container and particles. Motion requires overcoming friction. That is, at the contact the shear force, F_S, must be greater than μF_N, where μ is the friction coefficient, as discussed in Chapter 4. But the situation is complex because several forces act on a single particle.

Some qualitative factors influencing powder flow can be understood. First, since friction takes place at the particle-particle (and particle-container) contacts, reducing the number of such contacts should reduce the resistance to flow. Very fine powders have a large number of particle-particle contacts per unit volume and therefore a greater resistance to sliding and hence flow. Spherical particles tend to flow more easily than plate-like or irregular particles due to their tendency to have point contacts rather than contacts

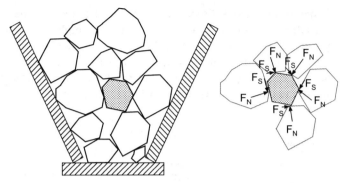

FIGURE 5.4 Schematic of powder in a funnel with a closed end. Each particle is under a complex load from gravity. When the end of the funnel is opened, the powder will flow if the shear forces are able to collectively overcome the frictional resistances at particle-particle contacts and between the particle and the walls. *Adapted from Reed (1995).*

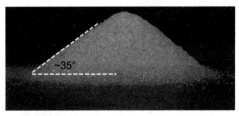

FIGURE 5.5 A pile granulated sugar created by flowing the sugar through a funnel onto a flat surface. The approximate angle of repose is shown.

that have appreciable surface area. Further, there are steric factors that cause asymmetric particles to become "stuck." Reducing the friction coefficient at interfaces, with a lubricant for example, is another important strategy for improving flow. Additionally, flow is encouraged with an entrained gas (i.e., fluidized bed transport).

Powder flow is a measurable quantity. One easy measure of powder flow is known as the angle of repose. In this approach, powder is allowed to flow from a funnel onto a horizontal surface. The powder naturally heaps up and develops an angle, which is characteristic of the friction and flow behavior. See Figure 5.5. Steeper angles correlate with poorer flow properties. For more quantitative information, there are apparatuses designed to shear powder and measure the resistance to shear. For example, the Jenike shear test involves placing two, packed powder beds in contact and recording the force needed to translate one relative to the other. See Figure 5.6a. These tests are useful for understanding not only the flow behavior an unconsolidated powder, but also they can be used to determine friction coefficient. More common are simpler tests of flow time. In powder metallurgy, powder is loaded into a standard funnel and then the time for a fixed quantity (e.g., 50 g) to pass through the funnel is found. See Figure 5.6b. This test is combined with a standard size cup in order to determine a bulk density of a powder.

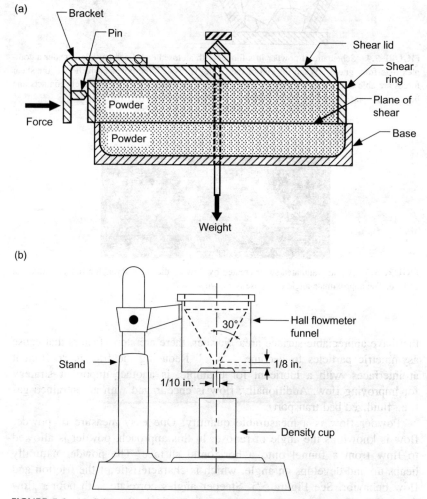

FIGURE 5.6 (a) Schematic of a Jenike shear apparatus. (b) Hall funnel with density cup. *Reprinted, with permission, from ASTM Standard D6128 (2014), Standard Test Method for Shear Testing of Bulk Solids Using the Jenike Shear Cell and ASTM Standard B213 (2013) Standard Test Methods for Flow Rate of Metal Powders Using the Hall Flowmeter Funnel, copyright ASTM International, 100 Barr Harbor Drive, West Conshohocken, PA 19428. A copy of the complete standard may be obtained from ASTM International, www.astm.org.*

$$\text{Bulk Density of a Powder} = \frac{\text{Mass of powder in the cup}}{\text{Volume of the cup}} \quad (5.1)$$

Frequently, a "tap density" is also found. In this case, the powder vessel is tapped gently and the volume remeasured. The above are more commonly used in the characterization of metal powders as compared.

Powder additives influence flow. Most notably, lubricants are designed to reduce friction between particles and between the surfaces of process equipment and the powder. Lubricants are either used to treat process surfaces or they are mixed in with the powder. Binders, which are additives designed to increase strength after compaction, may reduce or increase friction, depending on the chemistry and conditions.

5.2.2 Sintering and Microstructure Development

Sintering Overview. Parts made by compacting powder at room temperature are porous. To decrease the amount of porosity in the powder compact and thereby improve its mechanical strength and enhance many other physical properties, the part is sintered. Sintering is a high temperature treatment that converts the porous assembly of particles into a denser, stronger solid with a polycrystalline microstructure. A schematic diagram of the changes that occur during sintering is shown in Figure 5.7. This diagram begins from a compacted powder in which there are distinct particles and particle-particle contacts. As the temperature increases and the microstructural transformation

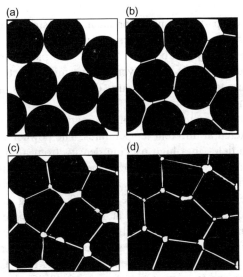

FIGURE 5.7 Cartoons of microstructure changes on sintering: (a) powder compact, (b) initial stage, (c) intermediate stage, and (d) final stage. *From German (1991), reproduced with permission from ASM International.*

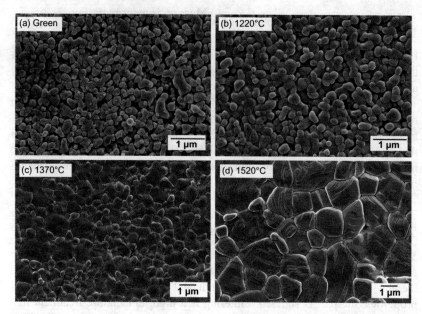

FIGURE 5.8 Scanning electron micrographs of the surfaces of α-alumina pellets in (a) the green powder compact state and after sintering for two hours at (b) 1220°C, (c) 1370°C, and (d) 1520°C. The arrows on the 1370°C image mark pores. The images of the green pellet and the pellet sintered at 1220°C are at a magnification of 20,000x and the images of the pellets sintered at 1370°C and 1520°C are at 10,000x. *SEM micrographs courtesy of H. Luo, University of Minnesota.*

begins, the particle-particle contacts broaden, increasing in area. The spaces between the particles, the pores, decrease in volume. The particles themselves lose their identity and become "grains," and particle-particle contacts become grain boundaries as the polycrystalline microstructure develops. As this microstructure transformation occurs, density increases, and the physical dimensions of the part shrink.

Figure 5.8 shows scanning electron micrographs of the surfaces of alumina pellets sintered at different temperatures. The powder compact consists of submicron particles in a high packing density from a compaction operation. Distinct particles with irregular shapes are clear. After sintering at 1220°C, the surfaces appear rounded and the boundaries between particles (now better referred to as grains) are flattened, forming grain boundaries. The porosity decreases after sintering at 1220°C. After sintering at 1370°C, the microstructure appears almost fully dense. Grains and grain boundaries are clear and a few gaps, pores in the surface image are apparent. The grain size is much larger than that in the pellet sintered at 1220°C. Facets on the grain surfaces are due to restructuring from surface diffusion, a process that only affects the surface microstructure. These facets are even more prominent in the specimen sintered at 1520°C. This high temperature sintering also results in larger grains. No pores are apparent in this surface image, but measurements of density reveal that the pellet still has about 4% porosity.

There are two main type of sintering: solid state sintering and liquid phase sintering. Solid state sintering is densification without the formation of a liquid phase during heating. The alumina sintering shown in Figure 5.8 is a good example solid state sintering. In this type of sintering, solid state diffusion is at the core of the structural change. Remarkably the massive change in dimensions and microstructure comes about as atoms diffuse, hopping from one vacant site in the crystalline lattice to the next. During sintering, solid/vapor interfaces in the compact are eliminated, grain boundaries are formed, and pores disappear. Competing processes that results in particle and pore shape change with little densification are also possible, and later in the sintering processes the grains may grow rapidly, which often prevents full densification. If the particles are amorphous, solid state sintering involves viscous flow rather than diffusion. However, this is a special case, treated separately at the end of this section. Liquid phase sintering, on the other hand, involves the formation of a liquid phase at high temperatures and the participation of that liquid phase in the densification process. The liquid allows rearrangement of particles and speeds the diffusion process. While the goal of sintering is densification, competing thermally activated changes in structure can interfere with densification. These processes are coarsening, which increases the particle size in the early stages of sintering, and grain growth, which increases in grain size in the later stages.

Solid State Sintering. During sintering, the microstructure changes dramatically from discrete particles in contact with pores interspersed to a dense microstructure consisting of grains, grain boundaries and some residual porosity. In this discussion, we consider the case in which the initial microstructure begins as a single phase powder compact with particles touching at point contacts so that there is an interconnected network of pores. During heating, changes take place on the microstructural level as shown in Figures 5.7 and 5.8, and also on the macrostructural level. That is, the physical dimensions of the part shrink as the porosity is removed. Therefore the progress in sintering can be tracked both by microscopy and by monitoring part dimensions or density changes.

Figure 5.9 shows an example of the density increase that accompanies sintering. The density rises gradually at lower temperatures and then more steeply before leveling off. These transitions together with characteristic microstructure changes mark the three stages of sintering. In the initial stage, there is very little macroscopic densification, but the microstructure begins to show signs of change. Notably, the particle-particle contacts begin to fill in to form "neck" regions and grain boundaries (i.e., areas of contact) form. Some coarsening or particle size increase may also occur. At this stage, porosity is interconnected. In the intermediate stage, rapid densification occurs with major changes in the microstructure. The grain boundary area increases and pore content drops. Pore size also decreases and at the end of this stage, the pores become isolated from one another (i.e., the interconnection is lost). Some increase in the grain size is also typical of the intermediate stage. In the final stage of sintering, only a small increase in density is observed. The pores are closed and isolated, and grain growth is rapid. Pores may become engulfed within grains. In the final stage, a limiting density is reached; longer hold times do not lead to increases in density but rather simply increases in grain size.

In trying to understand sintering, the first question to consider is why does sintering occur? The "driving force" for sintering is thermodynamic. That is, sintering lowers the total Gibbs free energy of system. The Gibbs free energy, G, has volume, boundary (solid-solid interface), and surface (solid-vapor interface) contributions.

$$G_{\text{total}} = G_{\text{volume}} + G_{\text{boundary}} + G_{\text{surface}} \tag{5.2}$$

Assuming that the volume of the material is constant then the change in the free energy on sintering is given by:

$$\Delta G_{\text{total}} = G_{\text{sintered}} - G_{\text{porous}} = \gamma_{GB}\Delta A_{GB} + \gamma_{SV}\Delta A_{SV} \tag{5.3}$$

where γ_{GB} is the solid-solid interfacial energy, or the grain boundary energy, per unit area, γ_{SV} is the solid-vapor interfacial energy per unit area, ΔA_{GB} is change in the area of the grain boundaries on sintering, and ΔA_{SV} is the

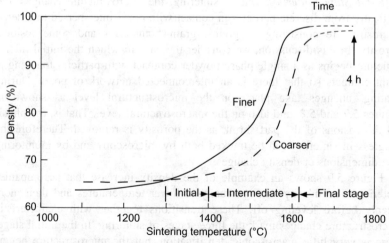

FIGURE 5.9 Increase in density of alumina (given as a % of theoretical, single crystal density) with sintering temperature. Data is shown for compacts pressed from alumina particles having two different averages particle sizes (1.3 μm and 0.8 μm). An extension shows little change with sintering time at the highest temperature. *Reproduced with permission of John Wiley & Sons from Reed (1995);* © *1995 by John Wiley & Sons, Inc.*

change in the surface (solid-vapor interface) area on sintering. Since the grain boundary area formed is less than the surface area removed and especially when γ_{GB} is lower than γ_{SV}, replacing the solid-vapor surfaces with grain boundaries is energetically favorable. The resulting negative ΔG_{total} is the thermodynamic driving force for sintering.

While this overall thermodynamic driving force is instructive, it does not provide the necessary understanding for the mechanistics of sintering. That is, it does not explain how the atoms rearrange themselves to accomplish this overall structural change. To dig down to this level, the microscopic, thermodynamic driving forces must be considered. That is, for an atom to move from point A to point B, the chemical potential (i.e., partial molar Gibbs free energy) of an atom at point B must be less than its chemical potential at point A. The chemical potentials of atoms at surfaces and boundaries are impacted by the local curvatures. We could base this description on the fundamental thermodynamic expressions for the chemical potential, but instead the mechanism is more easily understood by examining two quantities that are determined by the chemical potential: equilibrium vapor pressure and vacancy concentration.

First, surface curvature leads to changes in equilibrium vapor pressure over that surface. These vapor pressure differences are experienced by both liquids and solids, but the focus is on solids in this discussion. The Kelvin equation gives the relationship between vapor pressure and curvature. Here is the expression for the case of a spherical curvature:

$$\frac{P}{P_o} = \exp\left(\frac{2\gamma_{SV} V}{rRT}\right) \tag{5.4}$$

where P is the vapor pressure *over* a surface with radius of curvature r, P_o is the vapor pressure over a flat surface, and V is the molar volume of the liquid or solid material that is vaporizing. See Figure 5.10. A convex surface has a positive radius of curvature and therefore a higher vapor pressure as compared to a flat surface. Likewise, the concave surface has a negative

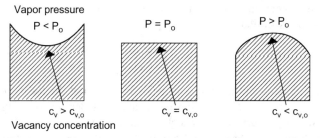

FIGURE 5.10 Effect of surface curvature (left: concave, center: flat and right: convex) on vapor pressure over a curved surface and vacancy concentration in the solid under the curved surface.

358 Materials Processing

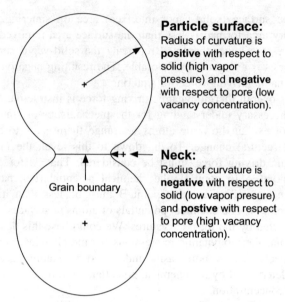

FIGURE 5.11 Curvatures in partially sintered particle compact.

radius of curvature and hence a lower vapor pressure. In the powder compact, therefore, one can expect evaporation from some surfaces in a powder compact and condensation onto others. The extreme would be the evaporation from the particle surface (positive curvature) and the condensation into the region between particles, which is known as the neck (negative curvature). See Figure 5.11. The mechanism of material motion in this case is evaporation—condensation rather than atomic diffusion. While this transport leads to the change in the shape of particles and pores, it does not lead to densification. That is, particle centers do not approach each other. For densification by solid state diffusion, it is more important to consider this problem from the viewpoint of vacancy concentration.

Surface curvature also changes the local vacancy concentration. The situation is somewhat similar to that of vapor pressure, and therefore the expression is similar.

$$\frac{c_v}{c_{v,o}} = \exp\left[\frac{2\gamma_{SV}\Omega}{(-r)kT}\right] \qquad (5.5)$$

Here, $c_{v,o}$ is the concentration of vacancies in a solid at the location *under* a flat surface, c_v is the concentration of vacancies under a surface that has a spherical radius of curvature r, which is defined in the same way as above, Ω is the volume of a vacancy, and k is Boltzmann's constant. In this expression the negative sign is used to convert the radius of curvature sign so that it represents the radius of curvature with respect to the pore or vacancy

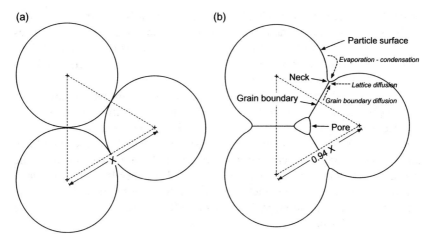

FIGURE 5.12 (a) Before sintering: particles just touching with their centers separated by a distance X and (b) After partial sintering: particles centers are closer (0.94X), indicating shrinkage. In (b) features (pore, grain boundaries, particle surfaces, necks) and local curvatures are shown along with pathways for material motion (dotted arrows); vacancy motion goes in the opposite direction. Lattice and grain boundary diffusion result in shrinkage, while evaporation-condensation changes particle shape and pore shape without changing the distance between particle centers. *Adapted from Ashby (1974).*

(rather than the solid). Vacancy concentrations in the neck region are high relative to a flat surface. See Figures 5.10 and 5.11. Therefore, a vacancy concentration gradient exists between the neck and grain boundary, as well as between the neck and the bulk or surface of the grain or particle.

Given the local concentration gradients in the powder compact due to curvature, densification occurs as diffusion proceeds. See Figure 5.12. In this example, we assume that the mechanism for self-diffusion (i.e., atomic mobility in the absence of compositional gradients) is by a vacancy mechanism in which atoms jump from regular lattice sites onto adjacent vacant sites. Therefore, the concentration gradient of vacancies drives diffusion of vacancies away from the neck. Concurrently, atomic diffusion goes in the opposite direction. That is, atoms move from grain to the neck via the bulk of the grain or along the grain boundary. Particle centers approach each other because material is transported, usually along the grain boundary, to the neck. Material is displaced from between the particle centers, causing the particle centers to approach each other and the part to shrink. The mechanisms of lattice and grain boundary diffusion cause shrinkage. Diffusion along the surfaces, like evaporation-condensation, merely reshapes the microstructure and does not lead to shrinkage.

From both macroscopic and microscopic viewpoints, the size of the particles is a key factor controlling the thermodynamic driving force. Small particles create compacts with large amounts of solid-vapor surface area, and hence large decreases in Gibbs free energy are possible by replacing these surfaces with grain boundaries during densification. Likewise, fine particles also have small radii of curvature at their contact points or necks; therefore, high vacancy concentration gradients are created in the compact, creating a strong driving force for diffusion.

With this understanding of driving forces and mechanisms, the next question to consider is what controls the rate of sintering and from an engineering perspective how can that rate be increased? From the discussion above it is clear that diffusion plays a key role in densification and that the driving force for diffusion is greatly increased when fine particles are used. However, the rate at which diffusion actually takes place also depends on the diffusion coefficient. In simple terms, Fick's first law states that the flux, J, is proportional to the concentration gradient of the diffusing species, dc/dx, and the diffusion coefficient, D:

$$J = -D\left(\frac{dc}{dx}\right) \tag{5.6}$$

Diffusion is a thermally activation process and the diffusion coefficient is given in general by:

$$D = D_o \exp\left(\frac{-Q_d}{kT}\right) \tag{5.7}$$

D_o is a preexponential factor, k is Boltzmann's constant, and Q_d is the activation energy for diffusion. Both D_o and Q_d depend on the specific material and mechanism of diffusion. In this expression notice the importance of temperature—higher temperature increases the diffusion coefficient and hence the flux of atoms. In fact a common rule of thumb is that solid state sintering begins to go forward at an appreciable rate when the temperature (in Kelvin) is about 2/3 of the melting point, also expressed in K. So a refractory metal like tungsten with its high melting point requires a much higher temperature for sintering than a lower melting point metal such as copper.

Grain boundaries are key to the densification process in part because diffusion along grain boundaries is faster than through the lattice. Initially grain boundaries form at the particle-particle contacts. Grain boundaries are interfacial zones that have high disorder and hence lower activation energies for diffusion and higher diffusion coefficients as compared to the bulk of the grain. The grain boundaries, therefore, are conduits for material transport. They are essential for the densification process. This point is important because over the course of sintering, grain growth typically takes place, lowering the overall content of grain boundaries.

Now, returning to the three stages in the sintering process. In the *initial stage*, grain boundaries form, pore shapes change and a small amount of densification takes place. Researchers have successfully modeled this initial stage by linking the linear shrinkage, $\Delta L/L_o$, for example, to parameters such as particle diameter, d, and temperature. For the case of a compact sintering by diffusion, the vacancy concentration gradient established due to the surface curvatures leads to a driving force for diffusion, as shown in Figure 5.11. With two parallel paths available for transport, the fastest path controls the process. In many cases, the grain boundary diffusion is faster than bulk diffusion and the linear shrinkage ($\Delta L/L_o$, see next section) depends on the parameters related to the grain boundary:

$$\frac{\Delta L}{L_o} = \left(\frac{48\gamma_{SV}\Omega\delta D_b t}{d^4 kT}\right)^{1/3} \qquad (5.8)$$

where the grain boundary has a width δ and a diffusion coefficient, D_b. Other variations of the shrinkage equation exist for different dominant pathways, but the same general dependencies on key variables hold.

Equation 5.8 shows the effects of key parameters on shrinkage and densification in the initial stage. First notice the importance of the particle size. Compacts composed of smaller particles shrink faster (Example 5.1). Here, the stronger driving force plays a role, but also the abundance of grain boundaries is important. The effect of the diffusion coefficient is also clear. If diffusivity can be boosted, for example by increasing the temperature, then shrinkage is faster. While temperature is in the denominator of the expression, shrinkage increases with temperature because the effect of temperature on D_b is more important due to the exponential dependence (see Eq. 5.7). The properties of the material, namely, the surface energy or solid-vapor interfacial energy, as well as the vacancy volume and boundary width also play roles though not variable over a wide range for a given material.

Most of the densification takes place in the next stage, *the intermediate stage*. In this stage, the pore content and size drops; pores intersecting grain boundaries shrink the most quickly. The pore can be thought of a vacancy "source" and the grain boundary a vacancy "sink." The pores are interconnected through most of this stage, which means that the gas in the pores has a pathway to escape as the pore volume drops. Energetically, pore removal depends on the grain boundary energy and the surface energy and the configuration of the pore relative to the grain boundaries of surrounding grains (i.e., coordination number of the pore).

As shown in Figure 5.13, the dihedral angle, Ψ, is defined from a force balance in a similar manner as the Young equation in Chapter 3, assuming interfacial energies can be expressed as interfacial tensions.

$$\gamma_{GB} = 2\gamma_{SV}\cos\left(\frac{\Psi}{2}\right) \qquad (5.9)$$

The example in Figure 5.13 shows a two-dimensional sketch of a material with a dihedral angle of $120°$ ($\gamma_{SV} = \gamma_{GB}$) and a pore that intersects six

EXAMPLE 5.1 (a) Show the effect of particle diameter, d, on the linear shrinkage as a function of time. Compare 0.1, 1.0 and 10 μm, keeping other parameters constant. (b) Find an expression for the rate of shrinkage as a function of time.

a. We can set all the parameters except for d and t in Equation 5.8 equal to one. This selection creates an arbitrary scale for shrinkage and time.

$$\frac{\Delta L}{L_o} \propto \left(\frac{t}{d^4}\right)^{1/3}$$

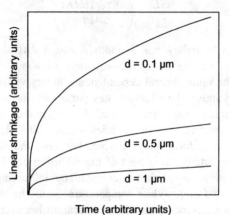

FIGURE E5.1

The effect of particle diameter is clear from this plot. The initial stage shrinkage at any particular scales with $d^{-4/3}$.

b. To find an expression for shrinkage rate, we take the derivative of Equation 5.8.

$$\frac{d}{dt}\left[\frac{\Delta L}{L_o}\right] = \left(\frac{48\gamma_{SV}\Omega\delta D_b}{27t^2 d^4 kT}\right)^{1/3}$$

Time is now in the denominator, indicating that the rate drops off with an increase in time. The plot also shows this trend; the shrinkage rate (slope of the plot) is steep initially and then tails off.

---

adjacent grains or four adjacent grains. The flatness of the surfaces in the case of six adjacent grains indicates that such a pore would be stable. The lack of curvature means no vacancy concentration gradient exist and no driving force for vacancy transport to the neck. By contrast, few intersecting grains would tend to lead to pore removal due to the curvature. The situation is somewhat different in three dimensions but the principle is the same. Again considering Eq. 5.9, materials with a high grain boundary energy compared to surface energy ($\gamma_{SV} < \gamma_{GB}$) have a dihedral angle that is lower

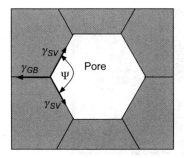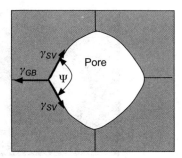

FIGURE 5.13 Sketch of a pore intersecting six grains (left) and four grains (right). The dihedral angle, Ψ, is defined and constant for the material. The pore intersecting fewer grains has a curvature that favors pore shrinkage.

($<120°$), and therefore the pore stability in the intermediate and final stages of sintering often limits the total final density.

The intermediate stage is complex with the possibility of multiple mechanisms. For example, local packing density variations influence shrinkage. Shrinkage proceeds more rapidly where the packing density is the highest (and hence pore size is the smallest). Defects can develop if these local gradients are severe. Another complication is grain growth, which begins in this state. At the end of the intermediate stage, the pores begin to pinch off such that they are no longer interconnected.

The *final stage* is one of slow densification and typically concurrent grain growth. Porosity trapped within grains is nearly impossible to remove and a limiting value of density is reached. In this stage the gas trapped within the pores must dissolve into the solid in order for the pores to be removed. Grain growth must be inhibited if a full density, fine-grained microstructure is to be achieved. The rationale for this statement is based on the important role of grain boundaries in densification. If the conduits for fast material transport are removed, then the densification rate falls.

Grain Growth. Like sintering, grain growth is thermally activated and driven by thermodynamic factors. The macroscopic thermodynamic driving force is the elimination of the boundary contribution to the Gibbs free energy. Although more favorable than solid-vapor interfaces, the solid-solid grain boundary interfaces increase the free energy relative to a single crystal. Growing the grains and eliminating grain boundaries is energetically favorable. So there is a macroscopic thermodynamic driving force for grain growth. Microscopically, grain boundaries have curvature and hence the chemical potential of vacancies is affected by the curvature. As mentioned above, grain growth occurs in the intermediate and final stages of sintering after grain boundaries are established.

In the later stages of sintering, grains are three-dimensional polyhedral structures with faces that abut adjacent grains. A two dimensional slice

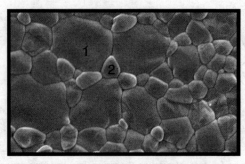

FIGURE 5.14 Scanning electron micrograph of an alumina pellet surface after sintering to a late stage. Notice that large grains (e.g., grain 1) have more sides and grain boundaries that bow inward, whereas small grains (e.g., grain 2) have fewer sides and grain boundaries that bow outward. This image is an enlargement of Figure 5.8c.

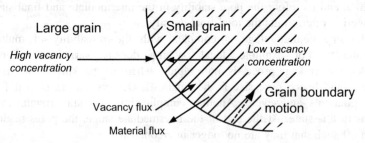

FIGURE 5.15 Effect of grain boundary curvature on vacancy concentration and grain boundary motion.

through a polycrystalline microstructure or a surface reveals how the number of such adjacent grains influences the grain boundary curvature. See Figure 5.14, for example. Large grains create more boundaries with surrounding grains and boundaries bow inward. Smaller grains have boundaries that bow outward.

A vacancy concentration difference is established across the grain boundary, based on the curvature. This concentration difference drives diffusion and boundary motion. See Figure 5.15. Based on the curvature, the vacancy concentration is high in the large grain adjacent to the boundary (compare to Figure 5.10) and low in the small grain on the other side. Vacancies diffuse across the boundary from high to low vacancy concentration and atoms diffuse in the opposite direction. This process causes atoms to be added to the larger grain and hence it grows, while the smaller grain shrinks. The boundary moves towards the center of the smaller grain. Small grains disappear so that overall the average grain size increases. Grain boundary velocities can be tracked experimentally. But it should be recognized the boundary structure is complex, and the movement is the consequence of atomic level diffusion.

The rate of grain growth is typically characterized by determining the grain size as a function of time at some temperature. The data show that the rate of increase in the average grain size is inversely proportional to that grain size. In other words, if we compare the grain growth rate for a polycrystalline material with fine grains with a similar material that has a larger average grain size, the fine grain material has the greater rate of grain growth. The macroscopic driving force provides a rationale that is easy to grasp—the finer grained microstructure has more grain boundary area per unit volume. From the microscopic viewpoint, the curvature of the grain boundary and hence the microscopic driving force lessens as grain size increases. This situation is unfortunate in ceramic processing because the ultimate goal is frequently a fine-grained, dense microstructure.

Grain growth interferes with achieving complete densification. Grain boundaries are the preferred pathways for material motion during densification, and grain growth not only decreases the quantity of grain boundaries per unit volume but it also can lead to pore isolation inside of grains. If a grain grows essentially around a pore and leaves it stranded inside of a grain, disconnected from a grain boundary, then removal of that pore is virtually impossible. Therefore, there has been considerable effort into developing strategies to limit grain growth and achieve high density.

Since grain growth and sintering (densification) are both thermally activated, the rates of both processes increase with temperature. Therefore, it is not possible to simply choose a process temperature so that densification occurs but grain growth does not. Engineering the microstructure is one way to limit grain growth. Second phase inclusions tend to inhibit grain growth. When a moving grain boundary intersects an inclusion, the grain boundary energetics are affected. For the boundary to move past an inclusion, it must "break away" from the inclusion and leave it stranded inside a grain. This change may not be energetically favorable. Additionally, if several inclusions are located along the boundary and locally "pin" the boundary, then the curvature of the boundary is limited, which slows down growth. Pores act in this fashion and therefore grain growth is slow in the intermediate stage but picks up significantly in the final stage of sintering after most of the pores have been eliminated.

The discussion thus far has focused on what is known as normal grain growth. In normal grain growth, the average grain size increases but the grain size distribution does not change much. Another type of grain growth is known as secondary or exaggerated grain growth. In this process, the grain size distribution changes significantly as some small fraction of abnormally large grains grow at the expense of a finer grained matrix. This sort of grain growth is particularly damaging to the microstructure development and physical properties. One method of lessening the likelihood for secondary grain growth is to ensure that the powder starting material does not contain abnormally large particles.

Liquid Phase Sintering. Formation of a small amount of liquid phase on heating typically enhances the densification rate of a powder compact

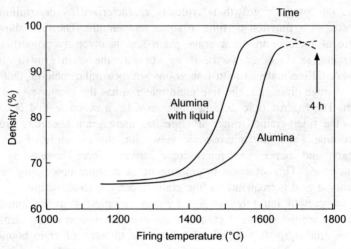

FIGURE 5.16 Effect of a liquid phase (5 wt% alkaline earth aluminosilicate glass) on the densification of alumina. *Reproduced with permission of John Wiley & Sons from Reed (1995);* © *1995 by John Wiley & Sons, Inc.*

significantly. The amount of liquid can be as little as a few volume %. See Figure 5.16. To be effective, the liquid phase should wet the powder solid surfaces well and preferably allow some dissolution of the powder phase into the liquid. The liquid phase plays several roles. First, if the liquid forms before significant solid-state sintering then it will facilitate the particle rearrangement to a denser state. Capillary forces from curved liquid menisci are primarily responsible. Next, the liquid provides a low diffusivity pathway for material transport. The powder phase dissolves in the liquid, diffuses through the liquid and then reprecipitates at the neck or elsewhere to increase the density. This mechanism is effective over a limited range of densities and hence solid state sintering typically takes over in the later stages of densification.

It is possible for pores to coarsen in the liquid phase and therefore to expand in the later stages of liquid phase sintering. This coarsening or growth in the size of pores comes about due to pore coalescence. The pressure in a pore increases as the size decreases. So small pores coming together to form a larger pore experience a lower pressure and hence a larger volume.

Another downside of liquid phase sintering is the fate of the liquid phase. The liquid phase solidifies in the microstructure either as a glassy phase, predominately at the grain boundaries, or as a crystalline phase. This second phase can affect the properties of the final part. Also important is control of the amount of liquid and temperature of firing. Nonetheless, the use of liquid phases to enhance sintering rates is a common practice that results in faster and lower temperature sintering processes.

Viscous Sintering. Viscous sintering is the process of densification of amorphous (glassy) powder compacts or mixtures of glassy and crystalline phases in which there is an appreciable quantity of the glassy phase. A good example would be traditional ceramics like porcelain. Viscous sintering is most easily discussed from the starting point of a compact of glassy particles. The macroscopic driving force for sintering is the same as solid state sintering of crystalline particles, but the microscopic mechanism is different. As the glassy particles are heated (e.g., $T > T_g$, glass transition temperature), their viscosity drops and they become "liquid." Hence the presence of curved surfaces leads to a pressure gradient that drives flow and the microscopic flows drive material to the neck regions (based on low pressure there), the compact shrinks and densifies. The initial rate of shrinkage follows:

$$\frac{\Delta L}{L_o} = \frac{3\gamma_{LV} t}{2\eta d} \qquad (5.10)$$

where γ_{LV} and η are the surface tension and the viscosity of the glass at the sintering temperature. Notice that shrinkage varies linearly with time and inversely with particle size. Since viscosity has a strong temperature dependence increasing the temperature leads to increased initial sintering rate. As sintering proceeds, glass particle fuse and eventually the pores become isolated.

The final stage of viscous sintering involves isolated spherical pores. As mentioned in Chapter 3, there is a pressure drop $\Delta P = 2\gamma_{LV}/r$, across the curved surface. The pressure inside the pore is lower than in the adjacent glass, which creates a hydrostatic pressure that drives the densification.

As mentioned, viscous sintering is active in the densification of traditional ceramics. In this way, viscous sintering could be considered as variation of liquid phase sintering. The glassy phase in traditional ceramics, like porcelain and stoneware, is an alkali aluminosilicate that develops on heating. The quantity of glass that forms influences the properties and appearance of the final ceramics. Control of temperature is very important due to the strong influence of temperature on viscosity of the glassy phase. Depending on size and geometry, the sintering part may slump under its own weight and have a final distorted structure. See Figure 5.17.

5.2.3 Dimensional Changes during Densification

Materials formed in the powder state contain pores after the forming is complete. For example, in uniaxial compaction, the pore content ranges from about 25% to 30% of the volume for most ceramics formed from powders down to around 5% for parts made from highly deformable metal powders. This section provides more information on the macroscopic changes that occur in powder-based materials during sintering, namely the changes in density and part dimensions (i.e., shrinkage).

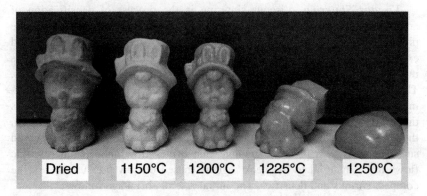

FIGURE 5.17 Effect of sintering temperature on densification and slumping of slip cast stoneware pieces. All pieces were held at peak temperature for one hour. A glassy liquid phase forms on heating; the viscosity of the liquid drops with temperature.

Shrinkage is expressed relative to some fixed starting point or initial state. For a part prepared by compaction of a powder, the starting point is the powder compact before it is heated for final densification. The linear shrinkage after some process step can be generally defined as a fraction or percent:

$$\frac{\Delta L}{L_o} = \frac{L_o - L_f}{L_o} \quad (5.11)$$

$$\% \text{ Linear Shrinkage} = \frac{\Delta L}{L_o} \times 100 \quad (5.12)$$

where L_o and L_f are the initial and final dimensions in a particular direction, respectively. By this convention, shrinkage is a positive number, although the final dimension is smaller than the initial. At this point, it is important to know if dimensional changes are isotropic or not. For isotropic changes, the percent linear shrinkage is the same in all directions. Shrinkage is also quantified on a volume basis:

$$\frac{\Delta V}{V_o} = \frac{V_o - V_f}{V_o} \quad (5.13)$$

$$\% \text{ Volume Shrinkage} = \frac{\Delta V}{V_o} \times 100 \quad (5.14)$$

where V_o and V_f are the initial and final volumes, respectively. If the linear shrinkage is isotropic, then the volume and linear shrinkages are related:

$$\frac{\Delta V}{V_o} = 1 - \left(1 - \frac{\Delta L}{L_o}\right)^3 \quad (5.15)$$

EXAMPLE 5.2 Find the relationship between the volume and linear shrinkages for a material that undergoes isotropic shrinkage. Consider the following general shape:

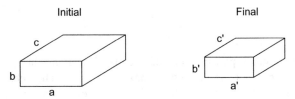

FIGURE E5.2

$$\frac{\Delta V}{V_o} = \frac{abc - a'b'c'}{abc} = 1 - \frac{a'b'c'}{abc}$$

Isotropic Shrinkage:

$$\frac{\Delta L}{L_o} = \frac{a-a'}{a} = \frac{b-b'}{b} = \frac{c-c'}{c}$$

$$\frac{\Delta L}{L_o} = 1 - \frac{a'}{a} = 1 - \frac{b'}{b} = 1 - \frac{c'}{c}$$

$$1 - \frac{\Delta L}{L_o} = \frac{a'}{a} = \frac{b'}{b} = \frac{c'}{c}$$

$$\left(1 - \frac{\Delta L}{L_o}\right)^3 = \frac{a'b'c'}{abc}$$

$$\frac{\Delta V}{V_o} = 1 - \left(1 - \frac{\Delta L}{L_o}\right)^3$$

This simple relationship is derived in Example 5.2

For many powder processes, the density of the part is tracked through forming and sintering. An as-formed ceramic or metal powder product is referred to as a *green* shape or green part. As discussed in the last section, green parts are porous and require a sintering treatment for densification. Typically, the goal of sintering is to reach a nearly pore-free state. In other words, during sintering the bulk density of the part should approach the theoretical density of the material.

$$\rho_{Bulk} = \frac{\text{Mass of part}}{\text{Geometric volume of part}} \quad (5.16)$$

The theoretical density of a material is either found experimentally from a single crystal, which is free or pores, or calculated from the crystal structure. That is, theoretical density is simply the mass of the atoms in a unit cell divided by the volume of the unit cell. Theoretical densities are also found in reference books and tables. Bulk densities are always found experimentally. The densities of green and sintered parts are often referred to in terms of their percent of theoretical densities:

$$\rho(\%) = \frac{\rho_{Bulk}}{\rho_{Theoretical}} \times 100 \tag{5.17}$$

For example, a green part may be "65% dense," meaning that the part has a density that is 65% of the theoretical maximum value. The percent theoretical density is generally more informative and more commonly reported in the literature, product specifications, and engineering designs because this value informs the reader about the porosity, which has significant effects on properties. Porosity, P, is given as a volume %. If a green part is 65% dense, then it is necessarily 35% porous. In equation form,

$$P(\%) = 100 - \rho(\%) \tag{5.18}$$

Bulk density measurements also give information on shrinkage. If the bulk density is measured before and after a processing step, such as sintering, shrinkage can be calculated. The relationships between density, porosity and shrinkage are best conveyed in a numerical example (Example 5.3).

Design of dies, tools and other process parameters requires knowledge of dimensional changes. Design calculations require turning the shrinkage data around. With the final part dimensions and density, and shrinkage data for a particular material and process, the required mold dimensions, for example, can be calculated.

5.3 PRESSING

5.3.1 Process Overview

Pressing or compaction processes convert a free-flowing powder into a "compact" with controlled shape and much higher density. The process is also known as dry pressing to emphasize the point that the powders themselves are "dry." There may be a small amount of moisture in some powders, but compared to the processes described in Chapter 6, these powders are indeed dry. The first step in the process is the preparation of the starting powder. The powder must be designed with all of the subsequent steps in mind. Then a controlled amount of powder is added to the die or mold and force is applied to compact the powder. After compaction and removal from the die or mold, the green part has integrity and strength. Even so, after compaction most parts are heated or sintered to attain an even higher density and improved properties. In some processes, heat and pressure are applied at the same time.

EXAMPLE 5.3 A disk-shaped green zirconia ceramic has dimensions of 1 cm diameter and 0.1 cm thickness and weighs 0.26 g. After a sintering treatment, the diameter of the disk is measured and found to be 0.83 cm. Find (a) the green density (bulk density of green ceramic) and the percent theoretical density of the green ceramic, (b) the linear and volume shrinkages (in %), and (c) the bulk density, percent theoretical density, and porosity of the final sintered ceramic. The theoretical density of zirconia is 6.0 g/cm^3. State assumptions.

a. Green Density = $\dfrac{\text{weight of part}}{\text{geometric volume of part}} = \dfrac{0.26 \text{ g}}{\pi(0.5)^2(0.1)} = 3.3 \text{ g/cm}^3$

% Theoretical Density = $\dfrac{3.3 \text{ g/cm}^3}{6.0 \text{ g/cm}^3} \times 100 = 55\%$

b. After sintering the diameter has shrunk to 0.83 cm. First the linear shrinkage can be calculated from this data:

Fractional Linear Shrinkage = $\dfrac{\Delta L}{L_i} = \dfrac{1.0 - 0.83}{1.0} = 0.17$

% Linear Shrinkage = 17%

Now, if the shrinkage is isotropic then the thickness of the sintered disk can be found. The fractional change in the thickness is also 0.17.

$$\dfrac{0.1 - t_f}{0.1} = 0.17$$

Sintered thickness = $t_f = 0.083$

The volume of the sintered ceramic can now be found as can the % volume shrinkage:

$$V_f = \pi \left(\dfrac{0.83}{2}\right)^2 (0.083) = 0.044 \text{ cm}^3$$

% Volume Shrinkage = $\dfrac{V_o - V_f}{V_o} \times 100 = \dfrac{0.079 - 0.044}{0.079} \times 100 = 44\%$

c. To find the bulk density, we have to make another assumption—that the weight is the same after firing.

Bulk Density = $\dfrac{\text{Mass of part}}{\text{Geometric volume of part}} = \dfrac{0.26 \text{ g}}{0.044 \text{ cm}^3} = 5.9 \text{ g/cm}^3$

%Theoretical Density = $\dfrac{5.9 \text{ g/cm}^3}{6.0 \text{ g/cm}^3} \times 100 = 98\%$

Porosity = 2%

Note: All 0.26 g are zirconia, which has a density of 6.0 g/cm^3. Therefore,

Volume of zirconia = $\dfrac{0.26 \text{ g}}{6.0 \text{ g/cm}^3} = 0.043 \text{ cm}^3$

Volume of pores = 0.044 − 0.043 = 0.001 cm^3

Volume% pores = $\dfrac{0.001}{0.044} \times 100 = 2\%$

There are several types of pressing processes based how the load is applied to the powder. See Figure 5.1a and b. First uniaxial pressing involves filling a cavity in a die with powder and then applying the load with punches that act along a single axis. The die and punches are high strength steel and do not plastically deform as the powder compacts. This process appears simple, but there is complexity in the deformation that occurs in the die cavity and a diversity of approaches to achieve both the desired shape and uniformity in the green part. For improved uniformity, isostatic pressing can be used. In this process, the mold is deformable and the pressure is applied via a liquid that surrounds the deformable mold. A different sort of geometry is accessed with isostatic pressing. Either of these two processes can be carried out at high temperature in processes called hot pressing and hot isostatic pressing, respectively.

A variety of shapes are made by pressing, ranging from very simple to fairly complex multilevel structures. Pressing is not used to create shapes from polymer powders. Therefore, the emphasis in this section is on the understanding of these pressing from the point of view of ceramic and metal powders. This section begins with a discussion of the steps needed to prepare powders for pressing and then continues with an exploration of uniaxial and isostatic pressing, and their high temperature variants. Post-processing operations, the science of which has already been introduced, conclude the section.

5.3.2 Powder Preparation

To attain the best performance and final properties, control over the starting material is important in pressing processes. All steps in the process, including sintering, need to be considered along with the properties of the final part. Therefore, the first step in any pressing process is powder selection and preparation. The requirements for ceramics and metals differ. The basis for the difference is the deformability of the individual particles. The particles in most metal powders plastically deform during compaction while ceramic particles as well as some brittle metal particles do not. Hence, the requirements and preparations are described separately.

Ceramic Powder Preparation. The brittle nature of ceramics enters into the pressing process and the design of powders for pressing in two ways. First, the particles are nondeformable and so the compaction process can, at best, achieve $\sim 60\%$ or 65% of the theoretical density of the material. This value would be achieved with a broad particle size distribution and fairly good space filling from the particles. Second, high density and fine grain size are typically desired for the final microstructure of ceramic products. Such a microstructure has high strength, because the strength of brittle materials is sensitive to imperfections associated with incomplete densification. Therefore, the starting powder must be designed with sintering in mind.

A reasonable route, therefore, is to use a powder with fine particles. Particles with diameters less than ~1 μm, for example, have enhanced sintering rates compared to larger particles. Herein lies the dilemma. While these fine particles are important to developing the desired dense, fine-grained microstructure, their flow properties are poor. To overcome this problem, fine primary particles are prepared into easy flowing, larger granules before pressing. This granulation also provides the opportunity to add organic binders, lubricants, and plasticizers to aid in compacting and developing high green strength.

Granules are controlled agglomerates containing a multitude of micron- or submicron-scale particles, organic binders, plasticizers including water, and pores. Granulation can be accomplished by one of several methods, but spray drying is the most common. In spray drying a dispersion of particles in a liquid is atomized into a heated chamber where the small droplets of dispersion dry into granules. Spray drying is an industrial process used to make a wide variety of materials, including food products, chemicals, and fertilizers, as well as powders for ceramic and powder metallurgy processes.

Spray drying begins with the preparation of a dispersion or suspension of the powder and additives in water or an organic solvent. For oxide ceramics, water is almost always used. The considerations in creating a stable and uniform suspension are the covered in Chapter 6. Key to controlling the suspension is ensuring that the particles do not aggregate and in fact, during the preparation of the suspension, steps are taken to break down agglomerates that were present in the dry powder. For example, the suspension can be ball milled. See Chapter 2. The role of the binder on the process is discussed in the subsection on compaction. About 0.5–5 wt% binder on a dry weight basis is common.

Figure 5.18 shows a schematic of a spray dryer. The dispersion is pumped to an atomizer, which creates a mist or droplets. Due to surface tension, these droplets are spherical or nearly spherical. The droplets are sprayed into a chamber containing heated air (or inert gas). As the droplets fall through the chamber they dry, leaving behind the solids—particles, binder, and plasticizer—in granule form. Granules have a range of particle sizes, reflecting the range of sizes in the droplets. See Figure 5.19.

The spray drying process is engineered so that the granules are well suited to pressing. The granule size should be on the order of 40–200 μm. This size flows easily and also with a range of granule sizes, the granules themselves can fill space well. Spherical granules are preferred for their lower frictional effects. As the mist droplets dry, the ceramic particle concentration increases to about 50–60 volume %. Only a very small amount of the remaining space is filled with binder; therefore, the granules are porous. They are also deformable; the binder and plasticizer are key to developing this property.

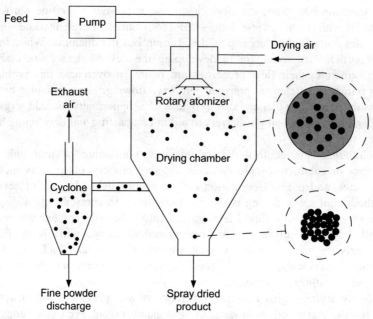

FIGURE 5.18 Schematic diagram of a spray dryer showing the structure change in droplets as they travel through the drying chamber. *Adapted from Reed (1995)*.

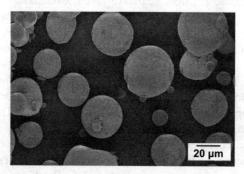

FIGURE 5.19 Scanning electron micrograph of alumina granules prepared by spray drying. Primary particles are 0.1–0.3 μm in size. *SEM courtesy of Y. Wu, University of Minnesota*.

Metal Powder Preparation. Metal powder preparation involves design of particle size distributions and additives to encourage flow, packing and deformation. Since the particles themselves are deformable, the green part densities are higher than those for ceramics. The final products themselves have higher fracture toughness, so larger grain sizes are acceptable and flaws from incomplete densification are less damaging as compared to ceramics.

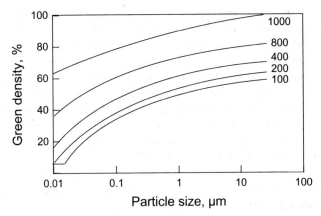

FIGURE 5.20 Model predictions of the effects of particle size and compaction pressure on green density (expressed as %theoretical density) of tungsten powder. Compaction pressure in MPa is labeled on the plot. *From German (2015), reproduced by permission of Elsevier.*

Therefore, granules are not used and the size distribution of the particles themselves is one of the more important features.

Metal powders designed for pressing processes tend to be composed of dense particles that are spherical or uniform in shape. As such the process of melt atomization, which in some sense parallels spray drying, is common in the primary synthesis of the powder. As described in Chapter 2, breaking up a metal melt stream creates droplets; the droplets adopt spherical shapes and then cool to create solid particles. The cooling time scales with the size of the particles so that large particles cool and solidify more slowly. Based on the kinetics on nucleation and growth, their slower cooling tends to lead to larger grain sizes in the particle. Larger particles, therefore are more deformable, than smaller, finer grained particles that result from the faster cooling of smaller metal melt droplets. See Figure 5.20.

For pressing, metal powders typically have a broad particle size distribution with particles ranging from 1 to 150 μm in diameter. The larger particles in the distribution are essential for improving the flow behavior and for their high deformability, which is helpful during compaction. The finer particles play two roles. First they assist in creating high packing density in the presence of the larger particles, and second, they improve the sintering rate during post-forming. To make sure that the fine particles do not interfere with the powder flow, the content of particles less than 44 μm in diameter is usually less than 30–40%.

Prior to compaction, the powder is formulated to improve the process. Tailored particle size distributions are prepared by mixing powders of differing size ranges. Sizing can be accomplished with the assistance of sieves. Lubricants are frequently added in a mixing step. Lubricants including steric acid, metal stearates, such as zinc stearate and calcium stearate, and waxes. These lubricants are added in small amounts, ~1 wt% and less. They decompose on heating, with the metal stearates leaving behind an oxide.

Alloy compositions deserve special mention. Steels are commonly used in powder metallurgy processes including pressing. They present a special challenge because their compositions are usually designed for high yield strength. This property is gained by the alloy composition and so fully prealloyed powders have low deformability. An alternative, as mentioned in Chapter 2, is mixing powders of the alloying elements with iron powder. The iron is deformable and the finally alloying is accomplished during the thermal post-forming treatment. The trade-offs are clear. Prealloyed powder is uniform but suffers from low deformability, whereas mixing in alloy elements separately creates deformability but relies on thermal treatment to achieve chemical uniformity. A host of other metal alloy parts are created by powder metallurgy methods, including aluminum alloys, magnesium alloys, refractory metal alloys, such as tungsten, and copper alloys.

5.3.3 Uniaxial Pressing

The goals for a uniaxial pressing process are speed, reproducibility and formation of a high-quality green compact. The desired features in the green compact are uniformity, lack of defects, high green density, and high green strength. Meeting these goals requires an understanding the compaction cycle, as well as the mechanics involved, so that the best pressing conditions can be chosen.

The pressing cycle for a simple uniaxial pressing operation is shown in Figure 5.21. The first step is filling the die cavity. In automated processes, a fixed amount of powder is sent to the die cavity by allowing the powder to

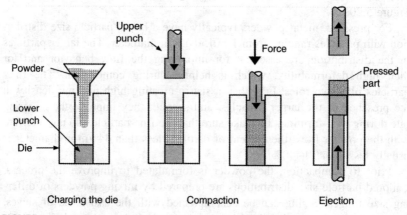

FIGURE 5.21 The compaction cycle used in uniaxial pressing.

flow from a hopper for a set time. This step can also be accomplished by a feedshoe, which is filled with powder, typically by gravity flow for a fixed time, and then shifted over the open die cavity to transfer the powder. The feedshoe then retracts. In the second step, the punch moves down with a controlled force to put pressure on the powder and compact it. In some processes both the bottom and the top punches move in this step. The choice of force and hence compaction pressure is key in the design of the pressing process. Lastly, the part is ejected. When a feedshoe is used the pressing cycle repeats with the feedshoe pushing the newly compacted part on to the post-processing step.

Figure 5.22 shows the density increase that accompanies compaction for several ceramics and metals. The fill density is the initial density before pressure is applied and the final green density is the value after the compaction is complete. Ceramics and metals differ in both these extremes with powdered metals typically filling to a higher fill density and compacting to a higher green density. The shape of the compaction curve is similar for both metals and ceramics. Initially, there is a steep increase in density and then a slower increase followed by a plateau. These density increases signal changes in the microstructure of the compact, which are discussed below. Again comparing metal compaction to ceramic compaction, it is clear that compaction pressures can vary significantly, but metals require higher pressures in general.

It is worth noting at this point that the air in the spaces inside the powder requires a means to escape from the die cavity as the compaction proceeds. While it is possible to compress the air by the action of the punch, it is more effective to allow the air to escape through the small gap between the punch and the die. The pore space in the compact is interconnected during a good fraction of the metal powder compaction process and throughout compaction

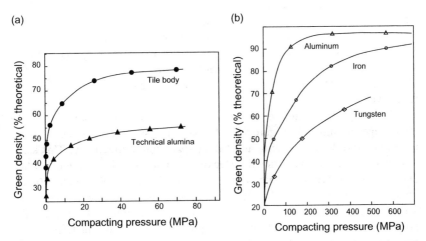

FIGURE 5.22 Examples of compact density as a function of pressure for (a) ceramics and (b) metals. *Data from (a) Reed (1995) (b) Davidson (1984).*

of a ceramic powder. Therefore, the air can be forced out of the compact via the pore space.

One of the most important parameters in the design of a pressing process is the maximum pressure. Pressing should be carried out to a pressure that results in the highest density and no more. Beyond the point where density stops increasing, the compact deforms elastically. This deformation is constrained by the die walls both along the axis of punch motion and perpendicular to it. So, the ejection force or pressure needed to remove the part increases if this elastic deformation occurs. Also, when the part is ejected, there is elastic recovery, which is known as *springback*. Springback not only changes the dimensions of the part but it can also cause defects, cracks that run perpendicular to the axis of punch travel. These cracks occur as the part is ejected and the top portion springs back while the bottom potion is still in the die cavity. Understanding how to choose the correct conditions for compaction and develop a fast, effective process depends on knowledge of the microscopic changes that occur during compaction.

Compaction Sequence for Ceramic Powders. Ceramic powders are granulated and the granules feature strongly in the compaction process. A single granule contains ceramic particles, binder, plasticizer, and pores. The packing of the particles in the granule results in about 50–60% of the theoretical density, depending on the particle size distribution in the initial ceramic powder. Since the binder and plasticizer are minor components with lower density than the ceramic, these materials do not play a significant role in the green density (i.e., they do not add much to the mass of the pressed compact). Moreover, since they are removed during sintering, they are virtual pores and so the quantity of binder and plasticizer should be at the minimum needed for producing granules with the best flow behavior and deformability and for achieving the required green strength. Granules have a range of sizes, a factor that influences their initial filling of the die.

The fill density for a ceramic powder, therefore, is the consequence of packing porous particles in the die cavity. For example, if granules themselves have a density that is 50% of theoretical density and they occupy 50% of the volume (packing factor $= 0.5$), then the fill density is 25% of the theoretical value.

The compaction diagram for a ceramic powder can be broken down into three stages as shown in Figure 5.23. In the initial stage (I), granules respond to the pressure by rearranging and filling space more effectively. This stage is sometimes called translational restacking. A microscopic picture of a filled die reveals two types of pores—those that are between the granules (intergranular pores) and those that are within the granules (intragranular pores). The rearrangement in the first stage results in a decrease in intergranular pores. This process takes place at low pressure and results in fairly minor changes in the density. As pressure increases, at a critical point, the stage II, granule deformation, begins.

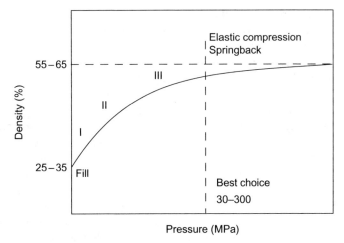

FIGURE 5.23 Schematic of punch pressure on the green density of a ceramic powder during compaction.

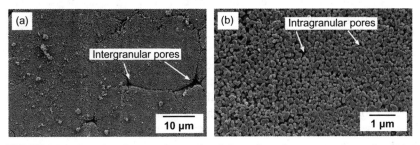

FIGURE 5.24 Scanning electron micrographs of the surface of a compacted, granulated alumina pellet. (a) Low magnification image showing the pores between the granules, intergranular pores, and (b) high magnification image showing the pores inside the granules, intragranular pores.

Granule deformation, stage II, results in a major increase in density. At a critical applied pressure, granules deform at granule-granule contact points and the deformation consumes intergranular pore space, resulting in a green density increase. Figure 5.24 shows SEM images of a compacted granulated powder with the two types of pores highlighted. This stage, including both the yield pressure and densification extent, is sensitive to the binder and plasticizer type and content. Common binders for compaction of ceramic powders are water-soluble polymers such as polyvinyl alcohol. In general, the granule yield strength increases with binder content and is in the neighborhood of 1 MPa. These binders are plasticized by water and so the relative humidity in the storage environment for the powder influences the plasticizer content. In general, the yield pressure decreases as the relative humidity decreases due to the plasticizing effect (i.e., lowering of the glass transition temperature of the binder). As compaction proceeds in this stage, the intergranular pore space is removed and microstructural images show the boundaries between the granules begin to disappear. The intergranular pore space is nearly removed at the point where this stage ends and the next begins, and the green density is the same as the granule density.

The final stage in compaction of a ceramic powder is *particle* rearrangement or translational restacking of particles. The particles in the granules may already have a high degree of packing with little room for rearrangement. Hence, the increase in density during this stage is less than that in the previous stage. The rearrangement of particles results in removal of some of the small intragranular pores (i.e., the pores between the particles themselves, which reside inside the remnants of the granules). Since binder and plasticizer are present inside the granules, the ease at which the particle rearrange follows the same trends as granule deformation. Depending on the characteristics of the particles and binder, fracture and fragmentation of the particles can occur, which leads to the opportunity for more rearrangement. This stage ends when there is no further density increase with increasing pressure. The final pressing pressure is therefore chosen based on compaction data to be at the end of this stage and before significant elastic compression. For ceramics, pressures in the 30–300 MPa range are typical.

The green strength of a pressed ceramic compact must be high enough for the part to withstand post-forming operations. For uniaxially pressed ceramic parts, post-forming is typically firing and so the green strength is needed for handling as the parts as they are transported to the kiln or furnace. The green strength in general increases with green density and with binder content.

Compaction Sequence for Metal Powders. Metal powders are not granulated because the particles themselves are deformable, and hence the fill density represents the ability of the particles to fill space, a property related mainly to the particle size distribution. However, some refractory metal powders are granulated because they have low deformability. There is a diversity in metal powders and therefore the fill densities vary. Powders consisting of particles with irregular shapes or some internal pores have lower fill densities than powders composed of dense spherical particles with a broad particle size distribution.

Figure 5.25 shows the compaction diagram for a metal powder. It has three stages, appearing very similar at first glance to the compaction diagram for ceramics. The first stage is the rearrangement of the particles. This stage results in small increases in the density as the particles move relative to one another into more favorable positions. The next stage starts as the stresses at the particle-particle contacts reach the yield strength of the particles and cause flattening at the contact zone. The deformation decreases the pore

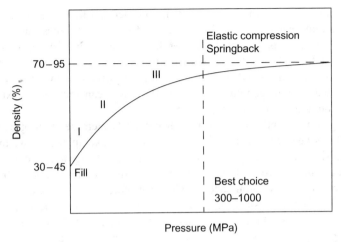

FIGURE 5.25 Schematic of punch pressure on the green density of a metal powder during compaction.

space and the density increases. The particle yield strength is a function of the same variables that affect the yield strength of bulk metals. For example, pure metals have lower yield strengths than alloys. Also, as mentioned previously, the grain size in particles prepared by melt atomization decreases as the particle size decreases and therefore finer particles are less deformable. As the local deformation at contacts proceeds, work hardening occurs, which makes it necessary to increase the punch pressure even more to attain more deformation and more densification. The final stage is linked to homogeneous plastic deformation that is not limited to the particle-particle contact region. As in ceramics, there is a pressure beyond which little densification is possible and hence elastic compression takes over. This is pressure is the best for a compaction process.

Due to the wide range of metal powders, it is more difficult to define typical values for compact green density and optimum compaction pressures. In general, green densities range from ~60% to 70% for refractory metals to over 95% for metals with a low yield strength. The pressure required to reach these valued varies over a wide range from the 100's of MPa to ~1 GPa. Achieving green densities as close to theoretical as possible is a goal for powder metallurgy as high green density translates to low shrinkage on sintering. It is possible to raise the temperature of die to ~150°C to process parts by "warm compaction" and achieve improved green density over room temperatures operations. In these processes, however, thermally stable lubricants are needed and sometimes other additives are used to assist the process.

The speed of compaction is a factor in some pressing operations. Increasing the rate at which the punch or punches are pressed into the powder can influence the response of the powder and the extent of compaction. The rate dependence is also important for some ceramics based due to the viscoelastic properties of the binder, but it is arguably more of a factor in the compaction of metal powders due to strain rate sensitivity in plastic deformation. As discussed in Chapter 4, this sensitivity is low at room temperature but at high compaction speeds the effect can reduce the extent of compaction at a given applied pressure.

The green strength of metal powder compacts increases with green density and has been linked to other phenomena. For example, cold welding or interdiffusion at particle-particle contacts has been suggested and linked to an increase in the green strength of compacts that contain a higher fraction of fine particles. Such compacts would have more interparticle contacts per unit volume and more sites for the cold welding. Others link the shape of the particles to green strength, noting that more irregular particles develop higher green strength. There is no consensus on a universal mechanism aside from the clear link between density and strength, and hence it is likely that different factors influence the green strength for different powders.

Die Wall Friction and Compaction Uniformity. The friction between the die wall and the powder compact has a significant effect on the compaction and compact uniformity. This friction results in a decrease in the pressure transmitted to the compact as a function of distance away from the punch. That is, a nominal pressure of 100 MPa may be established by placing a specific load on the known area of the punch, but only the powder immediately beneath the punch experiences this pressure. The frictional force diminishes the pressure so that the powder deeper into the compact may only experience 80 MPa, for example. Moreover, different local pressures cause different local green densities, leading to gradients that can cause defects during sintering. Additionally, frictional resistance at the die wall must be overcome during ejection of the compact. So, higher friction necessitates higher ejection forces. Lastly, friction can cause wear of components, such as the punches and interior walls of the die.

The effect of friction on pressure transmission and compact uniformly can be understood using a force balance analysis similar to that used in metal deformation processes in Chapter 4. The goal of this analysis is to develop an expression that describes how the pressure varies from the pressing face downward into the compact. Figure 5.26 shows the configuration for the analysis. Powder in a cylindrical die cavity is acted on by a top punch with the bottom punch stationary beneath. The application of pressure is uniaxial and from one direction only. The applied force, F_a, is applied to the top punch, resulting in an applied pressure, p_a, given by:

$$p_a = \frac{F_a}{A_p} \tag{5.18}$$

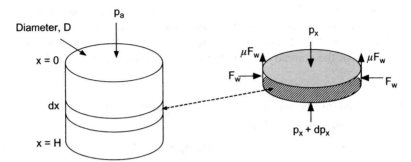

FIGURE 5.26 Schematic diagram for one dimensional analysis of pressure distribution in a uniaxially pressed compact. The pressure is applied on the top surface of the compact. The die and punches are not shown. The surface of the pellet is taken as the position $x = 0$ and x increases downward to the bottom of the pellet at $x = H$. A volume element of thickness dx is shown with the forces and pressures. *Adapted from German (1984)*.

The area pressed, A_p, for the cylindrical compact is given by:

$$A_p = \frac{\pi}{4} D^2 \quad (5.19)$$

The applied pressure in the vertical or axial direction causes a stress in the horizontal or radial direction. That is, as the powder is compacted downward it pushes out on the walls of the die. This action is similar to a Poisson effect in an elastic solid. So, there is a force, F_w, pushing out on the walls of the die and hence the die responds with an equal and opposite force pushing back. This normal force on the interface between the die wall and the powder is related to the frictional force, F_f, by the friction coefficient, μ:

$$F_f = \mu F_w \quad (5.20)$$

The ratio of the horizontal stress to vertical stress (pressure, p) is given by a factor, K_{hv}. This factor, which is also used in the field of soil mechanics, is a property of the powder and can be found experimentally with special fixtures. For powdered engineering materials, it is typically on the order of ~ 0.4 and does not vary over a wide range from powder to powder (e.g., 0.4 ± 0.2). Therefore, the horizontal stress and friction force can be expressed in terms of the pressure and K_{hv}.

$$\text{Horizontal stress} = p K_{hv} = \frac{F_w}{A_f} = \frac{F_f}{\mu A_f} \quad (5.21)$$

$$F_f = \mu p K_{hv} A_f \quad (5.22)$$

where A_f is the area of contact between the powder and the die wall, the friction area.

Now consider a force balance carried out on the small volume element shown in Figure 5.26. The vertical forces include those imposed by the applied force and the frictional response at the walls. The friction force acts on the area of the volume element that is in contact with the die wall.

$$A_f = \pi D dx \tag{5.23}$$

Therefore, noting the forces acting down as positive and the vertical stress, pressure, at any position x as p_x:

$$p_x A_p - (p_x + dp_x)A_p - F_f = 0 \tag{5.24}$$

Using Equations 5.19, 5.22, and 5.23, and simplifying:

$$\frac{dp_x}{p_x} = -\frac{4\mu K_{hv}}{D} dx \tag{5.25}$$

This differential equation is solved by integrating from the top of the pellet where $p_x = p_a$ and $x = 0$ to some distance x where the pressure is p_x

$$\int_{p_a}^{p_x} \frac{dp_x}{p_x} = \int_0^x -\frac{4\mu K_{hv}}{D} dx$$

$$\ln\left(\frac{p_x}{p_a}\right) = -\frac{4\mu K_{hv}}{D} x$$

$$\frac{p_x}{p_a} = \exp\left(-\frac{4\mu K_{hv}}{D} x\right) \tag{5.26}$$

The ratio p_x/p_a is the pressure transmission ratio. In an ideal situation, p_x/p_a would be close to one throughout the pellet so that the pressure would be uniform throughout.

Figure 5.27 shows a plot of the pressure transmission ratio p_x/p_a over a range of conditions. The x-axis on the plot is the distance from the pressing surface, x, normalized to the diameter of the pellet. In this way, we can see how the pressure drops as the distance from the pressing surface increases regardless of the size of the pellet. Clearly the vertical position relative to the diameter points to the role of pellet geometry. The importance of the friction coefficient is also clear. Increasing friction leads to more significant pressure gradients.

It is also instructive to consider the maximum pressure reduction by setting x equal to H, the height of the pellet:

$$\frac{p_H}{p_a} = \exp\left(-4\mu K_{hv} \frac{H}{D}\right) \tag{5.27}$$

Here the role of pellet geometry is apparent. The aspect ratio, H/D, is a key factor determining the extent of pressure gradient in the pellet. Keeping the

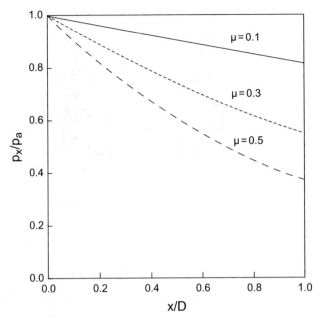

FIGURE 5.27 Pressure transmission ratio in a cylindrical powder compact (Eq. 5.26) as a function of distance from the pressing surface x normalized to the pellet diameter. The effect of friction coefficient is shown for $K_{hv} = 0.5$.

height of the pellet small relative to the diameter minimizes the frictional effects at the walls. Using the expressions for A_f and A_p, the pressure transmission ratio becomes:

$$\frac{p_H}{p_a} = \exp\left(-\mu K_{hv} \frac{A_f}{A_p}\right) \qquad (5.28)$$

This expression can be used to analyze different geometries and modes of pressure application. For example, double action pressing involves application of force via the upper and lower punches, effectively doubling A_p and decreasing the amount of pressure transmission loss. See Figure 5.28.

Gradients in pressure lead to gradients in compact density. Figure 5.29 shows calculated density distributions in compacts pressed with single action and double action uniaxial pressing (see Figure 5.29). In these results, there are two-dimensional effects that are not included in the one dimensional analysis above. The benefits of double action pressing are shown. The overall density is higher and gradients are less steep. Gradients in green density are a problem because the amount of shrinkage to reach a given endpoint density increases as the green density decreases. Therefore, in an inhomogeneous part, differences in shrinkage during sintering can cause defects and warping.

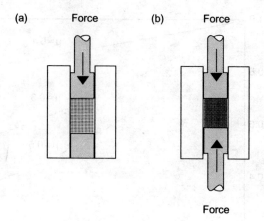

FIGURE 5.28 Comparison between (a) single action pressing and (b) double action pressing.

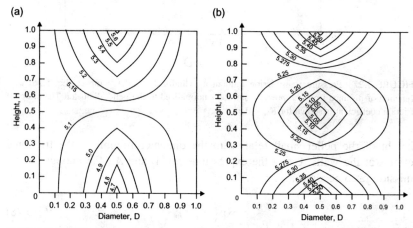

FIGURE 5.29 Calculations of density gradients a cylindrical compact after compaction with $K_{hv} = 0.5$, $\mu = 0.3$ and H/D = 1. *From Thompson (1981). Reproduced by permission from the American Ceramic Society.*

Using a lubricant to lower friction at the die walls and also between particles is an effective strategy. Lubricant reduces the friction coefficient, which leads to more effective pressure transmission and a higher and more uniform green density. In addition, lubricant lowers the ejection force and reduces wear of the die and punch system. There are two strategies for including lubricant: (i) admixing it into the powder and (ii) lubricating the die and punch surfaces. The former leads to some potential draw back. Figure 5.30 shows that there is an optimum amount of lubricant. Adding more beyond this point results in a lower green density. The increased lubricant content fills space between the particles and interferes with particle rearrangement and densification.

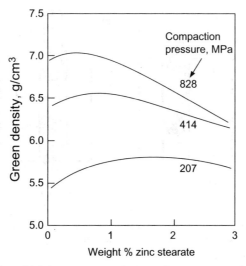

FIGURE 5.30 Effect of lubricant amount and compaction pressure on the green density of compacts made with sponge iron powder. *Reprinted with permission from* Powder Metallurgy Science *(1984) German R. M. (Figure 4.6 Page 107), Metal Powder Industries Federation, 105 College Road East, Princeton, New Jersey, USA.*

Uniaxial Tooling. Uniaxial presses are automated to provide an efficient continuous sequence of filling, compacting and ejecting. The force for compaction is applied by a mechanical or hydraulic system. A mechanical press consists of a series of crankshafts and linkages that supply the force and move the rams while a hydraulic press is based on pumps that move pressurized fluid to push the rams. The die and punch tooling must have high strength to resist plastic deformation during operation as well as features to create the desired shape. Hardened steel is commonly used. Double action and single action uniaxial tooling are two of the most common alternatives. Another is the floating die. As the name suggests, a floating die and punch system includes a die that is able to adjust its position during pressing. For example, the die could be affixed to a spring-loaded platform with the lower punch stationary and the upper punch applying the pressure. As the upper punch descends and the die moves down when a certain amount of frictional resistance is encountered. In this way, there is a virtual double pressing action from the bottom punch because the die position relative to the punch changes.

For uniaxial pressing of complex, multilevel parts, the tooling must also have some complexity. Core rods for internal central holes in gear-like object are used along with the die and punches. Further, uniformity in a complex, multilevel part requires tooling that allows the same extent of compaction to be achieved in each level. Figure 5.31 illustrates the complication with compaction of multilevel parts and the solution offered by a more intricate control of the punch motion. Simply compacting with a punch that has two

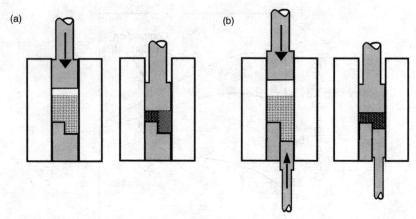

FIGURE 5.31 Compaction of a simple two-level part using a (a) monolithic lower punch with fill state (left) and pressed state (right) and (b) a two-part lower punch with fill state (left) and pressed state (right).

FIGURE 5.32 Example of a steel gear (∼13 cm diameter) created with a multilevel die and punch system. *Photo courtesy of W. Suszynski.*

levels creates a part that has different extents of compaction and therefore considerable inhomogeneity. A two-part punch provides the means to compact both levels of the part to the same degree, resulting in a uniform density compact. Figure 5.32 shows a two-level part. For more complex parts, several levels of punches are needed to create uniform states of compaction over different sections of the part. This sort of multilevel complexity is demanded of powder metallurgy components but less frequently engineered into pressed ceramic parts.

Advantages and Disadvantages. Uniaxial pressing is a relatively simple process that is easy to automate and produces high quality green parts quickly from powders. In ceramic processing it is an important forming operation for geometries with the appropriate aspect ratio for uniform compaction. For metals, powder metallurgy processes offer a distinct advantage over casting and forging in that the parts at completion of the process require very little if any machining or surface finishing. In this way the process is "near net shape." Of course, the dimensions of the green part and hence the tooling geometry must include allowances for densification during a firing or sintering process. Some overall disadvantages include limitations in shape complexity, issues with achieving complete densification, and density gradients. Some of these disadvantages can be overcome by improving compaction by uniform application of pressure.

5.3.4 Isostatic Pressing

Isostatic pressing is similar to uniaxial pressing in the requirements for the powder and in the general steps of the process, but there are several important differences. First, the compaction takes place under hydrostatic conditions. That is, the pressure is transmitted to the part equally in all directions, or very nearly equally. In this way, the die wall friction is significantly reduced or eliminated entirely. Second, the tooling consists of elastomeric molds rather than rigid dies. The powder is loaded into the flexible mold, the mold is sealed and the pressure is applied in a pressure vessel via a liquid. Isostatic pressing is also called cold isostatic pressing or CIP so that it can be distinguished from hot isostatic pressing or HIP, a similar process carried out at high temperature.

There are two general types of isostatic pressing operations: wet bag and dry bag. See Figure 5.33. The wet bag variation involves a separate elastomeric mold that is loaded outside of the press and then submerged in the pressure vessel. After pressurization and compaction, the mold is removed from the vessel, the part is retrieved, and the process repeated. Multiple molds can be loaded into the vessel for a single pressurization run. The dry bag variation circumvents the immersion step by creating a mold that is integrated into the pressure vessel. In the dry bag process, the powder is added to the mold, the mold sealed, pressure applied and then the part ejected. The integrated mold in this process makes automation easier than the wet bag process.

Powder preparation for isostatic pressing is typically the same or very similar to that for uniaxial pressing. The basic powder requirements—free flowing, easily compacted, good sintering performance—are the same. However, the particle size distribution and binder content are adjusted if the compaction is followed by green machining, a step that is used in high volume operations that produce the ceramic bodies for spark plugs and sensors. In some cases, parts are precompacted using uniaxial pressing operations and then further compacted with isostatic pressing. In this case the elastomeric mold is not involved in the shaping but only in the transmitting the pressure and isolating the part from the fluid in the pressure vessel.

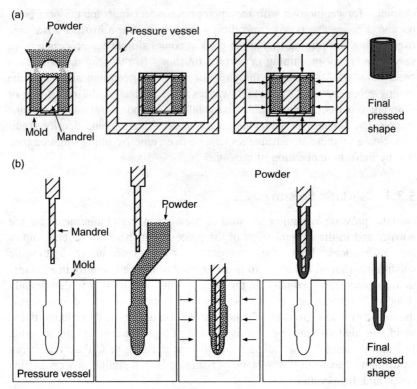

FIGURE 5.33 Schematic diagrams for (a) wet bag isostatic pressing with collapsing bag tooling, including a mandrel, to make a tube and (b) dry bag isostatic pressing. *From Kennard (1991), reproduced with permission from ASM International.*

The wet bag isostatic pressing process is a versatile option for achieving high compact densities from metal and ceramic powders. A common shape for this type of pressing is an axisymmetric open or closed end, hollow shape, such as a tube. Figure 5.33a shows a collapsing bag type tooling with an elastomeric mold, a rigid mandrel insert and elastomeric end caps. The geometry and volume of the mold cavity are designed to achieve a target final dimension. The inner diameter of such a pressed shape would have higher precision than the outer diameter. Another option is a dilating tool that has the pressurized fluid accessing the inner surface of the object; powder is pressed outward from an inner elastomeric mold wall against an outer rigid casing. The process provides a means to produce high and uniform pressure and high compact density due to the elimination of die wall friction. Enhanced density is only one feature of interest. Perhaps more important is the access to much different shapes than are possible with uniaxial pressing.

Dry bag isostatic pressing is an efficient production method for small parts with axisymmetric shapes. It is used to produce high-quality ceramic bodies for spark plugs. Since the rubber tooling is integrated into the apparatus, separate immersion and removal steps are not needed, which leads to ease in automation and fast production rates. However, there is friction on the side of the mold that does not experience compression from the pressurizing liquid. Likewise there are more shape and size limitations in the dry bag process. For example, spark plugs require green machining to develop the surface features needed.

Advantages and Disadvantages. Isostatic pressing is frequently chosen to achieve high compact densities and also to access shapes that cannot be compacted in uniaxial presses. Somewhat complex shapes can be engineered into the elastomeric molds if desired. The wet bag variation is better suited to production of large parts as compared to the dry bag process. However, loading and unloading of the molds decreases the productivity and limits the automation for wet bag pressing. Likewise somewhat higher densities are possible with wet bag processes as they involve very little friction. The dry bag version has the edge on automation and production rate. For both types of isostatic pressing, the tooling cost and complexity of the process are higher than for uniaxial pressing.

5.3.5 Post-Forming Processes for Green Parts

Post-forming operations are essential for green parts. The most important and ubiquitous post-forming operation is heating in a furnace to remove binders, plasticizers and lubricants, and sinter the part to high density. The thermal processing for metal and ceramic green parts have similarities as well as differences as described below. This section provides more of an overview of the technology, building on the science of sintering presented earlier in the chapter. While pressing operations are near net shape, some machining may be necessary. For ceramics, green machining is used to add detail to some isostatically pressed parts before thermal treatment. For metal parts, details might be added after thermal treatment by traditional machining operations.

Green parts are fired or sintered at elevated temperature in either batch or continuous operations. The terms kiln, furnace and oven are often used interchangeably for equipment used for the heat treatment. Kiln is more frequently used in the context of ceramics. In a batch operation, parts are loaded into a furnace, heated according to a specific time-temperature schedule and then unloaded. To fire large quantities of parts, a shuttle or "car" is loaded with parts and rolled into a large kiln for firing. For continuous processes, tunnel kilns or furnaces are used. Green parts are conveyed on a belt down the length of the furnace continuously. Zones within the tunnel furnace provide the appropriate temperature and atmosphere conditions. In both routes control of the thermal schedule (time, temperature) and atmosphere are essential to developing the desired microstructure and properties.

In ceramic processing, heating of green parts is known as firing. Firing encompasses heating to remove binders and plasticizers as well as to densify or sinter. Additive contents are fairly low in pressing operations, but the heating rate and atmosphere in the furnace are designed to encourage volatilization of the additives. Namely, temperatures in the range of 300–500°C with oxidizing conditions are typical for binder removal. Since removal of the binder leaves the green parts fragile, parts are not handled after this "binder burn out" step. In the next step in the firing schedule, the temperature is increased to the sintering temperature for densification. The sintering atmosphere for oxide ceramics is typical air or oxygen. For nonoxide ceramics, such as aluminum nitride, inert gases are used to prevent oxidation.

Thermal treatment of green powder metal parts requires attention to removal of lubricants and controlling the temperature and atmosphere for the sintering process. The initial temperature and atmosphere in the furnace is designed for removing lubricants. Complete removal is possible with hydrocarbon lubricants, but some residue is left behind for lubricants such as metal stearates. As the temperature increases to the sintering temperature, a key concern is preventing oxidation of the metal. Inert or reducing conditions are necessary in this stage. Also important for steel is the control of carbon content, which is also sensitive to the atmospheric conditions. Finally, the temperature and time required for the sintering process are chosen based on the metal composition and mechanism of sintering.

5.3.6 Hot Pressing and Hot Isostatic Pressing

Hot pressing and hot isostatic pressing (HIP) use a combination of pressure and temperature to produce high density parts. These methods can produce ceramic and metal parts with nearly 100% density and with fine grained microstructures. For both processes, powder compaction into a shape is sometimes carried out before the hot pressing or HIP process. In this case, hot pressing and HIP are used to enhance sintering rather than as a method of shaping.

Pressure increases the densification rate significantly. The mechanism for enhanced densification varies with the material and the conditions of temperature and pressure used. The compressive stress from hot pressing drives diffusion and/or dislocation motion in a manner analogous to creep. Deformation maps are available for some materials and allow prediction of the mechanisms under particular conditions of temperature and pressure. Figure 5.34 shows an example of a deformation map tuned for pressure densification in the HIP process. The pressure effect is superimposed on the sintering process, which makes a complete description of densification a challenge.

Hot Pressing. Hot pressing is the simultaneous application of heat and uniaxial pressure. Due to uniaxial nature of the process, only simple shapes are created by this method. Further, the application is much more common

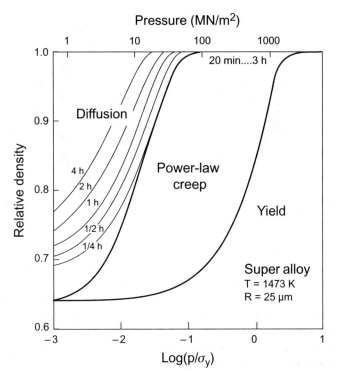

FIGURE 5.34 Schematic diagram showing the effect of pressure on densification mechanism at constant temperature for a superalloy powder with an initial particle radius, R, of 25 μm. *From Arzt et al., (1983) with kind permission of Springer Science + Business Media.*

in ceramic processing, perhaps due to the greater diversity of processing options open to metals or due to the specific applications, such as transparent polycrystalline ceramics, that require high density and fine grain size.

Hot pressing equipment integrates mechanical loading from a simple die and punch system with heating from a furnace that surrounds the die and punches, as shown in Figure 5.35a. The die and punches are typically made of graphite, which has the appropriate thermal and mechanical properties and is machinable. Use of graphite, however, requires nonoxidizing conditions (e.g., vacuum, inert), and hence some ceramic oxides are reduced during hot pressing and must be post-annealed in oxygen to regain their appropriate stoichiometry. Alternatives to graphite include aluminum oxide and silicon carbide. Liners, such as foils of tungsten and molybdenum, are used to prevent reaction for some materials with the punch and die. The material that is loaded into a hot press is frequently already compacted by a room temperature uniaxial pressing process. The hot press in this case is used for densification. Operating temperatures are up to 2500°C and pressures are typically 10–70 MPa.

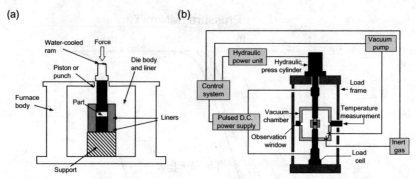

FIGURE 5.35 (a) Schematic diagram of a standard hot press with a furnace containing resistive elements or coils for inductive heating. (b) Schematic diagram of a spark plasma sintering apparatus. *(b) Courtesy of GT Advanced Technologies, www.GTAT.com.*

For hot pressing, a controlled sequence of pressure and temperature is used. Frequently, the pressure is applied after some heating has occurred because applying pressure at lower temperatures could have adverse effects on the part and tooling. Hot pressing temperatures are several hundred degrees lower than regular sintering temperatures. And nearly complete densification occurs rapidly. The speed of the process as well as the lower temperature required naturally limits the amount of grain growth.

A related method, spark plasma sintering (SPS), provides an alternative to external resistive and inductive modes of heating. In SPS, a sample, typically powder or a precompacted green part, is loaded in a graphite die with graphite punches in a vacuum chamber and a pulsed DC current is applied across the punches, as shown in Figure 5.35b, while pressure is applied. The current causes Joule heating, which raises the temperature of the specimen rapidly. The current is also believed to trigger the formation of a plasma or spark discharge in the pore space between particles, which has the effect of cleaning particle surfaces and enhancing sintering. The plasma formation is difficult to verify experimentally and is topic under debate. The SPS method has been shown to be very effective for densification of a wide variety of materials, including metals and ceramics. Densification occurs at lower temperature and is completed more rapidly than other methods, frequently resulting in fine grain microstructures.

Hot Isostatic Pressing (HIP). Hot isostatic pressing is the simultaneous application of heat and hydrostatic pressure to compact and densify a powder compact or part. The process is analogous to cold isostatic pressing, but with elevated temperature and a gas transmitting the pressure to the part. Inert gases such as argon are common. Powder is densified in a container or can, which acts as a deformable barrier between the pressurized gas and the part. Alternatively, a part that has been compacted and presintered to the point of pore closure can be HIPed in a "containerless" process. HIP is used to achieve complete densification in powder metallurgy and ceramic processing, as well as some application in the densification of castings. The method is particularly important for hard to densify materials, such as refractory alloys, superalloys, and nonoxide ceramics.

Container and encapsulation technology is essential to the HIP process. Simple containers, such as cylindrical metal cans, are used to density billets of alloy powder. Complex shapes are created using containers that mirror the final part geometries. The container material is chosen to be leak-tight and deformable under the pressure and temperature conditions of the HIP process. Container materials should also be nonreactive with the powder and easy to remove. For powder metallurgy, containers fashioned from steel sheets are common. Other options include glass and porous ceramics that are embedded in a secondary metal can. Glass encapsulation of powders and preformed parts is common in ceramic HIP processes. Filling and evacuation of container is an important step that usually requires special fixtures on the container itself. Some evacuation processes take place at elevated temperature.

The key components of a system for HIP are the pressure vessel with heaters, gas pressurizing and handing equipment, and control electronics. Figure 5.36 shows an example schematic of a HIP set-up. There are two basic modes of operation for a HIP process. In the hot loading mode, the container is preheated outside of the pressure vessel and then loaded, heated to the required temperature and pressurized. In the cold loading mode, the container is placed into the pressure vessel at room temperature; then the heating and pressurizing cycle begins. Pressure in the range of 20–300 MPa and temperature in the range of 500–2000°C are common.

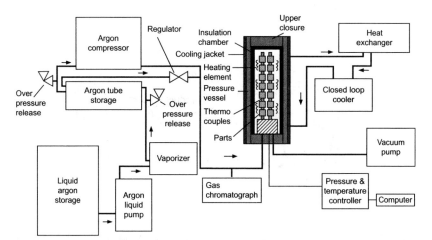

FIGURE 5.36 Schematic diagram of hot isostatic pressing equipment. *Courtesy of Pressure Technology, Inc., www.pressuretechnology.com.*

396 Materials Processing

Advantages and Disadvantages. Hot pressing and hot isostatic pressing are the methods of choice of achieving high density in parts designed for demanding applications or prepared from materials that are difficult to densify by other means. Pressure enhances the rate of densification at a given temperature and so densification can be completed in shorter times and at lower temperatures than conventional sintering. A benefit of the enhanced densification kinetics is final materials with lower grain size, because pressure does not affect the rate of grain growth. However, the equipment and tooling are more complex, the operation is inherently batch rather than continuous, and the processes overall are more expensive than the sequential approach of compaction followed by conventional sintering.

5.4 ROTATIONAL MOLDING

5.4.1 Process Overview

Rotational molding or rotomolding converts polymer powder into a hollow shape. The sequence of process steps is shown schematically in Figure 5.37. Powder is loaded into a mold, which is closed and translated into an oven. The mold is then heated and rotated. Polymer particles tumbling inside the

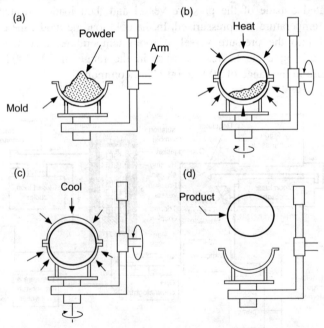

FIGURE 5.37 Schematic diagram of rotational molding steps: (a) charging the powder into the mold, (b) heating and rotating on two axes, (c) cooling and rotating on two axes, and (d) removing the part. *From Crawford (1996), reproduced with permission from Wiley-Interscience.*

mold adhere to the mold surface and melt to form a uniform polymer layer. Then the mold, still rotating, is cooled, and lastly the part is removed. The process is widely used to make large hollow shapes, such as tanks, bins, and silos as well as smaller objects such as traffic cones and trash cans. A key feature of the process, and a contrast to blow molding (see Chapter 3), is a very uniform thickness. Control of the process requires attention to the starting material characteristics, mold design, and structure development during the process steps.

Rotational molding is used to make a variety of thermoplastic polymers with the most popular being the various forms of polyethylene, including low−density polyethylene (LDPE), high-density polyethylene (HDPE), and linear low-density polyethylene (LLDPE). The popularity of PE polymers is based on their ability to be ground into powders, their thermal characteristics and their final properties and cost. As a semicrystalline polymer, PE transitions to a low viscosity melt over a narrow temperature range, which facilities the change from powder to molten layer during the process. Also, PE has good mechanical properties as well as chemical resistance, which is important for many of the applications. Other polymers that are rotationally molded include Nylon, polyvinyl chloride (PVC), and polycarbonate (PC).

5.4.2 Powder Preparation

The polymer powder must meet several requirements for rotational molding. First, it should flow readily so that it is easy to handle and add to the mold. Good flow properties are also needed in the initial stage of the molding process, when the particles are tumbling inside of the mold. During this phase, the particles adhere to the mold surface and ideally form a uniform particle layer. For flow characteristics, larger particle sizes are advantageous. However, for adherence to the mold and fast melting, a smaller particle size is better. Therefore, the particle size range for rotational molding is a compromise of these two requirements, typically in the range of 150−500 μm for PE polymers. Particle shape is also important to the flow and performance of the polymer in the process. Based on their flow behavior, symmetrically shaped particles are preferred over fibular or irregular shapes.

Thermoplastic polymers are routinely made into millimeter size pellets (which are also known as granules). Extra steps are needed to convert these pellets to powder. Mechanical grinding is commonly used, but care is needed to keep the polymer from overheating and forming filaments and irregular shapes. One grinding method involves feeding the granules between two closely spaced circular, grinding plates, one of which is rotating at a high speed. The plate design has a taper so that the spacing between the plates decreases from the center axis outward. Granules are fed into center region with its larger gap; then under action of the rotating plates they are size reduced and driven into the narrow gap periphery, where eventually they exit. A sizing system with sieves and a cyclone, which captures and removes very fine particles, is used to tailor the final particle size distribution. For some polymers, grinding at low temperature, cryogenic grinding, is advantageous as the low temperature promotes brittle fracture during grinding.

5.4.3 Rotational Molding Process Steps

Before the rotational molding process begins, molds are designed and fabricated. The mold is designed in two parts so that it can be opened for charging and part removal at the beginning and end of the process, respectively. Because rotation occurs on multiple axes, asymmetric shapes are easily accommodated. The interior of the mold, however, is free of features such as fins or other projections so that the powder is able to flow freely over all the interior surfaces. Molds are fabricated with thin walls ($\sim$6–10 mm) to speed heat up and cool down. This lightweight construction is in contrast to other polymer molding processes that involve pressure (e.g., injection molding). Aluminum molds made by melt casting and steel mold fabricated from sheet metal are examples.

The four steps of rotational molding are outlined in Figure 5.37. In the first step, the powder charge is added to the mold. The design of the part thickness is made prior to this step as the thickness is proportional to the powder mass. The wall thickness requirements vary with the type of polymer, the size of the part and the application requirements. Typical wall thicknesses are in the range of 1.5–25 mm. With knowledge of the mold geometry, particularly the area of the inner surface, the polymer density and the target part thickness, the weight of polymer powder is calculated. In the first step, the mold is charged with powder and closed. At this point, it is outside of the oven, which is usually kept at the temperature of interest; the next step begins as the mold is transferred into the oven and begins to rotate.

In the second step of the cycle, the polymer powder is transformed to a layer of polymer melt. One way to monitor this transformation is to follow the temperature on the inside of the mold. Figure 5.38 shows an example of this internal temperature during rotational molding of PE. As the mold shuttles into the oven, the temperature inside the mold is low. The oven temperature for this example is set at $\sim$250°C and so in this initial stage the interior temperature rises fairly steeply at the beginning (point A). The heat transfer from the hot air in the oven through the metal mold is fairly fast. Hence the air and the freely tumbling polymer heat up. As time goes on, however, the rate of temperature rise slows. This change is concurrent with the adherence of a layer of polymer on the mold surface. The low thermal conductivity of the polymer slows the heat transfer. Further, the heat needed

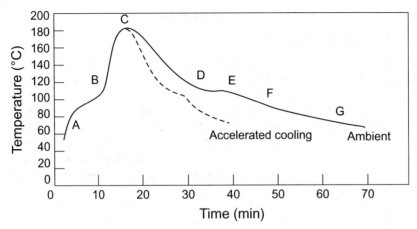

FIGURE 5.38 Temperature inside of a mold during a rotational molding cycle. A combination of air and water was used for the accelerated cooling data. *From Nugent & Crawford (1996), reproduced with permission from Wiley-Interscience.*

to melt the polymer itself is supplied from the air and so this process too slows the internal temperature rise. At the end of this melting phase (point B), the rate of internal mold temperature rise increases again.

The transformation from powder to a uniform melt layer requires time. The phase transformation is one aspect, as noted above, but also the powder layer as it is deposited contains pores filled with air and so as the particles melt, air filled bubbles are retained in the layer. These rise to the surface and are removed; the increasing temperature and hence decreasing viscosity of the polymer layer facilitates this stage in the process. The second step in the rotational molding process is complete when a uniform polymer layer is formed. Next (point C), the rotating mold is transferred out of the oven.

During the cooling step the polymer melt layer solidifies as it continues to rotate. The mold is out of the oven and now the external temperature is lower than the internal temperature. The rate of cooling is dictated by the conditions as demonstrated by comparing the curve for ambient cooling with that for accelerated cooling that involves air and water directed at the mold surface. Considering the ambient cooling data, the rate of temperature drop is moderate for some time and then reaches a plateau (point D). The reason the cooling slows is crystallization, which is exothermic. When crystallization is complete (point E), the cooling continues. The now solid polymer layer is also shrinking due to thermal contraction and separates from the mold (point F). At some point after solidification (i.e., when the temperature is lower than the heat distortion temperature of the polymer), the part can be removed and cooled more efficiently outside of the mold.

5.5 POWDER-BASED ADDITIVE PROCESSES

5.5.1 Process Overview

Powders provide a versatile starting point for additive processes. In fact, two of the oldest additive methods use powders. The selective laser sintering (SLS) process was invented at the University of Texas in the late 1980s. In this process, a laser beam scans over the surface of a bed of powder, locally fusing the powder particles together. The original process produced plastic parts, but the technique has been adapted to create metal and ceramic parts as well. The second early process uses an inkjet printing head to deposit a 2D pattern of binder droplets onto a powder bed. These droplets link particles locally. This method, the first to be called "3D Printing," was invented at Massachusetts Institute of Technology as a method to produce complex ceramic molds for metal casting. These two methods of using powder in additive processes are explored in this section.

Powder-based additive processes share a series of common processing steps. See Figure 5.39, a schematic of the SLS process, for example. The first step is to create the CAD file for the part. As in all additive processes, the CAD file defines the size, shape and features of the final object. The next step is the preparation of the powder bed in the feed chamber. Powder is chosen based on its flow properties as well as its characteristics related to the subsequent fusion. Powder is loaded into a feed chamber or sometimes into two feed chambers that flank the build chamber. The powder bed and the chamber are typically preheated; control of the atmosphere is also key in many processes. The platform in the build chamber is initially at a small distance below the bottom surface of the apparatus; this offset establishes the thickness of the powder layer. The platform on the feed chamber is then

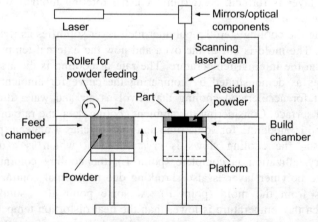

FIGURE 5.39 Schematic of a selective laser sintering apparatus that uses a feed chamber and counter-rotating roller to distribute powder. *Adapted from Gibson et al. (2010).*

notched up and the powder is pushed over to the build chamber where it is deposited as a uniform layer. Next, the powder is fused selectively into a 2D layer (e.g., by a scanning laser beam). Once the layer is established, the platform in the build chamber notches down and the process repeats. Unlike melt-based additive processes, powder bed methods do not require a support material for overhangs, because unfused powder effectively acts as support. In some instances, however, supports are used to stabilize the part. When the layer-by-layer building is complete, the part is removed from the powder bed, dusted off and in some cases. sent on to a post-process, such as sintering. A portion of the unfused powder is recycled.

5.5.2 Selective Laser Sintering (Melting)

Selective laser sintering (SLS) and selective laser melting (SLM) techniques use a scanning laser beam to transform powder particles into a fused solid. There are several variations on this method. In general, the laser light is absorbed by the powder bed, which leads to local heating. The local heating can sinter particles, either by solid state or liquid phase mechanisms. Also, the heat can melt or partially melt the particles. The ambiguity or overlap between sintering and melting is exemplified in partial melting. In this case, a chemically homogeneous powder can be sintered by a liquid phase mechanism. Further, powders are designed to fuse in a *direct* mode in which the particles themselves sinter or melt, or in an *indirect* mode in which a second phase is present with the main particle phase and this second phase melts to cause fusion. For example, a polymer coating on metal or ceramic powder melts as the laser beam passes over the bed, fusing the solid metal or ceramic particles, and a post-process heat treatment is used to burn out the polymer and sinter the metal or ceramic.

In this section, the fundamentals of the process steps in these selective laser fusion techniques are explored. To simplify the nomenclature, the acronym SLS is used in this discussion with the understanding that the laser scan frequently causes melting and the structure development is more complex than simple sintering. Lastly, while there are a large number of powder materials and approaches used in these additive methods, the focus here is on the direct mode.

Powder Preparation and Flow. Powders for SLS are designed to meet several requirements. Powders for SLS are free flowing so that they are easily distributed on the build plate. Achieving a high packing density on the build plate is also a goal so that shrinkage during laser fusing is minimized. Additionally a smooth surface on the powder bed is also desired. The interaction between the powder and the laser, which is a function of the powder chemistry as well as the particle size and powder bed density, is still another consideration. Lastly, the powder thermal properties, such as heat capacity and thermal conductivity, are important as are the melting, sintering and solidification characteristics.

Particle size distribution in SLS powders is designed to best meet the requirements mentioned above. To achieve good flow characteristics fine particles (e.g., <20 μm) are eliminated. A distribution in sizes is useful for packing and the maximum particle size is kept as small as possible so that surface smoothness and melting or sintering rates, which typically increase as particle size decreases, are kept as high as possible. For many materials, particle sizes in the range of 20–100 μm are common. Particle shapes tend to be spherical to encourage flow.

Powders are distributed onto the build plate by one of two main routes: a counter-rotating roller and a blade. The roller mechanism is depicted in Figure 5.39. The powder layer exposed by ratcheting up the feed platform is pushed by the roller over the build chamber, which is recessed by a fixed amount (e.g., 0.1 mm). The counter-rotation entrains air to fluidize the powder and distributes it such that there is little shear and disturbance to the part and powder already in the build chamber. In the other mode, a fixed amount of powder is fed from a hopper to a blade mechanism, which likewise spreads the powder. Additionally, heaters in the apparatus preheat the powder to prepare it for laser fusion.

Laser Fusion. A laser is an intense source of monochromatic light. The interaction between the laser beam and the material leads to heating. To understand this effect more, consider a laser spot incident on a material. The laser spot is assumed to have a Gaussian distribution in intensity (in Watts/m^2, for example), which is also known as the laser irradiance, H. For a circular spot, this distribution follows:

$$H(r) = H_o \exp\left(\frac{-2r^2}{W_o^2}\right) \tag{5.29}$$

where r is the radial distance and $H = H_o/e^2$ at $r = W_o$, making W_o a benchmark for the size of the spot. Hence, the laser spot diameter, D_L, is approximately $2W_o$ and the surface area illuminated by the laser is $\pi(D_L/2)^2$. See Figure 5.40a and b. The power, P_L, (in Watts) delivered by the beam is found by integrating the intensity distribution:

$$P_L = \int_0^\infty 2\pi r \bullet H(r) dr = \frac{\pi}{2} W_o^2 H_o \tag{5.30}$$

In other words, as the laser rests on the spot, it delivers P_L Watts or Joules per second to the material. To find the average total energy delivered per unit area, E, for a stationary spot exposing the surface for time t, the power is divided by the area of the beam and multiplied by the time:

$$E = \frac{4P_L}{\pi D_L^2} \bullet t \quad \text{(stationary spot)} \tag{5.31}$$

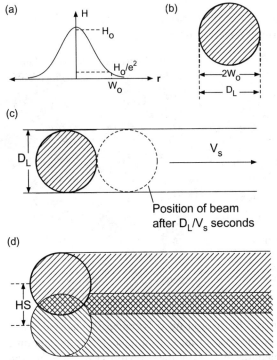

FIGURE 5.40 Schematic diagrams of (a) Gaussian distribution in incident power intensity of a circular laser beam, (b) laser spot size defined, (c) a single scan of a laser spot, showing the origin of the exposure time and energy density, and (d) multiple scans, showing overlap and hatch spacing, HS. *Adapted from Venuvinod and Ma (2004).*

E is known as the "energy density." Now, consider a beam scanning over the surface with velocity V_s. See Figure 5.40c. The beam spends a time D_L/V_s on each $\pi(D_L/2)^2$ area, so the energy density is:

$$E = \frac{4P_L}{\pi D_L^2} \cdot \left(\frac{D_L}{V_S}\right) = \frac{4P_L}{\pi D_L V_S} \approx \frac{P_L}{D_L V_S} \quad \text{(scanning beam)} \quad (5.32)$$

In an SLS process, the laser scan pattern includes overlap and so some parts of the powder bed experience a higher energy density. See Figure 5.40d.

The incident laser light is either reflected or absorbed. (The powder beds are deep and hence no light is transmitted.) For heating to occur, the powder bed must absorb the light. The absorption (or absorbance) characteristics of the powder bed depend on the chemistry of the particles as well as the particle size and bed packing density. Table 5.2 shows data for powders used in SLS processing at the wavelengths of two common lasers used in SLS. This

TABLE 5.2 Absorbance of Powders Used in SLS at Two Different Wavelengths[a]

Powder	Absorbance at 1.06 μm (%)	Absorbance at 10.6 μm (%)
Copper	59	26
Iron	64	45
Titanium	77	59
Alumina	3	96
Silica	4	96
Silicon carbide	78	66
Polytetrafluoroethylene	5	73
Polymethyl methacrylate	6	75

[a] Data from Tolochko et al. (2000).

data was gathered by measuring reflectance, R, and calculating absorbance as $1-R$ (i.e., for the case of no transmittance). The metals on the list absorb at both wavelengths but favor the 1.06 μm (Nd-YAG, neodymium yttrium aluminum garnet, laser). Oxide ceramic powders on the other hand absorb strongly at the higher wavelength of 10.6 μm (CO_2 laser) as do the polymers. Absorbance in these materials is related to the lattice and molecular vibrations. For all, absorbance is higher for powder beds as compared to the dense bulk material of the same composition.

Understanding the increase in temperature and the sintering or melting behavior as the laser scans over the surface is more challenging. It is important to first note that SLS is carried out on a preheated powder bed to limit the amount of heating required from the laser and to lessen thermal gradients. Figure 5.41 shows the various heat transfer events. The scanning laser irradiates the powder layer, some fraction of the light is absorbed and the temperature is raised locally, but the heat is also lost to radiation and convection in the adjacent gas phase as well as conduction to surrounding portions of the powder bed. The rates of these processes as well as the scanning rate and the properties of the material itself determine the local temperatures in the powder bed. In some materials and conditions, the temperatures exceed the melting point and a "melt pool" forms in the track. This pool cools and solidifies in the laser beam's wake. In other materials, there is partial melting or sintering.

Experimental data show correlation between the energy density (Eq. 5.32) and the extent of melting or sintering. Process maps, such as the one in

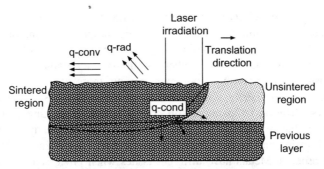

FIGURE 5.41 Schematic diagram showing heat transfer mechanisms at play in SLS as the laser scans over the powder bed, where q-conv, q-rad, and q-cond are the heat fluxes from convection, radiation, and conduction, respectively. *From Williams & Deckard (1998) with permission from Emerald Group Publishing.*

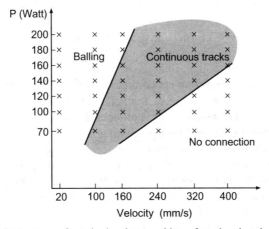

FIGURE 5.42 Process map for selective laser melting of an iron-based powder mixture, showing the effects of laser power and scan speed on morphology. *From Kruth et al. (2005) with permission from Emerald Group Publishing.*

Figure 5.42, show regions for achieving high density, continuous tracks without defects. In this diagram, low energy density is in the lower right (fast speed and low power) and here the particles are disconnected, unsintered. High energy density in the upper left (slow speed and high power). These excessively high energy densities lead to defects known as balling. "Balling" is the formation of spherical clusters; this defect is a sign of a large melt pool. At intermediate values of energy density, continuous tracks are formed. Under these conditions, high quality parts can be made. Achieving an overall cohesive structure is another goal. As mentioned, there is overlap in the scan tracks, which provides added laser energy input to those regions, improving the bonding and cohesion in-plane. The layer-to-layer bonding requires attention to the energy density so that adequate penetration can be achieved.

A variety of materials are prepared by SLS, each having its own set of considerations regarding the fusion process. For polymers, an important issue is melting behavior. Semicrystalline polymers have the advantage of melting into a low viscosity liquid, whereas the more gradual decrease in the viscosity of an amorphous polymer presents more of a challenge in the fusion process. Further, a semicrystalline polymer that has a considerably lower crystallization temperature (on cooling) than its melting temperature (on heating) has an advantage. On cooling, such a polymer has more time for interdiffusion and layer-to-layer bonding before crystallization effectively halts the process. Metals have different issues when processed by the direct route. On melting, their low viscosity and high surface tension lead to the "balling" defect. Further, there are issues with thermal stresses. Ceramic SLS processes are challenging and not as well-developed as polymers and metals. One problem is related to developing a powder bed with high packing density, and ultimately good sinterability. The use of granulated ceramic might seem attractive for this purpose, but the packing density in the bed is very low. So alternative powder distribution methods involving slurries (suspensions of ceramic powder in water) have been explored. Further, due to their high melting points, indirect methods are also attractive. Polymer additives lead to fusion at low temperatures, while additives that form liquid phases to assist sintering at higher temperatures are also employed. For all materials, the laser scanning may not be sufficient to reach the desired density or microstructure and post-processing is needed.

SLS Post-Processing. Post-processing for SLS begins with the removal of the part from the powder bed. Some of the unfused powder is recycled. Complete recycling is not possible due to degradation of the powder. Some parts are complete at this point and need no other post-processing. Others require thermal treatment for binder removal (i.e., for indirect processes) and sintering. Another route to increase the density of porous parts is infiltration. The porous SLS part is filled with a liquid thermosetting resin and cured, for example.

5.5.3 Inkjet Binder Printing ("3D Printing")

Inkjet binder printing (or 3D printing) processes have several steps in common with the SLS processes. Inkjet binder printing also begins with a powder bed and involves the same sort of initial powder spreading step. Therefore, the powder requirements are similar—the powder must flow evenly to make a fresh layer. And, the mechanism for controlling the layer thickness is the same; the build platform drops and a layer of powder is added onto the previously printed part. Like SLS, the powder bed can be heated to encourage the binding action and post-processing involves removal from the bed and thermal treatment. The method of fusing, however, is quite different.

In inkjet binder printing, an inkjet head translates across the plane of the powder layer (x-y plane), depositing "ink" droplets. The ink can take on several forms. As the process name implies, one type of ink is a binder solution. For example, a solution of polymer or a dispersion of nanosized particles. The liquid ink droplet is absorbed by the bed and binder is deposited at particle-particle contacts during evaporation. Another route is similar to indirect SLS. In this route, the binder is mixed with the powder so that it is uniformly distributed in the dry powder bed and then the inkjet printer supplies a liquid medium that activates the binder. This route has some advantages because the liquid can be designed to better meet the requirement of the inkjet process (see Chapter 6).

The method was originally developed to make ceramic cores and shells for metal casting operations. In this example, a nanosized silica suspension is deposited onto a bed of alumina particles. This application is ideal for binder printing, because larger alumina particles are acceptable. Achieving high sintered densities in ceramics, however, ordinarily requires micron to submicron particle sizes. The same issue with SLS of ceramics is encountered—achieving a powder bed with high packing density. Therefore, another version of this method involves depositing a layer of slurry and allowing it to dry to create the layer in the powder bed.

Perhaps the most visible applications of binder inkjet printing are architectural modeling and prototyping. In these applications, gypsum-based or polymer powders are used along with polymer binder solutions. The binders can be colored and multinozzle inkjet print heads are employed so that a multicolor model or prototype can be prepared.

5.6 SUMMARY

Powder processes provide a means to create simple as well as complex shapes from powder starting materials. The engineering of powder processing methods requires an understanding of how the characteristics of powders, including size, size distribution and shape, impact the flow and packing behaviors. Powders composed of fine particles (e.g., in the range of ~ 20 microns and below) have many more particle-particle contacts per unit volume than powders with larger sized particles and hence they do not flow well due to friction at these interfaces. Particle size distribution factors into the packing density that can be achieved without external pressure. Powders having a range of particle sizes are better able to form dense packings compared with powders with narrow size distributions. The small particles in the distribution can fill spaces between the large particles. Powder characteristics play vital roles in the forming of parts by compaction, rotational molding, and powder-based additive methods.

For most powder processes, a separate heat treatment is needed after a forming operation ("green" part) or additive process to create a dense polycrystalline microstructure. Densification occurs by a thermally activated process known as sintering and the final microstructure is affected by another thermally activated process—grain growth. The macroscopic thermodynamic driving force for sintering is the energy decrease from replacing high energy solid-vapor surfaces with lower energy solid-solid surfaces (grain boundaries). Sintering occurs by either solid state or liquid phase mechanisms. In solid state sintering, diffusion of material from between the particles to the "necks" at particle-particle contacts leads to the formation of grain boundaries. This process is enhanced in green parts composed of fine particles. Diffusion of atoms or ions along grain boundaries to the neck regions, one common solid state sintering mechanism, causes pore content to decrease and shrinkage to occur. Lastly, as the pore content decreases and pores become isolated, the rate of densification drops and a competing process, grain growth, accelerates. Since grain growth lowers the grain boundary content of the material, it interferes with densification. In liquid phase sintering, a high temperature liquid forms on heating, which provides a means for rearrangement of particles and for fast mass transport and shrinkage. Viscous sintering is a related mechanism in which amorphous particles fuse together by viscous flow.

Powder pressing operations, including uniaxial compaction and isostatic compaction, are effective methods for making shapes from powders. They are used in ceramic processing and are at the heart of "powder metallurgy." A compaction process requires attention to preparation of the starting powder material so that it is free flowing and deformable, and yet has the appropriate characteristics for sintering. In ceramics, these features are built into the powder by creating "granules," controlled clusters of fine particles with binder. In metals, the particles themselves are plastically deformable; a broad particle size distribution provides deformability and sintering characteristics. During uniaxial compaction the powder density increases gradually with an increase in pressure, eventually reaching a limit that is based on the mechanical properties and packing of the particles. Geometries of parts that can be uniaxially compacted are limited by friction at die wall interfaces, which tends to lead to pressure and density gradients. Isostatic pressing, which uses a pressure transmitting liquid and a deformable mold, overcomes some of the problems with uniaxial compaction. Full density can be achieved with the combination of heat and pressure in hot pressing and hot isostatic pressing.

Rotational molding is a forming operation that uses polymer particles. In this process, free flowing polymer particles are loaded inside a metal mold. On rotation and heating, the powder coats the inside of the mold and forms a molten layer. The mold is cooled with rotation to solidify the layer into a hollow polymer part. The thermal characteristics and final properties of polyethylene make it a common "rotomolded" material.

Powder-based additive processing methods, including selective laser sintering (SLS) and inkjet binder printing, have broad application across metals, ceramic and polymers. The powder material is built into a part in a layer-by-layer fashion. A 2D pattern is created in a layer of powder by a fusion technique, then another layer of powder is added on top of that, followed by fusion into the next 2D layer, and so on. SLS uses a scanning laser to fuse particles, most commonly by melting or partially melting them (and hence selective laser melting, SLM, is also used as a name for this process) or by sintering. SLS can create high density parts or porous parts that need to be sintered in a separate step. Another option is to locally fuse powder with drops of binder "ink" in the binder inkjet printing process, which is also known as 3D printing. One advantage of this process is that inkjet technology affords the option of multiple inks and hence multiple colors can be created in some versions of the method.

BIBLIOGRAPHY AND RECOMMENDED READING

Powder Processing—Ceramics

Barsoum, M.W., 1997. Fundamentals of Ceramics, second ed. Institute of Physics Publishing, London, UK.

Ewsuk, K.G., 2001. Powder granulation and compaction. In: Buschow, K.H., Cahn, R.W., Flemings, M.C., Ilschner, B. (Eds.), Encyclopedia of Materials—Science and Technology. Elsevier, New York, NY, pp. 7788—7800.

Eziz, A., 1991. Hot pressing. In: Schneider Jr., S.J. (Ed.), Engineered Materials Handbook, Vol. 4: Ceramics and Glasses. ASM International, Materials Park, OH, pp. 186—193.

German, R.M., 1991. Fundamentals of sintering. In: Schneider Jr., S.J. (Ed.), Engineered Materials Handbook, Vol. 4: Ceramics and Glasses. ASM International, Materials Park, OH, pp. 260—269.

Kingery, W.D., Bowen, H.K., Uhlmann, D.R., 1976. Introduction to Ceramics, second ed. John Wiley & Sons, New York, NY.

Quinn, D.B., Bedford, R.E., Kennard, F.L., 1984. Dry-bag isostatic pressing and contour grinding of technical ceramics. In: Mangles, J.A., Messing, G.L. (Eds.), Advances in Ceramics, Vol 9: Forming of Ceramics. American Ceramic Society, Columbus, OH, pp. 4—15.

Rahman, M.N., 1995. Ceramic Processing and Sintering. Marcel Dekker, New York, NY.

Reed, J.S., 1995. Principles of Ceramic Processing, second ed. Wiley Interscience, New York, NY.

Ring, T.A., 1996. Fundamentals of Ceramic Powder Processing and Synthesis. Academic Press, New York, NY.

Powder Processing—Metals

Alpien, D., 2001. Particulate processing (powder metallurgy). In: Buschow, K.H., Cahn, R.W., Flemings, M.C., Ilschner, B. (Eds.), Encyclopedia of Materials—Science and Technology. Elsevier, New York, NY, pp. 6769—6776.

German, R.M., 1984. Powder Metallurgy Science. Metal Powder Industries Federation, Princeton, NJ.

German, R.M., 1996. Sintering Theory and Practice. Wiley Interscience, New York, NY.

German, R.M., 1998. Powder Metallurgy of Iron and Steel. John Wiley & Sons, New York, NY.

Klar, E. (Ed.), 1984. Metals Handbook, Vol. 7: Powder Metallurgy. ninth ed. ASM International, Materials Park, OH.

Lenel, F.V., 1980. Powder Metallurgy: Principles and Applications. Metal Powder Industries Federation, Princeton, NJ.

Rotational Molding

Crawford, R.J., 1996. Rotational Moulding of Plastics. John Wiley & Sons Inc., New York, NY.

Osswald, T.A., 1998. Polymer Processing Fundamentals. Hanser/Gardner Publishing, Cincinnati, OH.

Rao, M.A., Throne, J.L., 1972. Principles of rotational molding. Polymer Engineering and Science 12 (4), 237–264.

Rees, R.L., 1988. Rotational molding. In: Dostal, C. (Ed.), Engineered Materials Handbook, Vol. 2: Engineering Plastics. ASM International, Materials Park, OH, pp. 360–367.

Strong, A.B., 2000. Plastics Materials and Processing, second ed Prentice-Hall, Upper Saddle River, NJ.

Powder-Based Additive Processes

Gibson, I., Rosen, D.W., Stucker, B., 2010. Additive Technologies. Springer, New York, NY.

Kruth, J.-P., Levy, G., Klocke, F., Childs, T.H.C., 2007. Consolidation phenomena in laser and powder-bed based layered manufacturing. CIRP Ann. Manuf. Techn. 56, 730–759.

Liou, F.W., 2008. Rapid Prototyping and Engineering Applications: A Toolbox for Prototype Development. CRC Press, New York, NY.

Venuvinod, P., Ma, W., 2004. Rapid Prototyping: Laser-Based and Other Technologies. Kluwer Academic Publishers, New York, NY.

Sachs, E., Cima, M., Bredt, J., Curodeau, A., Fan, T., Brancazio, D., 1992. CAD-casting: direct fabrication of ceramic shells and cores by three dimensional printing. Manuf. Rev. 5 (2), 117–126.

CITED REFERENCES

Arzt, E., Ashby, M.E., Easterling, K.E., 1983. Practical applications of hot-isostatic pressing diagrams: four case studies. Metall. Trans. A 14 (2), 211–221.

Ashby, M.F., 1974. A first report on sintering diagrams. Acta Metall. 22, 275–289.

ASTM Standard B213, 2013. Standard Test Methods for Flow Rate of Metal Powders Using the Hall Flowmeter Funnel. ASTM International, West Conshohocken, PA.

ASTM Standard D6128, 2014. Standard Test Method for Shear Testing of Bulk Solids Using the Jenike Shear Cell. ASTM International, West Conshohocken, PA.

Crawford, R.J., 1996. Introduction to rotational moulding. In: Crawford, R.J. (Ed.), Rotational Moulding of Plastics. John Wiley & Sons Inc., New York, NY, pp. 1–31.

Davidson, J.E., 1984. Compressibility of metal powders. In: Klar, E. (Ed.), Metals Handbook, Vol. 7: Powder Metallurgy, ninth ed. ASM International, Materials Park, OH, pp. 286–287.

Gauthier, F.G.R., Danforth, S.C., 1991. Packing of bimodal mixtures of colloidal silica. J. Mater. Sci. 26, 6035–6043.

German, R.M., 1991. Fundamentals of sintering. In: Schneider Jr., S.J. (Ed.), Engineered Materials Handbook, Vol. 4: Ceramics and Glasses. ASM International, Materials Park, OH, pp. 260–269.

German, R.M., 2015. Sintering: From Empirical Observations to Scientific Principles. Butterworth-Heinemann, New York, NY.

Gibson, I., Rosen, D.W., Stucker, B., 2010. Additive Technologies. Springer, New York, NY.

Kennard, F., 1991. Cold isostatic pressing. In: Schneider Jr., S.J. (Ed.), Engineered Materials Handbook, Vol. 4: Ceramics and Glasses. ASM International, Materials Park, OH, pp. 147–152.

Kruth, J.-P., Mercelis, P., Van Vaerenbergh, J., Froyen, L., Rombouts, M., 2005. Binding mechanisms in selective laser sintering and selective laser melting. Rapid Prototyping J. 11, 26–36.

McGeary, R.K., 1961. Mechanical packing of spherical particles. J. Am. Ceram. Soc. 44, 513–522.

Nugent, P.J., Crawford, R.J., 1996. Process control for rotational moulding. In: Crawford, R.J. (Ed.), Rotational Moulding of Plastics. John Wiley & Sons Inc., New York, NY, pp. 196–216.

Reed, J.S., 1995. Principles of Ceramic Processing, second ed. Wiley Interscience, New York, NY.

Svarovsky, L., 1990. Characterization of powders. In: Rhodes, M. (Ed.), Principles of Powder Technology. John Wiley & Sons, New York, NY, pp. 35–69.

Thompson, R.A., 1981. Mechanics of powder pressing: I, model for powder densification. Am. Ceram. Soc. Bull. 60, 237–243.

Tolochko, N.K., Laoui, T., Khlopkov, Y.V., Mozzharov, S.E., Titov, V.I., Ignatiev, M.B., 2000. Absorptance of powder materials suitable for laser sintering. Rapid Prototyping J. 6 (3), 155–160.

Venuvinod, P., Ma, W., 2004. Rapid Prototyping: Laser-Based and Other Technologies. Kluwer Academic Publishers, New York, NY.

Williams, J.D., Deckard, C.R., 1998. Advances in modeling the effects of selected parameters on the SLS process. Rapid Prototyping J. 4, 90–100.

QUESTIONS AND PROBLEMS

Questions

1. What is the difference between a particle and powder?
2. Describe how the characteristics of individual particles are similar to and different from the characteristics of a powder.
3. How is the packing fraction of powder mass related to its porosity?
4. What factors influence powder flow and how can a powder be engineered for the best flow characteristics?
5. What is sintering? Give some characteristics of the three stages of sintering. What stages of sintering are represented in Figure 5.8?
6. Does the particle size influence the thermodynamics driving force for sintering or the kinetics of sintering?
7. Why are grain boundaries so important to densification?
8. What is the difference between normal and abnormal grain growth?
9. What are the applications, advantages, and disadvantages of liquid phase sintering?
10. How can impact of grain growth be lessened?
11. Why are ceramic particles made in granules by spray drying before pressing?

12. What key features are engineered into a powdered metal for the best compaction?
13. Describe at least three ways that friction impacts a uniaxial pressing operation.
14. How are complex shapes made in uniaxial pressing?
15. What are the advantages of compaction and sintering (powder metallurgy) as compared to casting? Forging?
16. Why would dry bag isostatic pressing be chosen over the wet bag version?
17. Why are polymer parts not made by pressing?
18. What are the key steps in the rotational molding process?
19. How does rotational molding compare to injection molding? List similarities and differences.
20. In powder-based additive processes, how is powder flow controlled and why is this step important? What is the typical particle size range for these processes?
21. The powder bed should have high packing density before SLS or binder inkjet printing. Why?
22. What is meant by energy density in SLS processes? How does part density change with energy density?
23. What are some applications of SLS? What are some limitations?

Problems

1. Consider monodisperse spherical particles packed on a cubic lattice arrangement. (a) Find the expected packing density and porosity (in volume %). (b) If the sphere radius is R_1 and smaller spheres (with radius R_2, $R_2 < R_1$) are added such that the cubic arrangement of the large spheres is retained, find the size R_2 that results in the highest packing fraction that can be achieved? At what volume fraction of smaller particle does this condition occur?
2. Find an expression for grain size as function of time at constant temperature for normal grain growth. Assume that the change in grain size, G, with time is equal to a rate constant, k, divided by time (i.e., dG/dt = k/t). What form do you expect the rate constant to take with respect to temperature?
3. Mackenzie and Shuttleworth (Proc. Phys Soc. (1949) B62, 833) showed that the density in the final stage of viscous sintering follows:

$$\frac{d\rho_{th}}{dt} = \frac{3\gamma_{LV}}{2r\eta}(1 - \rho_{th})$$

where ρ_{th} is the density of the compact divided by the density of the glass (with no pores) and r is the initial radius of the glass particles, which approximates the size of the isolated pores. Plot the density of soda lime silica glass prepared from 1 μm radius particles as a function of time at 650°C. Assume an initial packing fraction of 0.6, use data from Chapter 3 and state assumptions.

4. Grain growth and sintering are both thermally activated, following an Arrhenius form [Rate = Aexp(−Q/RT)]. If the activation energy (Q_{GG}) for grain growth is higher than the activation energy for sintering/densification (Q_S), can a thermal treatment be engineered to densify with limited grain growth? Or densify and then grow grains? Make a sketch and explain.

5. Consider fabrication of parts by uniaxial pressing of Ancorsteel 150 HP powder and sintering. Use the data sheet in Figure 2.12 to answer the following questions. (a) The manufacturer recommends compaction at 620 MPa. Using the data sheet, find the green density at this condition and discuss the recommendation. Include the general changes that occur during the compaction as the pressure increases to 620 MPa. (b) When the powder is compacted at 620 MPa and then sintered, a 2% linear shrinkage is noted. What is the density of the compact after sintering? Assume no mass change on sintering. (c) The manufacturer also gives information on adding graphite to the powder before compaction. Why would this addition be done? What is the strategy?

6. An MgO powder compact is prepared by dry pressing of an granulated powder with an average particle size of 0.5 μm. The green density is 62% of theoretical. The compact is sintered in air at a temperature of 1500°C to produce a ceramic that has a 99% theoretical density and an average final grain size of 2 μm. (a) If the green compact has a diameter of 2 cm and a thickness of 1 mm, predict the fired geometry. (b) Predict how the following changes in the processing will affect the microstructure and the porosity the polycrystalline ceramic; assume all other process steps are unchanged. (i) The powder has an average particle size of 3 μm. (ii) The sintering temperature is reduced to 1300°C.

7. Consider the fabrication of metal parts by uniaxial pressing of a metal powders and sintering. The metal powder particles have a yield strength of 70 MPa. (a) The desired dimensions of the green part are 7 cm × 2 cm × 1 cm. Sketch the die and show how pressure should be applied. Give a reason for your design. (b) What factors would you consider in choosing a sintering temperature and conditions? (c) If the pressed part described in part b has a green density of 80% of theoretical and experiences 7% linear shrinkage on sintering, find the final dimensions. (d) Based on your answers to a–c would you consider isostatic pressing or hot isostatic pressing as an alternative? State your reasoning.

8. Derive Equation 5.5.

9. Consider pressing of a metal slab with dimensions of 11 × 11 × 2.5 cm. Compare single action, double action and dry bag isostatic pressing in

terms of (a) pressure transmission ratio assuming that $\mu K_{hv} = 0.12$, (b) green density (qualitative comparison).

10. A ceramic part with final dimensions of 7 cm × 2 cm × 1 cm is desired. Design a punch and die system for pressing this part, assuming that the compacted density is 55% of theoretical and the fired density is 99% of theoretical. Be clear on how pressure should be applied.

11. Consider the processing of a copper powder into an automotive part using uniaxial pressing followed by sintering. The final part should be produced as rapidly as possible to a high density. Two suppliers of powder have provided information on their powders:

Supplier	Purity	Size Distribution	Other Features
1	99.9%	90% < 50 μm, 30% < 10 μm	Spherical shape, lubricant admixed, yield strength of 300 MPa
2	99.9%	90% < 150 μm, 20% < 44 μm	Lubricant admixed, irregular powder shape, yield strength of 185 MPa

(a) Which supplier would you choose? Explain your choice in terms of all the steps in the processing (forming and post-forming).

(b) Sketch the compact density (in % theoretical density) as a function of applied pressure during uniaxial pressing. Specify units and put typical values on both axes. Describe the processes occurring in the compact as the pressure is increased.

(c) Specify the sintering conditions that you would choose along with the rationale for your choices

12. Discuss the effect of friction on powder processing. Use examples and consider all the phases of a powder process.

13. Discuss the advantages and limitations of powder metallurgy (pressing methods) in creating relatively complex metal parts (e.g., gears) as compared with casting (Chapter 3) and forging (Chapter 4).

14. This question compares the rotational molding of high density polyethylene (HDPE) and low density polyethylene (LDPE). (a) Calculate the quantity of each powder that you would need to prepare a spherical part with a 20 cm diameter and a thickness of 0.5 cm. (b) Sketch what you would expect for the temperature of the air in the mold versus time for HDPE and LDPE parts. (c) What general property differences would the two parts have?

Chapter 6

Dispersion and Solution Processes

By Lorraine F. Francis and Christine C. Roberts

6.1 INTRODUCTION

Dispersions and solutions are liquids that contain the building blocks of solid engineering materials. As described in Chapter 3, liquids have advantages for creating detailed 3D shapes, but it is not always possible or convenient to make a melt. A dispersion is a suspension of fine ceramic, metal, or polymer particles in a liquid medium, such as water. A solution, on the other hand, consists of molecules dissolved into a solvent. While there are a variety of solution precursor methods used to process materials, "solution" in this chapter refers to a solution of the final desired material in a solvent. This definition naturally restricts this category to polymer processing. Dispersions and solutions are similar in that they both involve a transient liquid. That is, the liquid medium is used to make the particles or molecules flow, but then eventually that liquid is removed, usually by evaporation, to leave behind a solid. The solid shape or coating may require additional post-processing steps, such as sintering, to develop structure and properties.

Though not a dispersion or a solution, liquid monomers and oligomers are also useful in materials processing and share the strategy of employing a liquid precursor to the final solid. In this regard, liquid monomers and oligomers (i.e., low molecular weight polymers) are easily made into coatings and polymerized by heating or exposure to ultraviolet (UV) radiation. Likewise, monomers are formed into 3D shapes, by virtue of additive manufacturing processes that also involve photopolymerization.

Figure 6.1 shows some examples of processes that employ dispersions, solutions, and liquid monomers. Complex 3D shapes are created using casting processes, such as slip casting (Figure 6.1a). These processes make use of the ability of the liquid precursors to flow into complex shapes defined by molds. After removal from the mold, drying, and sintering are needed. Coating processes (such as slot coating, Figure 6.1b) involve distributing

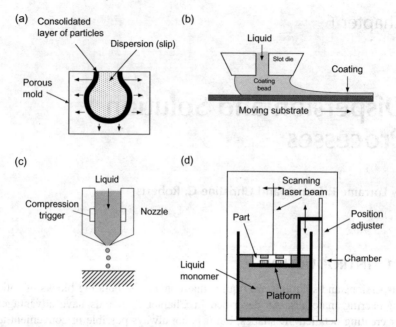

FIGURE 6.1 Schematics for processes that use dispersions, solutions and liquid monomers. (a) Slip casting—an aqueous dispersion is poured into a porous mold; water is extracted into the mold to create a layer of particles, which is the basis of the part. (b) Slot die coating—a dispersion, solution or monomer is forced under pressure through a die onto a moving substrate to create a coating. (c) Inkjet printing—the liquid flows into a nozzle where a compression trigger creates a pressure pulse in the liquid that results in droplets and printed structure. (d) Stereolithography—monomer is distributed on a build platform and converted to solid, layer-by-layer, using a scanning laser beam. *[adapted from Venuvinod and Ma (2004)].*

liquid uniformly onto a substrate and then solidifying by drying or curing. Printing operations, such as inkjet printing (Figure 6.1c), create 2D patterns on substrates. Such printing can be extended out of the plane of the substrate to make 3D parts, frequently with the use of photopolymerization. Another additive process is stereolithography (Figure 6.1d). Stereolithography, which is given the acronym SLA, uses a scanning laser to convert liquid monomer to a solid 3D part.

The fundamental steps of flow, shape definition, and shape retention are central to the processes in this chapter. Dispersions, solutions, and liquid monomers *flow* by gravity or under pressure into molds and onto substrate surfaces. In additive processes liquid flow is also required; for example, the monomer liquid must flow over the previous layer of the 3D part when the build platform is lowered. *Shape definition* in casting processes is clearly the function of the mold, while in coating processes the shape is essentially the thickness, which is defined by conditions of the process. For example, in slot coating the flow rate of liquid into the slot determines the thickness of the deposited liquid layer. As in other additive manufacturing techniques, shape is defined by the CAD file, which provides the part size and features. The CAD file is converted into the instructions that drive the position of the inkjet nozzle or laser, for example. *Shape retention* is accomplished in a variety of ways due to the diversity that spans dispersions, solutions and monomers. In some cases, shapes, including coating thicknesses, are stable after drying. In dispersion processes, particles form bonds during drying, sometimes with the aid of a binder, and in polymer solution processes drying leads to solidification by glass transition or a process of fusing latex particles called film formation. Shape retention in monomer processes is due to chemical reactions that increase the molecular weight and form crosslinks (i.e., curing). Likewise, a variety of post-processing operations, including continued drying and curing, binder removal, and sintering, are used.

Ceramic and metallic parts are made by a variety of processes that begin with a dispersion. The same sort of metal and ceramic particles that are discussed in Chapter 5 are used in dispersions. Particle sizes, however, are frequently tailored differently than they are for compaction of a dry powder. Ceramic and metal coatings are prepared by depositing dispersions onto substrates, followed by drying and sintering. Since these dispersions are made in simple liquids, such as water and organic solvents, fine particles are needed to prevent gravitational settling from dominating the dispersion characteristics. The particle size is frequently small enough that they are referred to as colloidal dispersions. However, in processes that involve the particles dispersed in polymer melts, which have a high viscosity, the sizes are tailored more to maximize the particle content as well as encourage subsequent sintering. Dispersions of metal and ceramic powders in polymer melts are injection molded into complex shapes. These shapes contain fairly high polymer content and require careful pyrolysis or "burnout" of the polymer before densification by sintering. For ceramics injection molding and slip casting, another dispersion-based process, are two common methods for creating complex shapes.

Polymeric parts and coatings are prepared from dispersions, solutions, or monomers. The polymer dispersion of most importance is latex. Latex is a generic term for submicron polymer particles created by emulsion polymerization in water. It is the starting material for making coatings, including adhesives and when combined with ceramic particles and additives, paints. Polymer solutions are frequently polymers dissolved in organic solvents that are deposited onto substrates and dried into solid polymer coatings. Lastly, monomers and oligomers are converted to polymer coatings or 3D shapes using thermal or more frequently energetic polymerization, such as UV curing.

The engineering of processes based on dispersions, solutions, and liquid monomers involves considerable attention to *formulation* in addition to the process steps themselves. In processes based on dispersions and solutions, the choice of the transient liquid and the formulation (i.e., the relative amounts of all the components) are critical. The rheology of the liquid or the microstructure of the solids within the liquid can impact final product qualities and process costs in positive or negative ways. For example, for a liquid to properly fill a mold or spread onto a substrate requires it to flow easily, yet diluting the liquid can be expensive in terms of initial starting materials and create extensive drying process steps. Choice of the transient liquid, aqueous or organic, impacts the stability of the dispersion or solution as well as the drying process. Consequently, understanding and optimizing the liquid formulation so that its behavior is acceptable throughout all of the process steps is a main challenge. For liquid monomers, the engineering involves formulating monomers with initiators and selecting the conditions, such as the wavelength and intensity of the UV source, for rapid curing.

In this chapter, there is a section on fundamentals followed by examples of processes. The fundamentals section begins with an introduction to the colloidal dispersions, with an emphasis on understanding how to create a stable dispersion that does not sediment or aggregate appreciably with time, followed by an introduction to polymer solutions. The rheology of dispersions and solutions is then discussed, building on the rheological principles presented in Chapter 3. The fundamentals section rounds out with introductions to drying and curing. Four types of processes are then explored: shape casting, coating and tape casting, extrusion and injection molding, and liquid monomer-based additive processes.

6.2 FUNDAMENTALS

6.2.1 Colloidal Dispersions

Solid particles are often suspended or dispersed in a liquid medium as a means of preparing them for shape casting or coating. Dispersions are also called "suspensions," and "slips" (i.e., ceramic dispersions that are used in the slip casting process). The phenomena that govern the interactions between particles and the behaviors of suspensions are independent of whether the particles are ceramic, polymer, or metal. However, there are substantial differences in behavior based on particle characteristics, such as density, electronic structure, and surface properties.

Dispersions are not thermodynamically stable, but dispersions of small particles can be made kinetically stable (i.e., the particles remain uniformly dispersed in the liquid medium without settling appreciably over time). In most cases, the particles are denser than the liquid that surrounds them and hence gravity opposes stability, especially for large particles or aggregates. Therefore, a stable suspension requires that the particles are small enough and that they do not aggregate. How small is small enough? The answer to this question depends on several factors, as described below. In general, the particles should be in the *colloidal* size range. Roughly speaking, a colloid is a particle with a diameter less than ~1 μm, small enough for Brownian motion to dominate over gravity (see below). Colloidal dispersions consist of colloids suspended in a liquid medium, usually, but not always, water.

This section provides background on the behavior of dispersions, focusing on the forces acting on particles and between particles. The simple forces acting on a single particle in a liquid are described first, followed by the attractive interactions between particles and the more complex repulsive interactions that are needed to make a suspension stable. The topic is presented as an introduction; the books on colloid science, listed at the end of the chapter, provide more details.

Forces Acting on a Single Particle in a Liquid. Consider the simplest suspension: a single, spherical particle in a liquid. This particle has a density of ρ_P, radius of R, and volume of V. The particle is suspended in a liquid of density ρ_L. The force of gravity, F_g, pulls the particle down and a buoyancy force, F_b, pulls the particle up. As shown in Figure 6.2, the net force depends on the densities and volume of the particle. In the most typical case, the density of the particle is greater than that of the liquid and the net force causes the particle to accelerate downward. As the particle velocity increases, the viscous force of resistance, F_v, opposing its movement also increases. In 1851, Stokes found that this viscous resistance is given by:

$$F_v = 6\pi\eta_L R v \quad (6.1)$$

where η_L is the viscosity of the liquid surrounding the sphere and v is the velocity. Equation (6.1) holds true only when the flow of the liquid around the particle is laminar, a condition conveniently expressed as a limit on the single particle Reynolds number, Re_p:

$$Re_p = \frac{2Rv\rho_L}{\eta_L} \quad (6.2)$$

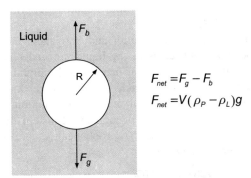

FIGURE 6.2 Forces acting on a spherical particle (density = ρ_P, radius = R, volume = V) suspended in a liquid (density = ρ_L, viscosity = η_L). The downward direction is taken as positive. When sphere begins to move, a viscous force is added in the direction opposite to the particle movement.

The Reynolds number is a ratio of the inertial force to the viscous force. For laminar flow, Re_p should be approximately less than one.

So, the velocity of the sphere in Figure 6.2 increases due to acceleration from gravity, but as its velocity increases so does the viscous drag force, which slows the particle. Eventually, the gravitation and viscous forces balance and a steady state or terminal velocity is reached:

$$v = \frac{2}{9}\frac{R^2(\rho_p - \rho_L)g}{\eta_L} \quad (6.3)$$

In Equation (6.3), g is the gravitational constant and ρ_p is the density of the particle. Note that this equation predicts the rate of downward motion of particles denser than the liquid (sedimentation) as well as the rate of upward motion of particles (or bubbles) that are less dense than the liquid. The rate of sedimentation of particles (denser than the liquid) increases with particle size and density. The influence of liquid viscosity is also apparent.

The particle size dependence of sedimentation rates is exploited in particle size measurement instruments such as an optical or X-ray sedigraph. In these techniques, the sedimentation of a dispersion is tracked, not a single particle. The dispersion is dilute enough to ensure that the presence of neighboring particles does not influence the liquid flow. Hydrodynamic and thermodynamic interactions become important at high particle concentrations, causing deviations from Stokes' law (Eq. (6.1)). The method has other limitations. The upper size limit is set by the restriction that Stokes' law is obeyed, which requires Re_p to be below 1. Large particles accelerate to velocities high enough that the inertia of the liquid surrounding the particle becomes important and the condition of laminar flow is not met. See Example 6.1. For these particles, Eq. (6.1) cannot be used. The lower size limit is based on the time that it takes for very small particles to settle. This size limit can be adjusted by using a centrifuge, which adds a force to accelerate the particles.

In addition to the effects of gravity and buoyancy, another force can affect particle position. Molecules in the liquid are under constant thermal motion. Therefore, the local molecular density in the liquid fluctuates, leading to unbalanced forces on the particle surface and random motion of the particle. This random motion is known as Brownian motion, after R. Brown, the scientist who first observed the phenomena. Einstein treated Brownian motion as a random walk problem. The root mean square displacement of the particle over time t is given by:

$$\bar{x} = \langle x^2 \rangle^{1/2} = \sqrt{2Dt} \quad \text{(one dimensional)}$$

$$\bar{x} = \langle x^2 \rangle^{1/2} = \sqrt{6Dt} \quad \text{(three dimensional)} \quad (6.4)$$

EXAMPLE 6.1 Calculate upper particle size limit for the use of sedimentation velocity for particle size measurement of the following particles suspended in water (20°C): (a) tungsten, (b) alumina, and (c) polystyrene. Assume spherical particles.

The use of sedimentation velocity (Eq. (6.3)) requires that Re_p is less than ~1. So, at the upper size limit Re_p is equal to 1.

$$Re_p = \frac{2Rv\rho_L}{\eta_L} = 1.0$$

The velocity is given by Eq. (6.3).

$$v = \frac{2}{9}\frac{R^2(\rho_p - \rho_L)g}{\eta_L}$$

Therefore,

$$1.0 = \frac{4}{9}\frac{R^3 \rho_L (\rho_p - \rho_L) g}{\eta_L^2}$$

At 20°C, the viscosity and density of water are 1.005×10^{-3} Pa·s and 998 kg/m³, respectively. The expression above can be solved for R, which in this case is the maximum size. The magnitude of R depends on the density of the particle. Density data and the calculated R_{max} are given in the following table.

Material	Density (kg/m³)	R_{max} (μm)	Max particle diameter (μm)
Tungsten (W)	19,300	23	46
Alumina (Al$_2$O$_3$)	3,980	43	86
Polystyrene (PS)	1,050	164	328

---

where t is time and D is the diffusion coefficient of the particle. Here, $\bar{x}$ is used for the average distance a particle will travel from its starting point by Brownian motion. The particle diffusion coefficient takes into account the viscous resistance discussed above and is given by the Stokes-Einstein relationship:

$$D = \frac{kT}{6\pi R \eta_L} \quad (6.5)$$

where k is Boltzmann's constant, R is the sphere radius, and T is temperature in Kelvin. See Example 6.2. In combination with Fick's laws, this diffusion coefficient can be used to predict how the concentration of particles varies in space and time. For example, when a drop of a concentrated dispersion of small particles is added to water, the particles diffuse from high concentration to low concentration, eventually erasing the concentration gradient.

Since the rate of Brownian motion depends on the particle diameter, it is also exploited in particle size measurement techniques. Light scattering, also called photon correlation spectroscopy, is one of the most common techniques for measuring particle size. As particles diffuse in and out of the path of a laser

EXAMPLE 6.2 Consider an alumina particle with a diameter of 0.1 μm suspended in water at 20°C for one minute. Calculate the mean distance the particle travels by Brownian motion in three dimensions. Repeat for a 1 μm particle and a 10 μm particle.

The diffusion coefficient of the 0.1 μm alumina particle is:

$$D = \frac{kT}{6\pi R \eta_L} = \frac{(1.38 \times 10^{-23} \text{J/K})(293 \text{ K})}{6\pi(5 \times 10^{-8}\text{m})(1.005 \times 10^{-3}\text{Pa} \cdot \text{s})} = 4.27 \times 10^{-12} \text{ m}^2/\text{s}$$

Therefore the mean distance traveled in 1 min is:

$$\bar{x} = \sqrt{6Dt} = \sqrt{(6)(4.27 \times 10^{-12} \text{m}^2/\text{s})(60\text{s})} = 3.9 \times 10^{-5} \text{ m} = 39 \text{ }\mu\text{m}$$

Repeating these calculations for 1 μm and 10 μm size particles reveals diffusion coefficients of 4.27×10^{-13} m²/s and 4.27×10^{-14} m²/s, respectively, and mean displacements of 12.3 μm and 3.9 μm, respectively.

---

beam, they create light intensity fluctuations that can be correlated to the diffusion rate. The inferred diffusion coefficient can then be used to calculate the particle size assuming the particles are only experiencing Brownian motion. Particle tracking is another common technique, whereby an optical microscope and a high-speed video camera are used to directly image diffusing particles. Particle tracking algorithms have been developed to identify the particles in the images and link their positions into tracks. Since the images are only two-dimensional projections of a three-dimensional diffusion process, a modified form of Eq. (6.4) is used to relate the distance traveled by the particle and the diffusion coefficient ($\bar{x} = \sqrt{4Dt}$). From the Brownian diffusion coefficient, the particle size can be calculated using Eq. (6.5). However, particle tracking is more commonly used to find the fluid viscosity when the particle size is known and the fluid viscosity is unknown. As with sedigraphy, if Eq. (6.5) is to be used, it must be valid; particles must be dilute in the suspension and of colloidal size. Furthermore, the fluid must be optically clear for both techniques.

Unlike this simple starting point of a single sphere is a liquid, dispersions are composed of many, many particles in a liquid medium. Each particle is simultaneously moving due to gravity, Brownian motion, and perhaps other forces. As a consequence, particles periodically approach one another, where they are subjected to particle-particle interaction forces. For example, van der Waals attractive forces act to draw particles nearer and nearer to each other so that agglomerates are formed. However, repulsive forces that come about due to particle surface charge or steric effects from adsorbed polymers or surfactants on the particle surfaces can counter this attraction.

Van der Waals Attraction. The origin of van der Waals attractive forces between particles is the interaction between atomic and molecular dipoles on the particle surfaces. There are different types of van der Waals forces, the most common of which are called the "London forces." London forces arise when an atomic dipole, created by a momentary fluctuation of the atom's electron cloud, induces a dipole in a nearby atom or molecule. The negative side of one dipole is attracted to the positive side of the other. Additional van der Waals forces arise if a molecule has a permanent dipole (e.g., HCl); neighboring permanent dipoles may attract each other (Keesom forces) and the permanent dipole can induce dipoles in neighboring atoms or molecules (Debye forces).

Van der Waals forces are considered weak in the context of interatomic bonding. However, the forces are additive, which means that the strength of the van der Waals attraction between surfaces is obtained by summing over the attractive interactions between many, many atoms or molecules on the neighboring surfaces. For interactions between colloidal particles, therefore, van der Waals forces are significant.

To find an expression for the van der Waals attractive force, the geometry of the surfaces must be specified. For two identical spherical particles of radius R, the van der Waals force of attraction, F_A, is given by:

$$F_A = \frac{AR}{12h^2} \tag{6.6}$$

where A is the Hamaker constant (see below) and h is the separation distance between the particles. This expression holds when the separation distance is small compared to the particle size ($h << R$). Equation (6.6) shows that the force of attraction becomes stronger as the separation distance between the particles becomes smaller.

The Hamaker constant depends not only on the material that makes up the colloid, but also on the liquid (or gas) that intervenes between the particles. The Hamaker constant required to predict the van der Waals force of attraction between two identical particles of material 1 with material 3 intervening is approximately:

$$A_{131} \approx \left(\sqrt{A_{11}} - \sqrt{A_{33}}\right)^2 \tag{6.7}$$

where A_{11} and A_{33} represent the constant for the material interacting with itself across a vacuum. Different equations are needed to find the attractive interactions between dissimilar particles. Hamaker constants for several materials with vacuum or water as the intervening material are given in Table 6.1.

The potential energy of attraction, U_A, is found by integrating Eq. (6.6) with respect to separation distance:

$$U_A = \frac{-AR}{12h} \tag{6.8}$$

This equation, like the force equation, is valid only for the condition that the separation distance is much less than the size of the particle ($h << R$).

TABLE 6.1 Hamaker Constants for Metals, Ceramics and Polymers (in units of 10^{-20} J)[a]

Material	With Vacuum Intervening	With Water Intervening
Silver	50	40
Copper	40	30
Fused Silica (SiO$_2$)	6.5	0.83
Sapphire (Al$_2$O$_3$)	15.6	5.32
Polystyrene (PS)	7.9	1.3
Polytetrafluoroethylene (PTFE)	3.8	0.33

[a]Data tabulated in Russel et al. (1989)

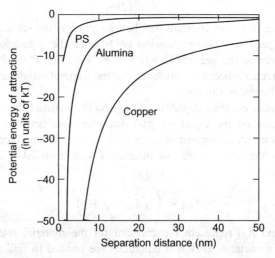

FIGURE 6.3 The effect of separation distance on the van der Waals attraction between 0.1 μm diameter particles of polystyrene (PS), alumina and copper at 25°C. Note at 25°C, $kT = 4.11 \times 10^{-21}$ J.

Figure 6.3 shows the effect of separation distance on the potential energy of attraction for 0.1 μm diameter colloids of different materials in water. As particles approach, attraction becomes stronger and stronger. The strength of the attraction also depends on the type of material; materials with higher Hamaker constant are more strongly attracted to each other than those with lower Hamaker constants.

While van der Waals attraction pulls particles close together, repulsive interactions between the particles prevent the separation distance from reaching zero. For bare, neutral particles, this repulsion arises from the overlap of core shell electrons of the atoms on the surface of the particles (i.e., the same repulsion that is considered in interatomic bonding). This repulsion is very short range (i.e., less than 1 nm) and the resulting total interaction energy attained by summing the attractive and repulsive components has a very deep, primary minimum. This minimum means that it is energetically favorable for particles to approach each other and remain at the separation distance defined by the position of this deep potential energy minimum. The depth of this minimum is many times the thermal energy (kT). Intense mechanical agitation is needed to break apart particles drawn together in such a primary minimum.

Repulsive forces, beyond those provided by inner shell electrons, are needed to prepare a stable suspension. When van der Waals attraction dominates, particles that approach one another are drawn together, forming agglomerates. Over time, more particles join the agglomerates and eventually, the size of the agglomerates makes them susceptible to sedimentation. Or if the particle concentration is high then the particles can actually interconnect into a gelled network. To prevent this spiraling process, repulsive interactions must be deliberately built into the system. One approach is to design the suspension so that the particles gain a surface charge and electrostatic repulsion goes into effect. Another route is to adsorb a polymer or surfactant on the particle surfaces, creating a steric layer that presents a repulsive interaction.

Electrostatic Repulsion. Particles that have the same type of surface charge (positive or negative) repel each other. The effectiveness of this electrostatic repulsion depends on the magnitude of the surface charge and also the characteristics of the liquid in which the particles are dispersed. Below, mechanisms for gaining surface charge are outlined, followed by a description of the response of the liquid medium to the charged surface (the electrical double layer) and, lastly, expressions for the potential energy of electrostatic repulsion.

There are several ways in which solid particles acquire a charge in a liquid medium. The most common are adsorption of potential determining ions, ionization of surface groups, and ion dissolution. The tendency of a material to acquire charge by any of these mechanisms depends on the material's chemistry and structure, and the nature of the liquid. Water, as a polar liquid, is an effective medium for developing a surface charge on a particle. The modes of surface charge development are discussed by way of examples, below.

A common mechanism is the adsorption of potential determining ions from the liquid. The term "potential determining ions" comes about because the adsorption of these ions onto the surface of the material sets up a charge on the surface and in turn, the charge determines the electric potential at the surface. For example, hydrogen and hydroxyl ions are potential determining ions for ceramic metal oxide particles dispersed in water. In water, the surfaces of metal oxides tend to be hydrated (i.e., covered with hydroxyl groups). Lowering the pH of the water results in adsorption of hydrogen ions and an increase in the number of positive surface sites (OH_2^+), while raising the pH leads to adsorption of hydroxyl ions and an increase in the number of negative surface sites (O^-). For a generic metal oxide (MO) with surface hydroxyls (MOH), these adsorption reactions are given as:

$$MOH + OH^- \leftrightarrow MO^- + H_2O$$

$$MOH + H^+ \leftrightarrow MOH_2^+ + H_2O \tag{6.9}$$

Each reaction has an equilibrium constant and the tendency for one reaction over the other depends on the metal oxide structure, particularly the nature of the metal-oxygen bond. At a certain pH, there are equal numbers of positive and negative surface charges and hence a neutral surface. This special pH is known as the point of zero charge (PZC). See Figure 6.4. The PZC

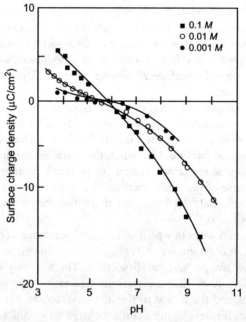

FIGURE 6.4 Change in the surface charge density of titania particles with pH in aqueous KNO_3 solutions. At pH < point of zero charge (PZC), the net surface charge is positive and at pH > PZC the net surface charge is negative. PZC is pH 5.8. Note that the presence of the K^+ and NO_3^- ions does not change the position of the PZC, but does affect the magnitude of the surface charge density away from the PZC. K^+ and NO_3^- are called indifferent ions. *Adapted from Yates and Healy (1980).*

depends on the chemistry of the oxide. At pH values well above the PZC or well below the PZC, surface charge density is greatest.

Non-oxide ceramics, such as silicon nitride, and some metals develop a thin layer of metal oxide on their surfaces when placed in water and therefore develop surface charge in a similar way as their the complementary metal oxide.

Dissociation or ionization of surface groups is a common means by which polymer latex particles acquire charge. The emulsion polymerization technique that generates the latex often leaves behind initiators that bear charge or stabilizers (e.g., surfactants) that are ionic or contain ionizable groups (e.g., carboxylic acid groups, COOH). Carboxylic acid groups, for example, ionize as the pH increases, forming COO^- groups, leading to a negative surface charge. Likewise, polyelectrolytes (polymers that have ionizable groups) behave in a similar way when they are adsorbed onto particle surfaces.

Ion dissolution is a means of developing surface charges on clay particles. Clays have a layered crystal structure with positive cations between the layers. These cations dissolve into water when the particles are dispersed, leaving the particles deficient in positive charge. This interaction results in a net negative charge on clay platelet faces, while the charge on the edges of the platelets responds to the pH-dependent ion adsorption.

Surface charge density can be quantified using titration methods; the data in Figure 6.4 is an example. However, expressions for the repulsive force between charged particle surfaces require knowledge of the electric potential, which is the consequence of the charged surface. There is a distribution in electric potential away from the charged surface into the electrolyte liquid. The sign of the surface potential is the same as that of the surface charge, but uncovering the relationship between the two requires first an understanding of the electric potential distribution.

The Electrical Double Layer. Charged particle surfaces influence the distribution of ions in the adjacent liquid. Oppositely charged ions (counter-ions) are attracted to the surface and those of the same charge (co-ions) are repelled. Electrical neutrality is achieved by the accumulation counterions near the surface, but thermal motion results in a diffuse region of varying ion concentration rather than a discrete layer of ions. (See Figure 6.5.) Hence, the electric potential varies continuously as a function of the distance away from the surface. As might be imagined, the nature of this variation depends on properties of the liquid, namely its dielectric constant and its ionic composition. The combination of the charged surface and the diffuse charged layer in the liquid is commonly referred as a *double layer*.

In the early 1900's, Gouy and Chapman derived an expression for the variation in potential away from the charged surface. The starting point for the derivation is Poisson's equation, which is given here for the

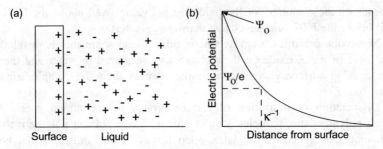

FIGURE 6.5 Schematic diagrams showing (a) a charged surface and the distribution of ions in the adjacent liquid and (b) the variation in electric potential with distance from the surface, based on a functional form derived in the text.

one-dimensional variation in electric potential Ψ as a function of distance (x) away from the charged surface:

$$\frac{d^2\Psi}{dx^2} = -\frac{\rho}{\varepsilon_o \varepsilon_r} \tag{6.10}$$

where ε_r is the relative permittivity or dielectric constant of the liquid, ε_o is the permittivity of free space (8.85×10^{-12} F/m), and ρ is the charge density of the liquid. A solution to this equation provides the functional form of the potential distribution away from the surface, but it does not give the magnitude of the potential at the surface (Ψ_o). Relating this value to the surface charge density is the second step.

The charge density of the liquid is found by summing the contributions of all free ions in the liquid.

$$\rho = \sum_i Z_i e N_i \tag{6.11}$$

where Z_i is the sign and valence of the ith ion (e.g., for Na$^+$, $Z = +1$ and for Cl$^-$, $Z = -1$), e is the electron charge (1.602×10^{-19} C), and N_i is the concentration of the ith ion in number per unit volume. It is instructive to consider the charge density of an aqueous NaCl solution in the absence of any charged particles. The charge density of this solution is 0, because there are equal numbers of positive and negative ions (electrical neutrality). Due to the presence of ions, the solution is called an electrolyte. The electrical neutrality condition holds for all electrolytes, even pure water. The charge density of an electrolyte, however, changes locally near a charged particle surface and this variation gives rise to the variation in potential.

The concentration of ions in the electrolyte at some position near a charged surface depends on the value of Ψ at that position and the effects of thermal disorder. The Boltzmann equation provides a distribution based on these two effects:

$$N_i = N_{io}\exp\left(\frac{-Z_i e\Psi}{kT}\right) \quad (6.12)$$

where k is Boltzmann's constant (1.38×10^{-23} J/K) and T is absolute temperature. N_{io} is the bulk concentration of ion i, the concentration at some distance away from the charged surface where Ψ drops to zero.

With this information, Poisson's equation becomes:

$$\frac{d^2\Psi}{dx^2} = -\frac{1}{\varepsilon_o \varepsilon_r}\sum_i Z_i e N_{io}\exp\left(\frac{-Z_i e\Psi}{kT}\right) \quad (6.13)$$

The solution to this equation is the electric potential as a function of position away from the charged surface. To get to a solution easily, a simplifying assumption is helpful. If $Z_i e\Psi$ is much less than kT (which amounts to assuming that the electrostatic energy of the ions is much less than their thermal energy), then the higher order terms in the expansion of the exponential in Eq. (6.13) can be ignored (i.e., $\exp(y) = 1 + y$). This assumption is called the Debye-Huckel approximation and results in:

$$\frac{d^2\Psi}{dx^2} = -\frac{1}{\varepsilon_o \varepsilon_r}\sum_i Z_i e N_{io}\left(1 - \frac{Z_i e\Psi}{kT}\right) = -\frac{1}{\varepsilon_o \varepsilon_r}\left[\sum_i Z_i e N_{io} - \sum \frac{(Z_i e)^2 N_{io}\Psi}{kT}\right]$$

(6.14)

Electrical neutrality requires the bulk solution to have no net charge (i.e., $\sum_i Z_i e N_{io} = 0$); therefore:

$$\frac{d^2\Psi}{dx^2} = \frac{\sum_i (Z_i e)^2 N_{io}}{\varepsilon_o \varepsilon_r kT}\Psi = \kappa^2 \Psi \quad (6.15)$$

The term κ lumps together a number of parameters that are constant for a given electrolyte liquid. The solution to Eq. (6.15) then can be found, using the boundary conditions of $\Psi = \Psi_0$ at $x = 0$ (the position of the charged surface) and $\Psi = 0$ at a distance far from the surface ($x = \infty$).

$$\Psi = \Psi_o \exp(-\kappa x)$$

or

$$\Psi = \Psi_o \exp\left(-\frac{x}{\kappa^{-1}}\right) \quad (6.16)$$

The term κ^{-1} is a benchmark distance known as the double layer thickness or Debye length. Numerically, κ^{-1} is the distance at which the potential drops to $1/e$ of its value at the surface, Ψ_o (the surface potential).

Based on Eqs. (6.15) and (6.16) the double layer thickness is given by:

$$\kappa^{-1} = \left(\frac{\varepsilon_0 \varepsilon_r kT}{\sum_i (Z_i e)^2 N_{io}} \right)^{1/2} \qquad (6.17)$$

The double layer thickness decreases as the concentration of ions in the liquid increases. This trend is intuitive as the presence of more ions allows the surface charge to be screened more effectively. Likewise, ions of higher valence (Z_i) compress the double layer more than those of lower valence. The dielectric constant of the liquid also plays an important role in the potential variation: liquids with higher dielectric constants create larger electrical double layers if all other parameters are equal. Note that the double layer thickness characterizes the potential distribution; it is not affected by the magnitude of the surface charge. Example 6.3 shows how the double layer thickness is calculated.

EXAMPLE 6.3 (a) Calculate the double layer thickness for a 0.1 M aqueous solution of potassium nitrate (KNO_3) with a pH 4. (b) Repeat for a 0.01 M KNO_3 solution also with pH 4. (c) Plot the variation in potential as a function of distance away from a charged surface for the two electrolytes in (a) and (b). Assume T = 25°C for all calculations.

a. The concentrations of K^+ and NO_3^- are given in moles per liter and therefore must first be converted to number per m^3.

$$[K^+] = [NO_3^-] = 0.1 \text{ mol/L}$$

$$\frac{0.1 \text{ mol}}{L} \times \frac{1000 \text{ L}}{m^3} \times \frac{6.023 \times 10^{23} \text{ ions}}{\text{mole}} = 6.023 \times 10^{25} \text{ ions}/m^3$$

$$N_{K,o} = N_{NO_3,o} = 6.023 \times 10^{25} \text{ ions}/m^3$$

The pH of the solution is given and so the concentrations of H^+ and OH^- can also be found.

$$[H^+] = 10^{-pH} = 10^{-4} \text{ mol/L} \text{ and } [OH^-] = 10^{-(14-pH)} = 10^{-10} \text{ mol/L}$$

The concentrations of these ions are negligible compared with those of K^+ and NO_3^-.

Given or known: ε of water is 80, $\varepsilon_o = 8.85 \times 10^{-12}$ F/m, $k = 1.38 \times 10^{-23}$ J/K, $T = 298$K, $e = 1.602 \times 10^{-19}$ C. [Note that F/m = C/(V•m) and J = V•C.]

$$\kappa^{-1} = \left(\frac{\varepsilon_0 \varepsilon kT}{\sum_i (Z_i e)^2 N_{io}} \right)^{1/2}$$

$$= \left(\frac{(8.85 \times 10^{-12} F/m)(80)(1.38 \times 10^{-23} J/K)(298K)}{(1.602 \times 10^{-19} C)^2 [(+1)^2 (6.023 \times 10^{25} \text{ ions}/m^3) + (-1)^2 (6.023 \times 10^{25} \text{ ions}/m^3)]} \right)^{1/2}$$

$$= 9.7 \times 10^{-10} m = 0.97 \text{ nm}$$

b. For the electrolyte with lower salt concentration, the calculation is repeated using the same methods as in (a). The result is $\kappa^{-1} = 3.07 \times 10^{-9}$ m = 3.07 nm.
c. The potential distribution is found by plotted Ψ/Ψ_o as a function of x from Eq. (6.16). Note that the thin horizontal line represents the condition of $\Psi = \Psi_o/e$ and therefore the intersection of the horizontal line with the data gives the double layer thickness.

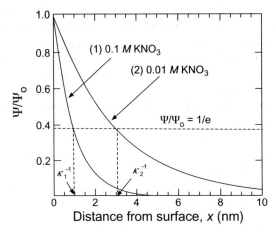

FIGURE E6.3

Equation (6.16) shows how the potential varies as a function of distance away from the surface, but it does not give any information on the value of surface potential, Ψ_o. The surface potential depends on the surface charge density, σ_o, which is established by one of the mechanisms described above, and the properties of the liquid medium. For electrical neutrality, the surface charge density is balanced by an equal and opposite number of charges per unit area in the electrolyte as it extends from the surface.

$$\sigma_o = -\int_0^\infty \rho dx \qquad (6.18)$$

Again, x is the position coordinate with $x = 0$ at the charged surface and the electrolyte at $x > 0$. Using Poisson's equation (Eq. (6.10)), the surface charge density can be shown to be directly proportional to the gradient in potential at the surface.

$$\sigma_o = \varepsilon_r \varepsilon_o \int_0^\infty \frac{d^2\Psi}{dx^2} dx = \varepsilon_r \varepsilon_o \frac{d\Psi}{dx}\bigg|_0^\infty = -\varepsilon_r \varepsilon_o \frac{d\Psi}{dx}\bigg|_0 \qquad (6.19)$$

Then, using the approximate solution for $\Psi(x)$ in Eq. (6.16), the surface charge density is found to be proportional to the surface potential.

$$\sigma_o = \frac{\varepsilon_r \varepsilon_o}{\kappa^{-1}} \Psi_o \text{ or } \Psi_o = \frac{\kappa^{-1}}{\varepsilon_r \varepsilon_o} \sigma_o \tag{6.20}$$

Equation (6.20) shows that surface potential has the same sign as the surface charge density and is proportional to it, but the characteristics of the electrolyte are also important. For example, consider the data in Figure 6.4, which shows surface charge density of titania particles in water at a pH less than the PZC increases with an increase in salt concentration (decrease in κ^{-1}). For titania (and most other metal oxides), H^+ and OH^- are the potential determining ions and their concentration determines the surface potential; the salt ions are indifferent (located in the diffuse layer only). At a given pH, the surface potential remains constant as the salt is added; to maintain electrical neutrality, the surface charge density increases. That is, more H^+ ions are adsorbed onto the surface as salt is added. Example 6.4 explores local ionic concentrations. On the other hand, if the charge on the particles is determined by ion dissolution (e.g., clays), then that surface charge density remains constant as salt is added and the surface potential adjusts.

The repulsion that develops between the charged surfaces depends on the magnitude of the surface potential and the nature of the potential distribution away from the surface (i.e., the double layer thickness). The effect of the surface potential is easy to understand—the higher the surface potential, the higher is the repulsive force and the potential energy of repulsive interaction. The effect of the double layer thickness takes some thought. If we imagine two charged surfaces approaching one another, they begin to "feel" repulsion when the outermost portions of their diffuse layers overlap. Therefore, if the surface potential is constant, a larger electrical double thickness results in repulsion appearing at a point further from the charged surface. Having this sort of far-reaching repulsive effect is important for stability, which is essentially combating the van der Waals attraction as discussed more below.

Deriving an expression for the potential energy of repulsion between particles requires assumptions about the particle size, distance over which the expression is accurate, and the double layer thickness relative to the particle size. Here, two expressions for the potential energy of repulsion between spherical particles that have constant surface potential are presented. When the particles are large relative to the double layer thickness ($\kappa R > 10$), the potential energy of repulsion, U_R, as a function of separation distance, h, is:

$$U_R = 2\pi \varepsilon_r \varepsilon_o \Psi_o^2 R \ln[1 + \exp(-\kappa h)] \tag{6.21}$$

EXAMPLE 6.4 Consider a suspension of titania particles in an aqueous KNO_3 solution at pH 4. (a) Estimate the surface potential of the titania particles in a suspension 0.1 M KNO_3. Use data in **Figure 6.4** and **Example 6.3**. (b) Plot the potential as a function. (c) Plot the variation in ion concentrations (K^+ and NO_3^-) as a function of distance from the titania surface. Assume $T = 25°C$ for all calculations. Assume that the Debye-Huckel approximation holds.

a. The surface potential can be calculated using Eq. (6.20) and data from Figure 6.4 and Example 6.3.

From Figure 6.4, the surface charge density of titania at pH 4 and 0.1 M KNO_3 is $\sim 5\ \mu C/cm^2$. From Example 6.3, κ^{-1} for this electrolyte is 9.7×10^{-10} m.

$$\Psi_o = \frac{\kappa^{-1}}{\varepsilon_r \varepsilon_o} \sigma_o = \frac{9.7 \times 10^{-10} m}{(80)\left(8.85 \times 10^{-12}\ \frac{C}{Vm}\right)} (5 \times 10^{-2} C/m^2) = 6.85 \times 10^{-2} V$$

b. The potential distribution is plotted using the answer to (a) and Eq. (6.16).

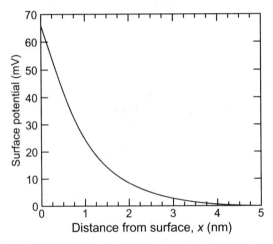

FIGURE E6.4A

c. The ion concentrations are found using the answers above and Eq. (6.12). The bulk ion concentrations are given in Example 6.3.

$$N_{K,o} = N_{NO_3,o} = 6.02 \times 10^{25}\ ions/m^3$$

$$N_K = N_{K,o} \exp\left(\frac{-(1)e\Psi}{kT}\right) \text{ and } N_{NO_3} = N_{NO_3,o} \exp\left(\frac{-(-1)e\Psi}{kT}\right)$$

$$\frac{N_K}{N_{K,o}} = \exp\left(\frac{-e\Psi}{kT}\right) \text{ and } \frac{N_{NO_3}}{N_{NO_3,o}} = \exp\left(\frac{e\Psi}{kT}\right)$$

The ion concentrations as a function of position are calculated using the potential distribution found in part (b), being careful of units (recall that $J = V \cdot C$).

434 Materials Processing

The plots above show the higher concentration of anions near the positively charged surface. The concentrations plotted are relative to the bulk concentration.

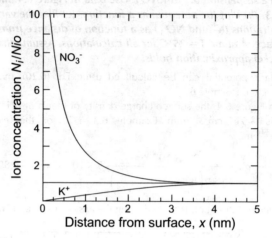

FIGURE E6.4B

When the particles are smaller relative to the double layer thickness ($\kappa R < 5$), the potential energy of repulsion as a function of separation distance is:

$$U_R = 2\pi\varepsilon_r\varepsilon_o\Psi_o^2 R \exp(-\kappa h) \qquad (6.22)$$

These expressions are approximations that are best used with particles having a low surface potential (<25 mV). In spite of this limitation, they are useful in illustrating the effect of key parameters on repulsive interaction. Both show that regardless of the position h, U_R increases with an increase in the dielectric constant of the liquid as well as with an increase in the size or surface potential of the particles. The spatial variations depend on the double layer thickness: both $\exp(-\kappa h)$ and $[1 + \exp(-\kappa h)]$ decrease with h and that decrease is steeper when κ is large (κ^{-1} small). For more information on the origin of these expressions and for additional information, see Hunter (1989).

Clearly, the surface potential is an important characteristic; however, it is difficult to quantify exactly. The most practical way to estimate the surface potential is to set the particle into motion by applying an electric field and determine the potential at the shear plane between the moving particle and the liquid. This shear plane is close to the particle surface and so the potential there, a quantity known as the zeta potential, can be considered as an approximation for the surface potential. Consider a nonconducting particle of radius R in a liquid with a viscosity of η_L and dielectric constant of ε_r. Applying an electric field of strength E causes the particle to move with a velocity v_E, which can be measured and the zeta potential is computed by the following expression:

$$\zeta = \frac{C\eta_L v_E}{\varepsilon_o \varepsilon_r E} \qquad (6.23)$$

where C is a constant that depends on κR. The value of C ranges from $C = 2/3$ at $\kappa R < 1$ to $C = 1$ at $\kappa R > 100$.

The value of the zeta potential is an important factor determining the extent of the electrostatic repulsive interaction between particles. The zeta potential is used in the calculation of the potential energy of repulsion (Eqs. (6.21) and (6.22)). Like surface potential, the zeta potential of an oxide particle changes with pH. See Figure 6.6. The pH value at which the zeta potential is zero is called the isoelectric point, *IEP*. Suspensions at pH values appreciably higher or lower than the *IEP* have higher zeta potentials and hence higher surface potentials and greater repulsive interactions. Because the magnitude of the zeta potential is close to that of the surface potential,

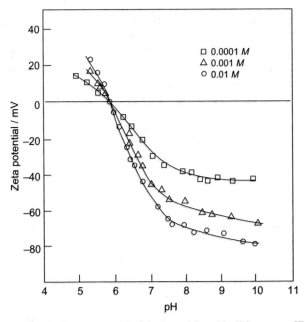

FIGURE 6.6 Change in the zeta potential of titania particles with pH in aqueous KNO_3 solutions (particle loading is 0.05 g/L). At pH < isoelectric point (IEP), the zeta potential is positive and at pH >PZC the zeta potential is negative. The IEP is 5.9, a value very close to the PZC in Figure 6.4. Note that the presence of the K^+ and NO_3^- ions does not change the position of the IEP, but does affect the magnitude of the zeta potential away from the IEP. *Reprinted from J. Colloid Interface Sci., 51, Weise, G. R., & Healy, T. W., Coagulation and electrokinetic behavior of TiO_2 and Al_2O_3 colloidal dispersions, 427−442, Copyright (1975), with permission from Elsevier.*

the *IEP* is close to the point of zero charge. At constant surface potential, however, the zeta potential changes with the characteristics of the liquid medium (i.e., the electrical double layer thickness). When κ^{-1} is large (e.g., ion concentration low), potential drops more gradually with distance away from the particle surface, and hence the zeta potential is closer to the surface potential. See Figure 6.7. Table 6.2 gives some examples of isoelectric points of common ceramics.

Steric Repulsion. Polymers adsorbed on particle surfaces also cause a repulsive interaction. A complete discussion of the factors that encourage adsorption on surfaces is beyond the scope of this section, but some principles can be simply stated. A polymer adsorbs when it is energetically favorable. Therefore, one does not expect much adsorption when the liquid phase of the dispersion is a very good solvent for the polymer, because the polymer has a lower free energy in the solvent than on the particle surface. Further, adsorption is favored when there is some chemical interaction (i.e., hydrogen bonding) between sites on the polymer chain and the particle surface. Another strategy is to use diblock copolymers, which consist of two dissimilar polymer chains covalently bonded together. For example, a PS-PEO block copolymer consists of a segment of polystyrene (PS) connected to a segment of polyethylene oxide (PEO). One block can be designed to bond with the particle surface and the other to extend into the solvent.

Once the polymer molecules have adsorbed on the particle surfaces, repulsion between particles arises as particles approach and the polymer layers on adjacent particles interact. Mixing of these layers (in most solvents) results in repulsion due to an osmotic pressure effect related to the increase in

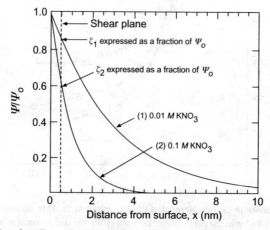

FIGURE 6.7 Data from Example 6.3 plotted to show the position of the shear plane, the definition of zeta potential (in this plot it is expressed as a fraction of the surface potential but ordinarily it is a number expressed in volts), and the variation in zeta potential with ionic concentration of the liquid. The higher concentration electrolyte has a smaller electrical double layer thickness and a lower zeta potential.

TABLE 6.2 Isoelectric Points (IEP) of Ceramic Oxides[a]

Material	pH at the IEP
Quartz (SiO_2)	2
Soda lime glass	2–3
Orthoclase ($K_2O \cdot Al_2O_3 \cdot 6SiO_2$)	3–5
Silica, Amorphous (SiO_2)	3–4
Zirconia (ZrO_2)	4–5
Rutile (TiO_2)	4–5
Tin oxide (SnO_2)	4–7
Kaolin, edges only ($Al_2O_3 \cdot 2SiO_2 \cdot 2H_2O$)	6–7
Mullite ($3Al_2O_3 \cdot 2SiO_2$)	7–8
Alumina ($\alpha\text{-}Al_2O_3$)	9–9.5
Zinc oxide (ZnO)	9
Yttira (Y_2O_3)	11
Magnesia (MgO)	12–13

[a]Data tabulated in Reed (1995).

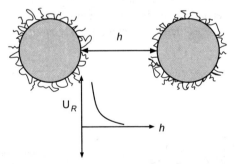

FIGURE 6.8 Schematic diagram of particles with steric layers approaching one another and the increase in potential energy of repulsion with decrease in separation distance, h.

concentration of polymer between the particles. In addition, intermingling of the adsorbed polymers is not favored because it decreases the configurational entropy of the polymer chains and thus results in a higher free energy. These factors lead to a repulsive interaction. Qualitatively, the U_R increases as separation distance approaches the thickness of the steric (polymer) layer itself. See Figure 6.8. The thickness of the layer as well as other factors, such as the amount of polymer adsorbed, influences the strength of the repulsion.

Steric repulsion is commonly used to stabilize suspensions in organic solvents. Electrostatic repulsion is not very effective in organic solvents because organic solvents have low dielectric constants compared to water and are not good media for creating surface charge on particles. On the other hand, there is a wide selection of polymers that are soluble to varying degrees in most organic solvents; hence one can find a polymer that is suitable for steric hindrance in most cases. Nonpolymeric additives, such as certain oils and surfactants, can also adsorb on particle surfaces, creating a steric layer and leading to repulsion.

Lastly, a very effective means of stabilization is known as electrosteric stabilization. As the name implies, it is the combination of electrostatic and steric. This type of stabilization comes about with the adsorption of polyelectrolytes on the particle surfaces. The polyelectrolytes ionize to create charge and electrostatic repulsion. At the same time, their polymeric nature results in a steric repulsion.

Total Potential Energy of Interaction. The total potential energy of interaction is the sum of the attractive and repulsive components (U_A and U_R). The idea of making this summation and using it as a predictor of the tendency for aggregation is attributed to four scientists: Derjaguin, Landau, Verwey, and Overbeek. Hence, it is called the DLVO theory. Figure 6.9 shows the features of the total potential energy of interaction schematically. One important feature is an energy barrier, the height of which is important to the stability of the suspension. The energy barrier prevents the particles from approaching close enough to fall into the deep primary minimum. Another feature of DLVO plots is a shallow secondary minimum that appears in some cases at a greater separation distances. Particles can become weakly agglomerated at a distance

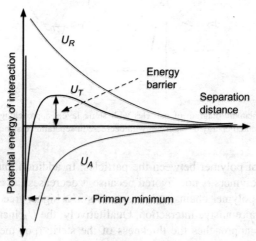

FIGURE 6.9 Schematic diagram showing total potential energy of interaction as the sum of the attractive and repulsive components.

defined by the secondary minimum; however, these agglomerates are easily broken in the suspension by stirring, for example.

Figure 6.10 shows the effect of varying surface potential and double layer thickness on the total potential energy of interaction. The potential barrier height is sensitive to both parameters: higher surface potentials and larger double layer thicknesses lead to stronger repulsion and hence higher barriers. One rule of thumb states that a barrier of about 25 kT is sufficient to prevent aggregation. Another factor to consider is the strength of attraction on the total potential energy of interaction. Particles with higher Hamaker constants have stronger attractive interactions, which brings down the height of the energy barrier (not shown).

The total potential energy of interaction for particles stabilized by a steric layer cannot be calculated easily. Experimental measurements of the force between flat surfaces covered with polymer as a function of separation have revealed that the shape of the total potential energy of interaction is a bit different than those plotted in Figure 6.10. The repulsive potential from the polymer steric layer dominates if the polymer covers the particle well such that the total energy of interaction increases dramatically at a separation distance close to the polymer layer thickness.

Overall Colloidal Stability. Achieving colloidal stability is a typical goal for a dispersion-based process. A stable suspension can be used to make coatings or parts that are consistent. The appearance and properties of a "stable" suspension do not change over the required processing time interval. This apparent stability is influenced by many factors, including the particle size and density, the particle-particle interactions, the mechanism of stabilization, and the particle concentration in the suspension.

The simplest issue to address is gravity-induced sedimentation, which is particularly important for larger and denser particles. The time scale for the development of a sediment is predicted using Eq. (6.3). This equation assumes that the particle is surrounded an expanse of liquid, and so it overpredicts the terminal velocity in suspensions containing more than a few volume percent. For example, one correlation for "hindered settling" in suspensions shows that the sedimentation velocity of a 30 vol% suspension of spherical particles has a terminal velocity one-tenth of that for a very dilute suspension. Nonetheless, a quick calculation with the terminal velocity equation can reveal the time frame over which a suspension might be expected to be stable to sedimentation.

The second consideration is aggregation, which is also called flocculation. The first step in aggregation is particle encounters or collisions due to Brownian motion. From Eqs. (6.4) and (6.5), the mean distance traveled by a particle in 3D Brownian motion in time t is given by:

$$\bar{x} = \sqrt{\frac{(kT)t}{\pi R \eta_L}} \qquad (6.24)$$

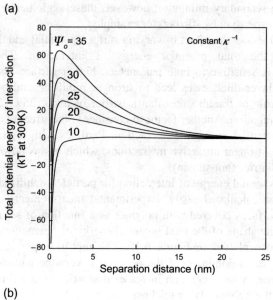

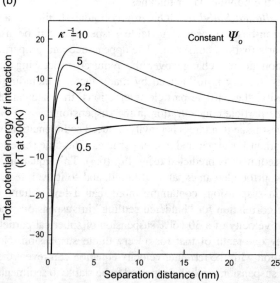

FIGURE 6.10 Calculated total potential energy of interaction (in units of kT at 300 K) as a function of separation distance for: (a) spherical silica particles ($R = 100$ nm) dispersed in water with double layer thickness of 5 nm and varying surface potential from 10 to 35 mV, as labeled and (b) spherical silica particles ($R = 100$ nm) with a surface potential of 20 mV dispersed in water with varying double layer thickness from 0.5 to 10, as labeled. The repulsive potential is calculated with Eq. (6.21).

From this equation, it is clear the distance traveled in time t is greater for smaller particles moving in liquid media with low viscosity. Now we can compare this length scale with the separation distance between particles to find a characteristic time to an encounter. Clearly as the concentration of particles in the suspension increases, the separation distance decreases and the time to an encounter becomes shorter. Less obvious is that for a given volume % of particles in a suspension, the time to an encounter gets shorter as the particle size gets smaller. Example 6.5 compares the time scale for micron-scale particles to nanoparticles.

The next consideration is the energy barrier—the magnitude of the barrier determines the probability that the encounter leads to aggregation. That is,

---

EXAMPLE 6.5 Consider two suspensions of spherical silica particles in water: one has a particle diameter of 10 nm and the other has a particle diameter of 1 μm. Each suspension has a concentration or loading of 10 volume %. For each, find the spacing between particles and the time between encounters based on Brownian motion (Eq. (6.24)). Assume particles are arranged on a cubic lattice for simplicity. The viscosity of water is ~1 mPa·s.

Let's start with the 10 nm particles. The suspension is 10 vol% so that is equivalent to 10 cm³ of particles in 100 cm³ total. Each 10 nm particle has a volume of $4/3 \pi R^3 = 5.236 \times 10^{-19}$ cm³. Therefore, the number of particles is:

$$\frac{10 \text{ cm}^3}{5.236 \times 10^{-19} \text{ cm}^3/\text{particle}} = 1.91 \times 10^{19} \text{ particles}$$

These particles are in 100 cm³ total volume, which we can envision as a box with an edge length of $(100 \text{ cm}^3)^{1/3} = 4.64$ cm. Each particle is a cube in the box and so there are 1.91×10^{19} little cubes inside the box. So if each particle is a cube, then there are $(1.91 \times 10^{19})^{1/3} = 2.67 \times 10^6$ boxes on edge and each box is $4.64 \text{ cm}/2.67 \times 10^6 = 1.736 \times 10^{-6}$ cm on its edge. Considering that there is one particle in the center of each little cube, the separation between particle centers is the same as the edge length, 17.36 nm. Given that the particle diameter is 10 nm, we can conclude that the surface-to-surface separation distance is 7 nm.

Now, let's find the distance traveled by Brownian motion:

$$\bar{x} = \sqrt{\frac{(kT)t}{\pi R \eta_L}} = \sqrt{\frac{(1.38 \times 10^{-23} \text{J/K} \cdot 300 \text{K})t}{\pi (5 \times 10^{-9} \text{m})(0.001 \text{ Pa} \cdot \text{s})}} = 1.62 \times 10^{-5} (t)^{1/2} \quad [\text{m}]$$

So, we can find the time between encounters as:

$$1.62 \times 10^{-5} (t)^{1/2} = 7 \times 10^{-9} \rightarrow t = 1.9 \times 10^{-7} \text{ s}$$

The time between collisions is extremely short for the 10 nm particles.

Repeating this exercise with the 1 μm diameter particles, we find the center-to-center separation distance is 1.74 μm and the surface to surface separation is 0.74 μm. The time between collisions is 0.21 s. This time is also quite short, but it is six orders of magnitude longer than the collision time for the nanoparticles.

without a barrier, when two particles that approach each other, van der Waals attraction pulls them close and prevents their separation. In other words, the particles are stuck in the primary minimum shown schematically in Figure 6.9. However, with a barrier the particles must approach each other with an energy greater than the energy barrier in order to contact one another and aggregate. With a higher energy barrier, the probability of such encounters decreases. Typically, barrier heights in excess of $\sim 25\,kT$ or so are best for reducing the aggregation rate.

When aggregation or flocculation occurs, it takes on different forms depending on the nature of the dispersion. If the particle concentration is low, then multiparticle aggregates tend to form. These aggregates remain suspended and continue growing until they sediment due to their larger size or interconnect to form a network or "gel" with other aggregates. Aggregate morphologies can be tightly packed or open and fractal-like. In concentrated suspensions, the tendency is for gelation, the formation of an interconnected network of particles. As can be imagined from Example 6.5, nanoparticle suspensions have a tendency to gel due to the close proximity of particles.

6.2.2 Polymer Solutions

Polymeric solutions consist of a polymer dissolved in a solvent, either water or an organic liquid. Unlike particulate dispersions, polymer solutions can be formulated to be at thermodynamic equilibrium and so they are stable. One key to developing a polymer solution for a process (i.e., to make a coating) is to find the appropriate solvent. Many factors, including cost, environmental effects, and boiling point, might be important for solvent selection for any given process. In this section, a short introduction to the structure of the polymers in solution and the thermodynamics of solubility is provided along with a general picture of how solution structure changes with polymer concentration.

Polymer Structure in Solution. To illustrate the behavior of a flexible polymer chain, like polyethylene, in a solvent, the beads-on-a-spring model is often used. Polymer chains have many flexible, elastic elements that are in constant motion, just like springs. Polymer chains, like particles, are also subject to hydrodynamic forces and undergo Brownian motion, represented by the beads. Additionally, the beads act as pivot points where the springs can freely rotate. The motion of each bead is often imagined as a random walk, where the physical conformation of each joint is equally likely at any given time. In reality, the accessible rotations of polymer segments are restricted by how flexible the bonds are in the polymer backbone. For example, a polymer that contains many double bonds or benzene rings in its backbone is much less flexible than a chain that contains only carbon-carbon single bonds. Chain branches can also hinder motion by sterically restricting the available conformations of the polymer chain. Furthermore, interactions between the polymer and the solvent bias the chain to interacting more with itself or with the surrounding fluid, causing the polymer to be either tightly coiled or more extended and flexible.

Because a polymer chain can adopt many different conformations in a solvent, the size of a chain and, consequently, its volumetric concentration, is difficult to define. The most common specification of size is the molecular weight, or average mass per individual chain. The molecular weight can either be a number average or a mass average, and can be measured by numerous ways including gel permeation chromatography, membrane osmometry, or light scattering. Because all polymer chains in a certain sample are not always the same length, the polydispersity index (PDI) is also cited. The PDI quantifies the broadness of the molecular weight distribution. A perfectly monodisperse polymer has a PDI of 1, a common free radically polymerized polymer has a PDI of about 2, and a polymer with a very broad molecular weight distribution has a PDI greater than 10.

Another description of size is the end-to-end length of the polymer chain, which can be estimated from its composition and molecular weight. First, the number of monomer repeat units in a polymer or the degree of polymerization, n, can be found by dividing the molecular weight by the monomer molecular weight. If a is the number of carbon-carbon bonds along the backbone of a single monomer, the total number of bonds in the polymer is $n_b = an$. Knowing that the length of a single carbon-carbon bond, l, is about 0.154 nm, the end-to-end distance of a polymer chain if it were stretched out completely, h, would be $h = n_b l$. For high molecular weight polymers, chain lengths can be longer than you would naively expect. In reality, diffusion causes polymers to coil into a messy sphere, as illustrated in Figure 6.11. For a freely jointed, flexible chain coiled in a solvent the mean square end-to-end distance is $<h^2> = n_b l^2$. This is a statistical quantity, calculated considering all of the possible chain confirmations.

A more useful measure of a polymer size in solution is the radius of gyration, R_g, which is the average radius of the polymer with respect to its center of mass. The radius of gyration is directly related to the chain length:

$$R_g^2 = \frac{C_\infty n_b l^2}{6} = \frac{nb^2}{6} \qquad (6.25)$$

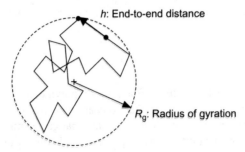

FIGURE 6.11 Polymer chain dimensions defined.

Here, n is the number of monomer repeat units in the chain (i.e., the degree of polymerization) and b is the statistical monomer length in the solvent. Recall that a monomer can contain many carbon-carbon bonds along its backbone, so b is often larger than the bond length l. This expression can also be extended by noting that $n = M/M_o$, where M_o is the monomer molecular weight. Hence $(R_g)^2$ is proportional to the molecular weight, M.

In a good solvent, the interactions between the solvent and the polymer are favorable, the chain avoids itself, and the coil expands. Steric effects become important, and the chains undergo what is termed a self-avoiding random walk instead of a pure random walk, which does not include steric effects. In this case, R_g is proportional to $n^{3/5}$ (rather than $n^{1/2}$). A special case is a polymer in what is called a theta solvent. Adding a polymer to a theta solvent requires no entropy of mixing (see below), and the attractions between polymer chain segments exactly cancel the steric exclusion effect. Polymers in theta solvents do undergo pure random walk motion with R_g is proportional to $n^{1/2}$ and have smaller radii of gyration than in good solvents.

Solvent Quality and Solution Thermodynamics. The ability of a solvent to dissolve a polymer can be predicted based on thermodynamic principles. Dissolution occurs when the change in Gibbs free energy on mixing, ΔG_m, is less than zero.

$$\Delta G_m = \Delta H_m - T\Delta S_m \tag{6.26}$$

where ΔH_m and ΔS_m are the change in the enthalpy on mixing and the change in the entropy on mixing, respectively. Entropy always increases on mixing ($\Delta S_m > 0$), which favors mixing. Therefore solubility is most closely connected with ΔH_m, which is sometimes referred to as the heat of mixing. If $\Delta H_m < 0$, then mixing the polymer and solvent together releases heat (exothermic) and dissolution occurs ($\Delta G_m < 0$). However, if $\Delta H_m > 0$, then the mixing is endothermic and dissolution may occur if $\Delta H_m < T\Delta S_m$. Flory-Huggins theory gives insights into this balance.

The Flory-Huggins theory considers the statistical distribution of n_s moles of solvent molecules and n_p moles of polymer segments on lattice sites. It predicts that the entropy change on mixing arises from configurational entropy change:

$$\Delta S_m = -R[n_p \ln\phi_p + n_s \ln\phi_s] \tag{6.27}$$

where ϕ_p and ϕ_s are the volume fractions of polymer and solvent, respectively ($\phi_p + \phi_s = 1$). Since volume fractions are less than one, the term in the brackets is a negative number. The enthalpy change on mixing depends on the number of contact points between polymer and solvent on the lattice, the interaction energies between solvent and solvent, polymer and polymer, and solvent and polymer, which define the energy polymer-solvent contact, and the composition of the solution.

$$\Delta H_m = n_s \phi_p \chi RT \tag{6.28}$$

where χ is the Flory-Huggins interaction parameter. This interaction parameter contains the interaction energies mentioned above. Therefore,

$$\Delta G_m = RT[n_s \ln \phi_s + n_p \ln \phi_p + n_s \phi_p \chi] \tag{6.29}$$

For solubility, $\Delta G_m < 0$, and hence smaller values of χ favor solubility. Values for the χ parameter can be found in tables in reference volumes.

The Flory-Huggins theory provides a good starting point for understanding the factors that affect solubility. The χ parameter decreases with temperature, which means that solubility is favored at increased temperature. A theta solvent is a solvent that has a χ of ½ at the temperature of interest. Likewise, the theta temperature of a solvent is the temperature at which χ is ½. Polymer solubility is also influenced by molecular weight. Intuitively, higher molecular weight polymers would seem to be more difficult to dissolve. This is indeed the case, as shown in Example 6.6.

A related route to understanding polymer solubility in various solvents is through the solubility parameter, which is derived from an empirical energetic analysis. The main premise of the analysis is that polymers dissolve in solvents that share chemical similarity. The solubility parameter, δ, is a measure of the cohesive energy density, which is related to the heat of vaporization and molar volume. The heat of mixing found from this approach scales with the square of the difference in solubility parameters between the solvent and polymer $(\delta_s - \delta_p)^2$. Hence, to reduce the heat of mixing and encourage dissolution, the solvent should have a solubility parameter that is similar to the polymer. Table 6.3 gives solubility parameters for common solvents and polymers. It is clear from this table that the common thermoplastic polymers on the list require organic solvents for dissolution.

Effect of Concentration on Structure. When a polymer solution is dilute, individual polymer chains are separate from each other. Each coil has a size equal to its equilibrium radius of gyration in that solvent. Adding more polymer to the solution causes the coils to become closer together until C^* is reached, the concentration at which the polymer chains begin to overlap. The structural changes are shown schematically in Figure 6.12. At concentrations greater than C^*, polymer chains interact with one another and with a further increase in concentration become entangled. As discussed in the next section, polymer concentration affects the viscosity considerably.

Since dilute and concentrated polymeric solutions act very differently from each other, C^* is an important concentration to know for a polymer/solvent system. The volume of one polymer coil in a solvent is:

$$V_C = \frac{4}{3} \pi R_g^3 \tag{6.30}$$

EXAMPLE 6.6 Calculate ΔG_m for a 5 vol% solution of polystyrene in toluene at 25°C for polymers with a molecular weight of 10,000 g/mol and 50,000 g/mol. The χ parameter for polystyrene – toluene is 0.37. The density of PS is 1.06 g/cm^3 and the density and molecular weight of toluene are 0.867 g/cm^3 and 94.14 g/mol, respectively.

For this solution, the volume fractions are $\phi_p = 0.05$ and $\phi_s = 0.95$. To solve this problem, it is convenient to assume that we have 100 cm^3 total volume—5 cm^3 of PS and 95 cm^3 of toluene. From there, the number of moles of each component can be found. Starting the toluene,

$$n_s = \left[\frac{0.867 \text{ g/cm}^3}{94.14 \text{ g/mol}}\right] \cdot 95 \text{ cm}^3 = 0.875 \text{ moles}$$

For the PS, the number of moles depends on the molecular weight. Starting with the lower molecular weight:

$$n_p = \left[\frac{1.06 \text{ g/cm}^3}{10000 \text{ g/mol}}\right] \cdot 5 \text{ cm}^3 = 5.3 \times 10^{-4} \text{ moles}$$

Now, ΔG_m is calculated:

$$\Delta G_m = RT[n_s \ln\phi_s + n_p \ln\phi_p + n_s \phi_p \chi]$$

$$= [8.314 \text{ J/(mol·K)} \cdot 298\text{K}][0.875\ln(0.95) + (5.3 \times 10^{-4})\ln(0.05)$$

$$+ 0.875(0.05)(0.37)]$$

$$= -442 \text{ J}$$

So dissolution is favored. Repeating the calculation with the higher molecular weight polymer, we find that there are fewer moles of polymer (1.06×10^{-4}) and ΔG_m is less negative (-87 J). Toluene is a very good solvent for PS and dissolves both molecular weights. We could continue and find the molecular weight that results in a positive ΔG_m.

---

and as a rough estimate we can assign C^* as the concentration for which the volume of the polymer coils equals the volume of the solution, V_T. The typical units of concentration are grams of polymer per unit volume of solutions (e.g., g/cm^3 or g/dL, where 1 dL = 100 mL). (Note that while the units appear to be the same, C is a concentration and not a density.) With this definition:

$$V_T = NV_C = N\left[\frac{4}{3}\pi R_g^3\right] \approx N\pi R_g^3 \quad \text{at} \quad C = C^* \quad (6.31)$$

where N is the number of polymer molecules in the solution. Rearranging and recognizing the connection between the N/V_T and C ($N/V_T = CN_A/M$), where M is the molecular weight and N_A is Avogadro's number:

$$C^* \approx \frac{M}{N_A \pi R_g^3} \quad (6.32)$$

TABLE 6.3 Solubility Parameters for Solvents and Polymers[a]

Solvent or polymer	Solubility parameter [$(J/m^3)^{1/2} \times 10^3$]
Cyclohexane	16.8
Carbon tetrachloride	17.6
Xylene	18.0
Toluene	18.2
Tetrahydrofuran	18.6
Chloroform	19.0
Ethanol	26.0
Methanol	29.7
Water	47.9
Polytetrafluoroethylene (PTFE)	12.6
Polyethylene (PE)	16.4
Polystyrene (PS)	18.5
Polypropylene (PP)	19.0
Polymethyl methacrylate (PMMA)	19.0
Polyvinyl chloride (PVC)	20.0

[a]Data from Brandrup et al. (1999).

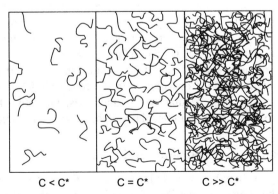

FIGURE 6.12 Schematic showing the changing structure of a polymer solution with concentration. *From Evans, D. F., & Wennerstrom, H.: The Colloidal Domain. Page 295. 1994. Copyright Wiley-VCH Verlag GmbH & Co. KGaA. Reproduced with permission.*

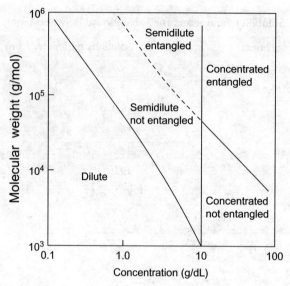

FIGURE 6.13 Effect of molecular weight and concentration on the structure (viscoelastic regimes) of polybutadiene in a good solvent. The regimes in the figure are based on rheological data. *Reprinted from Polymer, 21/3, Graessley, W. W., Polymer chain dimensions and the dependence of viscoelastic properties on concentration, molecular weight and solvent power, 258–262, Copyright (1980), with permission from Elsevier.*

According to Eq. (6.25), the radius of gyration is proportional to $(M)^{1/2}$, and so C^* is proportional to $M^{-1/2}$. Therefore, larger molecular weight polymers have lower C^* values.

Figure 6.13 shows how both molecular weight and concentration affect the solution structure. The dilute regime with separated coils occupies the lower left portion of the plot, where both concentration and molecular weight are below some limits. Considering moving from left to right on the diagram for a solution of a polymer of a certain molecular weight, the transition from dilute to semi-dilute occurs at C^*. With further increase in concentration, a concentrated regime is encountered along with entanglements if the molecular weight is high enough. Consistent with the rough model above, C^* gets smaller and smaller as molecular weight increases, indicating that making concentrated solutions of high molecular weight polymers is a challenge.

6.2.3 Rheology of Dispersions and Solutions

Fluid rheology, or flow behavior, is introduced in Chapter 3. As in melt processes, the rheological properties of solutions and dispersions are very important to the design of process steps. In initially formulating the dispersion or solution, rheology is often a rewarding method for characterizing particle and polymer interactions, and microstructure in the liquid. X-ray and light scattering as well as optical and confocal microscopy are other methods to obtain interaction information, but, unlike rheological methods, the solutions must often be diluted for these techniques to be possible. Rheology is also useful in designing the forming or additive process itself. For example, the fluid flow can dictate which coating dies are chosen, the speed that the process can run, or the feasible dimensions of the final part. Additionally, the viscosity changes by orders of magnitude during solidification. Controlling the rate at which solidification occurs can be crucial to the final microstructure of the product. Finally, once a process has been implemented, rheology is an excellent diagnostic tool for confirming that a process liquid behaves similarly between batches or lots.

The rheological behaviors of dispersions and solutions are complex. As these liquids become concentrated, interparticle forces and hydrodynamic interactions become important and more complicated behavior emerges. The dispersion or solution can act like a solid or have behavior that is time or shear rate dependent. The variety in the rheological responses is evident in familiar dispersions and solutions, such as toothpaste, shampoo, applesauce, mud, and paint. Some appear solid-like until a shear is applied and others flow readily. The qualitative behaviors of both particulate suspensions and polymeric solutions are predictable in most cases, even when they are not dilute. These behaviors, as well as the microstructures that create them, are discussed in this section.

Rheology of Particulate Suspensions. When solid particles are dispersed in a liquid, the viscosity and rheological properties of the mixture are altered from that of the liquid alone. The presence of the particles interrupts and disturbs the flow of the liquid; the liquid is forced to move around the particles, and thus the viscosity increases. The simplest case is the addition of particles to a Newtonian liquid under conditions where the particles do not interact with each other (i.e., in dilute suspension). Einstein showed that the following relationship holds for dilute suspensions of spherical particles:

$$\eta = \eta_o(1 + [\eta]\phi)$$

$$\eta = \eta_o(1 + 2.5\phi) \quad \text{for spherical particles} \quad (6.33)$$

where η_o is the viscosity of the liquid without any particles, ϕ is the volume fraction of particles, and $[\eta]$ is the Einstein coefficient or intrinsic viscosity. This relationship holds only for low volume fractions (less than around 0.10) of noninteracting particles.

The meaning of the intrinsic viscosity is discerned when the relative viscosity (η/η_o) is plotted as a function of ϕ. See Figure 6.14a. In this figure, the slope of the linear increase is the intrinsic viscosity, the increase in viscosity per unit of concentration. The intrinsic viscosity is formally defined as:

$$[\eta] = \lim_{\phi \to 0} \frac{\eta - \eta_o}{\eta_o \phi} \quad (6.34)$$

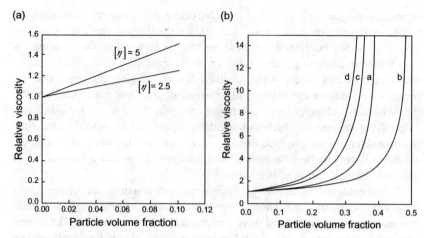

FIGURE 6.14 (a) Increase in relative viscosity (η/η_o) with addition of particles to a liquid. Data is plotted for $[\eta] = 2.5$ and 5 using Eq. (6.33). (b) Increase in relative viscosity (η/η_o) with addition of particles for dispersions with different maximum packing fractions ϕ_m and intrinsic viscosities: a. $\phi_m = 0.4$, $[\eta] = 2.5$ (spheres), b. $\phi_m = 0.5$, $[\eta] = 2.5$, c. $\phi_m = 0.4$, $[\eta] = 4$, and d. $\phi_m = 0.4$, $[\eta] = 5$, Results are calculated using the Doughtery-Kreiger relationship (Eq. 6.35).

A value of 2.5 is used for $[\eta]$ of spherical particles to recover Eq. (6.33). Higher values of $[\eta]$ represent asymmetric particles that, by virtue of rotation, appear to take up more volume and hence create an effect similar to a higher volume fraction of particles.

As more and more particles are added into the liquid, the distance between particles decreases. The chance that they physically collide increases as they crowd into the liquid. At some point, flow ceases because particles are unable to move around each other. This critical volume fraction of particles depends on particle packing, particle shape, and interparticle interactions. For now, we will ignore the last factor and assume that particles do not form agglomerates.

A useful concept in the rheology of more concentrated dispersions is the maximum packing fraction, ϕ_m, the highest volume fraction that the suspension can achieve and still flow. Monodisperse spheres, for example, can pack to a fraction of 0.74 if they are arranged in face-centered cubic or hexagonal close-packed geometries. Random close packing, on the other hand, leads to a maximum packing fraction of only 0.64. Particle size distribution plays an important role in packing, as discussed in Chapter 5. Mixing in fine particles increases the packing fraction as these fill interstitial spaces between larger particles. A broad particle size distribution typically results in a maximum packing fraction of 0.55–0.65.

Dougherty and Kreiger proposed the following expression for the effect of higher volume fractions of particles on viscosity.

$$\eta = \eta_o \left(1 - \frac{\phi}{\phi_m}\right)^{-[\eta]\phi_m} \quad (6.35)$$

Figure 6.14b shows calculated results from this equation, illustrating the effects of the maximum packing fraction and $[\eta]$ on the viscosity. The figure shows how the viscosity increases with ϕ as ϕ_m is approached and the fact that higher values of $[\eta]$, representing asymmetric particles, raise the viscosity more than spheres at a given ϕ. Note that in the limit of low dispersion concentration, Eq. (6.35) reduces to Einstein's equation (Eq. (6.33)).

Dilute suspensions of particles in a Newtonian liquid tend to display Newtonian behavior; however, at higher volume fractions these suspensions become non-Newtonian. Both shear thinning and shear thickening may be observed with increasing rates of shear. The shear thinning behavior results from particle alignment in the flow direction. Increasing the shear rate further can cause a shear thinning suspension to become shear thickening as higher shear rates disturb the particle alignment. Particle size and size distribution influence the shear thickening behavior. The shear rate at which shear thickening begins increases as the particle size decreases. That is, a larger particle size favors shear thickening. In addition, suspensions with narrower particle size distributions tend to shear thicken. In terms of processing ease in operations that require high shear rates, such as mixing and coating, shear thickening is not desired.

How do particle-particle interactions and agglomerates affect the viscosity and rheology? When attractive forces dominate, particles tend to agglomerate. The presence of agglomerates has the practical consequence of decreasing the suspension stability. That is, the larger agglomerates tend to settle over time. The tendency to agglomerate also reduces the maximum packing fraction, because particles cannot rearrange into favorable positions. In addition, large or oddly shaped agglomerates themselves trap liquid or immobilize liquid in their interiors. In effect, this results in the particle having a larger hydrodynamic size and the liquid in the agglomerate is therefore not available to assist flow. These effects lead to higher viscosity for agglomerated suspensions.

Flocculated suspensions have rheological behavior that is shear rate dependent. At low shear rates, the flocculated structure spans and connects large volumes of the fluid, causing the viscosity to be relatively high. As the shear rate increases, the flocculated structure breaks down and the viscosity drops. Therefore, flocculated suspensions are shear thinning. Increasing the shear rate on stable suspensions can also cause shear thickening. At these conditions, the particles do not have time to move relative one another and therefore, the particles "jam."

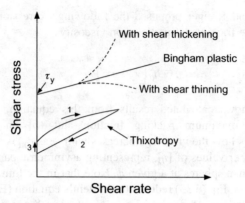

FIGURE 6.15 Schematic diagram of the shear stress versus shear rate behavior of several suspensions. The response of a suspension with a yield stress (Bingham plastic) is shown together with responses of those with a yield stress and shear thinning or shear thickening behavior. Thixotropy is also demonstrated in a suspension with a yield stress: 1: increase in shear rate, 2: response does not follow the same path on decrease in shear rate (hysteresis), and 3: with time the yield stress increases.

As particle concentration increases, suspensions rheology becomes more complex, including the appearance of yield stress, time dependence in the rheological response, and viscoelasticity. Figure 6.15 shows the response of a suspension with a yield stress, including possible responses that have shear thinning and shear thickening after the yield stress is surpassed. If the suspension flows as a Newtonian liquid after the yield stress is surpassed, then the suspension is called a Bingham plastic. The yield stress is sometimes convenient for processing methods as it assists in shape retention after the shear stress causing flow is removed (e.g., as in extrusion). An example of this behavior from everyday life is mayonnaise, which holds a shape with peaks and valleys in the jar, but flow easily under a knife. The high stresses applied by the knife break the microstructure of fats in the mayonnaise, allowing it to flow. Time dependence is also a possibility. A thixotropic suspension is one that is shear thinning, but experiences hysteresis and thickening as it sits undisturbed. For these materials, the history of shearing or stirring influences their rheological behavior.

Rheology of Polymer Solutions. Just as adding particles to a liquid causes the viscosity to rise, so does dissolving a polymer in a solvent. Factors such as polymer concentration, molecular weight, solvent/polymer interactions, and temperature all have a large effect on the viscosity of polymeric solutions.

When the solution is dilute, below the overlap concentration C^*, the separated chains can be modeled as noninteracting hard spheres of radius R_g. Just like particles, these spherical polymeric chains move in solution under Brownian motion and Einstein's equation (Eq. (6.33)) is valid to describe the relative viscosity of the solution. The volume fraction of polymer in solution is given by:

$$\phi = \frac{NV_C}{V_T} = \left(\frac{V_C N_A}{M}\right) C \qquad (6.36)$$

This relationship makes use of the connection between the number of molecules in the solution volume and the concentration in grams polymer per volume of solution ($N/V_T = CN_A/M$). Now, substituting into Eq. (6.33) for spherical particles reveals:

$$\eta = \eta_o\left(1 + 2.5\left(\frac{N_A V_C}{M}\right)C\right) \quad (6.37)$$

Using this, the intrinsic viscosity of a polymeric solution can be defined as:

$$[\eta] = \lim_{C \to 0} \frac{\eta - \eta_o}{C\eta_o} = \lim_{C \to 0} 2.5\frac{V_C N_A}{M} = 2.5\frac{4\pi R_g^3 N_A}{3M} \quad (6.38)$$

Therefore, this intrinsic viscosity, $[\eta]$, is different from the one used with particles; it has units of inverse concentration (e.g., cm^3/g). As in the particle case, $[\eta]$ gives the effectiveness of a single polymer chain in raising the solvent viscosity and therefore, the intrinsic viscosity can also be thought of as a measure of the size that a polymer chain takes up in a solvent, assuming that the dissolved polymer is a collection of non-interacting spheres.

Because the radius of gyration is proportional to $M^{1/2}$ (Eq. (6.25)), Eq. (6.38) can be re-written as:

$$[\eta] = \frac{A_2(A_1 M^{1/2})^3}{M} = A_3 M^{1/2} \quad (6.39)$$

Here, A_1, A_2, and A_3 are constants used to simplify the expression. Eq. (6.39) shows that the intrinsic viscosity depends on the molecular weight of the polymer. This makes intuitive sense; a greater molecular weight polymer chain is physically larger and impacts the polymer solution viscosity to a greater extent.

Equation (6.39) was derived assuming that the volume of a polymer chain in solution scales with $M^{3/2}$, which is not always true. A generalized version of Eq. (6.39) is the Mark-Houwink-Sakurada equation:

$$[\eta] = K_{MH} M^\alpha \quad (6.40)$$

where α and K_{MH} are the Mark-Houwink-Sakurada parameters. These parameters give some indication of the polymer conformation within the solvent, which is directly dependent on the affinity of the polymer to the solvent and also the polymer flexibility. They are also a function of temperature, with most polymers becoming more flexible with heat. If $\alpha = 0.5$, the polymer is dissolved in a theta solvent and Eq. (6.39) is followed exactly. In good solvents, polymer chains expand and contribute more to the rheology, and $0.5 < \alpha < 0.8$. Values of $\alpha > 0.8$ are possible for very rigid polymers. Mark-Houwink-Sakurada parameters are found in reference volumes, such as the *Polymer Handbook*, listed at the end of the chapter.

Polymer solution viscosity and rheological parameters change dramatically with concentration. In the limit of dilute solutions, the viscosity is linear with polymer concentration with a slope of the intrinsic viscosity:

$$\frac{\eta}{\eta_o} = 1 + [\eta]C \tag{6.41}$$

As the concentration of the polymer increases, polymer chains begin to interact in the solution. Equation (6.33) no longer holds as C^* is approached and the polymer chains overlap. Above C^*, polymer chains may also become entangled and their motion in the solvent becomes further restricted. The solution viscosity rises dramatically (see Figure 6.16 for an example). The viscosity of polymer solutions above C^* cannot quantitatively be predicted from polymer and solvent properties; instead, simple rheological experiments are conducted to properly characterize the polymer solution behavior. These data can then be fit to empirical models. For example, a power law of the type $\eta = BC^b$, where b and B are constants, is frequently successful.

Whereas dilute polymer solutions are relatively Newtonian like the solvents that are within them, the viscosity of concentrated polymer solutions is most often shear rate dependent. As a concentrated solution is sheared, polymer chains elongate in the velocity gradient direction and become disentangled. The elongated chains can move past each other more easily and the viscosity drops. Like polymer melts, solution viscosity is found to follow

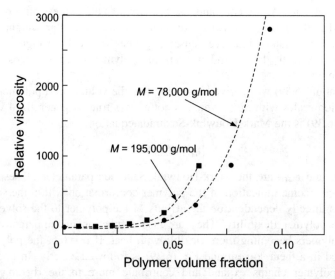

FIGURE 6.16 Relative viscosity of aqueous polyvinyl alcohol solutions made from polymer with $M = 195{,}000$ g/mol and $M = 78{,}000$ g/mol at 25°C. Empirical fits to the data are given as dotted lines. The relative viscosity is reported at zero shear but the solutions were only mildly shear thinning. *Data from Buss et al. (2011).*

power law behavior over a range of shear rates after initial Newtonian plateau (see Chapter 3). Like particulate suspension rheology, solution rheology is complex. Viscoelasticity is common in concentrated solutions and yield stresses and time-dependent responses can also be present.

6.2.4 Characteristics of Volatile Liquids for Dispersions and Solutions

Simple, volatile liquids play important roles in dispersion and solution processes. In a dispersion, the volatile liquid is a medium for dispersing particles, while in solutions, the volatile liquid is the solvent for polymer. As described above, these liquids play a central role in making the dispersions and solutions as well as in their rheology. Additionally, the liquid is removed by evaporation and so the volatility of the liquid is also important. Lastly, the surface tension and wetting behavior of dispersions and solutions are determined to a large extent by the surface tension of the liquids. Table 6.4 gives some useful parameters for common liquids. Not included in the table are environmental, health, and safety factors. Organic solvents have varying exposure limits and most require control conditions for handling, such as fumehoods. Materials safety data sheets (MSDS) should be consulted for each solvent.

6.2.5 Drying

Drying is a critical step for coatings and three-dimensional parts made from particulate suspensions and polymer solutions. Considering the amount of water or solvent present in a slip cast piece after removal from the mold or a polymer coating after deposition onto a substrate, the drying step is frequently both time consuming and energy intensive. Also, for continuous processes, such as roll-to-roll coating operations, the rate of the entire process can be limited by the drying step. Lastly, drying leads to shrinkage and if this shrinkage is not allowed to occur freely then stress and defects may result.

There are many common features in the drying process for ceramic particle suspensions, high T_g polymer latexes, and polymeric solutions. Low T_g latex suspensions are a special case, where the polymer latex particles actually deform into a nonporous polymer coating during the drying process. Although most of what follows is focused on drying of coatings, the same principles apply to drying of bulk shapes. The description below begins with an exploration of general phenomena and ends with comments on drying of coatings and pieces from dispersion and solution processes.

General Features of Drying. There are three steps in any drying process. See Figure 6.17. First, the volatile liquid is transported to the free surface of the coating or object. Of course, early in the drying process this step is

TABLE 6.4 Parameters for Some Volatile Liquids Used in Dispersion and Solution Processes[a]

Liquid	Boiling point (°C)	Evaporation rate[b]	Heat of vaporization (J/g)	Permittivity	Viscosity at 25°C (mPa·s)	Surface tension (mN/m)
Acetone	56	7.70	524	21	0.3	25
Butyl Acetate	127	1.00	310	5	0.5	26
Cyclohexane	807	5.60	392	NA	1.0	26
Ethanol	78	2.65	860	24	1.2	23
Ethylene Glycol	198	<0.01	800	37	20	48
Isopropyl Alcohol	82	2.08	578	18	2.4	22
Methanol	63	3.70	1100	33	0.6	23
Methyl Ethyl Ketone	80	4.59	444	18	0.4	25
Tetrahydrofuran	66	6.3	444	7.5	0.47	NA
Toluene	110	1.96	352	2	0.6	29
Water	100	0.16	2260	80	1	72
Xylene	140	0.55	327	2	0.7	28

[a] Data are from Hellebrand (1991) or Weast (1980).
[b] Relative to butyl acetate; NA = not available.

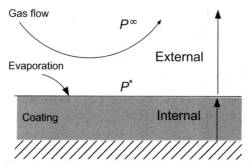

FIGURE 6.17 Schematic of a drying coating. The coating can be a polymer solution or a dispersion of particles in a volatile liquid.

instantaneous. Second, the liquid evaporates (i.e., it changes phase from liquid to vapor). Third, the vapor is transported away from the free surface. We can consider the first step as *internal*, dependent on transport processes inside the coating or object, and the second and third steps as *external*, dependent on conditions and transport processes in the surrounding gas phase. The three steps occur in series—the slowest step controls the rate of the drying process.

Initially, most drying processes are rate limited by external phenomena since most coatings are designed to be relatively dilute when deposited. So, as drying begins, the solvent or volatile liquid is plentiful at the free surface of the coating and the internal process that brings the volatile species to the surface (e.g., diffusion in the polymer/solvent system) is fast. Therefore, we do not expect internal factors to control the rate. Instead external factors, related to the driving force for the transport in the vapor phase or mass transfer coefficient, dominate. As drying continues, however, and the solvent content drops, eventually internal transport slows and becomes rate controlling until the end of the process.

External solvent vapor transport rates are affected by both the liquid-vapor equilibrium and gas phase conditions. An equilibrium between the liquid and the vapor is established at the free surface of the coating, and this equilibrium sets the partial pressure of the vapor adjacent to the free surface, P^*. An important point to remember here is that P^* increases with temperature. At some distance away from the surface, the partial pressure of the solvent vapor is set by the conditions far from the coating at P^∞. The "driving force" for the removal of solvent vapor away from the coating surface is therefore the difference of these two values, $P^* - P^\infty$. For any drying to occur, P^* must be greater than P^∞. The driving force is manipulated by temperature and solvent vapor concentration in the oven. Raising the temperature increases P^*; an obvious limitation is boiling, which should be avoided. Solvent vapor concentration in the oven is kept low by flowing solvent free air though the oven. In the special case of aqueous coatings and

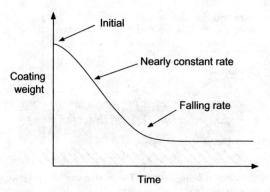

FIGURE 6.18 Schematic of weight loss from a drying coating. In the initial stage the drying rate (slope of the curve) is low as the coating or cast piece is heated to the temperature of the dryer. During the nearly constant rate period, the drying rate is nearly constant and external factors, such as airflow control the rate. In the last stage, the rate decreases as the supply of volatile liquid is depleted and internal factors, such as diffusion, limit the process.

water-saturated cast pieces, the drying rate is very much influenced by the relative humidity.

Drying is characterized by measuring coating weight as a function of time. Figure 6.18 shows a schematic representation. Usually, the data is normalized to the area of the coating, so that the slope of the curve, the drying rate is given in mass per area per time (e.g., kg/(m$^2 \cdot$s)). (*Note:* such normalization of the drying rate is not relevant in drying of 3D objects as the exposed surface area changes with time.) By convention, this drying rate is stated as a positive number. Considering the external control of drying, the drying rate represents the flux of the volatile species in the gas phase away from the coating surface, $J_{S,g}$:

$$J_{S,g} = k(P^* - P^\infty) \tag{6.42}$$

The flux is the product of the driving force and k, the mass transfer coefficient, an empirically derived parameter that increases with the rate of airflow. Increasing the airflow, with air jets or fans, for example, increases the mass transfer coefficient. Hot air flow with jets impinging on the coating surface is one of the most important modes of drying. The drying rate (i.e., $J_{S,g}$) is constant in the early stages of drying because the driving force is constant. That is, a coating or cast article with high volatile liquid content has a reasonably constant P^* so long as the temperature is constant.

Heat transfer is required for drying. Heat must be transferred to the drying material to supply the latent heat of vaporization and also to raise the temperature of the material in order to speed the drying process. If heat is not supplied or not supplied fast enough, then evaporation lowers the temperature of the material. Evaporative cooling is a familiar phenomenon; evaporation of sweat keeps our bodies cool. Heat may be carried to the surface by hot air (i.e., convection), transferred from hot solid that is in contact with the drying material (i.e., conduction), or absorbed from infrared radiation (IR) emitted by a hot solid nearby or an IR heater. Heat supplied to the surface of the drying material is then transferred to its interior by conduction.

Transient effects from evaporative cooling can also play a role in the drying rate. The heat of vaporization must be supplied for the phase change from liquid to vapor. So if the heat delivery to the drying coating or piece is not sufficient to raise the temperature of the drying material and supply this latent heat, then the temperature drops and P^* drops along with it. Typically, this phenomenon happens as the drying coating or piece first enters the drying oven. With time, temperature climbs to the temperature of the oven and P^* increases.

As drying continues, the content of the volatile liquid in the coating or part drops and the drying rate falls. The onset of falling rate period marks the beginning of the phase of drying that is controlled by internal factors. That is, the limiting step is getting the volatile liquid to the surface for evaporation and eventual removal in the gas phase. In this phase, drying is mainly controlled by diffusion. In the case of polymer-solvent coatings, the solvent must diffuse through the concentrated coating to the surface, and for coatings and cast pieces, the diffusion is of the vapor through the porous particulate material to the surface. More details are below.

Drying of Particle Dispersions. The course of events during the drying of a coating made from a particulate dispersion is shown in Figure 6.19. After the dispersion is applied to the substrate, evaporation begins at the liquid-air interface. As the volatile liquid evaporates, the liquid-air interface descends toward the substrate. As discussed above, it is likely that the coating liquid is relatively dilute so that it dries initially with a constant rate. The

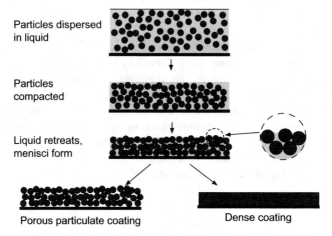

FIGURE 6.19 Stages of structure development in coatings made from particle dispersions.

evaporation rate (in grams per area of coating per time) found by a weight loss experiment can be converted to a rate of free surface descent (e.g., μm/s) during this phase. If the particles are colloidal in size or smaller, Brownian diffusion of the particles maintains a constant concentration of solids all of the way through the liquid. The decreasing amount of water brings the particles closer together and concentrates any soluble components that may be in the liquid phase. Increasing salt concentration can decrease the double layer thickness that surrounds particles, which may destabilize and possibly flocculate them.

Eventually, enough water is removed that a connected particle network is formed. At this point, the coating is quite similar in microstructure to a 3D part made by a slip casting process just as it is removed from the mold. Water evaporation and shrinkage continue, and the drying rate remains constant as long as particles remain submerged in water. The particles must rearrange into increasingly high packing fractions to remain completely saturated. At some point, the rearrangement stops and any further evaporation requires that the liquid-air interface retreat into the solids packing. Air invades into a particulate coating, creating curved menisci that, through surface tension and capillary forces, pull the particles toward each other. If the particles are colloidal in size, the pockets of air between them are sized for maximum light scattering and the coating turns from translucent to opaque.

As the air-liquid interface falls below the top surface of particles, the area for evaporation decreases and the effective evaporation rate is reduced. The "falling rate" drying period begins. The evaporation rate is reduced even further as the air-liquid interface recedes into the particle packing. In order for evaporation to occur, water-laden vapor must diffuse out through the pore space and fresh air must diffuse in. In the falling rate period, the evaporation rate is dominated by these diffusion processes, so later stage dryers often focus on heating the coatings to increase diffusion rates. Modifying convection (i.e., airflow), an external factor, is less effective. Likewise the same process occurs in a drying 3D part.

Just as the air-water interface is curved in a macroscopic glass capillary, the microscopic air-water interface is curved as it exists in the space between particles. The presence of curved liquid surfaces creates a pressure gradient in the packing, which is known as capillary pressure (see Section 6.3.2). When the particles are colloidal in size or smaller, the pressure difference across this interface can be an atmosphere or more. Even after most of the water has left the particle packing, some may be left behind at particle-particle contact points in what is termed a "pendular ring." Here, the low pressure in the liquid as well as van der Waals and surface tension forces cause the particles to be pulled toward each other. The resulting stress is measureable and can also manifest itself in cracking. In 3D parts, drying stresses can cause warping as well. In the special case of a low T_g polymer latex, the particles are soft polymer spheres that deform. During drying, particles flatten against each other at their contact points and the pore space shrinks until a dense coating is formed.

Drying Polymeric Solutions. Like particulate suspensions, polymeric solutions also dry with constant rate and falling rate periods. Initially, if the solution is below the concentration C^*, polymer chains distribute evenly in the suspension due to Brownian motion. The evaporation rate is constant, but it should be noted that P^* drops as the concentration of polymer near the free surface increases. Sometimes the drying rate still appears constant as this occurs because concurrently the coating is warming up to the oven temperature after experiencing evaporative cooling.

The falling rate period begins as further evaporation increases the polymer concentration to a critical level. Structurally, the polymer chains may start to entangle at this point and coating viscosity increases dramatically, The polymer solids now at the air-coating interface hinder evaporation. To complete the drying process, the remaining solvent must diffuse through the concentrated polymer-solvent network to the free surface. Again, the falling rate period is most affected by drying temperature and boosting the temperature of the dryer at this stage is effective at increasing the rate of evaporation.

Drying Stage Design Considerations. The drying stage takes up the most floor space and requires the most energy in a coating line, so mindful dryer design is essential for profitability. Since the drying process is multistage, dryers themselves are often sectioned into zones to target constant and falling rate drying differently. The atmosphere is most commonly air, but can be an inert gas, which can have the advantage of controlling explosive limits in the case of nonaqueous solvents. The gas flow can be either laminar or turbulent. Turbulence increases mass transfer but can affect surface flatness. The gas can also be heated, which is determined both for safety and energy considerations. Convection can be nonexistent, natural due to the evaporation process itself, induced (spin coating), or forced (tunnel dryers, flotation dryers, or impingement dryers). Finally, the gas can be recycled, which can decrease cost but also decrease the evaporation rate as compared to fresh air.

Although for economic reasons most drying processes are designed to dewater the product as quickly as possible, this can cause quality issues with the final part. Because the coating is effectively a solid during the falling rate period, stresses can build within it. In a particulate-based coating and slip-cast parts, stresses, as mentioned above, develop. If the coating dries unevenly the stresses are not balanced and cracks can develop. Gradients in the solvent concentration can develop both in the thickness direction and parallel to the substrate, and in various portions of a 3D part. Additionally, the constraint of the substrate prevents a coating from shrinking in the plane of the substrate and stress-induced curl or cracking is possible. One method to reduce cracking in 3D parts is to lower the evaporation rate, allowing the solvent time to redistribute when stresses develop.

Finally, drying multicomponent systems that contain both polymer and particles can often be a challenge; however, this is probably the most industrially common situation. Some polymer (called "binder") is often added to particulate suspensions to act as a soft glue to hold the particles together. Particles are often added to polymeric solutions to act as a pigment or to modify the final hardness of the coating. When drying particulate suspensions reach the maximum particle packing fraction, dissolved binder in the solvent is still mobile. Depending on the drying rate and the viscosity of the liquid medium, a gradient of polymer can develop in the particle packing. Likewise, in drying particulate suspensions large, dense particles can sediment toward the substrate if the coating is dried too slowly. The resulting layered composition can either be desirable or a process challenge.

6.2.6 Curing of Liquid Monomers

Liquid monomers and oligomers are converted to solid polymers by reactions driven by heat or exposure to UV radiation or other energetic sources, such as electron beams. The process of solidification by chemical reaction is typically called "curing". There are a host of chemistries designed for curing, including epoxies, polyurethanes, vinyl ethers, polyesters, and acrylates. Similarly the reaction mechanisms vary from condensation polymerization to free radical to cationic, for example. Given the richness and complexity of this topic and the processes that follow, the emphasis here is UV curing, a method used in the coating and 3D printing processes.

UV curing of liquid monomers and oligomers occurs in stages as the resin is exposed to UV light. Figure 6.20 shows an example for free radical curing of acrylates. The first step is the formation of free radicals from dissociation of UV initiators. The initiators are UV absorbing chemicals dissolved into the resin. UV light covers a portion of the electromagnetic spectrum from a wavelength of 10 to 400 nm, but most UV curing is carried out in the ~200–400 nm range where many UV initiators absorb energy. The primary free radicals react with monomers, forming reactive groups on the monomer, which then launch continued reactions. The chain length grows by these propagation reactions, and additionally cross-links form with other chains as well as within themselves to form cycles. An important distinguishing factor in the monomers and oligomers that are used in UV curing is the functionality. That is, each reacting unit—monomer or oligomer—is capable of extending in multiple directions. For example, a trifunctional monomer has three sites for reaction. This feature leads to highly cross-linked networks. The process ends with the termination of these reactive groups, for example, by combination or disproportionation. To understand how to design a process to cure a coating or use curing to make a 3D printed shape, the effect of variables on the reaction rate and conversion from monomer to polymer should be understood.

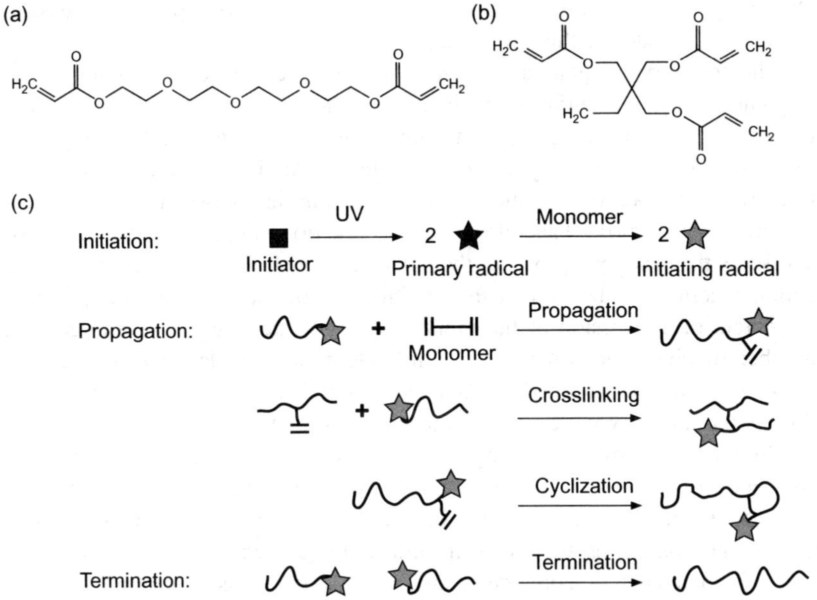

FIGURE 6.20 Acrylate monomer examples: (a) tetraethylene glycol diacrylate and (b) trimethylopropane triacrylate. (c) Stages of reaction in UV curing of a diacrylate, neglecting chain transfer. *Courtesy of the A. V. McCormick research group, University of Minnesota.*

The initiation step is affected by the intensity of UV light, the concentration of photoinitiator, *PI*, in the resin, and the system geometry. Initiators can dissociate into two primary free radicals as they absorb light of a specific wavelength. The rate of initiation is a function of the light intensity absorbed, I_A. The intensity of incident light, I_o (in W/m^2, for example), is a function of the specifications of the UV lamp and its proximity to the curing material. The light intensity transmitted, I_T, to a particular depth, z, is found with the Beer Lambert law:

$$I_T = I_o(10^{-\alpha[PI]z}) \quad \text{or} \quad I_T = I_o(e^{-\alpha[PI]z}) \tag{6.43}$$

where [*PI*] is the photoinitiator concentration and α is the molar absorption coefficient of the *PI* (at the wavelength of illumination). The choice of base 10 and base *e* depends on the field (chemistry vs. physics). The term $\alpha[PI]z$ is called the absorbance. The light intensity absorbed at a particular depth is:

$$I_A = I_o - I_T = I_o(1 - 10^{-\alpha[PI]z}) \quad \text{or} \quad I_A = I_o(1 - e^{-\alpha[PI]z}) \tag{6.44}$$

This relationship indicates that the gradient in intensity (and hence rate of initiation) is stronger in resins with high photoinitiator concentration. Absorbance increases with [*PI*] and hence lower photoinitiator concentration must be used if thicker coatings or deeper cure depths are needed.

Dropping [PI], however, slows the overall rate of conversion; an increase in I_o can compensate for this problem.

The free radicals generated in the initiation step react with monomers or oligomers to form initiating radicals and then these go on to react with monomers (or pendant groups on other chains) to propagate (or cross-link) the chains. Chains grow by monomer addition. Additionally, due to the multifunctional nature of the monomers, cross-linking between chains occurs and cycles can form. The rates of these reactions depend on the diffusion and so as the reactions proceed, their rates diminish. Further, there are termination reactions that result in the annihilation of the reacting sites. Overall the extent of conversion of the monomer to polymer can be monitored by a number of characterization tools. This extent of cure depends on the total amount of energy absorbed per unit area, which is known as the dose. The dose is the intensity multiplied by the illumination time.

Similar to drying, UV curing for coatings made on flexible webs is done continuously. After the resin is applied, the coated web passes under a UV lamp or several lamps. The speed of the process can be limited by this step and so the conditions for the resin and lamps are chosen for fast reactions and a high degree of conversion and cross-linking. Usually, high intensity lamps are used together with resins having an optimized photoinitiator content. For some resins, inert gas blankets the coating to prevent oxygen from inhibiting the cure by reacting with free radicals. This issue is experienced by acrylates, for example.

6.3 SHAPE CASTING

6.3.1 Process Overview

Shape casting from dispersions is a forming method to produce intricate and complex parts such as ceramic crucibles, toilets, and turbine blades. As a liquid, the dispersion is able to fill the fine features of porous and nonporous molds. Since ceramics are not made into complex shapes by melt casting, shape casting from a dispersion of ceramic particles is sometimes the only route to 3D ceramic shapes on an industrial scale. Since other, arguably more efficient methods of making metal and polymer shapes are available, shape casting from dispersions is almost exclusively a ceramic process.

There are several general steps in a shape casting process. The first step is the formulation and preparation of the dispersion. The dispersion must not only have the proper flow properties for casting, but also be composed of particles with the best characteristics for the process and other components to aid in the development of structure during and after solidification in the mold. The dispersions have a low enough viscosity to be poured under gravity into a mold. Next solidification occurs in the mold by extraction of the liquid medium of the dispersion into a porous mold (slip casting) or by reaction (gel casting). After this "casting" step, the part is removed from the mold, dried, heated to remove binder if necessary, and then sintered at high temperature to develop a dense microstructure.

Slip casting, which is the main topic of this section, begins with proper formulation and mixing of a high solids loading slip and the preparation of porous mold with the desired final shape. Several considerations go into the selection of the powder starting materials for slip, as discussed more below. Most molds are made from gypsum ($CaSO_4 \cdot 2\,H_2O$). Gypsum has a natural microstructure composed of needles and plates, which create a highly porous structure that is strong given the mold porosity. Molds are typically 50% porous, with a pore size of about 1–5 μm in diameter and up to 20 μm in length. Gypsum is also hydrophilic, so it takes up water easily. To protect the gypsum from acidic slips that degrade the material, the mold surface may be coated with a layer of inert material such as silica. In some high pressure casting operations, polymer molds are used.

After the slip is poured into the mold it is allowed to cast or solidify. See Figure 6.21. Capillary action draws water into the mold, leaving behind a consolidated layer of particles on the mold surface. This solid layer increases in thickness with time. Slip casting times vary greatly depending on the particle size, degree of flocculation, and part thickness. The casting time ranges from a few minutes for some parts to hours or even days. When the desired consolidated layer thickness is achieved, the excess liquid is drained from the center of the part, if a hollow part is desired. The part must then be removed from the mold. Separation from the mold may occur naturally as the part dries and shrinks. Parts are often removed as soon as they are robust enough to resist deformation and fracture during handling. Once the green part is removed, post-processing operations, including drying and sintering, are performed.

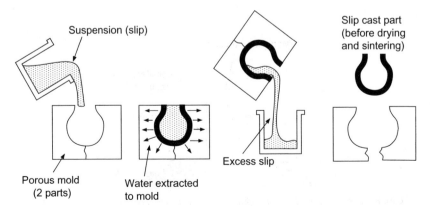

FIGURE 6.21 Schematic of the slip casting process. *Adapted from Richerson (1992).*

There are several variations of slip casting. Figure 6.21 shows "drain casting," the process variation for a hollow part; after a specific amount of time, slip is poured off and the hollow part is removed from the mold. Another variation is "solid casting," which produces a completely solid piece by filling the entire mold with slip. The mold is replenished with more slip as the liquid level falls until a completely solid piece is formed. After the part is removed from the mold, it is still "wet"; the pore space in the ceramic is filled with water. Therefore drying is required, followed by high temperature sintering for densification.

This section begins with a description of capillarity and an exploration of the casting rate, the rate at which solid ceramic particles accumulate on the mold wall. The casting rate is one of the important considerations when designing a slip composition and mold features. These and other process considerations are then discussed. The section concludes with comments on the post-processing of the parts made by shape casting.

6.3.2 Capillary Action

The flow of water into the porous mold is central to the slip casting. In Chapter 3, we learned that the liquid surface tension causes a pressure difference across curved liquid interfaces and that liquids wet surface based on a balance of interfacial energies. These phenomena are coupled in capillary action or capillary-driven flow into porous media.

Capillarity can be understood by considering a capillary tube plunged into a liquid, as shown in Figure 6.22. If the liquid wets the walls of the tube, then the liquid surface curves and a low pressure is established beneath the curved meniscus. The pressure difference for a spherical meniscus of radius, r, is:

$$\Delta P = \frac{2\gamma_{LV}}{r} \qquad (6.45)$$

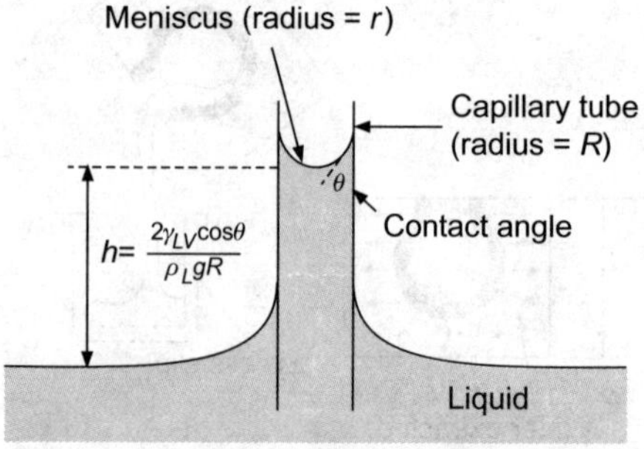

FIGURE 6.22 Schematic of rise of a wetting liquid up a capillary tube.

where γ_{LV} is the surface tension of the liquid and ΔP refers to difference in pressure between a flat liquid surface and the curved surface. The pressure causes flow of the liquid and the liquid rises until the hydrostatic pressure created by the column of liquid equals the capillary pressure given in Eq. (6.45). In slip casting, the pores are very tiny, commonly less than a micron in size and so the pressure difference is large across the meniscus. Therefore, the driving force for the liquid to enter the pores of the gypsum mold is very strong.

6.3.3 Predicting Cast Layer Thickness

After the slip or suspension is poured into the porous gypsum mold, water begins to move from the slip into the porous mold, leaving behind a layer of ceramic particles. Predicting how this cast layer thickness grows with time is essential to process design. In a drain casting operation, for example, this prediction is needed to select the time to pour off the excess slip. In general, the slip cast thickness increases with the square root of the casting time. In order to explain this observation and understand the factors that control the casting rate, a model is introduced for cast layer thickness.

There are two essential features of this model. The first is the motion of water from the slip into the mold. The driving force for this movement is a pressure difference set up by the curvature of menisci in the pores of the gypsum. One way to think of the mold is as a network of tiny capillary tubes, as described in the last section. Water wets the interior walls of the tube, forming a curved meniscus and the low pressure created by this curvature pulls water into the mold. This action is the same as a paper towel wicking up a spill. The second feature is accumulation of the ceramic particles; by tracking the water movement volumetrically the thickness of the stranded particles on the mold wall is found.

First, to understand the role of the mold alone, consider the simplified one-dimensional process illustrated in Figure 6.23, where water (with no particles) is drawn into a porous gypsum mold. At time $t = 0$, water is filled into a reservoir up to a level $x = L_s^o$, and contacts the mold. At the mold surface ($x = 0$), the pressure in the water is P_i and assumed to be independent of the water level (i.e., hydrostatic effects are not important). Water wets the hydrophilic gypsum and is drawn into the mold by capillary action. At time, t, some water has left the reservoir and entered the mold. In this model, there is no drying from the top surface so all of the water that leaves the reservoir enters the pores in the mold, saturating the mold to a level $x = -L_m$. The water within the mold is called the "filtrate."

Conservation of volume gives the relationship between the position of the wet/dry interface in the mold, $x = -L_m$, and the height of the remaining water in the reservoir, $x = L_s$. Essentially, all of the water missing from the reservoir is in the pores of the mold. If the volume fraction of pores in the

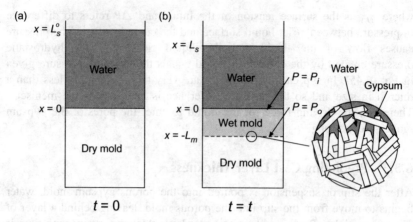

FIGURE 6.23 Schematic of the coordinate system for simple one dimensional water wicking from a reservoir into a porous mold. (a) At time $t = 0$, the water just make contact with the mold. (b) At time $t = t$, some water has moved from the reservoir into the pores of the mold. The inset shows water in the pores of the gypsum. Notice that the water wets the pore walls, creating a curved meniscus.

mold is ε_m and the area of contact between the mold and the water is A, then conservation of volume requires:

$$\int_{-L_m}^{0} A\varepsilon_m dx + \int_{0}^{L_s} A dx = AL_s^o$$

$$\varepsilon_m L_m = L_s^o - L_s \qquad (6.46)$$

The volume of filtrate V_F inside of the mold at time t is:

$$V_F = \varepsilon_m A L_m \qquad (6.47)$$

Where water fills a pore between gypsum particles, the water-air interface is shaped as a curved meniscus. The pressure in the liquid under the curved interface, P_o, is lower than atmospheric due to capillary effects. Capillarity is especially strong given that the pore size in the gypsum is on the order of a micron, so the radius of curvature of the meniscus is very small. A pressure gradient is therefore imposed between the bulk of the liquid and the filtrate, as illustrated in Figure 6.24. Under the influence of this pressure gradient, water flows into the mold. The flux of a liquid, J_F, through a porous network is modeled using Darcy's Law:

$$J_F = -\frac{1}{\alpha_m \eta_F}\left[\frac{dP}{dx}\right] \qquad (6.48)$$

Here, α_m is the resistance of the mold (units m^{-2}) and η_F is the viscosity of the filtrate. The mold resistance can be estimated using the Kozeny-Carmen approximation for the specific resistance

Dispersion and Solution Processes **Chapter | 6** 469

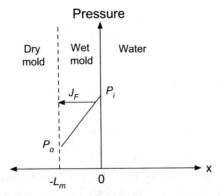

FIGURE 6.24 Pressure gradient established by the low pressure beneath the curved menisci in the porous mold.

$$\alpha = \frac{B(1-\varepsilon)^2 S^2}{\varepsilon^3} \qquad (6.49)$$

In this expression, B is a constant (e.g., ~ 5 for many porous materials) and S is the specific surface area of the gypsum, which is the total surface area per unit volume. Note that the specific resistance depends on both the size and quantity of pores in the porous material. As the void size increases, S decreases, and therefore α also decreases. With respect to the volume fraction of pores in the mold, α decreases as ε increases. So molds with high porosity and large voids allow faster flow of the filtrate. However the flux, J_F, also depends on the pressure gradient driving flow, and this gradient is also a function of pore size. The curvature of water menisci in small pores is extreme, and the capillary pressure driving the flow depends on the inverse square of the meniscus radius.

To derive an expression for the wet/dry interface L_m as it progresses in time, the water flux J_F is defined in terms of the change in the filtrate volume with time. A negative sign is added to account for the direction of the volume removal, which is in the $-x$ direction.

$$J_F = \frac{1}{A}\frac{dV_F}{dt} = \frac{1}{A}\frac{(\varepsilon_m A dL_m)}{dt} = \varepsilon_m \frac{dL_m}{dt}$$

$$J_F = -\varepsilon_m \frac{dL_m}{dt} \qquad (6.50)$$

By setting this expression and Darcy's law equal to each other, an expression for the position of the wet/dry interface as a function of time is found:

$$-\frac{1}{\alpha_m \eta_F}\left[\frac{P_i - P_o}{0 - (-L_m)}\right] = -\varepsilon_m \frac{dL_m}{dt}$$

$$\frac{P_i - P_o}{\varepsilon_m \alpha_m \eta_F} dt = L_m dL_m$$

$$\int_0^t \frac{P_i - P_o}{\varepsilon_m \alpha_m \eta_F} dt = \int_0^{-L_m} L_m dL_m$$

$$L_m = \left[\frac{2(P_i - P_o)t}{\varepsilon_m \alpha_m \eta_F}\right]^{1/2} \tag{6.51}$$

This expression for the capillary flux of water into the gypsum mold shows that the water interface position increases with $t^{1/2}$. Even without particles in the water, the $t^{1/2}$ dependence is apparent as are the roles of the pressure difference, specific resistance of the mold and viscosity of the filtrate (water) on the advance of the wet/dry front.

In slip casting, particles are suspended in the water phase. Therefore as water is pulled into the mold the particles suspended in that water are left behind on the mold surface (i.e., by design they are too large to be carried into the pores). The situation is shown in Figure 6.25. The cast layer becomes another porous layer that water must pass through before it reaches the mold. This layer has its own specific resistance, α_c and void fraction, ε_c. Because the cast layer resistance is not equal to the mold resistance, the pressure drop through the cast layer and mold are not equal. However, the minimum pressure in the system, P_o, that drives the water flux through both layers is the same capillary pressure that was used in the simplified case.

The derivation of the cast layer thickness, L_c, as a function of time follows the same general ideas as above, but is more complex. The volume

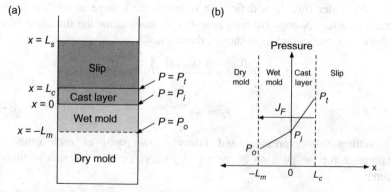

FIGURE 6.25 (a) Schematic of a one-dimensional slip casting process and (b) the pressure gradient established in the porous mold and altered by two resistive layers—the cast layer, which is a dense packing of ceramic particles, and the porous mold.

conservation must include not only the water but also the particles, and one more interface position must be tracked. Aksay and Schilling (1984) present the derivation in detail. The main result is the following equation for the cast layer thickness as a function of time:

$$L_c = \left[\frac{2(P_t - P_o)t}{\beta\eta_F\left(\alpha_C + \frac{\beta\alpha_m}{\varepsilon_m}\right)}\right]^{1/2} \quad \text{where} \quad \beta = \frac{1 - \phi_p - \varepsilon_c}{\phi_p} \quad (6.52)$$

where ϕ_p is the volume fraction of particles in the slip, which is assumed to be constant throughout the casting process, and β is a constant that arises from the conservation of volume that is carried out on the water and the particle phases. For a constant ε_c, β decreases as solids loading of the slip (ϕ_p) increases. The equation shows that the cast layer thickness grows with a $t^{1/2}$ dependence. In other words, to double the cast layer thickness, the casting time has to be increased by four times.

Typically, the cast layer resistance is greater than the mold resistance. The mold is designed to conduct the liquid away and is optimized for that purpose. The cast layer, on the other hand, is frequently composed of small particles packed with a small void fraction. Under these conditions, Eq. (6.52) simplifies to:

$$L_c = \left[\frac{2\Delta Pt}{\beta\eta_F\alpha_C}\right]^{1/2} \quad \text{when} \quad \alpha_C \gg \alpha_m \quad (6.53)$$

where ΔP is used for the pressure drop across the cast layer and the wet mold.

By analyzing Eqs. (6.52) and (6.53), the important design principles of slip casting are illustrated. The casting rate can be increased by increasing the pressure difference between the slip and the dry mold ($\Delta P = P_t - P_o$). This can be done by increasing the surface tension or decreasing mold pore size, which decreases the capillary pressure P_o, or by imposing an external pressure on the slip, which increases P_t. Another option is vacuum casting, which drops the pressure in the mold further. Increasing the solids loading of the slip also increases the casting rate by decreasing the parameter β. With higher solids, less water removal is necessary to deposit an equivalent cast layer as a lower solids slip. Also, decreasing the viscosity of the filtrate also speeds slip casting. (Note that the viscosity of importance here is the water viscosity and not the slip viscosity.) However, since the filtrate is most often water, this parameter is not adjustable. Finally, the specific resistance of the cast layer can be decreased. Such a change could be accomplished by increasing the particle size or by increasing the void fraction of the cast by introducing a destabilizer. Such changes, however, must be made in consideration of the entire process and the requirement for the final ceramic.

6.3.4 Slip Casting Process Considerations

In designing a slip casting process, multiple factors are considered and trade-offs made. A change that enhances the casting rate may lead to a slip that is not stable with time or too viscous to pour. Ultimately some compromises are necessary.

Successful design of a slip casting process relies on proper formulation of the ceramic slip. Casting rate, drying shrinkage, green strength, and sinterability all rely largely on the slip formulation. Unfortunately, optimizing some of these parameters means that others are detrimentally affected. Slip casting begins with proper formulation and mixing of a high solids loading slip. Viscous slurries of ceramic particles are challenging to mix and homogenize. Ball mills and vibratory mills are often used. Additives that might lead to foaming or degrade during the first mixing process are often incorporated to the mixed slurry in a second, less energy intensive process. If air bubbles are formed in the milling process, they are removed from the slip before the casting process.

One major consideration is the slip viscosity, which must be chosen so that the slip is easily poured into the mold and fills the detailed sections of the part. Typical slip viscosities range from 100 to 1000 mPa·s. High solids loading slips contain less water to remove into the mold, but they are also very viscous and possibly difficult to work with. Therefore, most slips contain 40–50 vol% particles. Water is the most common solvent, although in cases where the particles may become hydrolyzed or gelled (MgO, CaO) or form an undesirable oxide layer in water, other solvents such as an alcohol may be used. Binders (<1 vol%) such as carboxymethyl cellulose or low T_g latex can be necessary to give the green part enough strength so that it can be removed from the mold and handled when completely dry. The trade-off is that many binders increase the suspension viscosity and decrease the rate of the water flow rate into the mold.

Designing the particle size distribution is also a challenge that affects both dewatering of the slip as well as the sintering process. Smaller particles can be pulled into the pores of the mold, clogging the gypsum. Pore spaces between small particles are tortuous and constricting, leading to low dewatering rates. Larger particles, however, may sediment during the casting process and the resulting green ceramics sinter less efficiently. A typical slip contains powder particles ranging from 1 to 5 μm in size, although either smaller or larger particles can be used by making concessions in the process design.

One effective method for controlling dewatering rates is adjusting the stability of the slip. Highly flocculated slips form fractal, porous particle gel networks in the cast layer that allow water to pass through quickly, reducing casting time. Conversely, particles in highly deflocculated slips rearrange during the casting process into a dense packing. Water flows through the particle packing more slowly, increasing the casting time, yet the resulting green part is very dense, which is advantageous for the sintering process. Deflocculated slips pour and fill the mold easily, so one strategy is to add a coagulant just before casting so that the particles aggregate once the slip has filled the mold.

There are several ways to increase the casting rate without making trade-offs with the slip composition. "Vacuum casting" and "pressure casting" are other process variations that speed up the solidification process by applying a vacuum to the outside of the mold or pressure to the slip, respectively. In "centrifugal casting," the part is spun to create centripetal forces to consolidate and pack particles at the mold walls.

Slip casting is advantageous due to the high level of detail that can be produced in the green structure, relatively low expense of the process, and the level of homogeneity that is gained in the green part. Disadvantages include the time required to produce a single part, and the limits placed on the process by the porous mold and the capillary action. Namely, slip casting of submicron particles, which have considerable technical importance is a challenge due slow casting rate and clogging the mold.

6.3.5 Post-Processing Operations

Once the part is cast and the excess slip is poured from the mold, post-forming or post-processing operations are performed. The part is removed from the mold and dried completely. Although slip casting in general produces a green part with a smooth surface, there can be a ridge that is left behind at the seam between the two mold halves. This defect is sanded off in the green state. The organics in the part, including any binder, are then decomposed at a low temperature (300–600°C). Next, the part is sintered at a high temperature (900–1500°C) to form a dense, strong final product. Finally, surface coatings such as paints, glazes, or other decorations can be applied to complete the operation.

6.4 COATING AND TAPE CASTING

6.4.1 Process Overview

Thin, flat sheets of polymer, ceramic, or metal particles are produced from coating dispersions, solutions, and liquid monomers. Tape casting is a special type of coating process in which a dispersion containing ceramic particles with polymer binder is coated onto a substrate. In all cases, a liquid dispersion is deposited, spread onto a substrate, and subsequently solidified. In tape casting, the solidified sheet is stripped from the substrate and then heated to remove the polymer and sinter the ceramic sheet. Adhesive tapes, capacitor layers, displays, recording media, decorative varnishes, and

474 Materials Processing

protective barriers are all examples of functional coatings created from dispersions, solutions and liquid monomers.

There are a tremendous variety of coating operations, ranging from those designed to coat flexible substrates (webs) at high speeds to those that coat discrete objects like ophthalmic lenses or silicon wafers. The fundamental steps in a liquid coating process are: (i) preparing the liquid, (ii) feeding or transporting the liquid to the coating apparatus, (iii) distributing the coating liquid on the substrate, (iv) metering the coating liquid such that a uniform liquid layer is achieved, and (v) solidifying the liquid layer, typically by drying, curing, or chilling.

Figure 6.26 shows an example of a continuous operation for preparing polymer coatings on flexible substrates (webs). The flexible substrate comes in a roll and the apparatus unwinds it and controls its motion and tension during the process. Polymer solution is pumped at a controlled rate into a slot die, which is stationed over the web at a precise position beneath a back-up roll. The details of the die are shown in a later figure, but given that the solution enters the die in a central location a distribution chamber similar to the one shown in Chapter 3 for extrusion of polymer sheets would be used to assist the delivery. This chamber allows easy flow cross web so that the polymer solution exits the slot at a constant rate across the web. The next step is metering, creating a liquid layer with a uniform thickness. Slot coating is "pre-metered"—thickness is determined by the flow rate into the die and the substrate velocity. The coated substrate is carried through a dryer, where solvent is removed, and the final coated product is then collected in a roll at the end of the production line. This sort of process is frequently referred to as a "roll to roll" (R2R) process.

As discussed previously, the drying step often is rate limiting in designing a coating process. For example, if the coating is applied at 20 m/min and

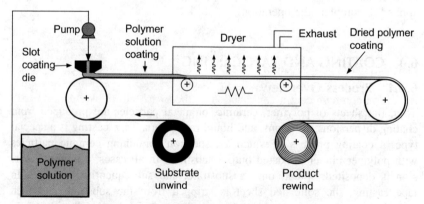

FIGURE 6.26 Schematic of a continuous process for coating a flexible substrate (web) slot coating of a polymer solution. This schematic shows some of the main features.

drying takes 2 minutes, a 40 meter long dryer section is required. If the dryer is impractically long and the drying process itself cannot be shortened, the coating or deposition speed must be decreased. Drying a coating too quickly can cause coating defects such as blisters and cracks. The drying section can also encompass annealing or curing for polymeric coatings.

Once a coating is dry, post-processing steps also occur at the speed of the web. For example, calendering is common in paper manufacturing. Most paper is coated to improve the surface characteristics and interaction with inks. This coating is calendered, or passed through a set of rollers that apply pressure to create a smoother, glossier surface finish. Laminates can also be applied in a separate step. In the case of ceramic tapes, the green tape must be sintered after it is dried of moisture and the binder removed.

Like all dispersion and solution processes, the liquid properties dictate the design of the coating process. Rheology is important for transporting, applying, metering, and leveling the coating liquid. High viscosity liquids are difficult to apply quickly to fast moving webs. Surface disturbances from coating liquid application may also take longer to level (by surface tension) and may be locked in by the drying stage. Low viscosity liquids may drip, pool, or be blown about in impinging air dryers. How the coating wets and spreads on the substrate is as important to understand as its rheology. In all cases, a coating liquid must replace air at a substrate interface without trapping bubbles or dewetting. Some advanced coating processes apply one liquid layer on top of another liquid layer before the drying stage. Here, the wetting of one liquid on another is important, as well as the surface tension between the adjacent liquids and their miscibility.

In this section, a brief overview of liquid coating methods followed by a discussion specific to how these processes are used to make polymer coatings and tape cast ceramic layers.

6.4.2 Coating Methods

Several coating application methods are available for selection based on considerations including coating liquid properties, production speed, and cost. Whether the substrate is a continuous web or an irregularly shaped object also may limit options for coating operations. Figure 6.27 shows schematics of several common coating application methods that all involve coating onto moving substrates. They range from methods that involve plunging a substrate into the coating and then pulling it out without (Figure 6.27a dip coating) and with post-metering (Figure 6.27b knife coating) to a fixed or adjustable gap applicator that feeds liquid from a reservoir (Figure 6.27c doctor blade coating). Precision high speed coatings are frequently prepared with pre-metered processes such as slot coating (Figure 6.27d), slide coating (Figure 6.27e), which involves liquid flowing down an inclined plane after it exits at a controlled flow rate from a slot die, and curtain coating

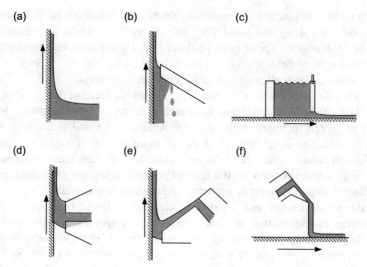

FIGURE 6.27 Different methods for applying liquid coatings onto moving substrates: (a) dip coating, (b) knife coating, (c) adjustable gap blade coating (doctor blade coating), (d) slot coating, (e) slide coating, and (f) curtain coating. In (a) and (b) the liquid is pulled from a vat or container; in (c) a liquid reservoir is kept full during coating and in (d)–(f) the liquid is pumped into the coating die. *Courtesy of M.S. Carvalho, Pontifical Catholic University of Rio de Janeiro.*

(Figure 6.27f), which is like a waterfall of coating liquid that impinges on a moving substrate.

Dip Coating. Perhaps the most obvious method of coating is dip coating. See Figure 6.28. A substrate is immersed in a pool of liquid and then removed at a fixed speed. The process can be conducted with a discrete substrate and or with a continuous substrate that plunges into the bath of coating liquid and then is pulled out and into a dryer.

Controlling the thickness of the as-deposited liquid entails a balance of forces. Liquid clings to the substrate as it exits the bath; a viscous drag pulls the liquid upward. This force is proportional to the product of the substrate speed and the viscosity and inversely proportional to the thickness, $\eta U/h$. At the same time, the liquid layer experiences the tug of gravity, which is proportional to $\rho g h$, where ρ is the density of the coating liquid. Considering only these two forces, a simple estimate is derived by setting these two forces equal:

$$h_{wet} = \sqrt{\frac{\eta U}{\rho g}} \qquad (6.54)$$

where the subscript "wet" is added to emphasize the thickness is that before drying or other solidification event. It represents the thickness of the liquid

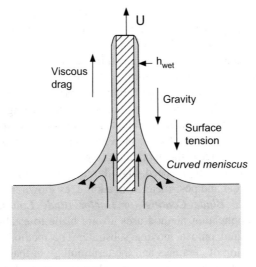

FIGURE 6.28 Schematic diagram of dip coating.

coating where it is flat, some distance away from the meniscus region near the bath. This simple relationship captures much of the action in dip coating. The viscous drag pulls liquid up and gravity forces it back down. The liquid attached to the substrate takes on the velocity of the substrate, moving upward, and the liquid at the free surface responds most to gravity, moving downward. Hence the velocity gradient between the substrate and the surface is complicated, especially near the bath. There is a stagnation point where the velocity is zero. Importantly, increasing the substrate withdrawal speed increases the thickness, making the prospect of depositing thin coatings time consuming.

In addition to viscous drag and gravity, yet another force comes from the curved liquid meniscus. The low pressure beneath this meniscus draws liquid from the newly deposited layer, which has a flat surface, downward back to the bath. This force is the result of surface tension. The relationship with all three forces included is:

$$h_{wet} = 0.944 \left(\frac{\eta U}{\gamma_{LV}} \right)^{1/6} \sqrt{\frac{\eta U}{\rho g}} \qquad (6.55)$$

where γ_{LV} is the surface tension of the coating liquid. Equation (6.55) is frequently referred to the Landau-Levich equation, after the scientists who derived it. The new term, the ratio raised to the 1/6 power, is called the capillary number. The Landau-Levich equation was designed for low capillary number, less than about 0.01. Notice that as the capillary number increases toward unity, a trend consistent with increasing coating speed or coating viscosity, Eq. (6.55) approaches the simpler Eq. (6.54). Also, the relationships for wet coating thickness do not account for non-Newtonian rheology or draining of the deposited liquid coating due to gravity. In practice, the second effect is usually not significant if thin coatings are prepared or if drying occurs immediately after deposition.

Dip coating has a number of applications, but the process also has a significant disadvantage. The technique is well-suited to coating discrete parts, exploring new coating materials and creating thin, uniform coatings. However, as Eqs. (6.52) and (6.53) show, the faster the substrate is drawn from the bath, the greater is the applied coating thickness. Therefore, the process is not well suited to high-speed, high-throughput coating operations. One solution is to dip coat and post-meter with a knife, as shown in Figure 6.27b.

Adjustable Gap Blade Coating (or Doctor Blade Coating). A second type of coating application method uses a rigid blade to establish a gap that meters coating liquid onto a moving substrate. The method is used extensively in tape casting of ceramics and the terminology in that field is "doctor blade coating." A schematic of an adjustable gap blade coater or doctor blade coater is shown in Figure 6.29. The two main features of the applicator are the reservoir and the blade. The reservoir holds the coating liquid, supplying it to the gap established by the blade. Coating liquid can be replenished periodically or fed into the reservoir continuously. The blade establishes the gap; applicators may have a fixed gap or a gap that is adjustable by micrometers. The gap is of central importance to this coating operation and so substrate flatness and rigidity is also key. Lastly, the applicator must be rigidly held in the coating apparatus to prevent it from moving as the substrate travels beneath.

In doctor blade coating, the moving substrate drags coating liquid and the blade meters the coating liquid to produce a controlled thickness. The reservoir also plays a role, because hydrostatic pressure drives flow beneath the blade as well. To develop an understanding for the factors that control the wet coating thickness, let's consider the flow rate, Q, in two regions, as

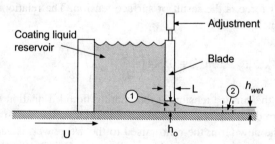

FIGURE 6.29 Schematic diagram showing the cross-section of an adjustable gap blade coating applicator. Two zones for flow analysis are noted: (1) directly under the blade and (2) the wet coating after exit.

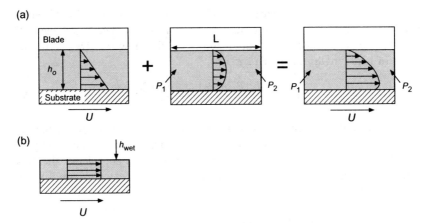

FIGURE 6.30 (a) Schematic diagram showing the velocity profiles under the blade in a doctor blade coating apparatus (region 1 in Figure 6.29). The drag portion of the flow (left) creates a linear velocity profile, the pressure-driven flow (middle) leads to a parabolic profile and flow with both drag and pressure-driven effects (right) is a summation of the two. (b) Schematic diagram of the coating after exiting the blade (region 2 in Figure 6.29), showing movement at the velocity of the substrate (plug flow).

noted in Figure 6.30. The first region is under the blade, where drag and pressure combine and the second region is the wet coating that is moving away from the blade with the substrate.

The action underneath the blade consists of two parts: a drag flow from the moving substrate and a pressure-driven flow from the hydrostatic pressure. Figure 6.30a shows the velocity profiles that develop from the drag and pressure effects separately and then added together. These two types of flow are explored in more detail in Chapter 3.

The drag from the moving substrate considered on its own (Figure 6.30a, left) produces a linear velocity profile in the liquid. Assuming that the liquid does not slip on any surface, the liquid in contact with the substrate moves with the velocity of the substrate, U, and the liquid in contact with the stationary blade is also stationary. Across the gap depth, the average velocity of the flow is $U/2$. Therefore, the volumetric flow rate in region 1 from the drag component, $Q_{1,d}$ is:

$$Q_{1,d} = \frac{U}{2} h_o w \qquad (6.56)$$

where h_o is the gap and w is the width of the blade and coating. In this analysis, the width is assumed to be large so that it does not affect the flow.

For a stationary substrate, the only driving force for flow would be the pressure difference through the gap. In this case, the velocity profile would be parabolic, as illustrated in Figure 6.30a, middle. Since in this situation the blade and the substrate are not moving, the no slip condition causes the flow to be stagnant at these two surfaces, and the greatest amount of flow is at the center of the gap. If the pressure in the liquid at the entrance to the gap is P_1 and the pressure at the exit of the blade is P_2, the flow rate $Q_{1,p}$ across the blade of land length L, is given by:

$$Q_{1,p} = \frac{h_o^3(P_1 - P_2)}{12\eta L} w = \frac{h_o^3 w \Delta P}{12\eta L} \qquad (6.57)$$

Notice that the pressure-driven flow has a stronger dependence on the gap height as compared to the drag flow component and is also influenced by the liquid viscosity η. Unless liquid curvature at the exit is considered, the pressure in the fluid exiting the die, P_2, is assumed to be close to atmospheric. Here, we have assumed that the liquid is Newtonian. A somewhat more complicated expression for pressure-driven flow of non-Newtonian power law liquids is given in Chapter 3 and could be used in this analysis.

The total flow rate under the blade, Q_1, is then found by adding the individual components:

$$Q_1 = \frac{U}{2} h_o w + \frac{h_o^3 w \Delta P}{12\eta L} \qquad (6.58)$$

The velocity profile (Figure 6.30a, right) is the addition of the two components, a distorted parabola.

In the deposited layer, region 2, all of the coating moves with the velocity of the substrate:

$$Q_2 = U h_{\text{wet}} w \qquad (6.59)$$

This type of flow is called plug flow. See Figure 6.30b.

Since all of the liquid flowing beneath the blade must also exit on the substrate (mass conservation), Q_1 and Q_2 are equal:

$$\frac{U}{2} h_o w + \frac{h_o^3 w \Delta P}{12\eta L} = U h_{\text{wet}} w \qquad (6.60)$$

Simplifying this expression, the coating thickness is:

$$h_{\text{wet}} = \frac{h_o}{2} \left(1 + \frac{h_o^2 \Delta P}{6\eta L U} \right) \qquad (6.61)$$

In many processes, the speed and viscosity of the coating liquid are high enough so that the second term in the parenthetical expression in Eq. (6.61) is small and the coating thickness can be approximated as half of the gap.

Doctor blade coating is commonly used for viscous suspensions, such as those developed for tape casting. The method creates controlled thickness layers with high precision. A downside of the method is the contact between the applicator and the web and the use of gravity for feeding the liquid to the gap.

Slot Coating. Slot coating (also called slot die coating) is a pre-metered process; coating liquid is pumped through a slot die, with a controlled flow rate, Q, onto a moving substrate. The output of the process is a liquid layer of thickness h_{wet} traveling with plug flow at the substrate speed, U; therefore, $h_{wet} = Q/U$ (on a per coating width basis). The ability to control the coating thickness by setting the flow rate and coating speed is a main advantage of slot coating as well other pre-metered methods, such slide and curtain coating.

Figure 6.31 shows a perspective schematic of a slot coating die as it applies coating liquid to a web. The coating liquid is pumped from a tank through a feed pipe located at the center of the die. To distribute the liquid across the length of the die and to ensure uniform output from the slot across the web, a distribution chamber with lower flow resistance is used. The die is spaced very close to the web, which is supported by a back up (or backing) roll. Because the process is very sensitive to the gap between the die and the substrate, this roll is designed to be smooth and have a highly circular cross-section so that no undulations occur as the roll rotates during the process. The coating "bead" is formed where the liquid meets the substrate.

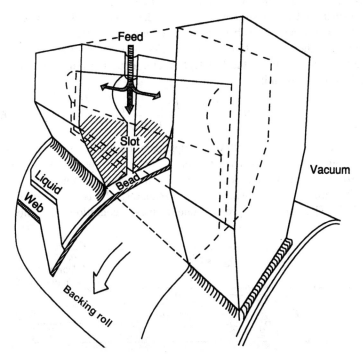

FIGURE 6.31 Schematic diagram of a slot coating die, web, applied liquid, and backing roll. The perspective of the drawing exaggerates the curvature of the roll. *From Stokes, R. J. & Evans, D. F.,: Fundamentals of Interfacial Engineering. Page 406. 1997. Copyright Wiley-VCH Verlag GmbH & Co. KGaA. Reproduced with permission.*

The shape of the bead depends on the process conditions. A vacuum box is placed on the upstream side of the die for extra control of the process, as discussed below.

The basic process analysis of slot coating has similarities to that of doctor blade coating; drag and pressure-driven flow combine to establish the conditions, but the fixed incoming flow rate leads to differences. A schematic of the lower part of the die, the coating bead, and the substrate is shown in Figure 6.32. Coating liquid is pumped in through a feed slot; the pressure in the bead at the entrance is noted as P_E. The shape of the die lips can vary considerably. In this example, the upstream and downstream die lips are parallel to the substrate surface and each is separated from the substrate by a gap, h_o. On the upstream side, the pressure in the gas phase, P_o, is either atmospheric or adjusted to be sub-atmospheric with a vacuum box. The downstream meniscus is bounded by gas at atmospheric pressure, P_{atm}. The pressures in the liquid at the upstream and downstream sides of the bead are noted as P_U and P_D. A first goal of the process analysis is to gain an understanding of the flow in the bead and the factors that control the coating thickness. To assist in this analysis, three zones are identified in Figure 6.32: 1. upstream, 2. downstream and 3. deposited coating.

Consider first the upstream side of the die (zone 1). Here, there is drag flow due to the moving substrate and pressure-driven flow. Again, we can add together the contributions of each. Combining the drag and pressure-driven terms as we did in the last section and assuming a unit coating width and a Newtonian liquid:

$$Q_1 = \frac{U}{2} h_o + \frac{h_o^3(P_U - P_E)}{12\eta L_U} \tag{6.62}$$

where L_U is the contact length of the upstream die lip. To maintain a steady-state location of the upstream meniscus, the net flow in zone 1 is zero ($Q_1 = 0$) and therefore the second term in this equation must be less than zero, indicating that P_U, which is approximately P_o, is less than P_E (i.e., $P_U < P_E$). The velocity profile under these conditions is shown in

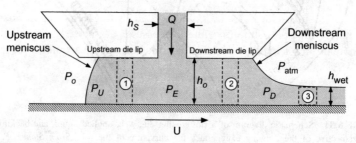

FIGURE 6.32 Schematic diagram of a slot coating die, coating bead and substrate. See text for description. *Adapted from Romero et al. (2004).*

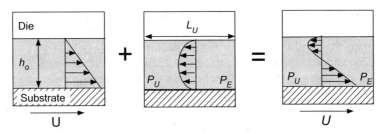

FIGURE 6.33 Schematic diagram showing the velocity profiles under the upstream die lip as a combination of drag and pressure-driven flow.

Figure 6.33. The direction of flow reverses across the gap and there is no net output.

In the downstream side of the die (zone 2), the liquid also experiences drag and pressure-driven flow, and last, the departing coating (zone 3) is transported by plug flow. By mass conservation, the flow rate in zone 2, Q_2, is equal to the flow rate in zone 3, Q_3, and both are equal to the flow rate from the feed slot. So, we can just use one "Q" for the process and set it equal to the plug flow output, assuming unit width:

$$Q = Q_2 = Q_3 = U h_{\text{wet}} \tag{6.63}$$

Analyzing now the flow rate in zone 2, we find that the wet film thickness follows the same expression as in doctor blade coating. Now, written with the variables of the slot die and on a unit width basis:

$$h_{\text{wet}} = \frac{h_o}{2} + \frac{h_o^3 (P_E - P_D)}{12 \eta L_D U} \tag{6.64}$$

where L_D is the contact length of the upstream die lip. Now the importance of the adjustable flow rate comes into play. If the flow rate Q is set (at fixed U) so that h_{wet} is greater than $h_o/2$ then P_E will be greater than P_D (i.e., the second term in Eq. (6.64) is greater than 0 and the pressure-driven flow is toward the deposited liquid layer). However, if the Q is decreased sufficiently so that h_{wet} is less than $h_o/2$, then P_E must be less than P_D ($P_E < P_D$) to maintain a stable coating bead. Since P_D is approximately atmospheric pressure, sub-atmospheric pressures are needed. To fulfill this requirement and achieve lower coating thicknesses, a vacuum is pulled on the upstream side of the die. The vacuum lowers P_U and P_E (but still maintains the requirement of $P_U < P_E$). The result of pursuing $h_{wet} < h_o/2$ and hence $P_E < P_D$ is that drag and pressure drive flows in opposite directions, as shown in Figure 6.33. When the thickness drops below a critical level ($h_o/3$), then recirculations appear in the downstream meniscus and these could have adverse effects as coating liquid is trapped in these circular pathways for some time.

As might be imagined, changing Q and the vacuum pressure (at a fixed U) results in different shapes for the coating bead. For example, starting from the conditions that lead the shape in Figure 6.32, increasing Q (and keeping all other conditions constant) would cause the bead to expand so that it contacts more of the upstream die lip. Likewise from the same starting point, lowering the pressure relative to ambient on the upstream side (i.e., increasing the vacuum) would lead to the same effect. Of course, variations in speed with Q and vacuum constant have similar effects. Under different conditions, the contact of the bead on the downstream die lip also changes. Based on these changing bead configurations and instabilities that come about in the exiting coating, a "coating window" can be defined.

Figure 6.34 shows a schematic of the coating window for slot coating. The coating window identifies regimes in process space that result in a stable coating bead and a high-quality liquid layer without defects. The axes on the plot are the main process variables. On the y-axis is the pressure on the upstream side of the beam represented as pressure below atmospheric. On the x-axis is $1/h_{wet} = U/Q$. This axis can be read in several of ways—increasing web speed or decreasing flow rate Q while keeping the other variable constant, for example. A good place to start is at the condition of no applied vacuum or $P_o = P_{atm}$ (y-axis is 0). Here, the coating window extends to a critical level of $1/h_{wet}$, where $h_{wet} = 1/2\ h_o$. This value lies on the lower limit line (labeled 2 in Figure 6.34) where the bead pulls away from the

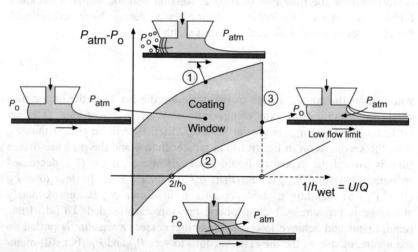

FIGURE 6.34 Schematic diagram of a slot coating window. Note that three limiting bounds are emphasized in this diagram: (1) an upper bound where coating liquid overflows the gap on the upstream side, (2) a lower bound where the coating liquid is depleted from the upstream side of the die, and (3) a low flow or high speed limit that occurs when attempting to coat very thin layers. *Reprinted from J. Non-Newtonian Fluid Mechanics, 118, Romero, O. J., Suszynski, W. J., Scriven, L. E. & Carvalho, M. S., Low-flow limit in slot coating of dilute solutions of high molecular weight polymer, 137–156, Copyright (2004), with permission from Elsevier.*

upstream die lip due to excessive drag relative to the rearward pressure-driven flow. The lower limit for the coating thickness is represented as the vertical line on the coating window (labeled 3 in Figure 6.34). This limit occurs in the extremes of low Q or high U, because the curvature at the downstream meniscus becomes extreme, causing to coating to breakup into rivulets or droplets. This limit is independent of vacuum. Lastly, the upper bound on the coating window (labeled 1 in Figure 6.34) is created by excessive vacuum, which leads coating liquid to be pulled out of the die and the loss of pre-metering action.

The analysis above takes into account only the viscous effects and neglects the role of capillarity (i.e., meniscus curvature) on the process. These effects are very important in slot coating and represent another process variable to contend with and adjust to achieve the desired coating thickness and uniformity. Hence the coating window is really a 3D coating space where a convenient third axis is the capillary number, $Ca = \eta U/\gamma_{LV}$.

Overall, slot coating is an excellent choice for coating thin liquid layers quickly. It finds application in the production of adhesives, magnetic tapes, and optical films to name a few. The process can be extended to two layers by including a second feed slot. Another variation is to remove the backup roll and use tension to support the web.

Curtain and Slide Coating. Curtain and slide coating methods are economical extensions of slot coating. Both methods have the possibility of coating multiple wet layers using only one die. In slide coating, the liquid is pushed through a slot and then slides down the die face and onto a substrate supported by a coating roll. Layers can be added sequentially to the bottom of the flow with the intention that the liquids remain distinct and unmixed even after encountering the coating web.

Curtain coating uses the same principles as slide coating. However, in curtain coating the liquid falls from the end of the inclined die onto the web. Although more difficult to design, curtain coating operations have many advantages including an extra gravity assist in wetting the substrate. Also, by changing the speed of the web, the thickness of the applied coating can be altered independently of the die design or liquid properties. These advantages apply whether the final product has single or multiple layers.

Multilayer coating applications were initially driven by the photographic industry to produce film that often had more than ten or more independent layers. By coating many liquids with one die, separate drying steps between each layer are unnecessary, resulting in great energy and space efficiency, and cost savings. However, a great deal of research and planning is necessary to implement a multilayer coating operation. The liquid surface tensions and miscibilities must be matched so that they can not only be coated but also dried without causing defects. Solvent migration between layers can cause precipitation of undesired solids and bare spots occur if one liquid does not wet another.

6.4.3 Polymer Coatings

Polymer coatings are thin layers of polymer applied to either flat substrates or irregular objects. Polymeric coatings can be functional (adhesives, photographic films), protective (anticorrosion), or decorative (paint). They are also used to modify surfaces (paper coatings, hydrophobic coatings). Although polymeric coatings are mostly organic, they can also include ceramic or metal particles to increase durability, functionality, or aesthetics.

Because polymer coating solutions are often expensive, coating methods are designed to produce as thin of a polymer layer as fast as possible. Polymeric coatings are typically on the order of 1–100 μm in thickness. Depending on the desired thickness of the coating, the rheology of the liquid and the speed of the web, polymeric coatings are applied many different coating methods, including those described in the last section. Multilayer coatings are also common.

Although all polymeric coatings are designed to be an impervious, homogeneous thin layers of polymer when dry, there are three unique starting liquid types that can be used to produce this outcome: polymer solutions, monomer liquids, and polymer latexes.

Polymer solutions. To reduce the viscosity of a polymer so that it becomes a coatable liquid, polymer is dissolved in a solvent. The rheology of the coating solution is tailored to the coating and drying method by altering the amount of solvent in the solution, as described in section 6.2.3. Once the liquid is deposited onto a substrate, the solvent must be removed in a drying process. As solvent is removed, the glass transition temperature of the solution increases, as shown in Figure 6.35. Nominally, when the glass transition temperature is greater than the drying temperature, the coating is solid. If the drying temperature is higher than room temperature, the coating may continue to solidify or harden upon cooling. Additionally, some polymers

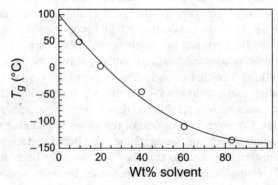

FIGURE 6.35 Glass transition temperature of polystyrene as a function of toluene content. *Data from Adachi et al. (1975).*

crystallize as solvent is removed or on cooling from the drying temperature, creating final polymer coatings that are semicrystalline.

Most polymers for functional coatings are not water soluble and so organic solvents are used. The solvent is chosen both for its ability to dissolve the polymer of interest but also for its cost and volatility, which influences the drying step. Frequently, the inclusion of additives or the need to reduce cost leads to the use of multiple volatile solvents. One potential issue with this practice is phase separation during drying if the more volatile solvent is also the better solvent for the polymers. Although usually undesirable, this possibility has been exploited to make porous polymer membranes.

Polymer solution coatings are advantageous because they can often be formulated to specifications at the processing site, applicable to a wide variety of polymers and tunable to create controlled thicknesses and properties in the final coating. Unfortunately, the amount of polymer that can be dissolved in a solvent and still maintain fluidity for coating is relatively small ($\sim$10−20 wt%) and hence a considerable amount of drying is required per unit of coating thickness. Complications related to solvent handling also present disadvantages, including environmental and safety concerns, solvent recycling, and the need for more expensive dryers that are capable of capturing flammable solvents.

Liquid Monomers. Many monomers are liquid at room temperature, so they can be coated directly often with no solvent or minimal solvent added to reduce their viscosity at the coating temperature. Oligomeric precursors are also in this category. Instead of a drying step, monomer liquids are solidified by curing reactions. As curing proceeds, the average molecular weight of the coating material increases until a solid polymer layer is formed. Curing reactions are initiated by heating or exposure to ultraviolet (UV) light or other energetic sources such as electron beams.

Compared to polymer solution coatings, a lesser variety of polymer chemistries are amenable to this route. One of the most popular curable coating materials is epoxy. Epoxies are not formed from monomers but rather by reaction of oligomeric resins with hardeners or curing agents. The coating liquids are made with solvent as well to improve the coatability. Among the liquid monomer alternatives, acrylates are common for UV curing.

Monomer liquids are attractive routes to coatings since they require little to no solvent and therefore a reduced or eliminated drying step. Final coating properties such as cross-link density can be controlled during the curing step by changes to the temperature and UV intensity as well as to the chemistry of the resin. Specialized, functional monomers and initiation agents may be more expensive than the materials used in polymer solution coatings, however. Also, achieving high degrees of cure and low residual monomer contents is sometimes a challenge in glassy polymers. The final product in many cases is brittle due to the high degree of cross-linking.

488 Materials Processing

Polymer latex. A latex is a dispersion of polymeric particles in water. Given the low solubility of functional polymers in water, latex routes provide an environmentally attractive approach for creating durable polymer coatings. Particles ranging in size from ∼10 nm to 1 μm can be produced from a variety of polymer chemistries by emulsion polymerization. Since they are synthesized as dispersions and stabilized in the process, they are particularly easy to use. For some application the latex is formulated with other phases, like ceramic particles.

The drying of latex particle suspensions is somewhat different from the drying of hard colloidal particles, which is described in an earlier section. The process known as "film formation" is shown in Figure 6.36. As water is removed, particles become more concentrated in the suspension during a "consolidation" stage. As the falling rate period ends and the water begins to invade into the particle packing, surface tension, capillary, and van der Waals forces start to pull the particles toward each other. These forces are strong enough that the particles start to flatten at particle-particle contact points. See Figure 6.37. Accordingly, the pore space between the particles shrinks. This stage is termed "compaction." Finally, in the "coalescence" stage, polymer chains diffuse across the boundaries between particles, welding the particles together. This process creates a final coating that is devoid of pores or boundaries between what were once individual particles.

Since water is the liquid medium for latex coatings, this route is an environmentally friendly alternative to monomer and solventborne coatings. Many consumer coatings such as varnishes and paints begin as latex dispersions. Complex polymer blends or structures can be polymerized into a latex particle, which retains its identity during the latex film formation process. However, latexes are expensive to transport and purchase as raw products in commercial scales, and the drying of water is an energy intensive process.

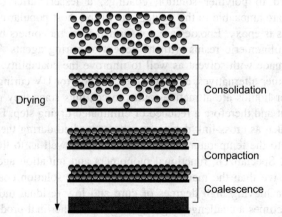

FIGURE 6.36 Latex film formation stages.

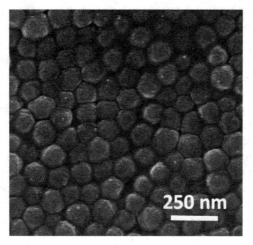

FIGURE 6.37 Top surface of a drying low T_g latex coating in the compaction stage. The coating was frozen and then imaged in a scanning electron microscope with a cryogenic stage to preserve the water in the pore space between the particles. Particles are flattened at their contact points. *From C.C. Roberts (2010) reproduced by permission.*

6.4.4 Tape Casting of Ceramics

Doctor blade coating of ceramic dispersions containing polymer binder is termed "tape casting." The end result of tape casting is a thin ceramic sheet or multilayer ceramic product. The steps in a tape casting process are very similar to any other coating process. The suspension (coating liquid) is formulated and mixed. Ball milling may be used to first disperse the ceramic particles, and then binder, plasticizer, or other additives are introduced in a second step. The suspension is then de-aired to remove bubbles. Finally, the liquid is coated onto the substrate, usually with doctor blade coating, but slot coating is also used. As in Figure 6.26, the as-deposited tape cast layer is sent immediately into an oven for drying. The layers are deposited onto a flexible plastic substrate that is wound up after drying and sent to the next stage of the process. Green tape thicknesses in the range of $\sim 10\,\mu m$ to over $200\,\mu m$ can be prepared. As thickness increases, the length of the dryer increases to provide for more drying time and/or the speed is reduced for the same reason. Based on the stringent applications in electronic devices, for example, tape casters may be located in clean rooms or have filters for their air supplies to reduce dust and other extraneous particulates. Once dry, green tape cast sheets go through a number of steps depending on the part geometry and application, including cutting to size, punching holes for electronic conductors, metalizing, and stacking. Next the parts are heated to remove binder and sinter.

Suspensions for tape casting are formulated for easy coating onto the substrate and flexibility in the green tape after drying. There are three main components in these dispersions: ceramic particles, a liquid medium (water or organic solvent), and additives. As in slip casting, achieving a high particle content is desired; however, the need to work with fine particles and a considerable amount of binder limits the solids loading. A particle concentration of 20–35 vol% is typical. Particle sizes are finer than those in slips for slip casting (i.e., much less than 1 µm in diameter). Large particles degrade the surface roughness of the tape cast layer, a feature that is especially important for multilayer and electronic applications. Since aggregates are problematic for the same reason and others, additives are used to ensure colloidal stability, usually by steric or electrosteric means. The ceramic particles are dispersed in the liquid medium before organic binders are added, because binders increase the viscosity and may interfere with achieving a highly dispersed state.

The choice of a water or an organic solvent for the liquid medium depends on a number of factors, including the suitability for dispersion of the particles and compatibility with the binder. Water is the best choice in terms of environmental, health, and safety factors. Most oxide particles are easily dispersed in water, and latex (i.e., emulsion) based binders are suitable. However, water also requires a lot of energy to evaporate in the drying stage due to its high heat of vaporization. Organic solvents have the advantage of fast drying and easy dissolution of binders and plasticizers, but they often have to be recaptured, contained, and disposed of in the drying step.

To increase flexibility, the green tape usually contains high concentrations of binders and plasticizers. Up to 15–25 vol% of the dry tape can be polymeric binders. Polymers, including polyvinyl alcohol, polyvinyl butyral, cellulosic polymers and acrylics, can either be dissolved into the liquid medium or added as latex. Plasticizers, such as phthalates and glycerol (for aqueous slips), can be employed to further increase the flexibility of the tape. Plasticizers are short chain molecules that are soluble in the binders and designed to reduce the modulus of the binders. Because of their low molecular weight, plasticizers are removed during thermal treatment at a lower temperature than the polymer binder. This sequence is helpful because pore channels are created as the plasticizer leaves, which allows easier volatilization of the binder. With the binders added, the viscosity of a suspension for tape casting is typically in the range of 0.5–20 Pa•s with shear thinning characteristics.

Once the coating is applied and dried, flexible green tape casts can optionally be removed from the substrate and fired in plates or sheets in a furnace. The heating step removes the binder and other organics remaining in the tape cast as well as sinters the particles together into a durable final product. High binder concentrations present challenges, such as large amounts of shrinkage during the sintering step, leading to possible curling or cracking.

Tape casting methods are also used to make multilayer devices, such as multilayer capacitors, fuel cells, batteries and multilayer ceramic substrates, and multilayer composite materials. These processes involve multiple depositions and multiple materials, including metals. Co-firing is frequently a challenge as the conditions for sintering of the ceramics and metals are not easily made compatible and thermal expansion mismatches lead to warping and other stress-induced defects.

6.5 EXTRUSION AND INJECTION MOLDING

6.5.1 Process Overview

In this section, extrusion and injection molding of concentrated dispersions is introduced. This brief section builds on techniques described primarily in Chapter 3 in the context of polymer melts. The main idea is to incorporate high loadings of ceramic or metal powder into a polymer melt, extrude or injection mold, and then remove the polymer and sinter the part. The "melts" are not designed to be functional polymers but rather convenient vehicles for the shaping process. As such they are designed to allow incorporation of ceramic or metal powder particles and for decomposition. Solvent can also be added. A main challenge is achieving a high solids loading and a low enough viscosity for processing. To this end, engineered particle size distributions are useful. In addition to polymer-based processes, traditional ceramic products are extruded from a highly loaded aqueous paste with little or no polymer added. These materials make use of the natural plasticity in clay-based ceramics.

After extrusion or molding is complete, removal of the volatiles and binders is typically a long process that scales with the size of the article. Ordinarily this "burnout" step is indeed a thermal process in which the polymers are pyrolyzed in controlled environments. Chemical debinding is also an option.

6.5.2 Extrusion of Concentrated Dispersions

In Chapters 3 and 4, extrusion was described as a process for polymer melts and solid metals. Extrusion is also a method to produce parts from a concentrated slurry or paste. This method is commonly used to produce ceramic brick, bathroom and roof tile, catalytic supports, and also tube-shaped ceramic parts such as furnace tubes and electrical insulators. Not exclusive to ceramics, metal powder can also be extruded to form high density metallic shapes. By starting with a metallic powder heterogeneous in composition, alloyed parts can be formed. Extrusion of pastes shares some features with melt extrusion (i.e., thermoplastic polymers) and others with solid extrusion (i.e., metals).

The stages of an extrusion forming process include first mixing the feed material and de-airing. The extruder feed material must be viscous and moldable when moving in the extruder, and yet when expelled as extrudate, it must hold its shape like a solid. Therefore, a major challenge of this type of extrusion is to achieve plasticity in a concentrated, particulate suspension or paste. There are at least two routes. The first is to use the natural plasticity that occurs in clay-water mixtures at appropriate moisture levels. Those familiar with pottery know that plasticity is achieved with a water content in between a fluid slip and a dry and crumbly mass. The "plastic" ceramic can be shaped by hand or on a potter's wheel. The plasticity is derived from thin layers of water on the clay platelets; under shear the platelets move relative to each other, but when the shear stress is removed, the water film and capillary effects hold the platelets in place. The second more universal route is to use polymer additives. In both routes, feed material can have up to 50 vol% particulates with the remainder water (or solvent) and binders. De-airing is carried out on its own, or the initial chamber in the extruder is designed for de-airing, possibly with a vacuum system. Because high concentration particulate slurries and pastes are often very viscous, removing bubbles is challenging but important for eliminating defects in the final part.

Once the extrudate is formulated, the material is then pressurized and forced through the extrusion die to make cylinders, hollow tubes, rectangular cross-sections, and even complex shapes like substrates and filters in automotive applications. See Figure 6.38. After exiting the extruder, the wet piece is dried, its surface optionally treated or smoothed, and, finally, organic binders are burned from the component and it is sintered to create the final form.

The most common extruder type for particulate slurries and pastes is the screw auger, which is essentially a single screw extruder (see Chapter 3). However, both piston press extruders, which forcefully ram material through a die opening, and twin screw extruders are used. Piston press extrusion is a slow, batch process, because the piston ram must be retracted between the formation of each part. However, the pressures that are possible with a piston press are much higher than those with auger extruders. Both co- and counter-rotating twin screw extruders have a higher capital cost than single screw augers, but require less power to move the material through the barrel. Also, as discussed in Chapter 3, the throughput of a single screw (auger) extruder is pressure dependent, whereas it is pressure independent in a twin screw extrusion system. Lastly, the extrusion process can be horizontal or vertical.

Metal and ceramic particles are typically hard and abrasive, which, when in concentrated slurries, create a lot of friction both on the extruder walls and between individual particles. When the auger applies pressure to such a slurry, the solid particles jam to form a stress-bearing network. The water or other fluid redistributes in this network, and can sometimes separate from the particle packing. This phenomenon is similar to what happens when you

FIGURE 6.38 Example of honeycomb shaped ceramic for environmental technology applications, such as diesel filters. *Courtesy of Corning, Inc. www.corning.com.*

step on moist sand at the beach. To prevent the migration of water, small hydroscopic clay particles are sometimes added to slurries to fill in spaces between larger ceramic materials to restrict flow through the pore space. Particulate fines are also useful for this purpose. Organic binders such as polyvinyl alcohol, methyl cellulose, or polyethylene glycol can also be added which serve to reduce interparticle friction, increase viscosity, and reduce liquid migration. Finally, coagulants like acids or bases can be helpful to provide the desired viscosity in the extruder and to encourage the formation of the desired stress-bearing solid network. These additives can occupy up to 20 vol% of the feed material.

As the high solids extrudate is pushed through the extruder, wall friction can increase the power requirements and cause heating and wear. Augers are made from high grades of steel or other suitably hard and durable metal. Sometimes small amounts of lubricant are injected at the walls of the barrel in order to better move the material. Finally, increasing the temperature of the material in the barrel may reduce friction by reducing the viscosity of the extrudate organics. Stress near the barrel walls can serve to align the particles, creating a unique microstructure at the edges of the part.

Flow profiles in the particulate dispersion extrusion formation process mirror those described in detail for polymer extrusion. However, most highly concentrated particulate systems have the added complication that the dispersions are shear thinning or have a measureable yield stress. If the die-entrance region is improperly designed for the material, a dead zone may appear where the shear stress during operation is not sufficient for the material to yield and flow forward. This problem can be corrected by increasing the taper of the die-entrance region, which increases the shear stress in the extrudate. Poiseuille flow through a rectangular or circular die is altered when the extrudate is shear thinning. Shear at the walls creates an effective lower viscosity there, which decreases the total pressure needed to push the material through the area.

6.5.3 Powder Injection Molding

Powder injection molding is the general term for the process of injection molding of polymer melt−based mixes containing a high loading of ceramic or metal powders. This process is similar to injection molding of thermoplastics (see Chapter 3) except that metal or ceramic particles are compounded with polymer binder into pellets before the process. The injection unit can be based on a reciprocating screw type or a piston. The molding process follows the same principles as thermoplastic injection molding. Parts are solidified by cooling in the mold, and then heated to remove binder and sinter.

Ceramic or metal powder incorporation into the polymer melt can be accomplished by high shear mixing in a twin screw extruder, for example. The powders used in this process have particles in the micron size range for ceramics and less than 20 μm or so for metals. Lubricants are also added to lessen wear of machine from contact with hard particles. At least two polymer binders are used. The major binder is a higher molecular weight polymer that is responsible for the solidification and strength of the molded part. The minor binder is a lower molecular weight polymer that enhances flow behavior and assists in the binder burnout process. The minor binder decomposes before the major binder, which clears out some pore channels and allows easier removal of the major binder. Plasticizer or solvent is also added to assist in flow. The mixes can have up to 50−70 vol% particles.

Post-processing is the rate limiting step in the powder injection molding. Decomposition and removal of large amounts of polymer from the parts requires time and ovens with atmosphere control. Of concern is retention of the shape of the parts considering that the polymers are thermoplastic. The time needed to remove the polymer increases with part size and hence powder injection molding applications are restricted to small parts. Given the high content of polymer there is considerable shrinkage during the polymer removal and then final sintering steps. Nonetheless, the advantage of creating complex shapes drives the use of this process.

6.6 LIQUID MONOMER-BASED ADDITIVE PROCESSES

6.6.1 Process Overview

Photocurable liquid monomers and resins naturally lend themselves to additive manufacturing. The conversion of liquid to solid is instigated by light and therefore one route is to selectively polymerize the liquid in a continuous liquid bath. The process of stereolithography (SLA—for stereolithography apparatus) is based on this premise. SLA uses a scanning laser to polymerize 2D layers; a moving platform and fresh layers of resin make it possible to build in the third dimension. Another route is to control the positioning of the liquid and illuminate broadly. This path is taken by an inkjet 3D printing method. Resin "ink" droplets are deposited and then illuminated to cure them into polymer. Again, a layer-by-layer approach creates the 3D object. This method lends itself to multiple materials and colors.

There are similarities and differences in the requirements for the photocurable liquid resins for these processes. The liquid itself must have a low enough viscosity so that it can be distributed onto the build platform or jetted into droplets easily. In addition, the inkjet process requires attention to the surface tension of the liquid. For both, fast curing is essential in order to increase the overall process speed. Moreover, for SLA in particular the curing process should not result in large amounts of shrinkage. For both, the resin and curing conditions should result in a final product with a required set of physical and mechanical properties. Other desirable features are insensitivity to atmospheric conditions, such as relative humidity and oxygen.

SLA and inkjet involve the same sort of preparation as other additive methods. The first step in is to create a CAD file of the part and convert the drawing to an STL format. The STL file is then used by the printer software to establish a sequence of scan patterns for the 2D slices of the 3D object.

It is worth noting here that there are additive manufacturing processes based on dispersions. In particular, lower viscosity dispersions can be inkjet printed directly and high viscosity dispersions can be extruded in processes that are similar to fused deposition modeling. In these processes, solidification is by drying, which can be encouraged using heated build plates. Of course, combining a curable polymer with particles is yet another variation. For the purposes of this section, however, the focus is on liquid monomer-based processes.

6.6.2 Stereolithography (SLA)

Stereolithography is a high-resolution additive process that uses photocurable liquid resin. The process begins with a thin layer of the liquid resin on a build platform. A laser scans over the layer to selectively polymerize a 2D pattern, a fresh layer of resin is added, and the process repeats to build a 3D object. SLA has a long history. The process was invented in 1986, in

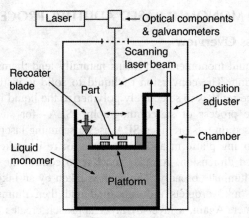

FIGURE 6.39 Schematic diagram of a stereolithography apparatus, showing the main components and analogies with selective laser sintering (see Chapter 5). Resin level height sensor is not shown. *Adapted from Venuvinod and Ma (2004).*

the same era as selective laser sintering (SLS) and fused deposition modeling (FDM).

Figure 6.39 is a schematic of an SLA system. The system consists of an enclosed chamber, containing a vat or bath of resin, as well a platform and positioning system. The enclosure prevents escape of resin fumes. Additionally, since photocurable polymers are used, the chamber is either opaque or tinted to prevent ambient light from starting the curing process prematurely. The other major pieces of equipment are the laser, and components that control the beam. Lasers are chosen based on the photoinitiator systems. Wavelengths in the 320–360 nm range are common.

In SLA, 3D parts are built layer-by-layer using a scanning laser beam. A build platform is immersed in a liquid resin bath; the height is adjusted so that it is very close to the free surface of the bath and only a thin layer of resin exists on the platform. The laser scans over the surface, creating a cured 2D pattern and adhering the part to the platform. The platform then moves down and a layer of resin floods the surface. To speed this process and establish a uniform layer thickness, a blade passes over the surface and meters the liquid in a manner similar to doctor blade coating (i.e., the layer thickness is ∼ ½ the gap between the blade and the previous layer). Other recoating strategies are also used. The scanning laser beam then creates the next 2D layer and adheres it to the previous. Supports are built at the same time. These extra pieces of the same polymer are needed to stabilize the part and for overhangs. An alternative setup locates the laser and optics beneath the platform and begins with the platform at the bottom of the vat. The laser scans through the transparent bottom, curing a layer and then the platform moves up and resin is pulled into the gap.

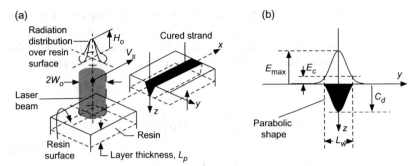

FIGURE 6.40 Schematic diagrams showing features of a laser line scan in a curing material. *From Venuvinod and Ma (2004) with kind permission from Springer Science and Business Media.*

The laser beam physics discussed in the last chapter as well as the introduction to UV curing are useful for understanding the SLA process. The laser beam spot has a Gaussian distribution in its intensity or irradiance, H. (In the context of UV curing, this parameter is equivalent to the intensity, I.) Considering the sketch of a scan track, the irradiance as a function of y is given by:

$$H = H_o \exp\left(\frac{-2y^2}{W_o^2}\right) \tag{6.65}$$

where H_o is the irradiance at the center of the scan ($y=0$) and W_o is the value of y at which $H = H_o/e^2$. See Figure 6.40a. W_o is a measure of the radius of the scanning beam. The power is found by integration. Considering an instant in time when the beam illuminates a circular cross-sectional area of radius y, the total power, P_L, provided is:

$$P_L = \frac{\pi}{2} W_o^2 H_o \tag{6.66}$$

The power at the resin surface is typically specified by the SLA manufacturer. Equation (6.66) can also be rearranged to provide a convenient expression for H_o.

To understand the depth and extent of cure, we need to find the exposure, which is the energy absorbed per unit area. (In the context of UV curing, this would be called the dose.) The exposure, E (also known as the energy density) at the surface ($z=0$) is found by integration of the irradiance:

$$E(y,0) = \sqrt{\frac{2}{\pi}} \frac{P_L}{W_o V_s} \exp\left(\frac{-2y^2}{W_o^2}\right) \tag{6.67}$$

where V_s is the scan velocity in the x-direction. The integration requires several steps and converting an integration over time to one over distance, using the definition of velocity. (Note that in the last chapter we did not consider the details of the Gaussian distribution in finding the energy density.) According to the Beer-Lambert law the irradiance, and hence the exposure, decreases with depth into the resin bath (i.e., with an increase in z).

$$E(y,z) = E(y,0)\exp\left(\frac{-z}{D_p}\right) = \sqrt{\frac{2}{\pi}}\frac{P_L}{W_o V_s}\exp\left(\frac{-2y^2}{W_o^2}\right)\exp\left(\frac{-z}{D_p}\right)$$

$$= E_{max}\exp\left(\frac{-2y^2}{W_o^2}\right)\exp\left(\frac{-z}{D_p}\right)$$

(6.68)

D_p, the depth of penetration, is the depth z at which $E(y,z) = E(y,0)/e$. Considering the terminology of UV curing, $1/D_p$ is equal to the product of the molar absorptivity of the photoinitiator and the concentration of the photoinitiator ($\alpha[PI]$). In other words, D_p is a constant for a given resin formulation. There is a critical exposure, E_c, needed for the liquid resin to solidify. This value is also a property of the resin material. The z and y values, where $E(y,z) = E_c$, define the dimensions of the solidified strand: C_d, cure depth, and L_w, line width, respectively:

$$C_d = D_p \ln\left[\sqrt{\frac{2}{\pi}}\frac{P_L}{W_o V_s E_c}\right] = D_p \ln\left[\frac{E_{max}}{E_c}\right]$$

(6.69)

$$L_w = W_o \sqrt{\frac{2 C_d}{D_p}}$$

(6.70)

From these relationships the required velocity for a scan to reach a certain cure depth can be found. See Figure 6.40b. Or, given information about the laser (P_L, W_o) and the resin (E_c, D_p), the geometry of the cured strand can be predicted for a given velocity. See Example 6.7. A complete derivation of these equations is found in Jacobs (1992).

The scan pattern involves overlap of adjacent tracks, resulting in increased exposure for the material in the overlap region. The spacing between adjacent strands or tracks is called the hatch spacing. The scan patterns are designed to create functional structures quickly. The cure depth is of particular importance as it relates to the layer of resin that is applied on the part. The layer thickness is less than the cure depth so that the laser penetrates to the previously deposited layer, creating a region of "overcure" that helps the layer-to-layer bonding. Due to the variable states of cure in the part, resins must have low shrinkage on cure to avoid strain and warping.

EXAMPLE 6.7 A resin designed for SLA has an E_c value of 10.1 mJ/cm², a D_p of 4.8 mil (0.122 mm), a viscosity of 370 mPa·s and density of 1.16 g/cm³. The SLA machine provides a power of 50 mW at the resin surface a beam diameter at $1/e^2$ of 250 μm. (a) From this information find the cure depth and line width as a function of scan speed as the speed varies from 0.1–1 m/s. (b) If the printer specifies a layer thickness of 0.10 mm, find the depth of the overcure region for a scan speed of 0.5 m/s

a. Using Eqs. (6.70) and (6.71) and recognizing that W_o is ½ of the $1/e^2$ diameter = 0.125 mm

$$C_d = D_p \ln\left[\sqrt{\frac{2}{\pi}} \frac{P_L}{W_o V_s E_c}\right]$$

$$= (0.122 \text{ mm}) \ln\left[\sqrt{\frac{2}{\pi}} \frac{0.1 \text{ W}}{(0.125 \text{ mm})(V_s[mm/s])(1.01 \times 10^{-4} \text{ J/mm}^2)}\right] =$$

$$= (0.122 \text{ mm}) \ln\left[\frac{6.32 \times 10^3}{V_s}\right]$$

$$L_w = W_o\sqrt{\frac{2C_d}{D_p}} = 0.125 \text{ mm} \left(\frac{2C_d[mm]}{0.122 mm}\right)^{1/2} = 0.506\sqrt{C_d}$$

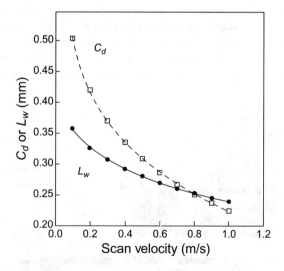

FIGURE E6.7

b. For a scan speed of 0.5 m/s, the C_d is 0.31 mm, which results in a overcure region that is ~0.2 mm deep.

SLA post-processing. Post-processing for SLA begins with raising the part out of the resin bath and allowing it to drain. Removal of uncured monomer then proceeds with physical wiping and rinsing with solvent. The part is solid at this point, but due to the fast scanning speed, it is not completely cured. Hence it is placed in a UV chamber to finish the process. Support structures are also removed at some point during this process.

SLA has some interesting features relative to other additive processes. The laser scanning technology combined with the precision of curing leads to high resolutions and high scan speeds. Layer thicknesses are on the order of 25–100 μm and laser spot location is precisely controlled on higher end systems (e.g., within 10 μm) though the spot sizes are larger than that 50–300 μm. Experimental microSLA is reducing sizes further. Moreover, the curing technology can also be adapted to exposing through a mask so that an entire 2D layer can cure at once. Other innovative techniques for ultrahigh resolution and fast speed are also under development.

6.6.3 Inkjet Printing with Liquid Monomers

Another liquid monomer-based additive process involves creating a 2D pattern by inkjet printing, curing the pattern and repeating the process to build a 3D object. Figure 6.41 shows an example of a inkjet printing setup designed for delivery of monomers and curing. Multiple printheads are mounted on a carriage that moves in the x-y plane. Each printhead contains multiple nozzles for ink delivery (i.e., the nozzle array extends in the y-direction in Figure 6.41 with a single ink reservoir feeding the set of nozzles so that each pass in the x-direction deliver ink over a wide area). An inline UV lamp cures the drops after they are deposited. Like other additive methods a 2D pattern is created and then the platform descends (z-direction) and the next

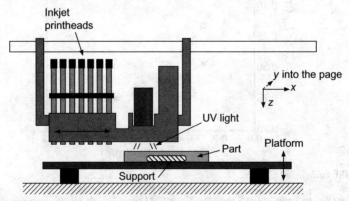

FIGURE 6.41 Schematic diagram of a multinozzle inkjet delivery system with inline UV curing. Cartoon show nozzles just after completing delivery with no ink dispensing. *Adapted Stratasys product information, www.stratasys.com.*

layer is printed. Supports are obviously needed in this process. The multiple printhead design accommodates this need as well as providing the means to print multiple materials and color in the final part. To understand more about this process, the fundamentals of inkjet printing are first explored.

Inkjet Printing Principles. Inkjet printing is a method for depositing small amounts of liquid onto surfaces at precise locations. In general, the printing process consists of generating discrete droplets and depositing them onto a surface. The printed liquid then solidifies, often by drying, but in the case of liquid monomer printing, by curing. This technology can create very small drops on the order of 10s of microns in diameter at a rate of 10s of kHz, with exact positioning of each drop. There are two main methods of inkjet printing. The first, termed "continuous inkjet printing (CIJ)," uses the Rayleigh instability of a liquid jet to generate drops of ink. The second method, drop on demand (DOD), produces single drops. Figure 6.42 shows schematics of both options.

Continuous inkjet printing produces a steady stream of drops through the Rayleigh instability of a falling stream of ink. The Rayleigh instability is the tendency of a perturbed stream of liquid to break up into drops. Any jet of liquid is slightly irregular in diameter and, as the liquid falls, the thin sections tend to become thinner and eventually pinch off, while the thick sections grow into drops. This instability is familiar in everyday life, for example in the breakup of a slow flow of water from a sink faucet. In CIJ,

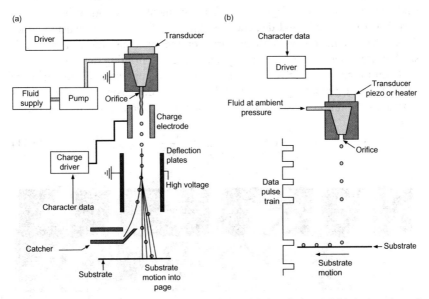

FIGURE 6.42 Schematic diagrams for (a) continuous ink jet printing and (b) drop on demand printing with the piezoelectric option. *Courtesy of MicroFab Technologies, www.microfab.com; MicroFab Technote 99-01 (1999).*

the liquid is perturbed with a sinusoidal pressure fluctuation while it is ejected from the nozzle to encourage the resulting drops to be regularly spaced and sized. Drops are slightly larger than the nozzle diameter, approximately 100 μm.

In continuous inkjet printing, drops are produced whether they are needed for printing or not. Therefore, drops are positioned either for printing or recycling as diagrammed in Figure 6.42a. This is done through electrostatics; each drop is charged as it falls and then its flight path is altered by deflector electrodes onto either the substrate or into a recycle stream. Drops that are not needed for the pattern or drops that are an abnormal size are deflected. Although drops are produced at high rates, typically only a small fraction are actually used for printing. The unused ink is recycled back to the nozzle, so as the printer operates the ink has the potential to become aged, contaminated, or dried, leading to printing defects.

By contrast, droplet on demand (DOD) printing creates only drops that are needed for the pattern. Here, drops are generated by a pressure pulse that forces just the desired amount of liquid from the nozzle when it is wanted. In bubblejet printers, the pressure pulse is created by locally heating and boiling a small amount of the ink solvent within the nozzle. The volume expansion of the vapor causes ink to be forced out of the nozzle. Pressure pulses can also be generated by using a piezoelectric transducer that mechanically deforms to squeeze the liquid out of the nozzle. The ejected drop, falling from the nozzle, remains connected to the fluid in the nozzle through a stretching ligament. See Figure 6.43. This ligament thins until it becomes so thin that it is unstable and detaches. As the drop falls, surface tension either draws the ligament back into the drop or the ligament forms a separate, smaller drop that falls independently and can cause printing defects.

Drop formation by DOD printers is bounded by operability limits that are defined by the Ohnesorge number, Oh, which is dimensionless number that represents the ratio of viscous forces to surface tension and inertial forces:

$$Oh = \frac{\eta}{\sqrt{\rho R \gamma_{LV}}} = \frac{\sqrt{We}}{Re} \tag{6.71}$$

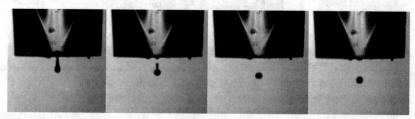

FIGURE 6.43 Images of a droplet emerging from a drop-on-demand nozzle captured at 25 microsecond intervals. The ink droplet is ~40 μm in diameter. *Courtesy of A. Mahajan and D. Barton, University of Minnesota.*

Here, η, ρ, and γ_{LV} are the ink viscosity, density, and surface tension, respectively, whereas R is the nozzle radius. The Ohnesorge number is a combination of the Weber number, We, where $We = v^2\rho R/\gamma_{LV}$ and the Reynolds number ($Re = v\rho R/\eta$). At high Oh, viscous dissipation prevents the drop from falling from the nozzle tip. At low Oh, small satellite drops tend to be present between the primary drops. For Newtonian inks, there is a process window that has been correlated to good inkjetting characteristics. If $1/Oh$ is less than ~1, then the viscosity of the liquid is too high, but if $1/Oh$ greater than ~10 then forming a single, distinct droplet is an issue. See Derby (2010). If the ink cannot be changed, this criterion requires that the drop size is the only variable parameter. Like in continuous inkjet printing, drop diameters naturally are sized on the same order as the nozzle diameter, with some variability that is possible by changing the pressure pulse magnitude and frequency. Typical nozzle diameters range from 30 to 100 μm. The highest resolution printers can deliver 10 picoliter drops (~13.2 μm drop diameter) and even 1 picoliter drops (~6.2 μm drop diameter).

DOD printing has a few advantages over CIJ printing. Because DOD printers do not require a recycle gutter to catch unwanted drops, the nozzle head can be placed more closely to the substrate (2–3 mm), which reduces the likelihood of air currents to misplace the drop. The lack of a recycle stream also reduces the chance of contamination or aging of the ink. Drop generation frequencies in DOD printers are an order of magnitude slower (1–20 kHz) than CIJ printers, but since every drop is used the two technologies actually are comparable in drop print rates. DOD drop sizes are smaller than those created with CIJ printing.

Another issue with inkjet printing is drop interaction with the substrate. Upon impact, a high energy, high velocity drop may recoil upwards before capillary forces bring the drop back to a spherical shape. This sort of splashing is discouraged by limiting the exit velocity. The impact event also starts the spreading of the drop on the substrate. Once the energy from the impact has dissipated, capillary forces remain, which bring the drop toward its equilibrium shape. This shape is determined by the equilibrium contact angle of the liquid with the surface, but other factors also influence the final feature size. In monomer printing, curing reactions increase viscosity and halt any shape changes. Multiple drops dispersed close together can merge if the curing is not fast enough. Other phenomena occur in droplets that solidify by drying.

Liquid Monomer Printing. Inkjet printing of liquid monomers is a synergistic combination of inkjet printing and photocurable resin technology. Starting with the requirement for inkjet printing first, we can estimate Oh using values typical of liquid monomers. As an example, trimethylolpropane triacrylate, has a surface tension of ~34 mN/m, viscosity of ~45 mPa·s, and a density of 1.1 g/cm^3. For a nozzle of radius 50 μm, Oh is ~1. Therefore $1/Oh$ is also ~1, which is right on the borderline for good droplet formation. With a smaller nozzle this monomer might not form droplets very well. Likewise, mild heating to reduce the viscosity would assist the process.

The next consideration is curing. As in the SLA process, the conditions should provide fast cure and an adequate cure depth. The effect of process parameters on curing and structure development can be inferred from the previous discussions of SLA and UV curing. The cure depth can be estimated in a similar way as that for SLA. In the case of monomer printing, however, the UV source is not a concentrated beam but rather a broader area and this broad UV source moves over the printed pattern at some fixed velocity. We expect the cure depth to increase as the velocity decreases and as the incident intensity from the UV lamp increases. Overcure for enhancing the layer-to-layer bonding is desired, as is printing patterns that creates good in-plane bonds. No post-curing is needed and the only additional step after a part is build is dissolution of the support.

One advantage of liquid monomer inkjet printing is the ease of printing multiple materials. At the simplest level, delivery of the monomer that forms the part and another material for the support. The supports also solidify by curing, but they are designed to be washed away in water after the printing is complete. More than two materials can be printed, making multiple colors and multiple physical properties possible. The method affords excellent resolution with layer heights $\sim 20\,\mu m$ and $\sim 40\,\mu m$ in plane resolution.

6.7 SUMMARY

Dispersions and solutions can be cast and molded into shapes and made into functional coatings. Liquid monomers are also useful starting materials for the fabrication of coatings and 3D objects by polymerization. The fundamental topics of importance to designing these processes include colloidal interactions and stability, polymer solution behavior, rheology, radiation curing, and drying. Creating a stable dispersion requires attention to particle-particle interactions. Van der Waals attraction tends to aggregate particles in dispersions, leading to gels and sediments; however, engineering repulsion between particles by electrostatic or steric phenomena allows suspensions to be stable over the time span needed for further processing. Creating a useful polymer solution requires attention to molecular weight and solvent choice. Solidification by drying and curing both involve kinetics and conditions such as temperature and intensity of UV light.

Shape casting from dispersions is an important process for creating ceramics in complex shapes. In slip casting a ceramic dispersion is poured into a porous mold. A layer of ceramic particles builds on the interior surface of the mold over time due to capillary action. The rate of casting depends on the solids loading and particle size in the dispersion as well as the mold characteristics. Predictions of casting time are necessary for engineering processes to create controlled thickness parts.

Coating processes create thin layers of dispersion, solution or liquid monomer that are then solidified by drying or curing. There are many coating application methods, including dip coating, blade coating, slot coating, and curtain coating. Coating processes require metering to ensure a controlled thickness. Pre-metered processes, such as slot coating, afford the best control as all of the coating liquid is deposited onto the substrate. Predictions of the effect of coating liquid properties, such as viscosity and surface tension, and coating speed on thickness allow processes to be engineered. In roll-to-roll processing, inline drying and curing occurs in series with the deposition. Hence, process speeds may be limited by the solidification equipment, such as the dryer length.

Ceramic and metal dispersions are also extruded and injection molded. Ceramic extrusion is carried out with thick pastes based on clays or dispersions containing polymers to assist in developing plasticity. Twin screw and single screw extruders operate on similar principles as polymer extruders to create extruded products. For injection molding, ceramic or metal particles are dispersed into polymer melts and then molded using techniques very similar to those used to mold thermoplastic polymers. Removal of the large amounts of polymer binder from these products is typically the rate-limiting step.

Liquid monomers are used in additive processes, including stereolithography and monomer inkjet printing. In stereolithography, the monomer is cured layer-by-layer with a scanning laser. Achieving high resolution and adequate bonding between layers requires attention to the laser energy density. The cure depth is a function of the laser specification, resin characteristics, and scanning speed. Liquid monomers have appropriate characteristics for inkjet printing. Scanning printheads deposit monomer droplets and UV light cures them. Multiple materials can be incorporated into the 3D printed object.

Chapter acknowledgment: Christine Roberts gratefully acknowledges helpful discussions with the late Professor L. E. Scriven as well as her coworkers at Sandia National Laboratories. Sandia is a multiprogram laboratory operated by Sandia Corporation, a Lockheed Martin Company, for the United States Department of Energy under contract DE-AC04-94AL85000.

BIBLIOGRAPHY AND RECOMMENDED READING

Colloid science
Evans, D.F., Wennerstrom, H., 1994. The Colloidal Domain. VCH Publishers, New York, NY.

Heimenz, P.C., 1977. Principles of Colloid and Surface Chemistry. Marcel Dekker, New York, NY.

Hunter, R.J., 1993. Introduction to Modern Colloid Science. Oxford Science Publications, New York, NY.

Larson, R.G., 1993. The Structure and Rheology of Complex Fluids. Oxford University Press, New York, NY.

Russel, W.B., Saville, D.A., Schowalter, W.R., 1989. Colloidal Dispersions. Cambridge University Press, Cambridge, UK.

Shaw, D.J., 1980. Introduction to Colloid and Surface Chemistry. Butterworth & Co, London, UK.

van Olphen, H., 1963. Introduction to Clay Colloid Chemistry. Wiley Interscience Publishers, New York, NY.

Polymer Solutions. Latex and Curing

Blackley, D.C., 1997. Polymer Latices: Science and Technology, second ed. Chapman & Hall, New York, NY.

Hiemenz, P.C., Lodge, T., 2007. Polymer Chemistry, second ed. CRC Press, Boca Raton, FL.

Keddie, J.L., 1997. Film formation of latex. Materials Science and Engineering 21, 101–170.

Macosko, C.W., 1994. Rheology: Principles, Measurements and Applications. Wiley - VCH, New York, NY.

Pappas, S.P., 1992. Radiation Curing: Science and Technology. Plenum Press, New York, NY.

Sperling, L.H., 1992. Introduction to Physical Polymer Science, second ed. Wiley Interscience, New York, NY.

Young, R.J., 1981. Introduction to Polymers. Chapman & Hall, New York, NY.

Slip Casting, Coating, and Tape Casting

Aksay, I.A., Schilling, C.H., 1984. Mechanics of colloidal filtration. In: Mangles, J.A., Messing, G.L. (Eds.), Advances in Ceramics, Vol. 9: Forming of Ceramics. American Ceramic Society, Columbus, OH, pp. 85–94.

Cohen, E., Gutoff, E., 1992. Modern Coating and Drying Technology. VCH Publishers, New York, NY.

Fries, R., Rand, B., 1991. Slip-casting and filter pressing. In: Brook, R.J. (Ed.), Concise Encyclopedia of Advanced Ceramic Materials. VCH Publishers, New York, NY, pp. 151–187.

Mistler, R.E., 1991. Tape casting. In: Schneider Jr., S.J. (Ed.), Engineered Materials Handbook, Vol. 4: Ceramics and Glasses. ASM International, Materials Park, OH, pp. 161–165.

Reed, J.S., 1995. Principles of Ceramic Processing, second ed. Wiley & Sons, New York, NY.

Schilling, C.H., Aksay, I.A., 1991. Slip casting. In: Schneider Jr., S.J. (Ed.), Engineered Materials Handbook, Vol. 4: Ceramics and Glasses. ASM International, Materials Park, OH, pp. 153–160.

Scriven, L.E., 2004. Coating Process Fundamentals Short Course Notes. University of Minnesota, Minneapolis, MN.

Stokes, R.J., Evans, D.F., 1997. Fundamentals of Interfacial Engineering. VCH Publishers, New York, NY.

Tormey, E.S., Pober, R.L., Bowen, H.K., Calvert, P.D., 1984. Tape casting – future developments. In: Mangles, J.A., Messing, G.L. (Eds.), Advances in Ceramics, Vol. 9: Forming of Ceramics. American Ceramic Society, Columbus, OH, pp. 140–149.

Extrusion and Injection Molding

Handle, F. (Ed.), 2007. Extrusion in Ceramics. Springer-Verlag Publishing, Berlin, Germany.

Mangels, J.A., Trela, W., 1984. Ceramic components by injection molding. In: Mangles, J.A., Messing, G.L. (Eds.), Advances in Ceramics, Vol. 9: Forming of Ceramics. American Ceramic Society, Columbus, OH, pp. 220–233.

Mutsuddy, B.C., 1991. Injection molding. In: Schneider Jr., S.J. (Ed.), Engineered Materials Handbook, Vol. 4: Ceramics and Glasses. ASM International, Materials Park, OH, pp. 173–180.

Ruppel, I., 1991. Extrusion. In: Schneider Jr., S.J. (Ed.), Engineered Materials Handbook, Vol. 4: Ceramics and Glasses. ASM International, Materials Park, OH, pp. 165–172.

Monomer-based Additive Processes

Basaran, O.A., Gao, H., Bhat, P.B., 2013. Nonstandard inkjets. Annu. Rev. Fluid Mech. 45, 85–113.

Derby, B., 2010. Inkjet printing of functional and structural materials: Fluid property requirements, feature stability, and resolution. Annu. Rev. Mater. Sci. 40, 395–414.

Gibson, I., Rosen, D.W., Stucker, B., 2010. Additive Technologies. Springer, New York, NY.

Jacobs, P.F., 1992. Rapid Prototyping and Manufacturing: Fundamentals of Stereolithography. McGraw Hill, New York, NY.

Martin, G.D., Hoath, S.D., Hutchings, I.M., 2008. Inkjet printing — the physics of manipulating liquid jets and drops. J. Phys. Conf. Ser. 105, 1.

Venuvinod, P., Ma, W., 2004. Rapid Prototyping: Laser-Based and Other Technologies. Kluwer Academic Publishers, New York, NY.

CITED REFERENCES

Adachi, K., Fujihara, I., Ishida, Y., 1975. Diluent effects on molecular motions and glass transition in polymers: I. Polystyrene-toluene. J. Polym. Sci. Polym. Phys. Ed. 13, 2155–2171.

Aksay, I.A., Schilling, C.H., 1984. Mechanics of colloidal filtration. In: Mangles, J.A., Messing, G.L. (Eds.), Advances in Ceramics, Vol. 9: Forming of Ceramics. American Ceramic Society, Columbus, OH, pp. 85–94.

Basaran, O.A., Gao, H., Bhat, P.B., 2013. Nonstandard Inkjets. Annu. Rev. Fluid Mech. 45, 85–113.

Brandrup, J., Immergut, E.H., Grulke, E.A., 1999. Polymer Handbook, fourth ed. Wiley Interscience, New York.

Buss, F.S., Crawford, K., Roberts, C.C., Peters, K., Francis, L.F., 2011. Effect of soluble polymer binder on particle distribution in a drying particulate coating. J. Colloid Interface Sci. 359, 112–120.

Derby, B., 2010. Inkjet printing of functional and structural materials: fluid property requirements, feature stability, and resolution. Annu. Rev. Mater. 40, 395–414.

Evans, D.F., Wennerstrom, H., 1994. The Colloidal Domain. VCH Publishers, New York, NY.

Graessley, W.W., 1980. Polymer chain dimensions and the dependence of viscoelastic properties on concentration, molecular weight and solvent power. Polymer 21.3, 261.

Hellebrand, H., 1991. Tape casting. In: Brook, R.J. (Ed.), Concise Encyclopedia of Advanced Ceramic Materials. VCH Publishers, New York, NY, pp. 190–265.

Hunter, R.J., 1989. Foundations of Colloid Science, vol. I. Oxford University Press, New York, NY.

MicroFab Technote 99-01, 1999. Background on ink-jet technology. <www.microfab.com>.

Richerson, D., 1992. Modern Ceramic Engineering. Marcel Dekker, New York, NY.

Roberts, C.C., 2010. *Understanding Particulate Coating Microstructure Development*, Ph.D. Thesis. University of Minnesota, Minneapolis, MN.

Romero, O.J., Suszynski, W.J., Scriven, L.E., Carvalho, M.S., 2004. Low-flow limit in slot coating of dilute solutions of high molecular weight polymer. J. Non-Newtonian Fluid Mech. 118, 137–156.

Russel, W.B., Saville, D.A., Schowalter, W.R., 1989. Colloidal Dispersions. Cambridge University Press, Cambridge, UK.

Stokes, R.J., Evans, D.F., 1997. Fundamentals of Interfacial Engineering. VCH Publishers, New York, NY.

Venuvinod, P., Ma, W., 2004. Rapid Prototyping: Laser-Based and Other Technologies. Kluwer Academic Publishers, New York, NY.

Weise, G.R., Healy, T.W., 1975. Coagulation and electrokinetic behavior of TiO_2 and Al_2O_3 colloidal dispersions. J. Colloid Interface Sci. 51, 427–442.

Yates, D.E., Healy, T.W., 1980. Titanium dioxide-electrolyte interface. J.C.S. Faraday I 76, 9–18.

QUESTIONS AND PROBLEMS

Questions

1. Why are van der Waals forces strong in particulate dispersions but only weak as a bonding mechanism in solids (i.e., compared to covalent and ionic bonding)?
2. Can electrostatic forces also lead to aggregation? Explain.
3. What factors related to the particles control the electrostatic repulsion? What factors related to the liquid medium control electrostatic repulsion?
4. What is an electrical double layer?
5. What is meant by "colloidal stability"?
6. What are the main factors that determine polymer solubility in a solvent?
7. How is the rheology of a colloidal dispersion similar to the rheology of a polymer solution?
8. What factors limit the solids loading in colloidal dispersions and solutions?
9. What are intrinsic viscosity and relative viscosity?
10. If a drying process is controlled by external factors, what does that mean? Internal factors?
11. How can the drying rate be increased early and late in the drying process?
12. What sorts of structure result from monomer curing? How are they the same and different from structure resulting from drying of a polymer solution?
13. Is the extent of cure always spatially uniform in UV cured materials?
14. List the steps in a slip casting process.
15. What factors control the thickness of the cast layer as function of time? Why is it frequently valid to ignore the resistance of the mold?
16. List several coating operations and determine if they are pre-metered.
17. How is a thin coating created in the dip coating process and why is the process not used widely in manufacturing of coating on continuous webs?
18. Why is slot coating a more controlled process compared to doctor blade coating?
19. What is latex film formation?

20. What are the chief advantages and disadvantages of powder injection molding?
21. In stereolithography, what is the importance of the Gaussian intensity distribution or irradiance?
22. What factors control the cure depth and line width in stereolighography?
23. What is the Ohnesorge number and how does it predict inkjet printing characteristics?
24. What are the advantages of monomer inkjet printing as a method of additive manufacturing?

Problems

1. (a) Consider a 1 μm diameter silica particle in water. Plot the distance traveled by sedimentation as a function of time and the distance traveled by Brownian motion as a function of time. Make your plot on the same graph. Assume T = 25°C. (b) Repeat for particles 10 μm and 0.1 μm in diameter. Comment on the effect of particle size on the relative influence of gravity and Brownian motion.
2. (a) Consider an aluminum oxide suspension in water at 20°C. Find a particle size below which Brownian diffusion dominates for a time scale of 1 day. State assumptions. (b) Repeat for a temperature of 50°C. Comment on differences. (c) Repeat for a suspension in ethylene glycol at 20° C. Comment on differences
3. Consider a suspension of particles (isoelectric point is at pH 6) in water at pH 2 and a NaCl concentration of 0.001 M. Describe how the strength of repulsion varies with the following changes, assuming all other conditions remain constant. Give a description (more than just increase or decrease) in terms of the effect on the double layer thickness and the zeta potential. (a) Change from 0.001 M NaCl to 0.1 M NaCl. (b) Change from 0.001 M NaCl to 0.001 M $CaCl_2$. (c) Change from pH = 2 to pH = 5. (d) Change from 25°C to 50°C. (e) Change from water to methanol.
4. A silica suspension (40 vol% silica, average particle size = 0.5 μm) in water is prepared at pH = 5 and 25°C. Assume 0.0001 M NaCl and a surface potential is 25 mV. (a) Plot the potential energy of attraction as a function of separation distance. (b) Calculate the double layer thickness. (c) Plot the potential energy of repulsion as a function of separation distance. (d) Plot the total interaction energy as a function of separation distance. (e) Do you think this suspension is stable (not prone to aggregation)?
5. A dispersion is made containing 50 g of alumina powder (average particle diameter = 0.5 μm) and 50 g of water. The pH of the dispersion is adjusted with HNO_3 to a pH of 3.5. (a) Calculate the double layer

thickness for dispersions having three different a concentrations of $CaCl_2$: (i) 0, (ii) 0.001 M and (b) 0.01 M. Plot the total interaction potential vs. separation distance for the alumina dispersions. Assume a surface potential of 50 mV. Comment on the stability of these dispersions.

6. Consider two silica particles (spherical, radius $R_p = 10$ μm) with water at their contacts, as shown in Figure P6.6. The radius R_n is 0.1 μm. The distance x is the distance between the center of the particle and the center of the radius of curvature at the ring of water at the neck. Drawing is not to scale. Calculate and compare the van der Waals force assuming a 1 nm separation distance (with water intervening) with the capillary force of adhesion. Assume room temperature and complete wetting ($\cos\theta = 1$). *Hint:* The principal radii of curvature are R_n and $x-R_n$. Also see Chapter 3. Comment.

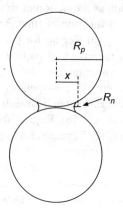

FIGURE P6.6

7. (a) Will the wetting of water on glass be improved as temperature increases? See Chapter 3 for discussion of wetting. State rationale for your answer and assumptions. (b) The capillary pressure pulling liquid A into a porous mold is twice that of liquid B. Liquid A has a surface tension 50 mJ/m^2 and a contact angle of 0° and liquid B has a surface tension of 73 mJ/m^2. What is the contact angle for liquid B?

8. Consider a stabilized aqueous suspension of alumina (pH << isoelectric point, IEP) has volume fraction of particles of 0.4, a maximum volume fraction of particles of 0.6 and intrinsic viscosity (shape factor) of 3.5, and an average particle size of 0.4 μm. (a) Calculate the viscosity of the suspension. (b) Graph the cast layer thickness as a function of time (use excel or other program). Assume the packing fraction in the cast is equal to the maximum packing fraction, Φ_m, the resistance of the cast layer is 10^{17} m^{-2}, the pressure drop from the mold is 140 kPa, and the mold resistance is much less than the cast layer resistance.

9. Consider your answer to problem 8. How would the time needed to form a 1 mm thick cast layer change (qualitatively relative to the original suspension) if: (a) the volume fraction of particle in the slip was 0.2, (b) the suspension has a pH near to the IEP, (c) 5 wt% polyvinyl alcohol is dissolved in the water, and (d) the average particle size of the alumina decreases to 0.1 μm.

10. An aqueous alumina slip containing 45 volume% alumina particles is cast into a gypsum mold. The ceramic particles are monodisperse with a size of 1 μm and they are well-stabilized. The as-cast layer is 65% of theoretical density. The gypsum mold has a specific resistance of 2×10^{10} cm^{-2} and a void fraction of 0.5. The total pressure drop from the slip to the saturation interface in the mold is 0.1 MPa. Plot the cast layer thickness as a function of time. Predict the time needed to cast a 1 mm thick layer and the position of the saturation front in the gypsum mold at that time

11. Polystyrene (PS) coatings are to be prepared on a polyester web using a blade coating apparatus like that used in tape casting. Two coating liquids are available: (1) polystyrene (MW = 100,000 g/mol) in toluene at a concentration of 15 wt% and (2) polystyrene latex particles in water at a concentration of 40 vol%. The coater has a land length $(L) = 1$ mm, a gap of 300 μm and the web speed is 100 m/min.
    (a) Estimate the viscosity of the suspension. State assumptions.
    (b) Estimate the as-deposited and final coating thickness. Assume that the latex coalesces completely. (c) Compare the advantages and disadvantages of making the coating from latex as compared with a solution of PS in an organic solvent.

12. (a) Find an expression for the wet thickness of a blade coated polymer solution, which is shear thinning and follows a power law expression with n and K as power law exponent and consistency index, respectively. (b) Find an expression for the dry thickness of a blade coated polymer assuming that the polymer solution has a polymer volume fraction of 0.1. State other assumptions.

13. Consider the preparation of a 5 μm thick polyvinyl alcohol (PVA) coating from deposition of a 7.5 vol% aqueous solution. Design processes based on dip coating and blade coating. Use the data in Figure 6.16 (lower M polymer) and assume that the surface tension of the solution is ½ that of pure water. Which method would you use and why?

14. (a) Using your answer to question 13, specify the length of dryer needed for the coating process you selected. Assume that the process is "roll to roll" with a drying rate of 20 μm of thickness loss per minute in the constant rate period.

15. Research the cell casting process used to make polymethylmethacrylate (PMMA, plexiglass) sheets. (a) List the advantages and disadvantages of cell casting and continuous cell casting. (b) Describe the typical

formulation of "syrup" used to make PMMA sheets. Why is it not just the monomer and initiator? (c) Polymers are rarely cast into thick parts or shapes. Why?

16. Compare ceramic and polymer extrusion. Include (a) the starting materials for the processes, (b) the feeding process, (c) the effect of rheological characteristics on flow, and (d) shape retention.

17. An acrylate monomer (trimethylopropane triacrylate) is mixed with a photoinitiator with a concentration of 0.01 mol/liter. At the UV wavelength of interest, the molar absorption coefficient is 1.1×10^4 liters/(mol·cm). (a) What is the fractional transmitted light intensity (Eq. (6.43)) through a coating that is 50 μm thick? 100 μm thick? (b) What is the total dose of UV delivered (J/m^2) for an incident intensity is 1 W/cm^2 and illumination time of 1 s?

18. Consider problem 17 from the perspective monomer inkjet printing. Assume the same monomer and photoinitiator are used, droplet create a monomer thickness of 20 μm and the conditions in 17(b) result in adequate curing. What intensity of UV source is needed to achieve a printhead speed of 100 m/min? State assumptions.

Chapter 7

Vapor Processes

By Bethanie Joyce Hills Stadler

7.1 INTRODUCTION

Vapor processes are used to make thin films that range in thickness from 1 atomic or molecular layer up to several microns. The starting materials for vapor processes are bulk solids, powders, or chemical precursors. These starting materials are also known as source materials or targets, depending on the process. These materials are vaporized, transported through the gas phase, and then deposited onto a substrate. Control of the gas phase is essential, and hence these processes are carried out in chambers with pressures controlled through the use of vacuum pumps. Vacuums are used to reduce undesired reactions and maintain purity. Substrates are typically any flat surface: a glass slide, Si wafer, computer hard disk, or an automobile window.

There are two broad categories of vacuum processes that depend on the atomization technique: physical vapor deposition (PVD), such as evaporation or sputtering, and chemical vapor deposition (CVD). Figure 7.1 shows schematics of the three vapor processes, illustrating how the vapor phase is created and deposited onto a substrate. In evaporation (Figure 7.1a), the source material is evaporated to form the vapor phase. By contrast, sputtering (Figure 7.1b) involves high-energy ions that knock atoms off of a "sputtering target," which is typically a disk about 0.25 inches thick. In chemical vapor deposition (Figure 7.1c), powders or other chemical precursors are evaporated or reacted to form vapors that flow into a vacuum chamber for subsequent chemical deposition. Central to all of these processes are four steps: (i) selection and preparation of starting materials and substrate, (ii) conversion of the starting materials to the vapor phase, (iii) transport of the vaporized materials and/or precursors to the substrate, and (iv) deposition onto the substrate. Additionally, the as-deposited thin films can be refined and structured by post-processing steps, such as annealing and patterning.

Thin films of ceramics, metals, and polymers typically are too small to see, but they enhance our lives in integrated electronic circuits, optical devices,

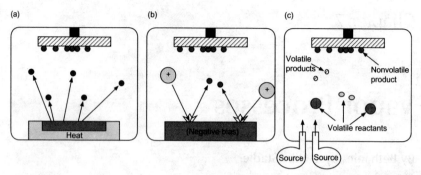

FIGURE 7.1 Schematic diagrams of the three most common vapor processes where atoms are collected on a substrate to obtain thin films: (a) evaporation, (b) sputtering, and (c) chemical vapor deposition. Vacuum components (pumps and gauges) are not shown.

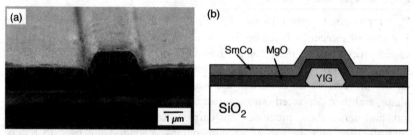

FIGURE 7.2 Yttrium iron garnet (YIG) waveguide device (photo courtesy S-Y. Sung) and schematic cross-section. A YIG film was sputtered onto a glass (SiO_2) substrate and then patterned to leave one line, which was to be used as an optical waveguide (trapezoidal cross-section). The line was annealed to ensure the garnet phase was present. Next, MgO, then Sm-Co films were sputtered onto the sample. Typical columnar grains are especially visible in the Sm-Co film.

magnetic sensors, and computer memories. Thin films or coatings can also be deposited as uniform layers that provide protection of functionality. For example, hard ceramic coatings are deposited on drill bits and other cutting tools. Alternatively, vapor deposited thin films are building blocks of complex circuits and devices. Figure 7.2 shows an example of how patterning and multiple depositions are used to create devices.

Ceramics have high melting points, which makes evaporation more difficult than sputtering. However, with recent advances in e-beam evaporation, it is possible to deposit most materials. In addition, ceramics are compounds with volatile components, such as oxygen or nitrogen, so it can be difficult to obtain stoichiometric films unless a separate gas source is used together with the bulk material source. Sputtering and CVD have been used extensively to produce oxide and nitride films that are too small (nm to μm thick) to make by bulk processing methods. A relatively new vacuum processing method, called atomic layer deposition (ALD), is particularly suited to deposition of atomically thin compounds, even over complex surfaces.

For metal films, evaporation and sputtering are equally appropriate choices. Many metals have low melting points and low binding energies, so they can usually be evaporated and sputtered 10 times faster than oxides. CVD, however, is not typically useful for metals, which tend to react to form compounds. Some refractory metals can be deposited using CVD for use as protective and wear-resistant coatings.

The deposition of polymer thin films by vapor phase processes is less practiced. High molecular weight polymers degrade on heating, which precludes evaporation as a method. However, low molecular weight organic semiconducting materials, such as pentacene ($C_{22}H_{14}$), are easily formed by evaporation. Sputtering processes have been developed for some polymers, such as Teflon, but these processes are complex, involving reactions in the plasma phase. Overall, traditional polymers are deposited as thin films or coatings by liquid phase processes involving solutions and dispersions and only rarely with vapor phase processes.

A key advantage of vapor processing is atomic control of the resulting material. Source materials are made in high purity and structure development is tuned through controlling conditions in the processing chamber. To form continuous thin films, the atoms or small clusters of reacted atoms, must reach the substrate with sufficient energy to diffuse. Also, the surface energies between the vacuum, substrate, and film must be balanced for "wetting" conditions. To form nanostructures, or islands, "non-wetting" conditions are needed so that the film "beads up" as atoms arrive at the substrate (similar to water droplets on a waxed car). In most cases, conditions are between these two extremes and small, columnar grains form as the film grows. This surface energy balance is important in all three vapor processes.

Before discussing the specifics of the three types of vapor processes, this chapter covers the fundamentals that all of the processes share. These include the transfer of atoms across a vacuum via the kinetic theory of gases and typical film microstructures. Under special conditions, single crystal films can be grown by epitaxy. Details on the specific processes of evaporation, sputtering, and chemical vapor deposition follow. Finally, post-processing procedures are discussed, including annealing to obtain either densification or a phase change and patterning, which removes part of the film to make device structures such as the one shown in Figure 7.2.

7.2 FUNDAMENTALS

7.2.1 Kinetic Theory of Gases and Its Relationship to Vapor Processes

This section demonstrates that there are two major regimes for vacuums: viscous and molecular. The *viscous* regime occurs at low vacuums, starting just below ambient (1 atm). Under viscous conditions, the vacuum behaves like a low density gas with molecules colliding with each other quite often. The *molecular* regime occurs at ultra-high vacuum (UHV) conditions, for example 10^{-9} atm. Here, molecules collide almost only with the walls of the processing chamber and gas interactions are rare. The delineation of regimes varies with the size of the processing chamber and other parameters, so pressure conditions at the boundary of viscous and molecular regimes are called "moderate" vacuums. Logically, processes that occur in these two regimes require very different control parameters. So, understanding the vacuums themselves is important to the materials that are made by vacuum processes.

Relating the Macroscopic to the Microscopic. Early observations of gases necessarily were macroscopic. From the combined work of Boyle, Charles, Gay Lussac, Boltzmann, and Avogadro, it was found that the pressure, P, volume, V, and temperature, T, of an ideal gas are related:

$$PV = nRT \tag{7.1}$$

where n is the number of moles in the sample and R is the gas constant (8.314 J/mol K). This expression is known as the ideal gas law. Relating these macroscopic observations to atomistics can begin by recognizing that

$$R = kN_A \tag{7.2}$$

where k is Boltzmann's constant (1.381×10^{-23} J/(molecule•K)) and N_A is Avogadro's number (6.02×10^{23} molecules per mole). In fact, atomistic models of gases often use a different form of the ideal gas law:

$$P = n_v kT \tag{7.3}$$

where n_v is the number of molecules per unit volume (nN_A/V) in the sample.

To connect the atomistic view of gases to the macroscopic parameters above, the kinetic theory of gases was developed. This theory provides an understanding of any gas in a processing chamber, providing a basis for evacuating air molecules from the chamber, measuring pressure, and transporting and collecting the desired atoms as the films grow. The basic assumptions involved in the kinetic theory of gases are threefold. First, gases are assumed to be a collection of atoms or molecules with diameters much smaller than distances between them. Second, gases involve a constant state of random motion with elastic collisions and no other interaction between molecules. Third, the number of molecule-molecule and molecule-wall collisions depends on concentration, n_v, as well as pressure (which depends on V and T).

To begin, consider a single molecule with mass, m, that has elastic collisions at the walls of a confining cube with dimensions, l. See Figure 7.3. The molecule is moving at a velocity, v, with respect to the walls. When it strikes the wall at $x = l$, it reverses its velocity in the x-direction v_x, but the

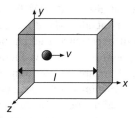

FIGURE 7.3 Schematic of a gas molecule as it transverses a cube between elastic collisions with the cube's walls.

components in the y and z directions, v_y and v_z, respectively, remain constant. The momentum change, Δp, perpendicular to the wall is:

$$\Delta p = p_{\text{final}} - p_{\text{initial}} = (-mv) - (mv) = -2mv \tag{7.4}$$

The momentum imparted to the wall is then $+2mv$. If the molecule crosses the cube without collisions in time $t = l/v_x$ and strikes the opposite wall, then it strikes the first wall again in twice that time. Therefore, the number of collisions with the wall at $x = l$ per unit time is:

$$\text{Number of collisions per unit time} = \frac{v_x}{2l} \tag{7.5}$$

and the rate of momentum transfer R_{momentum} to that wall is:

$$R_{\text{momentum}} = (2mv_x)\left(\frac{v_x}{2l}\right) = \frac{mv_x^2}{l} \tag{7.6}$$

Now consider a box with N gas molecules. The total force on a wall by a gas is the sum of the rate of momentum transfer for all molecules in that gas (1, 2, 3, ...) and pressure is force/area, where area (l). Therefore:

$$P = \frac{m(v_{x1}^2 + v_{x2}^2 + v_{x3}^2 + \ldots)}{l^3} = \frac{Nmv_{x,\text{ave}}^2}{l^3} \tag{7.7}$$

assuming that all of the molecules have the same mass. By defining density, ρ, as the total mass of molecules, $N \cdot m$, divided by volume, l^3, then

$$P = \rho v_{x,\text{ave}}^2 = \frac{1}{3}\rho v_{\text{ave}}^2 \tag{7.8}$$

since $v^2 = v_x^2 + v_y^2 + v_z^2$ and $v_{\text{ave}}^2 = 3v_{x,\text{ave}}^2$ if there is no preferred direction. It is now clear how the kinetic theory of gases can be used to relate a microscopic parameter, such as molecular velocity, to a macroscopic property, such as pressure.

Mean Free Path and Defining Vacuum Regimes. Now, one can ask: what is the result when gas molecules collide? Given that gases are defined as having elastic collisions, two colliding molecules would simply each have equal but opposite momentum transfer, and the resulting effect on the walls would be the same. This result also holds true for all shapes of containers. It is interesting to note that in a vacuum where the gas density is very low, the pressure on any surface will experience fluctuations. This is important, for example, when measuring pressures at high vacuum conditions.

So, the next question is: how are the vacuum regimes defined? Recall at the start of this section low vacuums were defined as the "viscous regime" and high vacuums were defined as the "molecular regime." Although it may seem like simple nomenclature, it is an important distinction for determining the necessary pumps, pressure gauges, and processing parameters for all vacuum processes. The key parameter is the mean free path of the gas molecules, or the average distance between collisions, l_{mfp}. The parameter l_{mfp} is proportional to the size of the molecule, as defined by the diameter, d, and the number of molecules per volume, n_v. Collisions occur when the sphere centers are separated by d, as shown in Figure 7.4a. To analyze atomistic collisions, we can consider one molecule having diameter of $2d$ and area πd^2 and allow all of the other molecules in the chamber to be points (see Figure 7.4b).

In time t, our "molecule" moves a distance vt, passing through a cylinder with volume $(\pi d^2)(vt)$. This cylinder contains $(\pi d^2 vt)n_v$ molecules, and therefore as many collisions occur in that time. The mean free path to a collision is the total distance traveled in time t divided by the number of collisions in that time:

$$l_{mfp} = \frac{vt}{\pi d^2 vt n_v} = \frac{1}{\sqrt{2}\pi d^2 n_v} \tag{7.9}$$

The factor $\sqrt{2}$ comes from the fact that the molecules in the cylinder are not stationary, so the velocity in the numerator, which is relative to the container, is not the same as the velocity in the denominator, which is relative to the other molecules. In fact, the latter is zero if the molecules are moving in the same direction and twice the v in the numerator if they are moving toward each other. The average difference in velocities is $\sqrt{2}$.

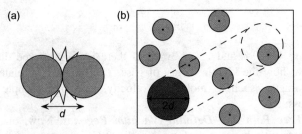

FIGURE 7.4 (a) Schematic of collisions between gas molecules with center to center separation of d. (b) A molecule with a diameter $2d$ collides with any other molecule whose center, noted by a dot, is inside the swept volume, shown with dashed lines.

One final piece of information that is important for solving problems in this field is a table of units. With diverse units, such as pressure in Pa or Torr or atmospheres (atm) and the gas constant in J/(mol•K), it is important to go back to the basic System International (SI) units and consult conversion tables, such as Table 7.1.

Pressures inside vacuum processing chambers, such as those shown in Figure 7.1, are typically around 10^{-6} Torr. At this pressure, the mean free path of air is on the order of meters. Since air is mostly nitrogen, a similar calculation as Example 7.1 can be used to verify this value, although more accurately air is a mixture of molecules with an average diameter of 4×10^{-8} cm. Such a long mean free path means that in a processing

TABLE 7.1 Pressure Conversions and Useful Constants

Conversions	Constants
1 atm = 760 Torr (mm Hg) = 14.7 psi = 101,323.2 Pa (N/m²) = 1013.232 mbar	$N_A = 6.022 \times 10^{23}$ molecules/mol
1 Pa = 7.5 mTorr	$R = 8.314$ J/(mol•K) = 0.08206 liter•atm/(mol•K)
1 Torr = 133.3 Pa	$k = 1.381 \times 10^{-23}$ J/K

EXAMPLE 7.1 Calculate the mean free path to a collision for nitrogen at room temperature and 1 atm pressure, assuming an average molecule diameter of 3.1 Å. What is the frequency of collisions if the average velocity of nitrogen at ambient conditions (room temperature and 1 atm) is 5.1×10^4 cm/s?

The ideal gas law can be used to find the number of molecules per volume and then the mean free path can be calculated:

$$n_v = \frac{P}{kT} = \frac{PN_A}{RT} = \frac{(1 \text{ atm})}{(0.08206 \text{ liter} \cdot \text{atm/mol} \cdot \text{K})(300 \text{ K})} \frac{6.02 \times 10^{23} \text{ molecules/mol}}{1000 \text{ cm}^3/\text{liter}}$$
$$= 2.45 \times 10^{19} \text{ molecules/cm}^3$$

$$l_{mfp} = \frac{1}{\sqrt{2}\pi d^2 n_v} = \frac{1}{\sqrt{2}\pi(3.1 \times 10^{-8} \text{ cm})^2(2.45 \times 10^{19} \text{ molecules/cm}^3)}$$
$$= 9.6 \times 10^{-6} \text{ cm}$$

Although this seems small, it is 310 diameters for the molecules. The collision frequency is $v_{ave}/l_{mfp} = 5.3 \times 10^9 \text{ s}^{-1}$, which means there are more than 5 billion collisions every second.

chamber with dimensions of m³, there are almost no intermolecular collisions despite having 300 billion air molecules present!

This is how we answer the question asked earlier: the difference between low and high vacuum conditions (viscous and molecular regimes) depends on the mean free path to collisions. There is a transition that occurs between 10^{-3} and 10^{-5} Torr where gases cease to flow like viscous fluids and rather behave as single molecules. A unitless number, called the Knudsen number, can be used to determine the vacuum regime of a processing chamber:

$$Kn = D/l_{mfp} \qquad (7.10)$$

where D is a characteristic dimension of the chamber. This definition, however, is not universal. Some texts prefer to invert the terms. However, in all texts, when the mean free path is smaller than the chamber dimensions ($l_{mfp} < D$), the chamber is in the viscous regime (low vacuum), which corresponds to $Kn > 1$ according to the definition in Eq. (7.10). By contrast, when the mean free path is longer than the chamber dimensions ($l_{mfp} > D$), the chamber is in the molecular regime UHV, which corresponds to $Kn < 0.01$ according to the definition in Eq. (7.10). Again, there is a transition region of "moderate" vacuums, this time $0.01 < Kn < 1$.

Distribution of Speeds and Molecular Flux. Various assumptions have been made so far about the speed of molecules in a gas. These molecules actually have a wide variety of speeds as they collide with each other and exchange kinetic energy. However, it is very unlikely that any molecule is sitting at rest ($v = 0$) or that it is infinitely fast. Rather the distribution of gas particle velocities can be expressed by the Maxwell-Boltzmann equation, which gives the number of molecules, N, at a given velocity relative to the total number of molecules in the system, N_{Total}:

$$\frac{N(v)}{N_{Total}} = 4\pi \left(\frac{m}{2\pi kT}\right)^{3/2} v^2 \exp\left(\frac{-mv^2}{2kT}\right) \qquad (7.11)$$

where m is the mass of each molecule. If molecular weight, M, and the gas constant, R, are used instead of the m and k, the same equation gives the *moles* of particles at a given velocity.

The distribution along each coordinate can similarly be expressed by a distribution, for example:

$$\frac{N(v_x)}{N_{Total}} = \left(\frac{m}{2\pi kT}\right)^{1/2} \exp\left(\frac{-mv_x^2}{2kT}\right) \quad \text{in the } x\text{-direction} \qquad (7.12)$$

In calculations requiring "typical" velocity, such as 5.1×10^4 cm/s for air in Example 7.1, the root mean square velocity of the Maxwell-Boltzmann equation is used:

$$v_{rms} = \sqrt{\frac{3RT}{M}} \qquad (7.13)$$

The ideal gas law (Eq. (7.3)) emerges when this velocity is substituted into Eq. (7.8):

$$P = \frac{1}{3}\rho v_{ave}^2 = \frac{1}{3}\left(\frac{n_v M}{N_A}\right)v_{ave}^2 = \frac{1}{3}\left(\frac{n_v M}{N_A}\right)\frac{3RT}{M} = n_v kT \quad (7.14)$$

Looking at the Maxwell-Boltzmann distribution for air in Example 7.2, it is clear that although a typical molecule has the velocity 510 m/s (or 5.1×10^4 cm/s), there is a wide distribution of velocities at room temperature. In fact, only 35% of the air molecules have velocities between 410 and 610 m/s (found by integrating the distribution over this range).

The Maxwell-Boltzmann distribution can be used in calculations of important parameters, such as the number of molecules that impinge on a surface perpendicular to any direction (x) per unit time, Φ, which can be found by integrating the Maxwell−Boltzmann distribution (fraction of particles at each velocity in that direction, v_x) over all velocities (m/s) times the total number of particles n_v (molecules/m^3).

$$\Phi = n_v\sqrt{\frac{RT}{2\pi M}} = \frac{N_A P}{\sqrt{2\pi MRT}} \quad [\text{molecules}/(\text{m}^2 \cdot \text{s})]$$

or
$$\frac{\Phi}{N_A} = \frac{P}{\sqrt{2\pi MRT}} \quad [\text{moles}/(\text{m}^2 \cdot \text{s})]$$

(7.15)

where M is molecular weight. This expression makes use of the ideal gas law in place of n_v. The term Φ is called molecular flux or impingement. We all experience molecular flux daily as molecules strike our skin (Example 7.3). Have you ever held your hand against the input of your vacuum cleaner? The "force" that seems to pull on your skin is actually the *lack* of impingement that your skin needs to hold you together!

This parameter of molecular flux, Φ, is critically important to vacuum processes as it represents the number of molecules that strike any surface, such as:

1. The inlet of a vacuum pump (Φ of air molecules determines evacuation speed)
2. The surface of a sample that is being made (Φ's of desired atoms and impurities determine both deposition rate and impurity content of a sample, respectively)

Table 7.2 summarizes the discussion of vacuum fundamentals thus far. The vacuum regimes have been discussed above with the viscous regime defined as low vacuum, where molecules collide with each other such that the vacuum acts as a low density gas with viscous flow. The molecular regime occurs at UHV. Here the molecules have long mean free paths such

EXAMPLE 7.2 Plot the Maxwell–Boltzmann distribution of molecular velocities for three gases at room temperature, including air which is a mixture of gases with an average molecular weight of 29 g/mol.

Choose velocities, for example 0–1000 m/s with increments of 10 m/s. Calculate and plot Eq. (7.11) at each velocity using a spreadsheet or MatLab. Be careful to use kg/mol for molecular weight if you are using R in J/(mol•K). For example,

$$\% \text{ air molecules at 510 m/s} = \frac{N(510)}{N_{Total}} \times 100\%$$

$$= 4\pi \left(\frac{(0.029 \text{ kg/mol})}{2\pi(8.314 \text{ J/mol•K})(300 \text{ K})} \right)^{3/2} (510 \text{ m/s})^2$$

$$\exp\left(\frac{-(0.029 \text{ kg/mol})(510 \text{ m/s})^2}{2(8.314 \text{ J/mol•K})(300 \text{ K})} \right) \times 100\%$$

$$= 0.181$$

We've plotted air, Ar and Xe (typical sputtering gases), and Au at room temperature. These curves demonstrate the dependence of molecular velocities on mass. Au is plotted at two temperatures to show the dependence of the molecular velocities on temperature. This example would be applicable to Au that has been atomized at 700 K by evaporation, but then the Au atoms slow down as they travel to the substrate, which is at room temperature.

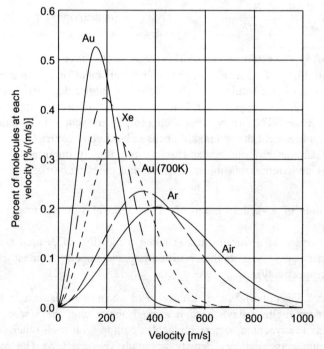

FIGURE E7.2 Distribution molecular velocities for ambient air and several elements, as well as Au at 700 K.

EXAMPLE 7.3 How many air molecules are striking the surface of an average human every second? (approximate an average human as a 5'9" tall, 1'4" diameter cylinder). Recall that air is a mixture of gases and has an average molecular weight of 29 g/mol.

First, we'll calculate the approximate area of this cylindrical human and convert inch^2 to m^2:

$$r = 16''/2 = 8'' = 0.203 \text{ m} \quad \text{and} \quad h = 69'' = 1.75 \text{ m}$$

$$A = 2\pi r h + 2\pi r^2 = [2\pi(0.203 \text{ m})(1.753 \text{ m}) + 2\pi(0.203 \text{ m})^2] = 2.49 \text{ m}^2$$

Next, Eq. (7.15) is used to find the impingement, making sure to use Pa for pressure. The unit Pascal "Pa" is equivalent to N/m^2 or kg/(m·s^2) and a Joule "J" is equivalent to N·m or kg·m^2/s^2.

$$\Phi = \frac{PN_A}{\sqrt{2\pi MRT}} = \frac{(101{,}323 \text{ Pa})(6.02 \times 10^{23} \text{ molecules/mol})}{\sqrt{2\pi(0.029 \text{ kg/mol})(8.314 \text{ J/mol}\cdot\text{K})(300 \text{ K})}}$$

$$= 2.86 \times 10^{27} \text{ molecules/m}^2\cdot\text{s}$$

$$\text{Units: } \frac{(\text{kg/m}\cdot\text{s}^2)(\text{molecules/mole})}{\sqrt{\frac{\text{kg}^2\cdot\text{m}^2}{\text{mol}^2\cdot\text{s}^2}}}$$

$$\frac{\text{Number of molecules}}{\text{s}} = \Phi A = (2.86 \times 10^{27} \text{ molecules/m}^2\cdot\text{s})(2.49 \text{ m}^2)$$

$$= 7.12 \times 10^{27} \text{ molecules/s}$$

So, there are $\sim 7 \times 10^{27}$ molecules of air bombarding you and I every second!

---

that in most processing chambers ($\sim$m^3 in dimension), the molecules collide mostly with the processing chamber walls and rarely with each other. Moderate vacuums are defined as transition pressures between the two regimes. Recall that the Knudsen number was also used to delineate these regimes according to processing chamber size, but for most chambers the delineation in the table applies. The molecular flux upon a surface perpendicular to any direction is distinct from mean free path in that it is calculated using the distribution of velocities present in a gas to find the number of molecules that strike the surface per area per time.

"Monolayer formation time" is often used to estimate impurity content in films as the average time for one monolayer of air to form on a surface. Impurities from air include nitrogen, oxygen, and water, which can be detrimental to metal films. Being volatile molecules, they do not "stick" to surfaces well, but if they react with a metal film, the chemical bond enables them to stick, and the metallic properties of the film are degraded. The deposition rate of a thin film should be orders of magnitude greater than the

TABLE 7.2 Characteristics of Vacuums[a]

Designation	P (Pa)	P (atm)	P (Torr)	Molecular density (#/cm^3)	Mean free path (cm)	Molecular flux (#/cm^{2}s)	Monolayer formation time (s)
Low vacuum (viscous)	101300	1	760	2.45×10^{19}	5.75×10^{-6}	2.86×10^{23}	2.18×10^{-9}
	1013	10^{-2}	7.6	2.45×10^{17}	5.75×10^{-4}	2.86×10^{21}	2.18×10^{-7}
	10.13	10^{-4}	0.076	2.45×10^{15}	5.75×10^{-2}	2.86×10^{19}	2.18×10^{-5}
	1.013	10^{-5}	0.0076	2.45×10^{14}	0.575	2.86×10^{18}	2.18×10^{-4}
Moderate vacuum	0.133	1.3×10^{-6}	10^{-3}	3.22×10^{13}	4.37	3.76×10^{17}	1.66×10^{-3}
	0.0133	1.3×10^{-7}	10^{-4}	3.22×10^{12}	43.7	3.76×10^{16}	1.66×10^{-2}
	0.00133	1.3×10^{-8}	10^{-5}	3.22×10^{11}	437	3.76×10^{15}	0.166
Ultra high vacuum (molecular)	0.00013	1.3×10^{-9}	10^{-6}	3.22×10^{10}	4370	3.76×10^{14}	1.66
	1.3×10^{-5}	1.3×10^{-10}	10^{-7}	3.22×10^{9}	4.37×10^{4}	3.76×10^{13}	16.6
	1.3×10^{-6}	1.3×10^{-11}	10^{-8}	3.22×10^{8}	4.37×10^{5}	3.76×10^{12}	166

[a] At room temperature, where air is considered a mixture of atoms with $M_{ave} = 29$ g/mol and $d_{ave} = 4$ Å.

monolayer formation time in a given process in order to obtain high purity films. The monolayer formation time is estimated as equal to $1/\Phi d^2$, where d is the average diameter of air molecules (4×10^{-8} cm). A more precise calculation could be done using the packing density of molecules, which will vary with material and temperature. Also, the atoms in air are volatile so that not all of the atoms that impinge on a surface will stick to it. We'll introduce a "sticking coefficient" later in the deposition processes, but the calculated monolayer formation time gives an order of magnitude estimate.

Another less obvious application of the molecular flux or impingement rate is the removal of atoms from a surface, such as the evaporation of atoms off a heated source. Vacuum processes are considered nonequilibrium processes because the vapor is not in equilibrium with the source material. Instead, the atoms that are driven from the source, especially in the case of evaporation, become gas molecules with velocities given by the Maxwell-Boltzmann distribution. If evaporation is done in a processing chamber that is evacuated to the molecular regime, the atoms will collide almost exclusively with the walls of the chamber, including the substrates, where they attach due to the relatively lower temperatures (e.g., room temperature).

Obtaining and Measuring Vacuum. In lay terms, any space that has less than atmospheric pressure ($\sim 10^{19}$ molecules/cm^3) is a vacuum because it will suck in air if an opening is introduced to ambient. And yet, a vacuum is also generically thought to be the total absence of air or molecules in a space. After the discussion above, it should be obvious that there is always a finite number of molecules present even in UHV.

Figure 7.5 shows a standard vacuum system, which is required for further description of vacuum pumps and gauges. In the simplest system, the source material and substrate on which the thin film is to be grown are placed in the main chamber before evacuation to high vacuum (Figure 7.5 (1)). To achieve high vacuum conditions, the system first has to reach a moderate vacuum ($Kn \sim 1$) via a positive displacement pump, such as a mechanical pump or roots pump (Figure 7.5 (2)). These pumps are designed to force the viscous flow of many molecules out of the system and are also called roughing pumps because they are used to obtain a rough (or moderate) vacuum. These pumps are also used to extract most of the air molecules out of the UHV pump. Next (Figure 7.5 (3)) the UHV pump is turned on and allowed to reach equilibrium. The mechanisms of various UHV pumps are described after this short overview of the pump-down process. After reaching equilibrium, most UHV pumps are ready to begin removing molecules from the main chamber, which takes 1–12 h, depending on the desired final base pressure of the system (Figure 7.5 (4)). A critical feature of this process is that the UHV pump cannot be exposed to atmospheric pressures once it is up and running at equilibrium, and valves at the input and output of the UHV pump are essential for protection of this pump against exposure.

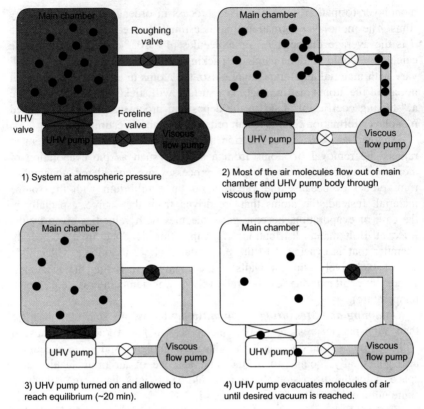

FIGURE 7.5 Schematic of a typical evacuation sequence for a processing chamber. At least two pumps and three valves are required. Open valves are white and closed valves dark gray. Most critical is to not expose UHV pump to atmosphere once it is running. Substrates and source materials (Figure 7.1) are not shown to help visualization of evacuation.

Various types of pumps and pressure gauges are listed in Figure 7.6. Viscous flow, or roughing, pumps are used in the viscous flow regime. They allow the particles to flow into the pump inlet, after which the gas molecules are compressed by mechanical motion into a smaller chamber, which is then rotated to an outlet. The compressed (high pressure) particles flow out and the pump continues to rotate to allow more air to flow in the inlet.

UHV pumps only work in the molecular flow regime because their molecule removal mechanisms cannot handle large molecular fluxes. First, diffusion pumps use exactly the same physical phenomenon as the large church-basement coffee brewers: percolation. Oil, rather than water, is heated at the bottom of the pump where a flat plate restricts the upward motion of the heated oil. Instead the oil evaporates, and the vapor is only allowed to rise through a tube in the center of the pump where it bounces off conical plates, causing it to jet downward. Any air molecules from the

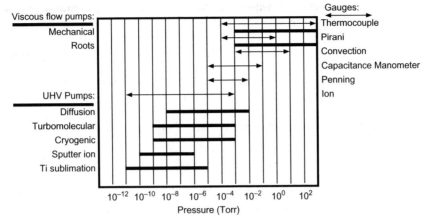

FIGURE 7.6 List of typical vacuum pumps and gauges, showing their useful pressure ranges as thick lines and arrows, respectively. As described by Figure 7.5, a processing chamber goes from atmosphere to moderate vacuum ($>10^{-3}$ Torr) on its way to ultra-high vacuum ($<10^{-6}$ Torr), so typically two pumps (one from each category) and several vacuum gauges are used in the evacuation process

processing chamber that have reached the pump inlet are caught in the downward motion by elastic scattering with the oil molecules. There is an effective pressure build up by the outlet, and the air escapes from the system through the viscous flow pump. The oil, however, condenses and drips back to the bottom of the diffusion pump where the process repeats. The viscous flow pump at the outlet of the UHV pump, called the foreline pump, can be the same pump that is used to "rough" the processing chamber, as shown in Figure 7.5, or it can be a separate viscous flow pump. The second type of UHV pump on the list in Figure 7.6 is also a positive displacement pump, the turbomolecular pump. This pump is much like a turbo engine on a jet, and in fact, it has a similar frequency, so it can sound like a quiet jet engine when starting and stopping. This pump simply spins, and air molecules that impinge on its opening are deflected through a series of spinning blades toward the pump outlet where a foreline pump again provides an escape. The next three UHV pumps work by entrapping the air molecules onto surfaces that are nanostructured to have large surface areas. These pumps differ in their mechanisms of entrapment. Cryogenic pumps freeze the molecules, and sputter ion and Ti sublimation pumps both deposit thin layers of titanium into which the air molecules become incorporated.

Pumping Speeds and Conductance. When it comes right down to it, time is money. Companies choose pumps that intrinsically make their processes run faster, but it is also essential that the processing chamber and plumbing are designed for optimal speed.

During evacuation, the pressure in a processing chamber will decrease exponentially with time. To see why, we will introduce a new parameter called "gas throughput," Q, with the following definition and relationship to the pump speed, S_p (in m³/s or liters/s).

$$Q = \frac{-d(VP)}{dt} = -V\frac{dP}{dt} = S_P P \quad [\text{Pa} \cdot \text{m}^3/\text{s}] \tag{7.16}$$

$$\frac{dP}{P} = -\frac{S_P}{V} dt$$

Integration of this relationship gives an "ideal" pressure at time, t, as

$$\frac{P(t)}{P_{\text{initial}}} = \exp\left(\frac{-S_P t}{V}\right) \tag{7.17}$$

However, in reality there is a limit to the minimum pressure that any pump can achieve. This can be accounted for as a minimum pressure, $P_p = Q/S_p$:

$$\frac{P(t) - \frac{Q}{S_P}}{P_{\text{initial}} - \frac{Q}{S_P}} = \frac{P(t) - P_p}{P_{\text{initial}} - P_p} = \exp\left(\frac{-S_P t}{V}\right) \tag{7.18}$$

Both P_p and S_p are specified by pump manufacturers. See Figure 7.7 as an example of pressure as a function of evacuation time in the viscous regime.

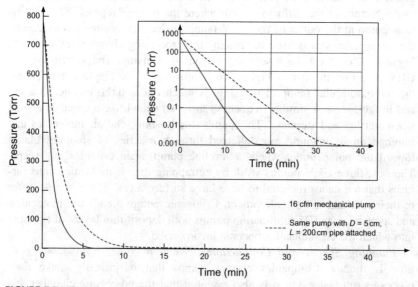

FIGURE 7.7 Typical evacuation times for the viscous flow regime using manufacturers specifications ($P_p = 1$ mTorr, $S_p = 16$ cfm) for the pump compared to evacuation times where plumbing is used to reduce noise in the processing lab. The conductance of the plumbing was equal to the pump speed ($C = S_p$).

Another parameter to consider when designing a vacuum processing chamber is the "Ohm's Law of Evacuation," where gas throughput, Q, is related to the pressure drop, ΔP, by a conductance, C:

$$Q = C\Delta P = C(P_{\text{final}} - P_{\text{initial}}) \tag{7.19}$$

Here the pressure drop is similar to an electrical voltage drop, Q is analogous to electrical current, and C is analogous to electrical conductance. The units on the conductance, C, are the same as pump speed, namely volume per time (e.g., m³/s). The conductance is a measure of the gas flow and is therefore be related to the molecular flux, Φ, which has unit of molecules per time per area (e.g., molecules/(m²·s)) by $C = \Phi A/n_v$.

If we consider an aperture, or opening, with area, A, between two vacuum chambers, the conductance is the molecular flux, Φ, impinging the aperture from the first chamber (Φ_1) minus that from the second chamber (Φ_2) multiplied by the area. If the first chamber is the processing chamber and the second "chamber" is a viscous flow pump, then Φ_2 is negligible when calculating C. Employing Eq. (7.15) the conductance in volume per second is:

$$C = \frac{(\Phi_1 - \Phi_2)A}{n_v} = \frac{PN_A}{\sqrt{2\pi MRT}} A \frac{RT}{N_A P} = A\sqrt{\frac{RT}{2\pi M}} \text{ (cm}^3\text{/s)} \tag{7.20}$$

See Example 7.4. The new parameter of gas throughput is most important to viscous pumping speeds. Viscous regime pumps are called "roughing pumps" because their intrinsic pump speeds are fast, and they evacuate most of the molecules from the processing chamber. For example, a 1 cm² aperture has $Q = A \cdot 11.7 \cdot (760 \text{ Torr}) = 8892$ Torr·liter/s.

However, once in the molecular flow regime of vacuum, the UHV pump speeds rely on less frequent impingement of molecules on their inlet (or aperture), and high vacuum pump speeds are intrinsically slow. In this regime, plumbing design is critical, or pump times would be prohibitively slow. Any additional components between the pump and the aperture on the chamber, such as the UHV valve shown in Figure 7.5, could substantially slow down the evacuation of the processing chamber. For example, the

---

EXAMPLE 7.4 Using Eq. (7.20), find the conductance of an aperture of area A for air at room temperature. Recall that air is a mixture of gases and has an average molecular weight of 29 g/mol.

$$C = A\sqrt{\frac{(8.314 \text{ J/mol} \cdot \text{K})(300 \text{ K})}{2\pi(0.029 \text{ kg/mol})}} (100 \text{ cm/m})$$

$$= A \cdot 11{,}700 \text{ cm}^3/\text{s} \quad \text{or} \quad A \cdot 11.7 \text{ liter/s} \quad (A \text{ in cm}^2)$$

aperture to a high-vacuum pump with area of 1 cm² only has a $Q = 11.7 \times (10^{-3} – 10^{-8})$ Torr•liter/s $= 11.7 \times 10^{-3}$ Torr•liter/s.

Since viscous pumps are typically loud, they will often be installed outside the laboratory walls in a chase with a long pipe connecting them to the processing chamber inside the lab. The conductance for a pipe is $C_{pipe} = 12.2 \, D^3/L$ [liter/s] where D and L are both given in cm. The conductances of an interconnected network of plumbing components (e.g., apertures, pipes) add in a similar manner as electrical conductances:

$$\text{Series:} \quad \frac{1}{C_{Total}} = \sum \frac{1}{C_i} \quad (7.21)$$

$$\text{Parallel:} \quad C_{Total} = \sum C_i \quad (7.22)$$

where i is each component between the vacuum chamber and the pumping mechanism. If a long pipe is used to connect the roughing pump to the processing chamber for sound control, the gas throughput could be cut substantially, but this roughing stage (#2 in Figure 7.5) is still significantly faster than the UHV evacuation stage (#4 in Figure 7.5) (Example 7.5).

For viscous pumps, which are relatively fast, this would make little difference. An example is plotted in Figure 7.7. Here, a 200 cm-long, 5 cm-diameter pipe is used to move the viscous pump into a chase outside the processing lab. The minimum pressure for the viscous pump (1 mTorr) is reached in less than 40 min with the pipe, instead of 20 min if the pump was directly attached to the processing chamber. However, a high vacuum pump with a low intrinsic gas throughput could take 5 h to reach a desired vacuum (from 1 mTorr to 10^{-6} Torr), but if it is connected to the chamber by a relatively short pipe (15 cm) it would take much longer. Figure 7.5 shows how a high

---

EXAMPLE 7.5 A processing chamber has an aperture with area 100 cm² connected directly to a pump. According to Example 7.4, this aperture has a conductance of 1,170,000 cm³/s (1170 liter/s). If the pump is attached to the chamber using a pipe with an area of 100 cm², determine what length pipe would cut the conductance to the pump by half.

The pipe would be in series with the aperture on the processing chamber. Therefore, the total conductance would be

$$\frac{1}{C_{Total}} = \frac{1}{1170} + \frac{1}{12.2 \cdot \frac{D^3}{L}} = \frac{2}{1170}$$

$$L = \frac{12.2 \left(\frac{400}{\pi}\right)^{3/2}}{1170} = 15 \text{ cm}$$

Therefore, a pipe with length 15 cm would cut the conductance in half.

vacuum pump is typically connected to the vacuum system—a direct connection with only a high vacuum valve between the pump and the system.

So, how can conductances be used to plot realistic evacuation times, as in Figure 7.7? Using Eq. (7.19) where P_p is the minimum pressure of the pump:

$$Q = C(P - P_p) = C\left(\frac{Q}{S} - \frac{Q}{S_P}\right) \quad (7.23)$$

then

$$\frac{1}{C} = \left(\frac{1}{S} - \frac{1}{S_P}\right) \text{ or } S = \frac{S_P}{1 + \frac{S_P}{C}} \quad (7.24)$$

This is the value of S that should be used to find a realistic evacuation time, as in Figure 7.7, including piping and other factors:

$$P(t) = (P_{\text{initial}} - P_{\text{final}})\exp\left(-\frac{St}{V}\right) + P_p \quad (7.25)$$

In the example of Figure 7.7, typical mechanical pump manufacturing specifications were used, namely $P_p = 1$ mTorr and $S_p = 16$ cfm (cubic foot per meter) = 7.5 (liter/s). The pump speed, S, was then cut in half by the pipe since $C = S_p$.

Once the molecular regime is reached, the evacuation time is dependent on many factors, such as adsorbed water slowly outgassing (desorbing) from the chamber walls into the vacuum and leaks (Figure 7.8). This water comes

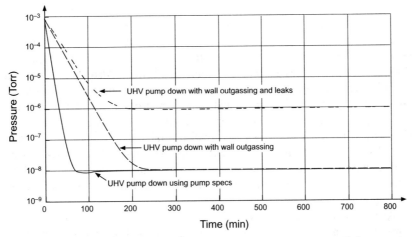

FIGURE 7.8 UHV pressure versus time curves showing three cases: an unrealistic case using Eq. (7.18) with typical S_p for a UHV pump (100 liter/s), a more realistic case where $S = 30$ liter/s to account for water outgassing from the chamber, and a pessimistic case in which the chamber has leaks (perhaps from O-rings being used rather than metal gaskets).

from the humidity in the air, and many processes will be behave differently in the summer than winter if the building climate is not well controlled. For this reason, the walls of a vacuum chamber are often heated to accelerate the evaporation of water and reach base pressures sooner. Each chamber also has leaks that occur with varying severity depending on the fixtures of the chamber and the desired base pressure. Special fixtures must be used if desired pressures are less than 10^{-6} Torr, such as metal gaskets instead of rubber O-rings. For pressures less than 10^{-8} Torr, even more specialized seals and gaskets are required. A liquid nitrogen cold trap can be added to a system to accelerate trapping of molecules, especially water from the air.

7.2.2 Thin Film Microstructures

Once the atoms arrive at the substrate, there are typical features in the microstructures of the resulting thin films made by any of the three techniques described in this chapter. There is a special case, called epitaxy, discussed in the next section, but otherwise most thin films have the microstructures discussed in this fundamental section due to the way they are "grown" from the bottom up. That is, atoms are collected on a substrate, and depending on the temperature, they move varying distances by surface diffusion before the next layer of atoms arrives to "freeze" them in place. Only if the temperature is quite high can atoms below the surface layer rearrange by bulk diffusion. Therefore, the parameters affecting the microstructure of the films involve a balance of shadowing, surface diffusion, and bulk diffusion.

In 1977, Thornton published a now famous diagram that relates the microstructure of polycrystalline films to this balance, Figure 7.9. Note that the temperature axis is normalized by the melting point of the film material. Therefore, the actual substrate temperatures needed to obtain specific film microstructures varies for each film material. For example, a substrate temperature of 700°C yields Zone 2 microstructures for a Cu film, but Zone 1 or Zone T microstructures for an aluminum oxide film. Also the "grains" in thin films are columns of atoms as they arrange on the substrate, but they are not always crystalline. In Zone 1 and Zone T microstructures especially, the dome-topped columns may have very small diameters, which appear amorphous by X-ray diffraction, but are small crystals when observed by transition electron microcsopy (TEM). The atoms in each column may also be void of any long-range atomic order, but the term "grain" is still used by most researchers as the columns resemble polycrystalline materials, even when amorphous.

At low pressures and temperatures, surface diffusion is too slow to overcome shadowing effects (Zone 1). Just like a shadow is cast behind an object in a directional light source, any roughness on a substrate is amplified by atoms sticking to peaks with "shadows" forming behind each grain. This

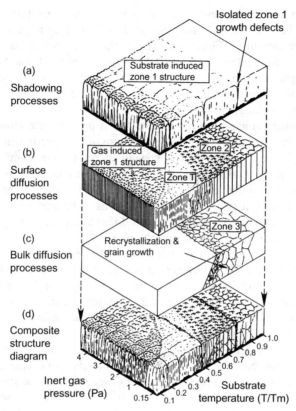

FIGURE 7.9 The Thornton diagram shows the film microstructures that are likely to result for given growth parameters, such as pressures and temperatures. *From Thornton (1977), reproduced by permission of Annual Reviews of Materials Science.*

leads to open boundaries between deposited islands of material and rough film surfaces.

With increased substrate temperatures, surface diffusion becomes possible because atoms have thermal energy to overcome the barrier to diffusion. Therefore, Zone T microstructures form on smooth substrates at low pressures and increased temperatures. These films are fibrous with dense boundaries, and they may possess more bulk-like properties than Zone 1 microstructures. For this reason, Zone T microstructures are often desirable, and methods such as substrate bombardment or radiation have been developed to encourage this structure. However, increased pressures (such as Ar in sputtering) increase the likelihood of voids due to gas adsorption.

At temperatures above approximately one-third to one-half of the melting point of the depositing material (Zone 2), surface diffusion dominates over shadowing, and the resulting microstructure is columnar with platelet- or whisker-like grains and dense grain boundaries. There are three ways that the balance between surface energies and the diffusion of the atoms impacts Zone 2 microstructures. First, simple nucleation and growth, including surface energy considerations, determine the grain size of the films. Second, the crystallographic orientation of the grains is determined by the faster growing surfaces. Third, columnar grains may have facetted tops with low-energy crystallographic planes.

Zone 3 microstructures occur in films when the substrate temperature is above three quarters of the melting temperature of the specific film material. At these high temperatures, there is enough energy for the atoms within the grains to move by bulk diffusion. Zone 3 microstructures, therefore, have equiaxed grains, similar to bulk, that are much larger than the grains found in the other Zones. It is more common for films to be post-annealed if bulk diffusion is desired, however, because it is difficult to provide very high temperatures inside a vacuum processing chamber.

Heat can be provided to a chamber by an electrical heater or by a lamp. As shown in Section 7.3.4, however, it is a good idea to rotate the substrate while the film is growing to minimize the shadowing that occurs in Zone 1, and to a lesser extent in Zones T and 2. Electrical feedthroughs into a vacuum chamber are difficult with rotating parts. When using heating lamps, there must be a window through which IR and other desired light can travel into the vacuum to heat the growing film. Or, the lamp can be located inside the chamber, but then it is likely to be coated at the same time as the substrate, so it needs to be cleaned often. For evaporation and sputtering, most substrate heaters are limited to 500°C. In these processes, sensors and process-related geometries (e.g., line-of-sight atomic deposition) make it impractical to heat the whole chamber just to have a heated substrate. In CVD, however, reaction temperatures must be very accurate and homogeneous, and the source materials can be evenly distributed over substrates with control of processing gas flow. So tube furnaces are often used where the processing chambers are quartz tubes and higher temperatures can be obtained.

The thermodynamics and kinetics of growth tie all of the Zone microstructures together. Atoms first arriving at the substrate are called "adatoms." Adatoms form nuclei, which can grow if the temperatures allow atoms to overcome the energy barriers of diffusion before the next monolayer of adatoms is deposited on the substrate. To begin, consider the *homogeneous* nucleation rate, I. In Chapter 3, we found a nucleation rate in number of nuclei per volume (of the parent phase) per time. This nucleation rate is the product of the concentration of critically sized nuclei, n_c, per volume and the frequency of atom addition to the nucleus, ν. Here, the concentration of critically sized nuclei is the concentration of sites available, n_s, multiplied by the probability that impinging atoms can thermally overcome the energy barrier of nucleation, ΔG^*. The frequency of atom addition to the nucleus is the product of the surface area of the spherical nucleus, $4\pi r^2$, a "sticking" coefficient, α, and the molecular flux (atoms per area per time), according to Eq. (7.15). The sticking coefficient varies from 0 to 1 depending on the volatility of the arriving atoms. For example, at high temperatures oxygen and nitrogen will have low sticking coefficients, and at low substrate temperatures evaporated metal atoms (from a hot source) will have $\alpha = 1$.

$$I = n_c \nu = n_c[A\alpha\,\Phi] = \left[n_s\exp\left(\frac{-\Delta G^*}{kT}\right)\right]\left[4\pi r^2 \alpha \frac{PN_A}{\sqrt{2\pi MRT}}\right] \quad (7.26)$$

Homogeneous nucleation can be engineered to occur if the desired product is nanoparticles (rather than thin films) by supersaturating the gas phase with a dense atomic flux, which makes ΔG^* very small. However, it is more often an undesired side effect during thin film growth (especially in CVD) where homogeneously nucleated particles would cause a "snow" precipitate in the system, and these precipitates become incorporated in the film as defects.

For the growth of thin films, *heterogeneous* nucleation occurs on the substrate. The new solid phase forms as a spherical cap, as shown in Figure 7.10. The relative interfacial energies determine the shape of the nucleus using a contact angle, θ, defined by the Young equation, as in the liquid wetting phenomena discussed in Chapter 3. Here, the interfaces of importance are between the substrate and the vapor, γ_{SV}, the substrate and the film nucleus, γ_{SF} and the film nucleus and the vapor, γ_{FV}.

$$\gamma_{SV} = \gamma_{FV}\cos\theta + \gamma_{SF} \quad (7.27)$$

The change in Gibbs free energy on forming this more complex nucleus shape can be derived using a similar approach to that used for homogeneous

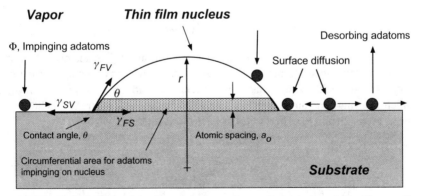

FIGURE 7.10 Schematic of the formation of a nucleus at the beginning of thin film growth. Adatoms impinge on the substrate at rate defined by the molecular flux; they travel to the nucleus via surface diffusion. There is a probability that they will desorb before reaching the circumferential area at the base of the nucleus.

nucleation (Section 3.2.5). Now, in addition to the volume Gibbs free energy change, ΔG_V, there are two surface or interfacial terms to include: one for the free energy change on creating the free surface of the nucleus and the other for creating the interface between the substrate and the nucleus. The critical size for the cap is found by differentiation, and then the barrier to nucleation is defined as the Gibbs free energy change associated with the formation of a critically sized spherical cap. Taking into account these multiple terms and the more complex geometry:

$$\Delta G^* = \frac{4\pi \gamma_{FV}^3}{3\Delta G_v^2} [2 - 3\cos\theta + \cos^3\theta] \qquad (7.28)$$

For a contact angle of 180°, ΔG^* is the same as it is for homogeneous nucleation. As θ decreases and the film phase wets the substrate, the barrier ΔG^* decreases.

For heterogeneous nucleation, the nucleation rate deviates from the homogeneous case not just because ΔG^* is reduced, but also because the rate at which the atoms are added to the critical nucleus changes. Nuclei are, by definition, very small so atoms are most likely to first impinge on the substrate and then travel by surface diffusion to the nuclei. The rate of surface diffusion in number of jumps per second is given by:

$$\nu' = \nu_o \exp\left(\frac{-Q_{sd}}{kT}\right) \qquad (7.29)$$

where ν_o is the jump or vibrational frequency (10^{13} Hz at 300 K) and Q_{sd} is the activation energy for surface diffusion. Since the adatoms land on a surface different than the nuclei, an adsorption/desorption thermodynamic balance is used rather than a simple sticking coefficient. This takes the form of a "lifetime to desorption," τ_{des}:

$$\tau_{des} = \frac{1}{\nu_o} \exp\left(\frac{Q_{des}}{kT}\right) \qquad (7.30)$$

where Q_{des} is the activation energy for desorption. If the adatoms do not desorb off the substrate before they reach a nucleus, they impinge onto the nucleus and stick there. Attachment occurs via an area of impingement, which is a ring, one atom high that encircles the nucleus where it touches the substrate ($a_0 2\pi r \sin\theta$). Therefore the nucleation rate is:

$$I = n_c \nu = n_c[A \cdot \tau_{des} \nu' \Phi]$$

$$= n_s \exp\left(\frac{-\Delta G^*}{kT}\right)\left[a_0 2\pi r \sin\theta \exp\left(\frac{Q_{des} - Q_{sd}}{kT}\right) \frac{PN_A}{\sqrt{2\pi MRT}}\right] \qquad (7.31)$$

This balance between thermodynamics, ΔG^*, and kinetics (the race between desorption and surface diffusion) determines the structural features of the growing film. If Q_{des} is large and Q_{sd} is small, the adatoms can diffuse a long way on the substrate before desorption, so the probability of nuclei capture is larger. If ΔG^* is small, nuclei density will be high, and the film will grow fast with small grains. These are the phenomena that determine film grain size. These parameters are fairly intrinsic to the materials chosen. For example, ΔG^* depends on gas supersaturation, but also on the surface energy balance between the vapor, substrate, and film. Another example is the growth of compounds, such as oxides. In vacuum processing chambers, oxygen has a low Q_{des} so it may not be incorporated into the film in the desired stoichiometry. Q_{sd} will also be lower for chemisorbed metals than oxides.

The nucleation rate is also directly dependent on the impingement rate, Φ, of atoms to the substrate. Microstructures can therefore be controlled via the temperature of the source in evaporation, the sputter gun power in sputtering, or the concentration of reactant gases in CVD. The other major external parameter of control is substrate temperature because the nucleation rate increases with decreasing temperature. Since more nuclei are likely to form at faster rates or lower temperatures, the film grain size will be smaller. To make a single crystal film, the rate of adatom impingement should be decreased and the temperature should be increased to allow adatoms to diffuse to a nucleus as it grows into a large grain.

Another thermodynamic factor is also important: does the film thermodynamically "wet" the substrate. The parameter, θ, shown in Figure 7.10, is important here because the single crystal film can only be grown if it has low surface energy with the substrate. θ determines if the film is stable as a monolayer ($\theta = 0°$) or as three-dimensional islands ($\theta > 0°$). For cases where film/substrate interfacial energies are high and θ approaches 180°, the film forms islands that are less stable. In the worst case, the film may even delaminate during or after growth, as shown in Figure 7.11. An intermediate mode of growth is called Stranski-Krastanov (SK). In this mode, the initial adatoms diffuse and nucleate in a monolayer or two, but after that islands

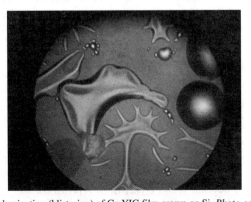

FIGURE 7.11 Delamination (blistering) of Ce:YIG film grown on Si. *Photo courtesy of A. Block.*

begin to form. In SK growth, the substrate/vacuum surface has higher energy than the film/vacuum surface, so the latter surface area is thermodynamically stable, or qualitatively speaking: $\gamma_{FV} < \gamma_{FS} < \gamma_{SV}$.

Surface energies also can vary within the film depending on the crystallography. In Zone 1 growth, most grains have dome-shaped tops and the grains themselves can even be amorphous, especially for oxide films, which have high melting points. Amorphous films result from a lack of thermal energy for atomic motion (surface diffusion). Because Zone 2 occurs at higher substrate temperatures, the adatoms can diffuse to low energy sites, and this means they move away from high energy surfaces and settle on low energy surfaces. Therefore, facets are formed. For example, the tops of Zone 2 grains may look like the corners of cubes instead of like domes. Wulff found a way of quantifying this effect using what we now call Wulff diagrams (Figure 7.12). These diagrams start at a point and arrows are drawn along crystallographic directions such that all of the arrows are proportional to the energy of the surface perpendicular to their crystallographic direction. In other words, $r_1/\gamma_1 = r_2/\gamma_2 = r_3/\gamma_3 =$ constant. This simple equality allows the expected crystalline facets to be simulated based on the respective surface energies of each crystallographic plane.

The final fundamental parameter in film microstructure is strain. Because films are grown onto substrates, they are automatically in a biaxial (in-plane) state of strain in most cases. As just mentioned, many films are grown at high temperatures. Due to thermal expansion difference between the substrate and film ($\alpha_s - \alpha_f$), the film may be in either tension or compression when the sample is cooled to room temperature after growth. Care must be taken to ensure that the stress, σ_{th}, resulting from thermal expansion mismatch does not exceed the fracture stress of the film. A quick estimate can be obtained using bulk values for Young's modulus and linear thermal expansion coefficient:

$$\sigma_{th} = E_f(\alpha_s - \alpha_f)\Delta T \tag{7.32}$$

where E_f is the Young's modulus of the film, and the film is assumed to be much thinner than the substrate. The equation assumes a starting point of near zero stress at elevated growth temperature and estimates the stress on cooling ($\Delta T < 0$).

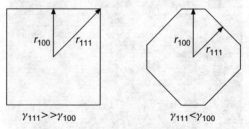

FIGURE 7.12 Wulff diagrams predict crystal facets expected due to differences in surface energies of various crystallographic planes.

7.2.3 Epitaxial Growth of Single Crystal Films

Epitaxy is a unique growth mode in which the crystalline structure of the growing film matches the substrate atom for atom as the layers are deposited. If there is a perfect match, the resulting film is a single crystal extension of the single crystal substrate, although there will always be some defects or strain at the interface so that the film is unlikely to be perfect. Even films with polycrystalline morphologies, such as those shown in the Thornton diagram (Figure 7.9) can have epitaxial lattice matching within the grains despite the grain boundaries between them.

From the Greek *epi* and *taxis*, the word epitaxy means "to arrange upon." As alluded to already, strain is the main determinant for whether a film grows epitaxially upon a certain substrate. Three cases of epitaxy are usually defined as shown in Figure 7.13. Matched films are grown on substrates with exactly the same lattice parameter (atomic spacing) as the film crystal. This is only truly the case in *homoepitaxy*, where a film is grown on a substrate of the same crystal, for example to obtain higher purities than that possible in the bulk growth process.

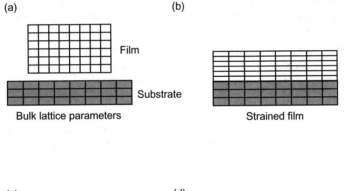

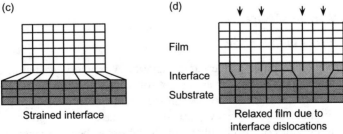

FIGURE 7.13 (a) Schematic of bulk lattice parameters, or atomic spacings, of the film and substrate materials if they were not constrained in any way. (b) Typical strain in a film, where the entire film is constrained by the substrate, which is much thicker than the film. (c) An unusual case where all of the strain is at the interface, for example when mechanically soft layers are used for compliance at the interface. d) A relaxed film where dislocations form between the film and substrate (below the arrows shown) to minimize the strain in the film.

Heteroepitaxy occurs when a film of different composition and/or crystal structure than the substrate is grown. In this case, the amount of strain in the film is determined by the lattice mismatch, f:

$$f = \frac{d_{hkl}(\text{substrate}) - d_{hkl}(\text{film})}{d_{hkl}(\text{film})} \qquad (7.33)$$

where d_{hkl} is the spacing between atomic planes in the substrate or film, respectively. The order of the terms in the numerator is important for nomenclature such that

$f = (+)$ means the film is in biaxial tension
$f = (-)$ means the film is in biaxial compression

In a first-order approximation, it is assumed that the film takes on all of the strain because the substrate is typically orders of magnitude thicker than the film (0.3–1.0 mm compared to 10–1000 nm for the film).

When could f be small, but the film still have large strain? The answer is found in the last equation of the previous section: thermal mismatch. The film and substrate could have similar lattice spacings but also have very different thermal expansion coefficients. If a film is then grown at a high temperature, then it can experience large strains upon cooling to room temperature. Also, subsequent device fabrication often involves thermal cycling.

In the Figure 7.13, the strain is overemphasized to clearly show the effects of mismatch in film and substrate lattices on strain at an interface. In reality, $f < \sim 5-10\%$ is necessary for obtaining epitaxy. If f is larger than that, the film experiences a volumetric strain that builds with each layer until a critical thickness is reached, d_c, is reached. With increased thickness ($d > d_c$) the elastic strain in the film is relieved by the formation of dislocations, each possessing local strain, but the sum of which is less than the volumetric elastic strain in the film prior to dislocation formation.

7.3 EVAPORATION

7.3.1 Process Overview

Evaporation is as simple as the name suggests. A source is heated in a vacuum until the atoms literally evaporate off of the surface. The atoms then fly through the chamber randomly, according to the kinetic theory of gases. Unlike air molecules, these evaporated atoms have a high probability of sticking to the sides of the chamber and the substrates because these surfaces are not heated, and the atoms are not volatile at room temperature. The sources are typically heated either using an electron beam or resistive heating, as shown in Figure 7.14. An electron beam (e-beam) is generated by a filament under the "hearth," which is the block holding the sources. A magnet in the hearth is oriented such that it steers the electrons around the side

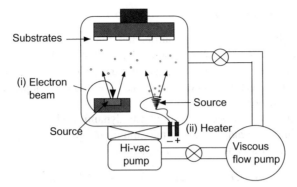

FIGURE. 7.14 A schematic diagram of an evaporation system showing two options for heating the source (i) an electron beam generated by filament directed to source by a magnetic field and (ii) a resistive heater. Source atoms evaporate and are collected on substrates as thin films.

of the hearth where they strike the source to heat it. A resistive heater is often simply a coil of refractive metal, such as tungsten, through which a current is applied to cause it to heat.

Evaporation is carried out at low pressures ($<10^{-6}$ Torr) such that the evaporated atoms have a long mean free path and do not undergo collisions. The evaporated atoms leave the source and travel unhindered in straight trajectories. Therefore, the entire chamber is coated with atoms, but only the atoms on the substrate are collected for subsequent device fabrication. Therefore, many vacuum chambers that are used for evaporation are either lined with metal, or are cleaned often between samples. On the other hand, some systems are made to hold substrates on curved surfaces that collect a higher fraction of the total number of atoms. The geometry of transport is very important in evaporation because films tend to grow by "line-of-sight," meaning that the atoms stick wherever they land and film uniformity can be difficult to obtain. The adatoms do not diffuse far on the surface because they arrive with only about 0.025 eV (300 K) compared to sputtered adatoms which can have energies of 0.04 eV (500 K).

Molecular beam epitaxy (MBE) is a special case where the source is heated in a holder called an effusion cell, and only a small opening is available for the evaporated atoms to flow out into the vacuum chamber. This slow rate of deposition, and heated substrates, allows the atoms to diffuse to low energy sites on the surface of the growing film, and epitaxial films can be achieved.

This section focuses on the fundamentals of producing thin films by evaporation. First, the pressure and temperature conditions needed for evaporation or sublimation from the source materials are established through thermodynamic relationships. Next, the challenges of evaporating alloys and compounds are covered. Lastly, the control of evaporation processes so as to produce uniform thin films is discussed.

7.3.2 Thermodynamics of Evaporation

To produce thin films by evaporation, the first step is to convert a solid source into a vapor. In many materials, the solid first melts and then vaporizes. In others, the solid directly sublimes. Understanding the thermodynamics of the change in state is necessary for choosing appropriate chamber conditions to form a thin film.

Evaporation is a classic study in thermodynamics where everything tends toward minimum enthalpy, H, and maximum entropy, S, and the total Gibbs free energy, G, of a system can be defined by:

$$G = H - TS \quad \text{and} \quad dG = dH - TdS - SdT \tag{7.34}$$

Enthalpy is the sum of internal energy, U, and work, which is mostly pressure•volume (PV) work in the case of vacuum processing.

$$H = U + PV \quad \text{and} \quad dH = dU + VdP + PdV \tag{7.35}$$

According to the first law of thermodynamics, an increase in heat, dq, must either increase the internal energy of the system or do work, dw, for the system to remain at equilibrium:

$$dq = dU + dw \tag{7.36}$$

Substituting in the second law of thermodynamics ($dq = TdS$, assuming reversibility) and assuming only PV work ($dw = PdV$):

$$dU = TdS - PdV \tag{7.37}$$

Combining the differential forms of Eqs. (7.34) and (7.35) with (7.37):

$$dG = VdP - SdT \tag{7.38}$$

This expression is a statement of equilibrium. At constant temperature and pressure, any process that lowers Gibbs free energy, $dG < 0$, is thermodynamically likely to occur.

Given the definition of equilibrium, we can consider the vaporizing of a material either by evaporation (from a melt) or by sublimation (from a solid). The reaction is given by:

$$C \rightarrow V \quad \text{(condensed phase} \rightarrow \text{vapor phase)}$$

The associated change in Gibbs free energy (per mole) is:

$$\Delta G_{vap} = [V_V dP - S_V dT]_{\text{product}} - [V_C dP - S_C dT]_{\text{reactant}} \tag{7.39}$$

where V_V and V_C are the molar volumes of vapor and condensed phases, respectively, and likewise S_V and S_C are the molar entropies of vapor and condensed phases, respectively. At equilibrium $\Delta G_{vap} = 0$, so

$$\frac{dP}{dT} = \frac{(S_V - S_C)}{(V_V - V_C)} = \frac{\Delta S_{vap}}{\Delta V_{vap}} \qquad (7.40)$$

Note that ΔS_{vap} is the change in molar entropy on vaporization, and ΔV_{vap} is the change in volume on vaporization.

Two important results come from this derivation. First, Eq. (7.40) is a statement of equilibrium; it provides pressure and temperature combinations where the liquid (or solid) is in equilibrium with a vapor of the same composition (i.e., the boiling or sublimation points). You may be used to thinking of a material as having a "known" boiling point. However, this standard value is only the boiling point at the specific pressure of 1 atm. The boiling point changes when a material is in a vacuum. Second is the fact that the elusive thermodynamic parameters of entropy and enthalpy can be measured using thin film vapor processes!

First, let's derive the pressure dependence of enthalpy of vaporization, ΔH_{vap}, also called the latent heat of vaporization. At equilibrium, $\Delta G_{vap} = 0$, so $\Delta H_{vap} = T\Delta S_{vap}$. Substituting into Eq. (7.40) brings us to the Clausius-Clapeyron equation:

$$\frac{dP}{dT} = \frac{\Delta H_{vap}}{T\Delta V_{vap}} \qquad (7.41)$$

As the molecules evaporate, they take up 1000x more volume than they occupied in the condensed phase, so ΔV_{vap} is essentially equal to the volume of the vapor, which can be related to pressure and temperature through the ideal gas law ($V_{vap} = \frac{RT}{P}$) as the molar volume.

$$\frac{dP}{dT} = \frac{P\Delta H_{vap}}{RT^2} \qquad (7.42)$$

To determine the enthalpy of vaporization per mole, we can use the following:

$$\frac{dP}{P} = \frac{\Delta H_{vap}}{RT^2} dT \qquad (7.43)$$

$$\ln P^* = 2.3 \log P^* = \frac{-\Delta H_{vap}}{RT} + P_0 \qquad (7.44)$$

where P_o is a constant and the asterisk is added to emphasize this pressure is the equilibrium vapor pressure, P^*, of the material being evaporated. By measuring the vapor pressure of a substance at a series of temperatures, one can find the heat of vaporization. But, it is difficult to measure the pressure of vapors at elevated temperatures because pressure gauges aren't made to withstand high temperatures. Therefore, the atoms would condense before being measured.

Instead, a film can be grown, and the deposition rates can be related to vapor pressure. The evaporation flux (in number of molecules per unit area per unit time) coming off an evaporating source into the vacuum is the same as the impingement rate for molecules in a vacuum, because the evaporation flux is also governed by the kinetic theory of gases:

$$\Phi_{evap} = \frac{\Delta P N_A}{\sqrt{2\pi MRT}} \tag{7.45}$$

This expression uses the definition of molecular flux (Eq. (7.15)), but accounts for the fact that the flux away from the surface of an evaporating substance depends on a driving force, ΔP. In a closed system that is entirely at a specific temperature, atoms evaporate and condense with a system pressure equal to the equilibrium vapor pressure of the substance P^*. The net flux off of the source would then be zero. In our processing chamber, however, only the source is heated, and it is open to the vacuum $P \sim 0$ (usually $< 10^{-6}$ Torr). So, the pressure differential driving flux off the source is $\Delta P = P^* - 0 = P^*$. If all of the atoms are collected from the source, the evaporation flux can be found. From this information, the vapor pressures at different temperatures can be plotted as $\log P^*$ versus $1/T$. Then, using Eq. (7.44), the slope will be equal to $-\Delta H_v/R$. Of course, collecting all of the atoms from the source is difficult, but Section 7.3.4 has a more practical solution.

Figure 7.15 shows the vapor pressure of various elemental source materials as a function of inverse temperature. This plot is used to

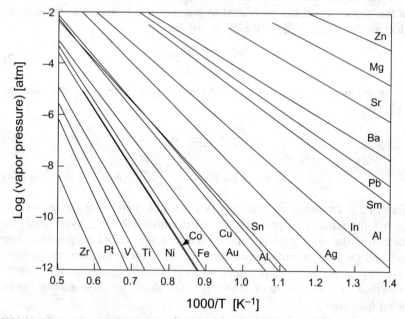

FIGURE 7.15 Vapor pressures (P^*) of many elements versus $1/T$. The equations for these metals and others can be found in *Alcock et al. (1984)*.

determine source temperatures based on the desired deposition rate of each material.

Example 7.6 shows how Figure 7.15 can be used to find molecular flux. However, not all of the Ag atoms can be collected onto a substrate because, as the kinetic theory of gases describes, the atoms will fly off of the source in all directions. Using Eq. (7.9), the mean free path of the atoms coming off of the source is 43 cm. Hence, evaporation occurs in the molecular vacuum regime. Section 7.3.4 will describe how to design the placement of substrates for optimized collection of atoms and for uniform thicknesses in the resulting films.

Some materials, such as carbon (not shown) and titanium, are similar to dry ice in that they sublime rather than evaporate. However, most materials melt and so they require a "boat" to hold them or a heating "hearth" if a large source is to be used. For small amounts of source material, a coil of resistive wire can be used as the heater, and

---

EXAMPLE 7.6 What molecular flux would evaporate from a silver (Ag) source if it was heated to $1000/T = 0.85$ K^{-1} ($T = 1176.47$ K)?

Using the line for Ag in Figure 7.15, the vapor pressure is found to be 10^{-6} atm, which is 0.76 mTorr, well above the typical base pressures for evaporation ($<10^{-6}$ Torr). So, $\Delta P = (0.76 \text{ mTorr} - 10^{-6} \text{ Torr}) \sim 0.76$ mTorr, and using Eq. (7.45):

$$\Phi = \frac{PN_A}{\sqrt{2\pi MRT}} = \frac{(10^{-6}\text{atm})(101,323\text{Pa/atm})\left(\dfrac{6.02 \times 10^{23}\text{molecules}}{\text{J/mol}\cdot\text{K}}\right)}{\sqrt{2\pi\left(0.1079\dfrac{\text{kg}}{\text{mol}}\right)\left(8.314\dfrac{\text{J}}{\text{molgK}}\right)(1176.47\text{K})}}$$

$$= 7.5 \times 10^{20} \frac{\text{atoms}}{\text{m}^2 \cdot \text{s}}$$

The atomic flux coming off of the Ag source would be 7.5×10^{16} atoms/cm²s. If all of the Ag atoms coming off of this source were collected onto a substrate, one monolayer would be deposited approximately every

$$t_{monolayer} = \frac{1}{(7.5 \times 10^{16}\text{atoms/cm}^2\text{s})(2.88 \times 10^{-8}\text{cm})^2} = 16 \text{ msec/atom}$$

assuming the diameter of a Ag atom is 288 pm. Recall from the discussion of Table 7.2, that this is an approximation and the actual monolayer packing density will vary with each material and temperature. Therefore, a 1 μm thick Ag film could be deposited in approximately:

$$t_{1\mu m} = \frac{10^{-4}\text{cm}}{2.88 \times 10^{-8}\text{cm}/16\text{ms}} = 56 \text{ s}$$

the surface energy of the molten liquid keeps it attached to the wire after it melts.

7.3.3 Evaporation of Alloys and Compounds

Metal alloys and ceramics, which are by definition compounds, contain more than one element. Vaporizing these source materials while maintaining stoichiometry is a challenge. The first issue is melting—not all alloys and compounds melt congruently (i.e., into a liquid of the same composition as the solid), which complicates the processing, requiring a higher temperature to reach a liquid solution state. Secondly, even after such a melt is formed, the vapor pressure of any given element in the melt is not the same as the others and hence the composition of the vapor emitted is different than the source. Nonetheless, there are methods to create thin films of these more complex materials.

Several methods have been proposed for the fabrication of metal alloy films by evaporation. Some involve a dynamic method, such as feeding wire stock into the source crucible during evaporation, to compensate for the preferential loss of more rapidly evaporating elements. However, the most practical solution is to have two sources with independent temperature control. By this strategy, the alloy composition can be controlled by tuning the pressure as well as the individual temperatures of the sources. Interestingly, this approach also allows atomic grading of composition, or multilayered films, which can provide properties that are not available in bulk materials.

Evaporation of compounds, such as ceramic oxides and nitrides, can also be challenging (Table 7.3). The trouble in this case is that compounds often dissociate in the vacuum, and the more volatile element has a significantly lower (or even zero) sticking factor at the substrate as in

$$MX \to M + \frac{1}{2}X_2(g) \qquad (7.46)$$

If X_2 is thermodynamically stable as a gas, such as O_2 or N_2, it elastically collides with the walls until it happens upon the aperture of the pump and is pulled out of the chamber. Some of the molecules may be trapped in the depositing metal film, similar to impurities being trapped from air as discussed above, but it may not be enough to make a stoichiometric film, such as MO (e.g., ZnO). The solution to this problem is to feed the reactive gas into the system separately. In fact, if carefully designed, a metal source can be used to allow a lower source temperature, while the reactive gas is fed into the system next to the substrates so that compound films form. However, the metal source most likely also reacts, unless it is inside an effusion cell (as discussed for MBE above). Reactive gas input is more effective

TABLE 7.3 Evaporation of Compounds[a]

Reaction type	Chemical reaction[b]	Examples	Comments
Evaporation without dissociation	MX(s or l) → MX(g)	SiO, B_2O_3, GeO, SnO, AlN, CaF_2, MgF_2	Compound stoichiometry maintained in deposit
Decomposition	MX(s) → M(s) + ½X_2(g) MX(s) → M(l) + (1/n)X_n(g)	Ag_2S, Ag_2Se III–V semiconductors	Separate sources are required to deposit these compounds
Evaporation with dissociation			Deposits are metal-rich
(a) Chalcogenides	MX(s) → M(g) + ½X_2(g) X = S, Se, Te	CdS, CdSe, CdTe	Separate sources usually required to deposit these compounds
(b) Oxides	MO_2(s) → MO(g) + ½O_2(g)	SiO_2, GeO_2, TiO_2, SnO_2, ZrO_2	O_2 partial pressure (reactive evaporation) recommended

[a]Tabulated by Ohring (2002), reproduced by permission of Academic Press.
[b]M = Metal, X = nonmetal

in sputtering (see Reactive Sputtering in the next section) because sputtering is a viscous regime process with inert processing gases that help to flow the reactive gas away from the source.

7.3.4 Transport Phenomenon and Film Uniformity

The example at the end of Section 7.3.2 required that all of the atoms evaporating from the source be collected and counted. This is unlikely given the random directions of their motion upon evaporating. More likely, one could measure the thickness of the deposited atoms around the chamber and calculate the moles evaporated:

moles evaporated = (thickness of film) (area of chamber) (density)/(M)

where M is the molecular weight of the atoms being evaporated. However, since the chamber is likely to have an odd shape, a fraction of the atoms could be collected on a substrate and its fraction of the total possible area could be determined (area fraction = substrate area/total area).

moles evaporated = (thickness of film)(substrate area)(density)/(area fraction)(M)

It turns out that this is very close to how important parameters, such as film uniformity, can be predicted based on the location and geometry of the substrate.

The simple calculation above works if the substrate is curved to be equal distance from the source at all points. One can imagine a "receiving sphere" around which an equal number of atoms are deposited (Figure 7.16). It is more likely that the substrate is flat, such as a Si wafer or a glass substrate, and there may be films grown on several substrates at the same time as shown in Figure 7.14. "Point sources," as in Figure 7.16, are typically resistive heaters, as in Figure 7.14 (ii). Let's consider one substrate at that is above the source at distance, h. The amount of the mass evaporated (dM) that is collected on the an elemental area (dA_{film}) of the substrate is

$$\frac{dM_{film}}{\cos\theta dA_{film}} = \frac{M_{evap}}{4\pi r_{sphere}^2} \quad (7.47)$$

where M_{evap} is the total evaporated mass, r_{sphere} is the radius of the receiving sphere at dA_{film}, and θ is the angle between the receiving sphere and dA_{film} (Figure 7.16). The factor $\cos\theta$ is used to "project" dA_{film} onto the receiving sphere to enable a proportional relationship to be expressed in Eq. (7.47), namely mass fraction is proportional to area fraction. By the definition of

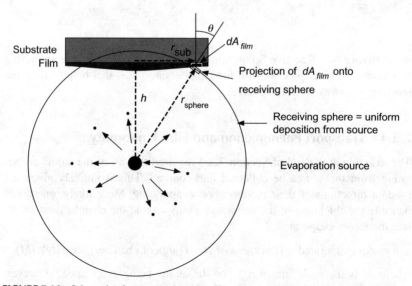

FIGURE 7.16 Schematic of atoms evaporating off a source and collecting onto a flat substrate.

density, ρ, the film thickness, d, is related to the mass by $d = \dfrac{dM_{film}}{\rho dA_{film}}$
Therefore, the thickness at each point along the film is

$$d = \frac{M_{evap}\cos\theta}{\rho 4\pi r_{sphere}^2} = t_{deposition} \cdot \frac{\Phi M}{N_A} \cdot \frac{\cos\theta}{\rho 4\pi r_{sphere}^2} \tag{7.48}$$

where $t_{deposition}$ is the deposition time, $\cos\theta$ is h divided by r_{sphere} (Figure 7.16), and Φ is the molecular flux from Eq. (7.45), M/N_A gives atomic weight (g/mole divided by atoms/mole). The center of the film shown in Figure 7.16 is the thickest point, d_o (Example 7.7).

By plotting d/d_o as a function of r_{sub}/h, the variation in thickness can be calculated, as shown in Figure 7.17. In determining uniformity along a substrate of interest, the size (l) of planar substrates is clearly important. In the manufacturing of thin film devices, uniformity of film thickness will be required to meet device specifications. For example, if waveguides as in Figure 7.2 were to be manufactured from a thin garnet film grown on a Si wafer, the waveguides on the outside of the wafer would need to have "similar" thicknesses as those at the center of the wafer. Specifications could be set, such as: "1 µm-thick" waveguides made on a 4″ Si wafer must have actual thickness is within 0.95–1.00 µm. In this case, using Figure 7.17, the r_{sub}/h ratio is ~0.18. In other words, for all of the waveguides on the 4″ wafer to be acceptable (100% yield), the substrate must located be (4″/0.18) 20″ above the

---

EXAMPLE 7.7 Find d_o in terms of M_{evap} and h. Also find d/d_o in terms of processing parameters: source-substrate distance, h, and substrate size, r_{sub}.

$$d_o = \frac{M_{evap}}{\rho 4\pi h^2}$$

$(r = h$ and $\cos\theta = 1)$

$$\frac{d}{d_o} = \frac{\dfrac{M_{evap}\cos\theta}{\rho 4\pi r_{sphere}^2}}{\dfrac{M_{evap}}{\rho 4\pi h^2}} = \frac{\cos\theta\, h^2}{r_{sphere}^2} = \frac{h^3}{r_{sphere}^3}$$

Using the geometry of Figure 7.16, $r_{sphere}^2 = h^2 + r_{sub}^2$, so
$\dfrac{d}{d_o} = \dfrac{h^3}{\left(\sqrt{h^2+r_{sub}^2}\right)^3} = \dfrac{1}{\left(1+\left(\frac{r_{sub}}{h}\right)^2\right)^{3/2}}$ **uniformity relationship for point sources**
(Figure 7.14 **(ii)**)

Although outside the scope of this chapter, the coating uniformity from a surface source, such as in e-beam evaporation, "i" in Figure 7.14(i), is

$\dfrac{d}{d_o} = \dfrac{1}{\left(1+\left(\frac{r_{sub}}{h}\right)^2\right)^2}$ **uniformity relationship for surface sources** (Figure 7.14 **(i)**)

550 Materials Processing

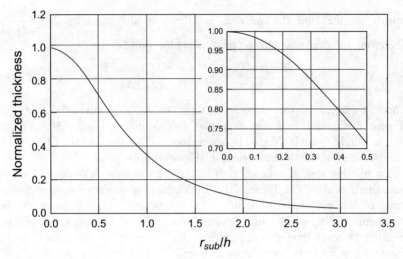

FIGURE 7.17 Film thickness (normalized to center thickness, d_o) versus distance on substrate (normalized by the height of substrate from source).

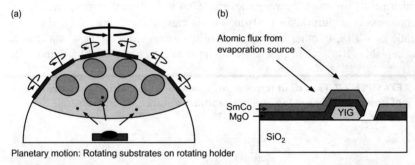

FIGURE 7.18 (a) Film uniformity can be improved using planetary rotation of substrates that are all located on the receiving sphere to maximize yield from the source. (b) Without planetary rotation, evaporation of films can result in poor coverage, especially if substrates are rough or have steps, such as the waveguide shown.

source. To avoid wasting atoms and also for better step coverage, many wafers could be placed on a spherical substrate holder so that they are all 20" away from the source (Figure 7.18).

Besides film uniformity, many devices require specific microstructures. As discussed in Fundamentals section, the films are likely to have voids and columnar grains with dimensions that vary with substrate temperature. Since evaporation occurs in the molecular regime, pressure is not typically a control variable, except that it should be as low as possible to avoid contamination. Therefore, substrate temperature is the main parameter of control. The atoms arrive at the substrate with little energy compared to sputtering (discussed next) so surface roughness plays a larger role in the microstructure.

Extreme case of roughness, such as steps in the substrate and/or pits, result in poor coverage (non-conformal). To use the waveguides in Figure 7.2 as an example again, assume that the garnet film uniformity was good (as calculated with substrates 20″ away from the evaporation source). These films could be removed and patterned (discussed last in this chapter), and then they could be put inside another processing chamber to evaporate the final two layers (MgO and SmCo). Although the waveguides in the center of the wafer are likely to be evenly coated, the waveguides on the edge of the wafer may instead experience poor step coverage (Figure 7.18b). Very high substrate temperatures may encourage better coating since atoms will have the energy to overcome surface diffusion barriers. Another option is to allow the substrates to rotate, ideally in a "planetary" motion, where each substrate would rotate around its own axis and also the entire substrate holder would be rotated (Figure 7.18a).

Summary. Table 7.4 shows experimental evaporation details for many materials. Also, the previous example of waveguides (Figure 7.2) is useful to facilitate a summary discussion of what we've learned about evaporation, even though these actual films were made by sputtering. First, consider the garnet ($Y_3Fe_5O_{12}$) waveguide. Garnet has a high melting point with a specific stoichiometry, so it would be difficult to evaporate. Sputtering is actually a better process for this layer. With e-beam evaporation, many ceramics can be evaporated, but they may disassociate. In rare cases, only metal films are obtained as the anion forms a volatile byproduct (e.g., O_2 or N_2) that is pumped out of the processing chamber. In less extreme cases of ceramic evaporation, the film is on the metal-rich side of stoichiometry. After patterning the waveguide (Figure 7.2), MgO could be evaporated by reactive evaporation where an Mg source would be heated to ∼700 K (P_{evap} ∼10^{-5} atm, Figure 7.15) and O_2 would be fed into the processing chamber next to the substrates. However, Mg is very reactive, and the source is likely to oxidize with or before the film. So, again sputtering is probably a better choice. Other ceramics are successfully made by reactive evaporation. Most ceramics require post-annealing because Zone 1 amorphous microstructures are common. These high melting temperature materials (see Figure 7.9) have high energy barriers for surface diffusion.

Finally, metallic SmCo could be evaporated from two separate sources, one with Sm and one with Co. Why two separate sources? Figure 7.15 shows that an alloy source of SmCo would be quickly depleted of Sm, which has about 1 million times higher vapor pressure than Co at any given temperature. Separate sources would be required to obtain say Sm:Co = 1:5, where example temperatures could be 850 K for the Sm source and 1600 K for the Co source. Metal films will most likely be crystalline, but post-annealing may be needed if another crystalline phase is desired rather than the as-grown phase. The final point we've mentioned in this chapter is substrate motion, which may be necessary to obtain better coverage in all layers and less material waste.

TABLE 7.4 Evaporation Characteristics of Materials[a]

Material	Min. source temperature[b]	State of vaporization	Recommended crucible material	Deposition rate (Å/s)	Power (kW) (e-beam[c])
Aluminum	1010	Melts	BN	20	5
Al₂O₃	1325	Semimelts		10	0.5
Antimony	425	Melts	BN, Al₂O₃	50	0.5
Arsenic	210	Sublimes	Al₂O₃	100	0.1
Beryllium	1000	Melts	Graphite, BeO	100	1.5
BeO		Melts		40	1.0
Boron	1800	Melts	Graphite, WC	10	1.5
BC		Semimelts		35	1.0
Cadmium	180	Melts	Al₂O₃, quartz	30	0.3
CdS	250	Sublimes	Graphite	10	0.2
CaF₂		Semimelts		30	0.05
Carbon	2140	Sublimes		30	1.0
Chromium	1157	Sublimes	W	15	0.3
Cobalt	1200	Melts	Al₂O₃, BeO	20	2.0
Copper	1017	Melts	Graphite, Al₂O₃	50	0.2
Gallium	907	Melts	Al₂O₃, graphite		

Germanium	1167	Melts	Graphite	25	3.0
Gold	1132	Melts	Al$_2$O$_3$, BN	30	6.0
Indium	742	Melts	Al$_2$O$_3$	100	0.1
Iron	1180	Melts	Al$_2$O$_3$, BeO	50	2.5
Lead	497	Melts	Al$_2$O$_3$	30	0.1
LiF	1180	Melts	Mo, W	10	0.15
Magnesium	327	Sublimes	Graphite	100	0.04
MgF$_2$	1540	Semimelts	Al$_2$O$_3$	30	0.01
Molybdenum	2117	Melts		40	4.0
Nickel	1262	Melts	Al$_2$O$_3$	25	2.0
Permalloy	1300	Melts	Al$_2$O$_3$	30	2.0
Platinum	1747	Melts	Graphite	20	4.0
Silicon	1337	Melts	BeO	15	0.15
SiO$_2$	850	Semimelts	Ta	20	0.7
SiO	600	Sublimes	Ta	20	0.1
Tantalum	2590	Semimelts		100	5.0
Tin	997	Melts	Al$_2$O$_3$, graphite	10	2.0
Titanium	1453	Melts		20	1.5
TiO$_2$	1300	Melts	W	10	1.0

(Continued)

TABLE 7.4 (Continued)

Material	Min. source temperature[b]	State of vaporization	Recommended crucible material	Deposition rate (Å/s)	Power (kW) (e-beam[c])
Tungsten	2757	Melts		20	5.5
Zinc	250	Sublimes	Al_2O_3	50	0.25
ZnSe	660	Sublimes	Quartz		
ZnS	300	Sublimes	Mo		
Zirconium	1987	Melts	W	20	5.0

[a]Tabulated by Ohring (2002), reproduced by permission of Academic Press.
[b]Temperature (°C) at which vapor pressure is 10^{-4} Torr.
[c]For 10 kV, copper hearth, source-substrate distance of 40 cm.

7.4 SPUTTERING

7.4.1 Process Overview

Sputtering is basically atomic billiards. Physically, sputtering involves bombarding a target with high-energy ions (analogous to the cue ball) at sufficiently high energies that target atoms (analogous to billiard balls) and secondary electrons, are ejected. To create ions, a process gas, such as argon (Ar), is fed into the vacuum chamber, and an electric field is applied to "breakdown" the gas such that the atoms ionize. See Figure 7.19. Due to the electric field, the positively charged ions (Ar^+) accelerate toward the target. On impact, the ions collide with target atoms. The interaction is complex, but a main outcome of this impact is momentum transfer from the incoming ion to a target atom, and the ejection of atoms from the surface. The ejected atoms travel to the substrate, where they are collected to form a film.

Figure 7.20 shows a commercial sputtering chamber based on two targets inside what are known as sputtering "guns." Unlike evaporation, the source material must have a specific shape to match the sputtering gun. This is usually a disk that is 1/8" to 1/4" thick with diameters ranging from 2" to 12". In some industrial, high-yield applications, sputter targets can be several feet in diameter. Also, the processing pressure is closer to the viscous than molecular vacuum regime because many ions (e.g., Ar^+ or Xe^+) are needed. However, it is very important that the processing chamber first be evacuated

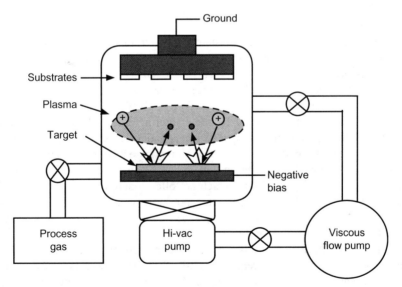

FIGURE 7.19 A schematic diagram of a sputtering system showing vacuum and gas handling. The potential difference set up by the negative bias on the sputtering target and the grounded chamber sets up conditions for ionizing an inert gas. Ionized gas molecules strike the target, knocking off the target atoms. The target atoms travel to the substrates and deposit there.

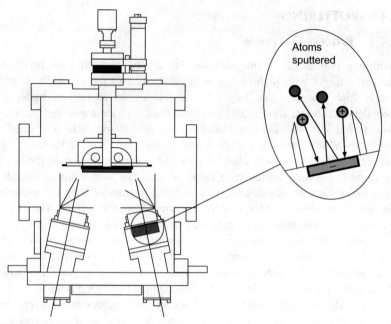

FIGURE 7.20 This sputtering chamber shows two sputter guns tilted toward the substrate with shutters that are shown in multiple stages of opening to allow the guns to keep running during multilayer growth. *Schematic courtesy of AJA International, www.ajaint.com.*

to the lowest pressure that the system can offer in order to minimize the impurities from the air (e.g., oxygen, nitrogen and water). After a decent base pressure is obtained ($<10^{-6}$ Torr), the UHV pump is protected, for example by throttling (partially-closing) the gate valve, and the process gas is fed into the chamber until 1–100 mTorr is obtained. The gas is ionized. Typically, the target is cleaned by pre-sputtering for 5–30 min before the shutter of the sputter gun is opened. Pneumatic shutters are quicker than turning on and off sputter guns, so these shutters are also used when growing multilayered films. Sputter systems in industry can have dozens of sputter guns for complex layered films, which are often required for microscale and nanoscale devices. The guns are often powered on for the whole process, and the shutters open and close in sequence to allow the atomic fluxes necessary for the required film multilayers.

The power applied to the target can be either direct current (DC) with a negative voltage or radio-frequency (RF). The RF power is required for oxide targets and for reactive sputtering of metal targets. Sputter guns also typically contain magnetrons, which help contain the plasma against the targets, increasing deposition rates and limiting accidental sputtering of other components in the chamber.

This section begins with an introduction to plasmas, which are vital to the sputtering process. Next, three variants on sputtering are introduced: magnetron, DC, and reactive. Finally, the focus shifts to understanding how to optimize sputtering rates and produce high-quality thin films.

7.4.2 Plasma Physics

Sputtering requires a "plasma", which is an ionized gas. If a DC voltage is applied across a gas using two electrodes (a cathode and an anode), the gas will be ionized if the voltage is higher than the breakdown voltage, V_B, of the gas. For example,

$$Ar + e^- \rightarrow Ar^+ + 2e^- \tag{7.49}$$

In a sputtering process chamber, the cathode is the sputtering target and the anode is the grounded chamber itself. After initial ionization, the newly created ion is accelerated toward the target, which has a negative bias applied to it, and the electron is accelerated away from the cathode. If ions strike the target with enough energy, atoms and secondary electrons are ejected from the target. The ejected atoms then travel to the substrate according to the kinetic theory of gases as discussed in Fundamentals section. There, they will be collected to form a film.

The breakdown voltage of the gas depends on the pressure of the processing gas, P, and the spacing between the electrodes, d. See Figure 7.21. Breakdown occurs when the electrons (produced by initial Ar^+ ionization and impact) have enough energy as they accelerate away from the target that they have ionizing collisions with other gas atoms. This produces more ions and electrons, which leads to more ionizing collisions, and so on. However, at low pressures or short distances, there are too few electron collisions. But, at very high pressures or very long distances, there are too many total collisions to allow the electrons to gain enough energy from the electric field to produce ionizing collisions. The initial ionization requires high voltages and optimized pressures that depend on the chamber geometry. Electrons or UV light can help "light" a plasma if injected into the gas from a filament somewhere in the processing chamber (like a light bulb filament which emits electrons as well as light). However, if the filament is turned off, the plasma will also be extinguished unless the applied voltage is $>V_B$.

Thankfully, V_B is lower than expected once the plasma is "lit," because the gas restructures into what is called a self-sustained discharge. The structure of such a discharge is shown in Figure 7.22a. From the target, the electrons travel quickly through Crooke's space, also called the cathode sheath. Then, at a distance known as the mean free path to an ionizing collision, $l_{mfp,ion}$, ion-electron pairs are created by colliding electrons in the negative glow. Importantly, $l_{mfp,ion}$ is 5–10 times longer than the mean free path to any collision because the electron has to accelerate and gain enough energy

558   Materials Processing

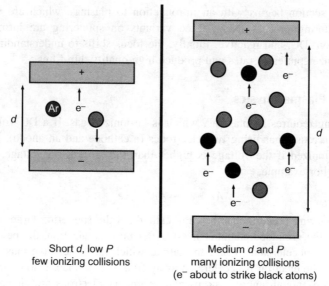

Short d, low P
few ionizing collisions

Medium d and P
many ionizing collisions
(e^- about to strike black atoms)

FIGURE 7.21 Schematic to show the influence of the pressure*distance product on the breakdown voltage of a gas. (a) Low P*d product gases are difficult to ionize. (b) Medium P*d product gases "breakdown" into a plasma at lower voltages, but high P*d product gases again have high breakdown voltages because there are more total collisions and electrons don't have the energy to ionize the atoms they impact.

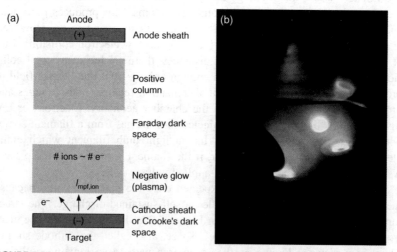

FIGURE 7.22 (a) Schematic of the structure of a self-sustained discharge. (b) Photograph of the inside of a sputtering chamber where three sputter guns are open at once. The ionization of Ar and some atoms from the targets gives the "glow" from each plasma a different color. The shutter of one gun is in the foreground as a dark circle. *Photo courtesy of Andrew Block.*

to ionize the process gas atoms. At this point, a negative glow or "plasma" is created. A plasma is a net neutral fluid in which there are equal numbers of negative electrons and positive ions. Here, the ionizing collisions produce characteristic colors associated with the ionic species in the plasma, Figure 7.22b. After the negative glow, the electrons have lost energy, and there is another dark space, called Faraday's dark space. Finally, the electrons accelerate again, and more ionization occurs in a weakly glowing positive column before the anode sheath.

In sputtering, ions strike a target with kinetic energies of 500–1000 eV which are the most effective energies for momentum transfer to target atoms. A substantial fraction of this kinetic energy is also transferred to the target as heat, so cooling water is very important to a sputter gun design. The gas atoms and ions are densely concentrated in the plasma, and there the ions and atoms thermalize each other via collisions, resulting in thermal energies of 0.04–0.08 eV (500–1000 K). This is high compared to the energies of evaporated atoms (0.025 eV = room temperature). Therefore, sputtered adatoms arriving at the substrate have energies between these values depending on the time of flight to the substrate and the number of collisions along the way, which is related to pressure. Since the thermal energy of the adatoms determines their ability to overcome the surface diffusion barrier, Q_{SD}, pressure is an important parameter in sputtering, unlike evaporation.

Sputtering involves an "asymmetric" discharge because the cathode (sputter target) is usually much smaller in area than the anode, which is the rest of the chamber maintained at ground. A sputter gun holds the target with a clamping ring, and a grounding shield surrounds the clamping ring with a very small gap separating it from the target. The target is electrically isolated from the rest of the gun, which also contains cooling water behind the target, an optional gas feed-through for the processing gas, and a shutter such as the one shown in Figure 7.20. Finally, it is likely that the sputter gun contains a magnetron, which is discussed next.

7.4.3 Magnetron Sputtering

Magnetrons create dense plasmas by confining electrons against the target for more efficient ionization and increased sputtering rates. Therefore, lower process gas pressures and voltages can be used during sputtering. Most magnetrons are a simple ring of magnets behind the outside of the target, with a magnet of opposite polarity behind the center of the target. This arrangement produces a magnetic field that extends over the surface of the target, in a ring that gathers electrons into a circle just above the target (Figure 7.23a) according to the Lorenz force, F_m

$$F_m = qv \times B \tag{7.50}$$

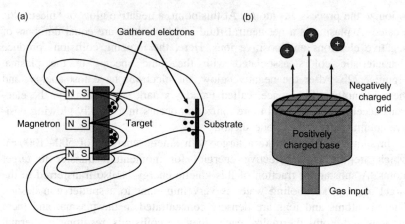

FIGURE 7.23 (a) Schematic of an unbalanced magnetron where the flux lines extend to the substrate to allow some ion bombardment of the film as it is depositing onto the substrate. The flux lines are most dense at the target, causing a racetrack to be worn into the target after a long time of sputtering, but electrons will also gather at the substrate surface. (b) Schematic of an ion beam that can be used to bombard films grown by any of the three processes discussed in this chapter. Bombardment is similar to annealing during growth as the sputter ions have energies greater than 0.04 eV ($\sim$500 K). Therefore, Zone T or Zone 2 microstructures could be produced, compared to Zone 1 for non-bombarded films.

This Lorentz force causes the electrons with charge, q, to move in a helical pattern around the magnetic flux, B, with velocity, v. Ions are slower than electrons, so they simply accelerate toward the net electron charge and sputter the target preferentially where the electrons are gathered. Since the electrons are gathered in a ring, a "racetrack" erosion pattern forms, and only 20–30% of the target is utilized before the pattern wears through it. Many target manufacturers will take the unused target material back to re-use in the next target (e.g., melting it to cast a new target).

A magnetron can be "unbalanced" to allow the magnetic flux lines to extend beyond the target, and even out to the substrate, Figure 7.23a. To do this, the center magnet is replaced with a magnet having different strength than the outer magnetic ring. The plasma will then extended out to the film, so it will be bombarded by energetic particles as it is deposited. This may cause some "resputtering," or removal of film atoms as they are forming the film. If the net flux of atoms to the substrate is positive, the film will be denser due to thermalization of the adatoms . The effect is similar to having an increased substrate temperature, without needing to have a heater behind the substrate. So, an otherwise Zone 1 film could be encouraged to form Zone T or even Zone 2 microstructures due to this bombardment. In many cases, a small negative "substrate bias" ($\sim$ tens of V) is applied directly to the substrate to further encourage ion bombardment of the film as it grows. Some processes use separate ion guns to bombard films during evaporation,

Figure 7.23b, and this is called "ion beam assisted deposition" (IBAD). It is also very common to use plasmas to enhance, or densify, films in CVD.

Given the popularity of magnetrons in sputtering, how can a magnetic material be used as a target? There are several configurations that work. In many cases, the sputter gun and magnetron are not altered, but the magnetic targets are just kept very thin to prevent the magnetic flux from shunting inside the material of the target. Shunting means that the magnetic flux is trapped inside the target, and it does not extend to the surface of the target. In this case, the plasma spreads out as if there were no magnetron. In other cases, the center magnet is replaced by a non-magnetic material of the same shape (for heat conduction) and the magnetic target itself acts as the center magnet.

7.4.4 Radio Frequency (RF) Sputtering

The DC sputtering techniques described so far require a conductive target because the target, plasma, and chamber walls (ground) make up a DC circuit. However, to sputter dielectric materials, such as ceramics and some semiconductors, a radio frequency (RF) alternating current can be used, typically 13.56 MHz because it is the approved scientific band according to FCC regulations. Under high frequency biasing, dielectrics respond by producing dipoles, and a high frequency oscillating charge appears at their surfaces.

Therefore, when RF power is applied to a dielectric target, the electrons gather at the surface of the target as the voltage sweeps positive. The processing gas ions are accelerated toward these electrons as a negative bias even though the target itself is not electrically conductive. The ions do not gather at the surface of the target when the applied RF voltage sweeps to negative values because they move much slower than electrons. So at very high frequencies, such as radio frequencies, the ions cannot fully compensate the electron saturation at the surface of the target. Therefore, the target surface gains a net negative bias called the bias voltage, V_{bias}.

A plasma is always more positive than its surroundings for this same reason. Electrons collect on surfaces, including the chamber walls, but the plasma-ground potential is small and spread out over a large surface area. The ceramic target, on the other hand, has a small surface area that is biased by roughly half of the peak-peak RF voltage. If a metal target is sputtered using RF sputtering, V_{bias} varies significantly depending on the state of the target's surface, for example, if the surface is oxidized in *reactive* sputtering (next section). Therefore, bias voltage (V_{bias}) is *the* key parameter to achieve repeatability between films.

7.4.5 Reactive Sputtering

Although ceramics can be directly sputtered using RF sputtering, the bonding energies of these materials are very high, and sputter rates can be very low.

For example, the sputter rate of a metal is usually an order of magnitude higher than the sputter rate of its oxide. The resulting low deposition rates for oxides may be useful when very thin films are needed. However, for films of 500–1000 nm, it may be preferable to sputter a metal target, while growing an oxide film by feeding oxygen gas into the system at the substrate. In this way, both metal and oxygen reach the growing film at the same time, but the target sputter rate is metallic in character.

The main concern in reactive sputtering is how to grow an oxide film without oxidizing the target. The answer, as suggested back in the evaporation section, is to design the gas flow inside the system such that minimal oxygen flows by the target. The processing gas (e.g., Ar) should be introduced at the sputtering target, a common feature in modern sputter guns. This keeps the partial pressure of the reactive gas low at the target. The reactive gas (oxygen or nitrogen) can be introduced at the substrate, which should be closer to the pump than the sputter guns. Several companies even make gas rings that introduce an even flow of reactive gas around the substrates. Recall that sputtering occurs close to the viscous vacuum regime, so there are many collisions that keep the gases flowing like low density fluids, and oxygen can be kept effectively away from the target in most processing chambers.

Even with a well-designed gas flow to the system, it is still recommended to use RF power with reactive sputtering so that an oxide can be removed from the surface of the target if one should form. In a sense, there is a competition between the rate of removal of metal atoms from the target and the rate of oxidation at the target. The best way to determine if the surface of the target has been altered by a reactive gas is to carefully record the bias voltage while doing RF sputtering. Once reacted, there is a hysteretic behavior as it is restored to its pure metallic state (Figure 7.24). This hysteresis is

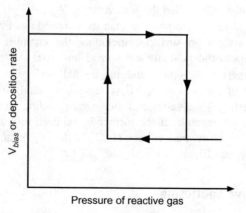

FIGURE 7.24 Reactive sputtering can have hysteretic behavior in that the target oxidizes and sputters back to metallic at different partial pressures of the reactive gas.

also exhibited in deposition rates. Starting with a metallic target surface, the partial pressure of the reactive gas (e.g., O_2) can be increased to produce the desired oxide film at the substrate. However, at a certain partial pressure, the reactive gas also reacts with (i.e., oxidizes) the target. At this point, the deposition rate decreases (often by an order of magnitude) and the bias voltage drops (often by a factor of 2 or more). The reactive gas then needs to be reduced substantially so that the process gas (e.g., Ar) can sputter-clean the target. The shutter on the sputter gun is closed first so that a pure metal is not deposited in the middle of the oxide film when the target is finally clean. Once the target surface is again metallic in nature, the sputter rate increases and the bias voltage goes back to its metallic value. In most cases, the reactive gas pressure can be increased again without inhibiting the deposition rate, as long as the partial pressure is below the value found in the hysteresis (down arrow in Figure 7.24). In some cases, the bias voltage changes slowly, rather than jumping as shown in Figure 7.24. In all cases, however, the bias voltage is an excellent way to monitor the surface condition of the target, and for the highest repeatability in films, this parameter should be maintained between samples by adjusting forward power and gas pressures accordingly.

7.4.6 Optimizing Sputtered Rates

It is important to control the sputtering rate and also the energy of the adatoms as they are collected on the substrate. In evaporation, the atoms are simply leaving a source due to vaporization, and they travel through the chamber in all directions with negligible collisions in the vacuum before landing on either the substrate or the chamber walls. They stick where they land because the substrate and walls are usually cool enough that they have a low vapor pressure (compared to the hot source from which they came).

In sputtering, the atoms coming off the source can have many collisions with gas atoms before they reach the substrate because the sputtering gas itself produces pressures of 1–500 mTorr. In addition, the gas molecules are colliding with each other, which can reduce their total energy. Sputter yield is defined by how many atoms can be ejected from the target per sputtering ion (S_y = target atom per sputtering ion). This yield depends on the respective masses (M_{target}, M_{ions}), the energy of the ion, E_{ion}, and the bonding energy of the target material, E_{bind}

$$S_y = \frac{3\alpha M_{target} M_{ion} E_{ion}}{\pi^2 (M_{target} + M_{ion})^2 E_{bind}} \qquad (7.51)$$

where α depends on the mass ratio (i.e., M_{target}/M_{ion}) and angle of impact, but usually has values of 0.2–0.4 in sputtering. This equation only applies when the E_{ion} is below 1 keV. Impacting ions with energies above this value begin to be implanted into the target and are therefore impractical for sputtering. The noble gases are preferred over other gases since they will not react with the target or film material which would greatly complicate the process. Some researchers use Xe or Kr as the process gas because these atoms have lower ionization energies and have more mass than Ar. However, Ar is orders of magnitude more abundant than Xe or Kr, and the resulting films produced by all noble gases are similar. Although the above equation is useful, there are many places where one can look for experimental and theoretical sputter yields (i.e. manufacturers such as Semicore, www.semicore.com).

From Eq. (7.51), it is clear that the sputter yields of various elements are different, which will affect the sputtering of alloy targets. Recall that evaporation rates between elements could vary by orders of magnitude, but differences in sputter yields are usually within 2–3 times of each other. In addition, sputtering targets can often be "conditioned" so that separate targets are not required. Conditioning involves sputtering an alloy target for a period of time prior to making films so that subsequent films are repeatable in composition. Following an example given by Ohring (2002), consider alloy AB with N surface atoms that have atomic concentrations of C_A and C_B, respectively. If element A has a higher sputter yield than element B ($S_{y,A} > S_{y,B}$), then the initial flux coming off the target surface is richer in A. But, then the surface becomes B-rich. So despite B atoms having a lower sputter yield, more of the surface is occupied with B atoms. After a time, the surface concentration is left rich in B according to

$$\frac{C'_A}{C'_B} = \frac{C_A S_{y,B}}{C_B S_{y,A}} \tag{7.52}$$

The enrichment of the surface in element B means that eventually, the flux coming off the target shifts away from being A-rich. This conditioned target will be able to produce films that have the desired composition of C_A and C_B, but the target surface will remain B-rich (Example 7.8).

These sputter yields (atoms per ion) indicate that at a certain energy (primarily determined by forward power to the sputtering gun), the sputter rate can be increased by linearly increasing the number of ions that strike the target. However, this means that the pressure in the chamber must be increased, and as you might have already guessed from your knowledge of kinetic theory of gas (Section 7.2.1), this means that gas phase collisions increase. In reality, these collisions are not strictly elastic, and the thermal energy of the ions and sputtered atoms is not uniform, especially in a plasma. Therefore, although increasing the process gas pressure increases the deposition rate of the film at lower pressures, there will be a maximum rate after which continued increasing of pressure actually reduces the deposition rate. See Figure 7.25. This phenomenon is very dependent on the vacuum chamber geometry, so empirical determination of ideal pressures is usually required.

EXAMPLE 7.8 *Calculate the composition of a conditioned target of permalloy ($Fe_{20}Ni_{80}$) sputtered at 20 mTorr with Ar ions at 500 eV. The sputter yields for Fe and Ni are 1.6 and 1.9, respectively.*

Use Eq. (7.52) with the sputter yields:

$$\frac{C'_{Fe}}{C'_{Ni}} = \frac{C_{Fe}S_{y,Ni}}{C_{Ni}S_{y,Fe}} = \frac{20 \cdot 1.9}{80 \cdot 1.6} = 0.30$$

So, the target surface changes from an Fe/Ni ratio of 0.25 to 0.30. Although this is fairly close, looking at Table 7.5, it is clear that the targets of other alloys will certainly require conditioning before repeatable film compositions can be obtained.

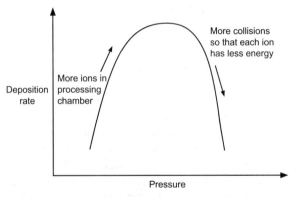

FIGURE 7.25 Deposition versus processing pressure. At low pressures, increasing the pressure means there are more ions that sputter the target (Ar^+ for example), but the sputter yield (atom/ion) is constant so film deposition rates increase. However at high pressures, ions lose energy before they strike the target due to collisions, so the sputter yield (atom/ion) decreases.

As with evaporation, thickness uniformity can be a problem with substrates that have complex textures. Figure 7.26 shows the top of a film that was sputtered onto a textured substrate. Shadowing prevented the filling of 500 nm-diameter holes in the substrate. However, the μm-wide waveguides shown in Figure 7.2 were made by sputtering, and the coverage of the guide is fairly conformal. In general, structures that are smaller than 1 μm, or that have aspect ratios greater than 2, tend to form uneven films like the one in 7.26.

Summary. In summary, the important parameters to optimize in sputtering are forward power and pressure. Deposition rates tend to be slower than evaporation rates (10–100 nm/min compared to 100–1000 nm/min for evaporation). If fast deposition rates are desired, it is best to calibrate a sputtering process as a function of pressure by growing several test films at various pressures between 1 and 500 mTorr. There will be a peak in deposition rate at a pressure between these values. Adatoms arriving at the growing film have more energy during sputtering than those during evaporation due to

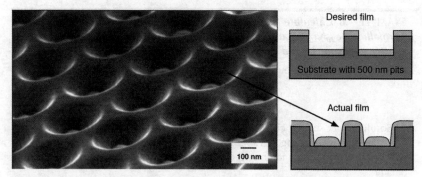

FIGURE 7.26 As in evaporation, shadowing in sputtering also leads to non-conformal coating of complex substrates. However, the waveguide in Figure 7.2 was made by sputtering, so uniform coating can be obtained with wide, shallow structures by utilizing substrate movement and high temperatures.

momentum transfer from ions and thermalization in the plasma. Therefore, Zone T and Zone 2 microstructures (Figure 7.9b) are more frequently obtained by sputtering, especially when using unbalanced magnetrons or substrate bias, unless an excess of Ar gas is used. Sputtered metal films also tend to have higher densities and better adhesion than evaporated metal films. Oxide deposition rates can be very slow (<1–10 nm/min) due to their high binding energies, and therefore low sputter yields. For faster deposition rates of oxide films, metal targets can be sputtered while oxygen is fed into the processing chamber next to the substrates. These oxide films tend to form amorphous "grains" as in evaporation, and annealing is required to obtain the desired crystalline phase. Finally, uniform coatings can be achieved via substrate rotation unless very small (<1 μm) or high aspect ratio (depth/width >2) features are present on the substrate

7.5 CHEMICAL VAPOR DEPOSITION

7.5.1 Process Overview

Chemical vapor deposition (CVD) is different than evaporation and sputtering in two important ways: the source atoms are removed by chemical, not physical means, and low vacuums are typically used. In CVD, volatile (or gaseous) reactants are flowed past the substrate by a carrier gas (e.g., H_2) in viscous regime vacuums (1 atm − 1 mTorr). If the thermodynamics are favorable for diffusion and reaction at the substrate, a solid film is formed and any volatile by-product flows away from the substrate. The reactors are therefore designed differently to allow a uniform viscous flow path, rather than the "line-of-sight" molecular motion that occurs in evaporation and sputtering.

The most common reactors are quartz tubes inside tube furnaces, or vacuum chambers that are configured for homogeneous gas flow. For CVD, the critical parameter for homogenous films is even gas flow, more than substrate position or motion. It is also important to consider that a viscous vapor can be expected to contact more of the substrate surface, even if the surface is rough. This means that CVD typically yields more "conformal" films, meaning uniform thickness, even inside deep holes.

Unlike the reactor in Figure 7.1c, which shows the general idea of CVD, gas flow is more uniform in reactors such as Figure 7.27. In other designs, a "showerhead" is used to distribute reactant gases evenly over substrates that are resting on holders, such as the reactor in Figure 7.28. Unlike evaporation and sputtering, substrate heating is not optional, but required since each reaction in CVD requires careful control of temperature. Some reactors, such as the quartz tubes shown in Figure 7.27, are heated by a tube furnace that is around the entire chamber to provide the heat of reaction. If the reaction is slow enough to enable only heterogeneous nucleation, the walls can be hotter than the substrates to improve volatility so that films are not deposited on the walls. In other cases, the substrate must be heated, but the walls of the chamber are kept cooler than the reaction temperature so they remain uncoated. The lamp heater in Figure 7.28, for example, shines through a window onto the back of the substrate to provide local heat only.

The source materials and precursors for CVD processes frequently present safety hazards. Some sources are flammable (silane), some carcinogenic (several metal-organics) and others toxic (arsenic-containing gases, such as arsine). The gases must be handled in a manner that prevents human

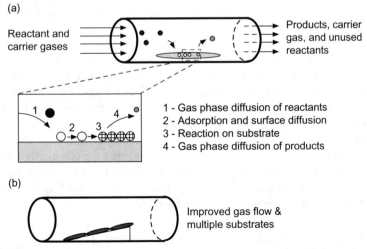

FIGURE 7.27 Schematic diagrams of tubular CVD reactors (a) simplified gas flow and (b) a design that allows multiple simultaneous depositions and improved gas flow (minimized boundary layers).

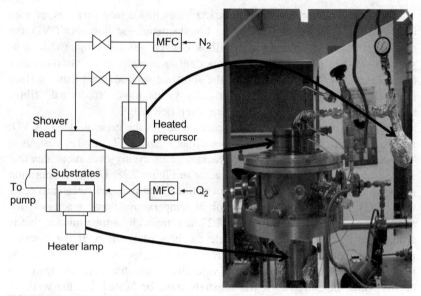

FIGURE 7.28 Schematic and photo of a CVD system where N_2 is flowed past a precursor that is heated such that the reactant molecules evaporate or sublime and N_2 carries the reactant to the chamber. (MFC = mass flow controller.) A showerhead is used to distribute the precursor gas over the substrate evenly. The substrate is heated to the temperature that drives the film-forming reaction (in this case oxidation). The bow-ties are valves. *Courtesy: Sun Sook Lee.*

exposure as they are flowed from the initial gas tank, through the reactor, and out a filtered vent. Leaks cannot be tolerated and many gas sensors and alarms have been created over the lifetime of CVD development. Materials Safety Data Sheets (MSDS) and exit procedures should be reviewed before one begins work with these materials.

A new thermodynamic versus kinetic balance is introduced when discussing CVD, namely reaction versus diffusion/convection. In viscous flow, a boundary layer of relatively stagnant vapor will exist at all surfaces. Reactants will need to diffuse through this boundary layer to react at the substrate, and products will need to diffuse back out of this boundary layer for the reaction to continue in the forward direction.

This section begins with general thermodynamics of formation reactions, and then it lists the six generic types of reactions that are commonly used in CVD. Once the reactions are explained, the viscous regime kinetics that are unique to CVD are discussed. Finally, surface coverage and film uniformity will be described as a major advantage for CVD.

7.5.2 Thermodynamics of Formation Reactions

CVD films are grown under conditions that are closer to equilibrium than sputtering and evaporation. In the preceding sections, atoms reached a substrate with varying degrees of energy with which they could diffuse along the surface before another layer of atoms landed and "froze" them in place. With CVD, a reaction that produces a solid phase must be thermodynamically favorable for the film to grow. In some cases, the walls can be heated or cooled to discourage growth except on the substrate.

To begin our discussion of chemical reactions, let's first discuss a generic example:

$$bB(g) + cC(g) \rightarrow dD(s) + eE(g) \tag{7.53}$$

Note that all of the reactants and products are gaseous (g) except D, which is the film. Also the lower case letters represent the number of molecules of each species taking part in the reaction. For example, using silane and oxygen to form SiO_2, the reaction will be

$$SiH_4(g) + O_2(g) \rightarrow SiO_2(s) + 2H_2(g) \tag{7.54}$$

one molecule of each is reacted, and the film plus two hydrogen molecules are produced. Although CVD is not strictly at equilibrium (gases are flowing in and out of the chamber), thermodynamics can still be a good indication of the likelihood of film formation and process analysis. For the generic equation above, we can find the Gibbs free energy of reaction (ΔG) by

$$\Delta G = dG_D + eG_E - bG_B - cG_C \tag{7.55}$$

where

$$G_i = G_i^\circ + RT \ln a_i \tag{7.56}$$

G_i is the free energy of species "i" in a reference state (e.g., 1 atm, 298 K), R and T are the gas constant and temperature, and a_i is the activity of "i" (e.g., partial pressure or concentration). Therefore the free energy of reaction is

$$\Delta G = \Delta G^\circ + RT \ln \left[\frac{a_D^d a_E^e}{a_B^b a_C^c} \right] \tag{7.57}$$

At equilibrium $\Delta G = 0$ so

$$-\Delta G^\circ = RT \ln k \quad \text{and} \quad k = \exp\left(\frac{-\Delta G^\circ}{RT} \right) \tag{7.58}$$

where k is the equilibrium constant (equal to the ratio of activities above). A reaction must have a negative ΔG to be thermodynamically feasible, but ΔG also has to be close to zero for a film to form by heterogeneous nucleation. If ΔG is more negative than 10s of kJ/mol, homogeneous nucleation may also occur, leading to airborne precipitates that can land on the substrate causing defects in the film. Of course, processes that want to form nanoparticles for other applications may use such a reaction intentionally.

Although the *ratio* of product to reactant could be determined with the equations above, how can the actual partial pressures of each component be found? The partial pressures of all gaseous species must equal the total pressure in the chamber.

$$P_B + P_C + P_D + P_E = P_{total} \tag{7.59}$$

This relationship gives us another equation, which is usually enough to solve for equilibrium constants and activities of each component.

Other methods to monitor the intentional, as well as unintentional reactions in CVD begin by controlling the total moles of reactant using a mass flow controller at the input. Next, a mass spectrometer can be used to measure the amount of unused reactant, volatile products, and any unintentional products from unpredicted gas phase reactions. Finally, a microbalance can be used to measure the mass of the film being deposited. Taken together, a complete model of each CVD process can be created.

7.5.3 Types of Reactions

There are six general types of reactions used in CVD, shown in Table 7.5. Pyrolysis is simply the thermal decomposition of a reactant, for example silane (SiH_4) into Si and H_2 gas. Reduction and oxidation use reactive gases (such as H_2 or O_2, respectively) to decompose the reactant by forming more stable products, such as Si or oxides. These reactions are usually forward progressing at lower temperatures than thermal decomposition, and can be used in combination with thermal decomposition. Compound formation is similar in that reactive gases (such as ammonia or water) are used to decompose reactants, to form nitrides and oxides. Hydrolysis is a popular method to obtain high-quality ceramic coatings, but the volatile products are

TABLE 7.5 Types of CVD Reactions

Reaction name	Generic reaction or example
Pyrolysis	$AB(g) \rightarrow A(s) + B(g)$
Reduction	$AX(g) + 1/2\, H_2(g) \leftrightarrow A(s) + HX(g)$
Oxidation	$AX(g) + O_2(g) \rightarrow AO(s) + [O]X(g)$
Compound Formation	$AX(g) + H_2O(g) \rightarrow AO(s) + H_2X(g)$ $AX(g) + NH_3(g) \rightarrow AN(s) + H_3X(g)$
Hydrolysis	$ACl_2(g) + CO_2(g) + H_2(g) \leftrightarrow AO(s) + 2HCl(g) + CO(g)$
Metalorganic	$Ga(CH_3)_3(g) + AsH_3(g) \leftrightarrow GaAs(s) + 3CH_4(g)$
Disproportionation	$2AB_2(g) \leftrightarrow A(s) + AB_4(g)$
Reversible Transfer	$6In(s) + 2AsCl_3(g) + 3H_2(g) + As_4(g) \rightarrow$ $6InAs(s) + 6HCl$

corrosive and toxic (HCl and CO). For this reason, and also to find volatile precursors for a wider range of materials, metalorganic precursors have been slowly introduced to the CVD market. These precursors are chemical compounds with metal cations combined with organic anions. The example below also uses arsine which has many safety hazards, but other precursors are less hazardous. Disproportionation uses compounds of elements with multiple valence states to deliver that element to the substrate. For example, the gas AB_4 passed by a heated bulk source of element A and the backward reaction shown below would take place. The newly formed AB_2 would then be transported to cool substrates for the forward reaction to occur. Similarly, reversible transfer reactions also use a reactive compound to "pick up" an element (such as In in Table 7.5) at a hot source and "drop it off" at the substrate (as InAs). Please remember to look up the MSDS before purchasing precursors.

7.5.4 Kinetics of CVD

CVD involves the viscous flow regime where flow is caused by a pressure differential as discussed in Section 7.2.1. There is also a diffusion component to CVD that is different than the surface diffusion of adatoms discussed in Section 7.2.2. In CVD, the reactants and products must also execute gas phase diffusion through a boundary layer of stagnant gas that exists next to the substrate. The total steps required for film growth in CVD are:

1. Gas phase diffusion of reactants through boundary layer
2. Adsorption and surface diffusion of adatoms
3. Reactions on the substrate
4. Gas phase diffusion of products away through boundary layer

First we'll define a velocity profile for the gas flowing due to the pressure gradient. A boundary layer forms due to the boundary condition of

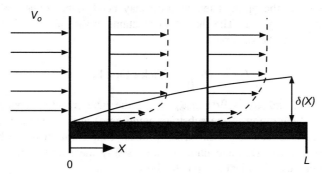

FIGURE 7.29 Velocity profile of gases flowing over a substrate in a CVD reactor (see Figure 7.27a).

(velocity = 0) at any surface, as shown in Figure 7.29. The thickness, δ, of the boundary layer depends on the density, ρ, and viscosity, η, of the gas and the length that the gas travels over the surface, x,

$$\delta(x) = \frac{5x}{(Re)^{\frac{1}{2}}} \quad Re = \frac{v_o \rho L}{\eta} \tag{7.60}$$

where Re, is the Reynolds number. Integrating with respect to x from 0 to L (length of the substrate), the average boundary layer will be

$$\delta_{ave} = \frac{10L}{3(Re)^{1/2}} \tag{7.61}$$

Reactants must diffuse through this stagnant boundary layer to reach the substrate and products must diffuse away to keep the reaction forward balanced. Therefore, it is desirable to have as thin a boundary layer as possible. Looking at the equations, the velocity, v_o, can be increased, but that would mean more wasted reactants. Also, if the flow becomes turbulent, the reactants will not reach the substrate homogeneously so the film will not be uniform. Another option is to choose a new carrier gas that has reduced viscosity or increased density. Finally, for the tubular reactors, quartz tubes are used that are usually long enough for the walls all act as surfaces with $v = 0$, so v_o is rarely reached at the center of the tube.

Diffusion through the boundary layer can be described by Fick's First Law. In a gas, diffusion is related to pressure (P) and temperature (T) as

$$D = D_o \left(\frac{P_o}{P}\right) \left(\frac{T}{T_o}\right)^{1.5} \tag{7.62}$$

where P_o and T_o are the standard state at which D_o is given. Since the diffusion coefficient increases with temperature and decreases with P, deposition rates can be increased by using higher temperatures and lower pressures. However, a balance must be struck because the reaction rate also depends on temperature, so the optimal temperature may need to be found experimentally for a given system. The flux, J, of reactants to the substrate and of products away from the substrate is

$$J_i = -D_i \frac{dc_i}{dx} \quad \text{(Fick's First Law)} \tag{7.63}$$

where $dx = \delta$, and dc_i is first $dc_{reactant}$, or the input reactant concentration (reactant concentration at the substrate being zero) and dc_i is next $dc_{product}$, the product produced at the substrate (product concentration in the bulk gas being zero). Since gas concentrations are usually measured as partial pressures, we can use the ideal gas law to show

$$J_{\text{diffusion},i} = -D_i \frac{P_i - P_{io}}{\delta RT} \tag{7.64}$$

where it is important to remember that a diffusive flux moves opposite to the concentration gradient (or in this case the partial pressure gradient $\frac{P_i - P_{io}}{\delta}$). Entropy drives motion from higher to lower concentrations, and this accounts for the negative sign in Eq. (7.64).

7.5.5 Deposition Rate and Uniformity

The end result of the varying δ as discussed above, is that substrates at the input side of a reactor (Figure 7.27) will have thicker films deposited on them than the substrates at the output side. As mentioned in Section 7.3.4, the devices that are being made using these thin films often have tight tolerances for layer thickness. The devices will not perform to specifications if the film thicknesses are outside their design. In early reactor designs (Figure 7.27a), the flux ($J_{\text{diffusion},i}$) films by the input could be 10 times thicker than films 10 inches into the reactor. To keep δ uniform for a given sample batch, the substrates are often tilted with respect to the input (Figure 7.27b).

Once the reactants arrive at the substrate, a reaction has to occur for the film to form in CVD. The flux of reactants that are removed from the gas due to this step is defined by

$$J_{rxn,i} = k \frac{P_{io}}{RT} \tag{7.65}$$

where k is the reaction constant defined by Eqs. (7.57) and (7.58), and P_{io} is the partial pressure of the reactant at the substrate. Surface adsorption and migration are often included in the kinetics of "reaction-limited," and the value of the measured activation energy distinguishes which step (#2 or #3 from Figure 7.27a) is responsible for limiting the deposition rate. At steady state, the flux of reactants to the substrate will be equal to the flux of reactant removal due to the CVD reaction, or $J_{\text{diffusion},i} = J_{rxn,i}$. Therefore,

$$\frac{kP_{io}}{RT} = -D_i \frac{(P_i - P_{io})}{RT} \quad \text{and} \quad P_{io} = \frac{P_i}{\left(1 + \frac{k}{D}\right)} \tag{7.66}$$

This equation indicates that the partial pressure of the reactant at the substrate will go to zero when the reaction is fast and the reactant is consumed as soon as it lands on the substrate ($k >> D$). Alternatively, if the reaction is slower than diffusion, the reactants will build up at the substrate ($D >> k$). As we've seen in Eq. (7.58), the reaction constant depends on temperature exponentially (Arrhenius behavior), but the diffusion coefficient in a gas is only a function of $T^{1.5}$, as shown in Eq. (7.62). Therefore, film uniformity is best achieved when the CVD process is diffusion-controlled because the deposition rate (μm/min) is equal to the reactant concentration at the substrate (atoms per area per time) divided by the atomic density (atoms per volume).

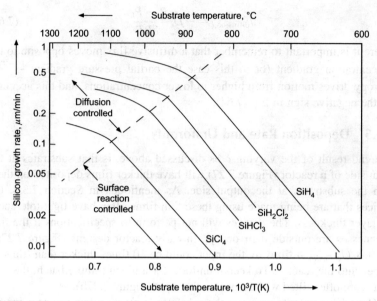

FIGURE 7.30 Deposition rate dependence on temperature for CVD grown Si. *Springer/ Microelectronic Materials and Processes 1989, page 230, Chemical vapor deposition, W. A. Kern, Figure 1, Copyright 1989]* With kind permission from Springer Science and Business Media.

An example of the temperature dependence of the deposition rate is seen in Figure 7.30. At high temperatures, gas phase diffusion controls the deposition rate, which is relatively constant with temperature. Therefore, for this Si growth process, high temperatures would yield more uniform films. At low temperatures (high 1000/T), the reaction controls the deposition rate as indicated by the Arrhenius behavior. The free energy change of the reaction is the slope/R. Interestingly, in this example, the slope and therefore ΔG^o was equal for many different reactants (also called precursors). Since the free energy of reaction is expected to vary with reactant composition, in this case, it appears that surface diffusion is actually the rate limiting step. The surface diffusion coefficient, D_s, has an Arrhenius dependence on temperature [$D_s \propto \exp(-E_s/kT)$] similar to k [$k \propto \exp(-\Delta G^o/kT)$]. The value of the energy barrier can help determine which step from Figure 7.27 is the rate-limiting step.

Pressure is another control variable if gas phase diffusion is the rate-limiting step. Atmospheric pressure CVD (APCVD) is common because it involves very simple equipment, and has the advantages of lower temperatures and higher deposition rates. But the rates can often be too high, resulting in homogeneous nucleation and defects. Equation (7.62) shows that diffusion in the gas is inversely proportional to P. For this reason, low pressure CVD (LPCVD) is often used for more conformal step coverage and larger wafer capacity compared to APCVD. In LPCVD, deposition rates at comparable

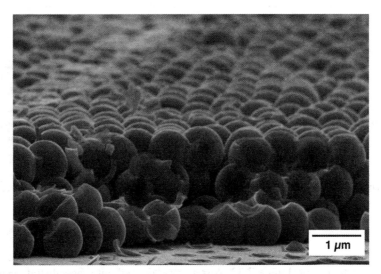

FIGURE 7.31 CVD coating of 500 nm diameter glass beads, demonstrating that carefully controlled CVD can result in conformal coating of very complex substrate structures. *Courtesy: Sun Sook Lee.*

reactant partial pressures are low because pressures are low. Plasmas can also be used to increase film densities, as with ion-beam-assisted evaporation. These processes are called plasma-enhanced CVD (PECVD).

Once engineered properly, transport in CVD can carry homogeneous reactants to the most complicated substrate surfaces, such as the nanospheres shown in Figure 7.31, for a completely uniform coating. Recall that evaporation and sputtering are more "line of sight" processes because they occur in molecular-regime vacuums. Atoms fly to substrates and stick, with some surface diffusion possible if substrate heaters or ion bombardment are used, so a complex coating, like the one in Figure 7.31, is not possible with PVD.

A particularly unique case of CVD is called "atomic layer deposition" (ALD). In this process, alternating reactant gases are passed over the sample sequentially. ALD reactants are very sensitive to surfaces, either by direct reaction or by surface-mediated catalysis, so the deposition stops after a monolayer of new growth. For example, in the case of aluminum oxide films, an Al-organic precursor would be passed over the sample in a pulse that is limited by special valves, called micro-valves. A monolayer of the Al-ligand would be deposited on the surface until all of the surface sites are full, then the rest of the precursor would be flushed out in the next pulse (called a purge) which is just the carrier gas by itself. Next, an oxygen precursor would be injected into the system to pass over the substrate, oxidizing the organic ligand to leave Al-O over the entire surface. In this way, very conformal films can be coated one atomic layer at a time, hence the name atomic layer deposition. ALD was specifically developed for dynamic

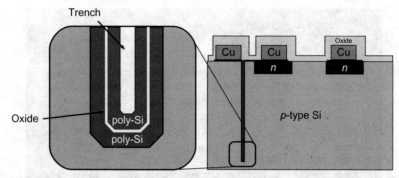

FIGURE 7.32 Simplified schematic of DRAM deep-trench capacitor with thin conformal coatings. Very high aspect ratios are possible with ALD.

random access memory (DRAM), where a deep-trench capacitor is needed next to a transistor, Figure 7.32. A trench is "drilled" into the substrate using a "focused ion beam" which is a nanometer wide ion beam that essentially sputters away the trench. Typical trench dimensions are 100s of nanometers in diameter and 10s of microns deep. Next these trenches are uniformly coated all the way down with metallic polycrystalline Si (poly-Si). Then, a very thin (<20 nm) dielectric material is deposited, and then poly-Si again. This way, a large area capacitor can be made while maintaining high areal densities for devices in memory applications.

Summary. DRAM is a great example to lead a discussion summarizing both the advantages and disadvantages of CVD. First, extremely conformal coatings are possible with CVD, especially the atomic layer deposition (ALD) variant of CVD, which are simply not possible with evaporation or sputtering. There are obviously many unique structures that can be made with this process. However, the deep-trench capacitor also demonstrates a disadvantage of CVD. The deep-trench capacitor uses poly-Si for contacts because metals are extremely difficult to make with CVD due to their reactivities. Semiconductors and compounds are typical CVD materials, and conveniently poly-Si is electrically metallic. However, it still has much higher resistivities (m$\Omega \cdot$cm) than pure metals ($\mu\Omega$ cm). Poly-Si is also used as a "Si-friendly metal," whereas many metals will destroy the device performance of Si. Therefore, poly-Si has a secondary purpose as a protective coating, which is deposited before the metallic contacts, which are coated in a separate processing chamber by sputtering or evaporation. Oxide coatings by either CVD or sputtering are used as protective coatings and as electrical insulators to isolate contact lines from each other in the DRAM example (Figure 7.32). CVD has a further advantage of large batch capability as long as even gas flow designs are used. A summary of other CVD processes and resulting crystallinity in the films is given in Table 7.6.

TABLE 7.6 Thermal CVD Films and Coatings[a]

Deposited material	Substrate	Input reactants	Deposition temperature (°C)	Crystallinity
Si	Single-crystal Si	$SiCl_2H_2$, $SiCl_3H$, or $SiCl_4 + H_2$	1050–1200	E (epitaxial)
Si		$SiH_4 + H_2$	600–700	P (polycrystalline)
Ge	Single-crystal Ge	$GeCl_4$ or $GeH_4 + H_2$	600–900	E
GaAs	Single-crystal GaAs	$(CH_3)_3Ga + AsH_3$	650–750	E
InP	Single-crystal InP	$(CH_3)_3In + PH_3$	725	E
SiC	Single-crystal Si	$SiCl_4$, toluene, H_2	1100	P
AlN	Sapphire	$AlCl_3$, NH_3, H_2	1000	E
In_2O_3:Sn	Glass	In-chelate, $(C_4H_9)_2Sn(OOCH_3)_2$, H_2O, O_2, H_2	500	A (amorphous)
ZnS	GaAs, GaP	Zn, H_2S, H_2	825	E
CdS	GaAs, sapphire	Cd, H_2S, H_2	690	E
Al_2O_3	Si, cemented carbide	$Al(CH_3)_3 + O_2$, $AlCl_3$, CO_2, H_2	275–475 850–1100	A A
SiO_2	Si	$SiH_4 + O_2$, $SiCl_2H_2 + N_2O$	450	A
Si_3N_4	SiO_2	$SiCl_2H_2 + NH_3$	750	A
TiO_2	Quartz	$Ti(OC_2H_5)_4 + O_2$	450	A
TiC	Steel	$TiCl_4$, CH_4, H_2	1000	P
TiN	Steel	$TiCl_4$, N_2, H_2	1000	P
BN	Steel	BCl_3, NH_3, H_2	1000	P
TiB_2	Steel	$TiCl_4$, BCl_3, H_2	>800	P

[a]Tabulated in Ohring (2002), reproduced by permission of Academic Press.

7.6 POST-PROCESSING OF FILMS AFTER DEPOSITION

7.6.1 Annealing

Many thin films do not have the properties that are expected based on bulk materials. In other words, a 1 μm thick film does not behave the same as if you had simply cut 1 μm thick slice of material from a chuck of the same material. What is different?

The main difference is that the film is formed by collecting atoms onto a substrate and "freezing" them in place. Although there is some surface diffusion as discussed above, the adatoms have a limited range of motion before they are locked in place by the next layer of depositing atoms. Therefore, films compared to bulk have typically more dislocations from strain and also smaller grains, so more grain boundaries. This increase in defects often means films exhibit degraded properties compared to bulk, but a device engineer can design around it.

Especially important is that the film may not be in the desired crystalline phase. A great example of this is aluminum oxide (or alumina). In bulk form, Al_2O_3 is extremely stable in the $\alpha - Al_2O_3$ crystalline phase. However, as a sputtered film, alumina is amorphous even when the substrate is heated over 500°C. Annealing can help transform a film into the desired phase. Alumina films must be heated to 1000°C to transform into γ-alumina (a metastable cubic phase), and only at 1200°C does it finally transform into the bulk equilibrium phase, α-alumina.

As mentioned in Section 7.2.2, slow deposition rates or high substrate temperatures can be used to obtain the desired crystalline phase. But if fast deposition is needed for large-scale batch processing or the required temperatures are too high for the vacuum system, then post-annealing will be needed. The downside is that more stress builds up. If conditions can meet Zone 3 of the Thornton diagram during deposition, the film might crystallize one layer at a time by a combination of surface diffusion and bulk diffusion. However after the film is made, the entire structure must be heated and cooled and thermal expansion mismatches may cause defects to form.

7.6.2 Patterning

This entire chapter dealt with the fabrication of thin films, or planar layers of atoms. In reality, these films are just the building blocks of integrated devices and each layer must be appropriately shaped before the next layer is fabricated. The waveguide isolator in Figure 7.2 is one example of the type of shaping, or patterning that is required. The layers are all grown as infinite (compared to the thickness) planes of atoms. The waveguide must be patterned from the film.

There are two popular types of patterning, Figure 7.33. One is photolithography, and the other is electron-beam (e-beam) lithography. In both

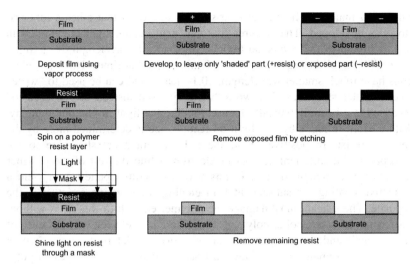

FIGURE 7.33 Patterning of thin films can be done by photolithography, involving either a positive or negative photoresist.

cases, a polymer layer is deposited onto the thin film layers. This polymer is then exposed to either light (photolithography) or an electron beam (e-beam lithography). The polymer layer is called "resist" because it will resist etching. For positive resist, the part of the polymer that is exposed to the light or e-beam is degraded by the incoming energy. Once the sample is exposed to a "developer," such as acetone, the degraded polymer dissolves away, any parts of the polymer that were shadowed from the light will not be removed. In the waveguide example, a line of polymer would be left over the place where the guide is to be located. Subsequent etching is then used to dissolve the thin film layers, except the line of film that is under the protective polymer resist. The end result is a line as shown in Figure 7.2.

If there is positive resist, you can imagine there is also negative resist. In this case, the structure remaining after etching will be the opposite of the mask. This happens when resists are designed to crosslink in the light so the exposed parts become stronger and therefore remain after the unexposed polymer regions are dissolved. The map below shows how these two types of resist work in making waveguide structures, such as that shown in Figure 7.2. By now, you may recognize from the figure that there are two subsequent steps for making the optical isolator shown. After patterning the waveguide, a top cladding (MgO) is deposited, followed by a permanent magnet (SmCo).

There is much more to post-processing than can be written in one section of one chapter. The main goals of this field have been (1) obtaining ever-smaller structures for higher density performance of systems like cell phones, hard drives, logic, etc., and (2) successful etching of a growing number of materials. It was for ever-smaller structures that e-beam lithography was developed. The features that can made using light are limited in size to structures that are equal to or larger than the wavelength of the light. Deep-UV photolithography (~100–200 nm) is used at times. However, electrons have much smaller wavelengths (0.1–1 nm) and can be used to "write" much smaller structures. The word "write" is used instead of "exposed to" because the e-beam is typically focused, unlike the light in photolithography. With e-beam lithography, the limit on feature size is determined more by the polymer resist. In positive resist, the polymer must crosslink due to the absorption of a sufficient number of electrons while the rest of the polymer remains weak enough to etch. In this way, the written pattern will yield a protective coating for subsequent film etching. In negative resist, it is the opposite. The e-beam must damage the polymer enough to weaken it without destroying the surrounding polymer. The difference between the etch rates of the "strong" and "weak" sections of the polymer determine how sharp a structure will remain after etching. In practice, it is difficult to design polymers that offer drastic differences with few enough electrons to yield a very small structure. For example, if the edges are fuzzy or rough for 1 nm, then the end devices must be larger than 2 nm or the whole structure is poorly defined.

The second goal (etching varied materials) has been accomplished by developing just as varied a number of etching procedures. For large features, such as those developed by photolithography, classic wet etchants can be used. However, these liquids often etch isotropically (the same in all directions), so the device structure will be smaller than the protective resist cap due to undercutting. Still, new techniques are constantly emerging and therefore a growing number of materials properties are available to thin film device designers.

7.7 SUMMARY

This chapter discussed ways of vapor processing techniques, which primarily yield thin films of desired materials. These films can be patterned and layered to build devices, such as integrated circuits, hard drive heads and media, and photonic lasers and circuits, to name just a few. The three most important processes are evaporation, sputtering and chemical vapor deposition. Hybrid methods are also used commercially, and were mentioned when appropriate in the chapter; for example ion-beam assisted evaporation combines evaporation with a mild sputtering of the film as it grows. Similarly, plasma-enhanced CVD also uses a plasma to densify the growing films. All three of these general processes involve atomizing a source material, and transporting the atoms to a substrate where they are collected to form a film. Post-processing steps, such as annealing and patterning, will be required for most of the devices listed above.

All of these vapor phase techniques take place in a vacuum processing chamber, so an understanding of vacuums and the kinetic theory of gases is central to understanding vapor processes. The kinetic theory of gases relates macroscopic (measureable) quantities, such as pressure, to microscopic quantities, such as gas molecule velocities and collisions. Two vacuum regimes are defined based on the mean free path to collisions. The viscous regime is defined as low vacuum (many molecules per volume) where gas molecules collide with each other much more than with the walls of the processing chamber. Alternatively, the molecular regime is defined as ultra high vacuum (UHV; few molecules per volume) where the molecular mean free path is larger than the chamber dimensions, so the molecules rarely collide with each other.

Within each regime, the gas molecules have a distribution in velocities, and by understanding this distribution, it is possible to determine impingement rates of gas molecules on any surface inside the processing chamber. Evacuation speeds are determined by impingement of molecules on vacuum pump inlets, film purity is determined by impingement of air remnants on substrates, and some of the processes (evaporation) have deposition speeds that are determined by the impingement of atoms off of a source. So, the kinetic theory of gas is required for accurate processing control.

This chapter also described typical film microstructures, using the Zone model from the Thornton diagram. The microstructural Zone of a given film is determined by the energy atoms have when they arrive at the substrate. These "adatoms" either stick where they land, or they move by surface or bulk diffusion. At low substrate temperatures (low adatom energies), Zone 1 columnar grains are formed with voids between them because shadowing dominates the microstructure. Adatoms gain energy with higher substrate temperatures (or in some cases from ion bombardment), and they can diffuse on the surface before the next layer of adatoms deposits. The microstructure becomes Zone T type with small grains and dense grain boundaries. With even higher temperatures, Zone 2 columnar grains are formed, larger and with faceted tops instead of rounded domes. Finally, at the highest substrate temperatures, bulk diffusion can become active and equiaxed grains are formed. Roughly speaking from evaporation to sputtering to CVD, the atoms landing on the substrates have increasing energy. The typical deposition rates for each are: evaporation $= 1-10$ μm/h, sputtering $= 0.1-1$ μm/h, and CVD $= 10-100$ μm/h. Under special conditions of low deposition rates and matching film−substrate pairs, single crystal films will form by epitaxy.

The first process described, evaporation, involves simple heating of a source such that atoms vaporize into the vacuum processing chamber. Low pressures ($\sim 10^{-6}$ Torr) can be used to avoid impurities and allow atoms to transfer to the substrate without collisions in the vapor. The rate of deposition can be fairly high (several μm in minutes to hours) depending on the temperature of the source. The vapor pressure increases exponentially with temperature, which enables a wide range of deposition rates. However, it also means that alloys and compounds can be difficult to evaporate due to non-stoichiometric evaporation. Ideal materials for evaporation are pure metals or combinations of these using several sources to obtain alloy films. Some oxides and compounds can be evaporated by e-beams, which heat small spots, so higher local source temperatures can be reached. Due to the low energies involved in evaporation, uniform thicknesses are more difficult to achieve than in sputtering or CVD, especially if the substrate has an uneven surface. The geometry of deposition is therefore critical during evaporation processes.

Sputtering involves the momentum transfer from a processing gas to the source (called a target) by ionization and electrostatic acceleration of the gas ions. The source atoms are then knocked off of the target and travel to the substrate for collection into a film. Higher pressures are used than in evaporation due to the need for a processing gas. Once ionized, the processing gas forms a plasma. The plasma can be either direct current (DC) for metal targets, or radio frequency (RF) for all targets including insulators. Magnets behind a target in a "magnetron" configuration are used to gather electrons, and therefore the plasma, close to the target surface so that higher sputter rates can be achieved. The sputter rates of different materials depend on the momentum transfer (e.g., atomic masses) and the binding energies of the material. For example, metals typically sputter 10 times faster than oxides. For faster deposition of oxides, reactive sputtering can be used in which metal targets are sputtered while oxygen is injected into the processing chamber to produce oxide films.

Chemical vapor deposition (CVD) involves the flow of volatile reactants over substrates using a carrier gas at higher pressures (low vacuum mTorr to atmosphere). These reactants then diffuse through the viscous vapor to the substrate to react and form a film. Reactions are designed so that any by-products are volatile, and they can diffuse away from the substrate to the vacuum pump. Reaction and transport thermodynamics and kinetics are therefore very important for understanding and controlling CVD processes. Although most metals are not compatible with CVD due to their reactivity, many recipes exist for semiconductors and insulators and the resulting coatings can be conformal even over very complex substrate geometries, especially in the special case of atomic layer deposition (ALD).

BIBLIOGRAPHY AND RECOMMENDED READING

Campbell, S.A., 2001. The Science and Engineering of Microelectronic Fabrication, second ed. Oxford University Press, New York, NY.

George, S.M., 2010. Atomic layer deposition: An overview. Chem. Rev. 110, 111–131.
Halliday, D., Resnick, R., 1974. Fundamentals of Physics. John Wiley & Sons, New York, NY.
Hoffman, D., Thomas, J., Singh, B., 1997. Handbook of Vacuum Technology. Academic Press, San Diego, CA.
Maissel, L., Glang, R., 1970. Handbook of Thin Film Technology. McGraw-Hill Professional Publishing, New York, NY.
Ohring, M., 2002. Materials Science of Thin Films, first and second ed. Academic Press, San Diego, CA.
Pierson, H., 1992. Handbook for chemical vapor deposition: Principles, technology, and applications. Materials Science and Process Technology Series. Electronic Materials and Process Technology. Noyes Publications, Norwich, NY.
Powell, R., Rossnagel, S., 1999. PVD for Microelectronics: Sputter Deposition Applied to Semiconductor Manufacturing Thin Films. Academic Press, San Diego, CA.
Smith, D., 1995. Thin Film Deposition. McGraw-Hill Professional Publishing, New York, NY.
Thornton, J.A., 1977. High rate thick film growth. Annu. Rev. Mater. Sci. 7, 239–260.
Vossen, J., Kern, W., 1979. Thin Film Processes I. Academic Press, San Diego, CA.
Vossen, J., Kern, W., 1991. Thin Film Processes II. Academic Press, San Diego, CA.

CITED REFERENCES

Alcock, C.B., Itkin, V.P., Horrigan, M.K., 1984. Vapour pressure equations for the metallic elements: 298–2500K. Can. Metallurg. Quart. 23, 309.
Kern, W., 1989. Chemical vapor deposition. In: Levy, R.A. (Ed.), Microelectronic Materials and Processes. Kluwer Academic Publishers, Boston, MA, pp. 203–246.
Ohring, M., 2002. Materials Science of Thin Films, second ed. Academic Press, San Diego, CA, p.101, p. 127, p. 286.
www.semicore.com for sputter yields.
Thornton, J.A., 1977. High rate thick film growth. Annu. Rev. Mater. Sci. 7, 239–260.
Yang, H.M., Flynn, C.P., 1989. Growth of alkali halides from molecular beams: global growth characteristics. Phys. Rev. Lett. 62, 2476.

QUESTIONS AND PROBLEMS

Questions

1. What are the two broad categories of vacuum processes?
2. What are the four steps involved in the making of thin films by vacuum processes?
3. Which processes are best for making metal thin films? Why?
4. What are the two regimes for vacuums and how are they different?
5. What are the basic assumptions involved in the kinetic theory of gases?
6. How is macroscopic pressure related to the microscopic average velocity of gas molecules?
7. Why can't UHV pumps be exposed to atmospheric pressures after they have been turned on?

8. Why are viscous flow pumps often connected to vacuum processing chambers by a long, thin pipes, but UHV pumps need to be connected directly onto the chamber with a large inlet as in Figure 7.5?
9. It can be more difficult to evacuate a processing chamber from ambient in the summer than in the winter if building climate is not well controlled. Why?
10. Why would a substrate temperature of 700°C yield films with large grains and faceted tops in sputtered Cu films, but yield films with small grains and voids between them in sputtered Al_2O_3 films? (*Hint*: See Figure 7.9.)
11. What are three ways surface energies affect Zone 3 thin film microstructures?
12. What role do surface energies play in varying the barrier to nucleation for formation of thin film nuclei compared to bulk nuclei?
13. In vacuum processes, nuclei form homogeneously on a substrate. From where does most of the atomic flux arrive as the nuclei grows?
14. What does it mean to say a growing film "wets" the substrate?
15. Why do some grains form facets on the top? Explain the Wulff diagram in your own words.
16. What does it mean to say a film is grown epitaxially onto a substrate?
17. For evaporation, processing chambers are evacuated to what vacuum regime? What maximum pressure is recommended? Why?
18. Evaporation source materials are heated to temperatures for which their equilibrium vapor pressures are 10^{-4} Torr or more. How does this value compare with the value of base pressures from the previous question? Why would lower base pressures be better?
19. Why are alloys and compounds difficult to evaporate from a single source? What is a solution to obtaining alloy films by evaporation? Compound films?
20. Why is uniformity of film thickness more of a challenge in evaporation than sputtering? How can this challenge be overcome?
21. What vacuum regime is typically used in sputtering?
22. In what cases is RF sputtering required rather than DC? What is the purpose of a magnetron?
23. Why is reactive sputtering often used to sputter oxide films?
24. Why is a magnetron sometimes unbalanced (*Hint:* It is the same reason that a small negative bias is sometimes applied to the substrate in sputtering)?
25. What is IBAD?
26. Is evaporation or sputtering usually preferred for making films from alloy source materials?
27. What does it mean to "condition" a sputtering target?
28. How does the pressure of the processing gas in sputtering affect the deposition rate of a film? Why?

29. What vacuum regime is used for most CVD processes?
30. What are the steps that occur during film deposition by CVD?
31. Why does a boundary layer form as the gases move over the substrates?
32. How can the boundary layer be minimized?
33. How can the CVD rate determining mechanism be found experimentally (e.g., gas phase diffusion vs. reaction rate)?

Problems

1. Calculate the mean free path to a collision for air molecules at room temperature in a vacuum of 10^{-3} Torr using the method in Example 7.1. Check your answer with Table 7.2.
2. If the pressure inside a full tank of propane is 200 psi on a hot summer day (100°F), what is the pressure of the same tank on a cold winter day in Minnesota (−20°F)? As a propane tank is filled, the store uses a scale to determine when it is full. What is the weight change when an empty 5-L tank is filled with propane (you may need to look up the chemical formula of propane)?
3. The world's largest vacuum chamber at NASA is 8000 cubic feet in volume. When it is evacuated, it takes 3 h. How many molecules are evacuated? (30 tons to 2 g)
4. Plot the distributions in velocities of vaporized atoms of Fe, Ni, and Co at 1000 K. What is the root mean square speed of each type of atom at this temperature?
5. At what temperature would Au molecules match the velocity distributions of Ar at room temperature? (*Hint:* Plot Maxwell Boltzmann distribution to match the figure in Example 7.2.)
6. Calculate the molecular flux of air into a pump (inlet area $= 10$ cm^2) at 300 K in mole/s versus pressure from 760 to 10^{-6} Torr.
7. a. A chamber used for deposition is evacuated to 6.4×10^{-6} Torr. Using the ideal gas law, calculate the number of molecules per unit volume (in cm^{-3} for air) in the chamber.
   b. If a 2″ diameter wafer is loaded into the chamber, calculate the rate at which atoms are impinging onto the surface of the wafer. You can check your answers with Table 7.2, but for this problem, show your work and show all of the units.
8. Calculate the impurity concentration in a film using Eq. (7.15) with a vacuum of 10^{-6} Torr and a sample that is being made with a deposition rate of 10 atoms/s.
9. Using Eq. (7.18), plot the evacuation time for an industrial processing chamber that has a volume of 10 m^3 where the pressure varies from atmosphere to 10^{-3} Torr. To find realistic values for S_p and P_p, search

the internet for two commercial mechanical pumps. Can the pumps you found also evacuate the chamber to 10^{-5} Torr?

10. Determine what aperture size would reduce the pump speed by an order of magnitude for a pump with $S_p = 800$ liter/s.
11. Choose one metal and one ceramic material. Determine what substrate temperature would produce Zone 1, Zone T, Zone 2 microstructures for each. (*Hint:* You will need to find the melting points of the materials.)
12. A metal film is being deposited, and it has grains that are 100 nm in diameter. Adatoms that have just been deposited onto the film will diffuse distances of $x = 2\sqrt{Dt}$ by surface diffusion. Once a new monolayer covers the current adatoms, these adatoms will be inside the grains (not on the surface any longer). At $T = 0.5\, T_M$, what deposition rate (monolayer formation time) is the maximum rate to allow these adatoms to reach the grain boundaries before being "frozen" in place. Use Figure P7.1 to estimate diffusion coefficients.
13. Repeat Problem 12 for a semiconductor film and for an alkali halide film.
14. After adatoms are "frozen" by the next monolayer arriving at the substrate, the atoms have to diffuse by bulk (or lattice) diffusion if they

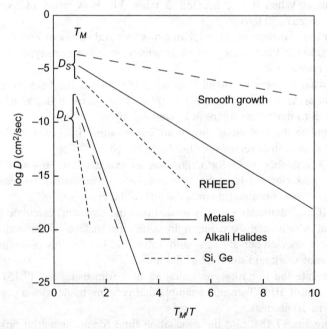

FIGURE P7.1 Diffusion coefficients versus T_M/T for bulk (lattice, D_L) and surface (D_s) diffusion for metals, semiconductors, and alkali halides. *From Yang (1989) — Reprinted Figure 6 with permission from Yang and Flynn, 1989 Copyright (1989) by the American Physical Society.*

move at all. Using Figure P7.1, find the ratio in the distances that an atom can move as an adatom (surface diffusion) compared to a lattice atom. Use $x_i = 2\sqrt{D_i t}$.

15. Choose a material to deposit as a thin film onto glass (SiO_2). Search the internet for the mechanical properties of that material and use Eq. (7.32) to determine how large a processing temperature it could withstand before its thermal stress would be greater than its yield stress. ($\alpha_{th,SiO_2} = 0.5$ ppm)

16. Determine the misfit for Ge ($d_{100} = 5.657$ A) grown onto Si ($d_{100} = 5.431$ A). Is epitaxy possible?

17. Choose an element from Figure 7.15. Determine to what temperature the source for this element should be heated to create a vapor pressure of 10^{-5} atm. How long will it take to evaporate a film that is 2 μm thick?

18. Evaporation can occur within a vacuum at 10^{-5} Torr. Find the monolayer formation time of air at room temperature at this pressure (quantitative). How would film purity be affected (qualitative)?

19. How would you plan to evaporate alloy films of equal (50:50) compositions of NiFe, FePt, InTi? Be specific regarding source materials and temperatures to which they would be heated.

20. Two point sources are to be used to evaporate an alloy film. The processing chamber configuration has one source at a distance of 5 cm and the other is at a distance of 7 cm from a flat plate that is holding the substrates. The element in the source that is 5 cm away deposits at 10 Å/min and the element in the source that is 7 cm away deposits at 15 Å/min. If the sources are placed 10 cm apart horizontally, where should the substrate be positioned between the sources to obtain an equal composition (50:50) of the two source elements? (*Hint:* Plot the d/d_o for each source on the same *x*-axis.) If the film is deposited for 30 min, plot the relative compositions of the two elements in the alloy as well as the total film thickness (normalized by the center thickness) as a function of distance along the substrate.

21. To meet device specifications, an Al film that is evaporated onto a 4″ Si wafer must be 2 μm thick. Only devices with thicknesses that are between 2 μm and +/− 10% can be used. What substrate-source distance is required for 100% yield (i.e., all devices will be useful) if a point source is used? Calculate the same for a surface source (as in e-beam evaporation).

22. Look up either one paper that compares sputtered and evaporated films or two papers (one discussing sputtered films and one discussing evaporated films). Report the differences and similarities between the microstructures and properties of films for the following materials:
    a. Alloys of Fe-Pt

b. Indium Tin Oxide

Submit a copy of the papers you read, and show what you found in tabular form (columns for differences and similarities).

23. Substrates are often bombarded during vacuum processes (e.g., ion-beam-assisted deposition, IBAD). Evaporated atoms arrive at a substrate with room temperature energies (0.026 eV). When they are bombarded with ions, they can be heated to 0.04 eV. Find five materials that would deposit with Zone 1 or Zone T microstructures (Figure 7.9) without bombardment, but that would deposit Zone 2 if bombarded. (*Hint:* You will need to find the melting points of the materials.)

24. Yttrium iron garnet can be made by reactive sputtering of a Y:Fe metal alloy target. After conditioning, what is the composition of the surface of the target if 500 eV Ar is used.

25. Find a journal article that describes a process for CVD of the films below. Next, find the MSDS for each chemical in the process and describe the precautions you would take before buying and using such chemicals.

    a. Si
    b. Tin oxide
    c. W

26. Calculate the "energy of reaction," ΔG_{rxn}, for Si films deposited using the precursors shown in Figure 7.30. How does this energy compare to the value for surface diffusion E_s according to Figure P7.1?

27. In the "diffusion controlled" section of Figure 7.30, there are four curves, each for a different precursor. Determine if the gas diffusion coefficient (D) varies with temperature according to the 1.5 power dependence as shown in Eq. (7.62). Assume the boundary layer thickness and partial pressure of reactant in the bulk gas do not change with temperature.

28. If a gas diffusion coefficient (D) were 0.2 cm^2/s in an atmospheric pressure reactor (APCVD), what diffusion coefficient would that gas have at 1 mTorr in an LPCVD reactor?

29. Choose any three deposited materials in Table 7.6 and, from the input reactants, determine which type of reaction was being used from Table 7.5.

Appendix A

TABLE A1 Selected Physical Constants

Quantity	Symbol	SI Units
Avogadro's number		6.022×10^{23} molecules/mol
Boltzmann's constant	k	1.38×10^{-23} J/(atom·K)
Electron charge	e	1.602×10^{-19} C
Electron mass	—	9.11×10^{-31} kg
Gas constant	R	8.31 J/(mol·K)
Acceleration due to gravity (Earth)	g	9.81 m/s^2
Permittivity of a vacuum	ε_o	8.85×10^{-12} farad/m
Planck's constant	h	6.63×10^{-34} J·s
Velocity of light in a vacuum	c	3×10^8 m/s

TABLE A2 Selected SI Units and Symbols

	Unit	Formula	Symbol
Elementary units			
Length	Meter	—	m
Mass	Kilogram	—	kg
Time	Second	—	s
Electric current	Ampere	—	A
Temperature	Kelvin	—	K
Plane angle	Radian	—	rad

(Continued)

TABLE A2 (Continued)

	Unit	Formula	Symbol
Derived units			
Acceleration	Meter per second squared	$m \cdot s^{-2}$	—
Area	Square meter	m^2	—
Capacitance	Farad	C/V	F
Charge	Coulomb	$A \cdot s$	C
Density	Kilogram per cubic meter	kg/m^3	—
Electric field strength	Volt per meter	V/m	—
Energy	Joule	$N \cdot m$ or $V \cdot C$	J
Force	Newton	$kg \cdot m/s^2$	N
Frequency	Hertz	s^{-1}	Hz
Power	Watt	J/s	W
Pressure	Pascal	N/m^2	Pa
Resistance	Ohm	V/A	Ω
Stress	Pascal	N/m^2	Pa
Velocity	Meter per second	m/s	—
Viscosity	Pascal second	$Pa \cdot s$	—
Voltage	Volt	W/A	V
Volume	Cubic meter	m^3	—

Index

Note: Page numbers followed by "*f*," "*t*," and "*b*" refer to figures, tables, and boxes, respectively.

A

Adatoms, 534–535, 535*f*, 541, 563, 581
Additive processes, 6, 416–417
   liquid monomer-based, 418, 495–504
   melt-based, 232–238
   powder-based, 345, 400–407
Additives, 21–22, 47–48, 59–61, 74, 84–85, 84*f*, 86*t*, 94, 170–171, 183
Agglomerates and aggregates, 348, 348*f*
Alumina, 31–32, 55
   origin of starting material, 51
Aluminum, 31–33
   origin of starting material, 51
Angle of repose, 351–353, 351*f*
Anisotropy, 235–237, 278–279
Annealing, 191, 578
Atomic layer deposition (ALD), 514, 575–576, 582
Atomization, 11–12, 44–45, 513

B

Ball milling, 41–44, 58, 65, 489
Barreling, 193–196, 205, 493
Batch calculation, 65
Bayer process, 31–32, 55–56, 56*f*
Bending, 325–328
Bernoulli's equation, 158–159
Binder, 344–345, 462
Binder burnout, 494
Bingham plastic, 452
Biot number, 454*f*
Blow molding, 106, 226–231, 239
   glass, 227–228
   polymer, 229–231
Brownian diffusion, 421–422, 459–460

C

Calcination, 54–58
Capillary action, 14–15, 14*f*, 465–467, 473, 504–505
Capillary number, 477–478, 485
Ceramics – categories, 50
Characterization, 86–87, 94
   powders, 347*t*
   starting materials, 7, 22, 95–96
Chemical vapor deposition (CVD), 15–16, 513, 566–578, 582
   deposition rate and uniformity, 573–578
   kinetics, 571–573
   thermodynamics, 566, 568, 582
   types of reactions, 568, 570–571
Chvorinov's rule, 164
Closed die forging, 308, 317–318
Coating, 473–491
   curtain coating, 476*f*, 485
   dip coating, 476, 476*f*, 477*f*, 478
   doctor blade coating, 476*f*, 478–480, 489
   methods, 475–485
   slide coating, 476*f*, 485
   slot coating, 474, 474*f*, 476*f*, 481–482, 481*f*, 484–485
Coke, 26–28
Cold working, 268, 299–300
Colloidal dispersions, 417–442
   aggregation, 438–442
   Brownian diffusion, 421–422, 459–460
   charging mechanisms, 425, 431
   electrostatic repulsion, 425, 438
   electrosteric repulsion, 438
   rheology, 448–455, 475, 486–487
   sedimentation, 420
   steric repulsion, 436, 438
   van der Waals attraction, 422–423, 424*f*, 425, 432, 441, 504
Communition, 57–59
Compaction, 14–15, 256, 315, 378–382, 387, 408, 488
Consistency index, 119
Constant rate period, 458*f*
Constitutional undercooling, 169–170, 170*f*

592   Index

Contact angle, 110–111, 535–536
Continuous casting, 30–31, 30f, 31f, 106
Continuous inkjet printing, 501–503
Cope, 156f, 157f
Crooke's space, 557
Crosslinking, 71, 90, 93, 463f
Crystal growth, 146, 149–150, 149f, 167–168
Crystallization, 105, 146, 150, 166–167
Curing, 18f, 89–90, 225–226, 462–464, 487, 495–496, 500, 504

D

Darcy's equation, 468–470
Data sheets, 22, 33, 48, 60, 65, 85–86, 94
Deborah number, 210
Deflocculation, 472–473
Dendrites, 164–166, 170, 179
Die casting, 9t, 117t, 175–178, 176f
Die swell, 207f, 208–210, 230, 233–234
Dielectric constant of liquids, 427–428, 430, 434, 438
Diffusion, 105, 267–268, 355, 359–360, 407–408, 571–573
Dihedral angle, 361–363, 363f
Dimensional changes during densification, 367–370
Dip coating, 475–476, 476f, 477f, 478
Direct extrusion, 300–301, 300f, 303–304
Dispersions. *See* Colloidal dispersions
DLVO theory, 438–439
Double layer thickness, 429–430, 432, 434, 439, 440f, 459–460
Dougherty-Kreiger equation, 451
Drag flow (Couette flow), 132–133, 133f, 134f, 194–195, 197–199, 238, 479–480, 482–483
Drop–on-demand inkjet printing, 237
Dry pressing, 370
Drying, 455–462
  particle dispersions, 459–460
  polymer sol, 461
Ductility, 179–181, 256, 267

E

Effective stress and effective strain, 279–282
Einstein's equation, 451–453
Electrical double layer, 425, 427, 430, 435–436
Engineering stress – engineering strain, 252–258, 255f, 265–266, 266f, 333

Equiaxed, 164–166, 170–171, 259, 267–268, 534, 581
Error function, 143–144, 144f
Evaporation (thin film), 540–554
  alloys and compounds, 546–547
  characteristics of materials, 552t
  thermodynamics, 542–546
  uniformity, 547–554
  vapor pressure vs. 1/T, 543–544, 544f
Extractive metallurgy, 21, 25–26, 96, 106
Extrusion, 9t, 192–213, 229, 299–308, 334, 491–494
  ceramic. *See* Powder extrusion
  metal. *See* Metal extrusion
  polymer. *See* Polymer extrusion
Extrusion blow molding, 105, 229–231, 230f
Extrusion die, 492
Extrusion limit diagram, 307–308, 308f
Extrusion ratio, 301, 303–308

F

Falling rate period, 459–461, 488
Flash, 308, 317–318, 318f
Flask, 156, 157f
Float glass, 15f, 106, 184–190
Flocculation, 439–442, 465
Flory-Huggins theory, 444–445
Flow stress, 251, 262, 288f, 289, 307–308, 318, 332–334
  average, 261f, 263–264, 289, 303–304, 334
  nomenclature, 263t
  plane strain, 263t, 310, 322–323
Forging, 308–318, 334
  forging force, ideal, 309–311, 313
  forging force, with friction, 282, 311–312, 312f, 317
  sticking friction, 317
  types, 308, 479
Forming, 4–6, 5f, 48, 60, 84–85
Fourier number, 142–143, 229
Friction, 282–284, 292, 301, 315, 350, 382
Friction coefficients, 282, 283t, 284, 297f, 314f, 350
Friction hill, 313, 322
Furnace, 26–29, 183
Fused deposition modeling, 5f, 6, 22, 85–86, 106–107, 232–237, 233f, 239–240, 495–496
Fusion downdraw process, 183, 189–190, 190f, 192, 239

G

Gas pressure, 44–45, 114b, 174–175, 559–560, 562–564
Gas pressure unit conversions, 519t
Gas throughput, 527–531
Gel casting, 464–466
Gelation, 54–55, 442
Glass
  batch calculations, 65
  frit, 60–61, 65, 96
  raw materials, 51, 60, 63t, 64–67, 66b
  reference point temperatures, 69t, 117
Glass codes, 69f
Glass melt preparation, 183–184
Glass plant layout, 62f
Glass sheet casting, 106–107
  float glass process, 9t, 15f, 106, 184–189, 239
  fusion downdraw process, 189–190, 190f, 192, 239
Glass transition temperature, 2–4, 85–86, 89t, 150–153, 188, 191, 235–237, 266, 271, 328, 329t, 367, 379–380, 486–487, 486f
Glow discharge, 557–559, 558f
Grain boundary, 355–361, 358f, 359f, 363–365, 407–408
Grain boundary diffusion, 359–361, 359f
Grain growth, 267f, 355–356, 360, 363, 365, 394, 396, 407–408
  abnormal, 365
  normal, 365
Granule, 14–17, 74, 85, 96–97, 192–195, 373–375, 374f, 378–380, 379f, 397–398, 408
Green density, 371b, 375f, 376–382, 377f, 379f, 381f, 385–386, 387f
Green machining, 389–391
Green strength, 380, 382

H

Hamaker constant, 348, 423–424, 424t, 439
Heat capacity, 141, 150–152, 161–163, 197b, 285
Heat distortion temperature, 89t, 145b, 399
Heat equation, 141, 143–144, 145f, 161–163, 172–173, 238–239
Heat transfer, 139, 143f, 238–239
  conduction, 142–143
  convection, 141–143

Heat transfer coefficient, 141–142
Heat treatment, 10t, 35, 178–179, 181–182
Hooke's Law, 115–116, 254, 255f, 259, 265–266
  triaxial, 275
  uniaxial, 254
Hot isostatic pressing (HIP), 389, 392–396, 395f
Hot pressing, 372, 392–396
Hot working, 268–269, 299–300

I

Ideal gas law, 516, 519–521
Ingot, 30, 269
Injection blow molding, 230–231, 231f
Injection molding, 213–226
  machine components, 214–217
  mold features, 214
  mold flow, 217–221
  molding cycle, 214–217
  packing and solidification, 221–223
  runner and cavity design, 220
Inkjet printing, 501
  binder-based, 406–407
  liquid-monomer based, 500–504
  melt-based, 237–238
Interfacial energy, 109–110, 146–147, 149
Intrinsic viscosity, 88t, 449–450, 453–454
Isoelectric point (IEP), 435–436, 435f, 437t
Isostatic pressing, 372, 389–391

J

Jenike shear apparatus, 352f
Johnson-Mehl-Avarmi equation, 150

K

Kelvin equation, 357–358
Kinetic theory of gases, 515–532, 543–545, 581

L

Ladle, 28–30, 30f, 153–154, 156–158
Ladle metallurgy, 29–30
Laminar flow, 131–132, 137b, 225–226, 420
Lasers, 191, 403–404, 496, 580–581
Latent heat, 161, 168–169, 173
Latent heat of vaporization of liquids, 458–459, 543
Latex coatings, 488, 489f
Life cycle assessment, 1–2, 3f

Liquid phase sintering, 355, 365–367, 407–408
Lubrication, 386

M

Machining, 6, 10t, 178
Magnetron sputtering, 559–561
Mass conservation, 129–131, 483
Materials cycle, 2f
Materials processing, 1–4
Maxwell-Boltzman equation, 520–521
Mean free path, 517–518, 524t, 557–559, 581
Mechanical properties
 ceramics, 265t
 metals, 257t, 267
 polymers, 265t, 270
Melt atomization, 44–46, 375, 380–381
Melt casting. *See* Glass sheet casting; Metal melt casting
Melt fracture, 210–211, 213
Melt pool, 196, 404–405
Melt processes, 105
Melts, 491
 density, 108, 112t
 rheology, 115–129
 structure, 107–114
 surface tension, 107–114
Metal alloy classifications, 10
Metal alloy codes, 23
Metal extrusion, 299–308
 die shape effects, 301–302
 extrusion pressure, 302–307, 303f, 305b, 306b
 limit diagrams, 307–308, 308f
Metal injection molding, 213
Metal melt casting
 continuous casting, 30–31, 30f
 die casting, 175–178
 ingot casting, 27f, 30
 microstructure development, 164–166, 353–367
 permanent mold casting, 171–175, 172f, 175f
 post-processing, 178–182
 sand casting, 156–171, 156f
 solidification time prediction, 161
Metal melt preparation, 155
Metering zone, 197, 200b
Mohr's circle, 273–275, 274f, 276f, 278b
Molecular beam epitaxy (MBE), 541

Molecular flux, 520, 524t, 529, 534–535, 535f, 545b
Molecular weight, 129
 average, 74–77, 78b, 88t
 characterization, 86–87
 distribution, 74, 77f
Momentum conservation, 129–130
Monolayer formation time, 523–525, 524t

N

Necking, 255f, 256, 262b
Neutral axis, 326–327, 326f
Neutral point, 319–320, 322
Newton's law of viscosity, 117, 132–134
Newtonian and non-Newtonian behavior, 119, 120f, 451
Newtonian cooling, 141–143, 141f
Newtonian plateau, 126–127, 454–455
Normal stresses, 272–273, 276f
Nucleation
 heterogeneous, 149, 166–167, 170–171, 200, 536–537, 567, 569
 homogeneous, 146–147, 149f, 150, 534–535, 569, 574–575

O

Ohm's law of evacuation, 529
Ohnesorge number, 502–503
Open die forging, 287f, 308–312, 309f, 312f, 334

P

Parison, 227, 231f, 239
Particle charging, 396f, 398
Particle packing, 348, 450, 460–462, 472–473
Particle size
 average, 48
 characterization, 36
 distribution, 36–38, 40b, 347–348, 380, 389, 402, 407, 450
Pattern, 578–580, 579f
Permanent mold casting, 171–175
Phenolics
 origin of starting material, 90–91
Photolithography, 513, 579–580, 579f
Plane strain, 281, 309–312, 322
Plane stress, 273–274, 274f, 276f, 277f, 281
Plasma physics, 557–559

Plastic deformation, 251–252
  effect of strain rate, 267–271
  effect of temperature, 267–271
  metals, 269
  polymers, 333
  temperature rise, 284–285
  uniaxial tension, 252–266
  work done per volume, 281–282
Plastic flow rules, 252–266
Plasticizers, 7, 85, 86t, 378, 490
Plug-assist thermoforming, 330f
Point of zero charge (PZC), 426–427, 426f, 435–436, 435f
Polyethylene, 72t, 78b, 81
  origin of starting material, 1, 81
  steps in reaction, 82f
  types, 81t
Polymer coatings, 486–488
Polymer extrusion, 192–213
  die exit effects, 207–211
  die flow, 201
  extruded products, 211–213
  melting and flow, 192–200
  metering model, 193f
  operating diagram, 201–204
  solidification, 211–213
  twin screw extrusion, 205–207
  viscous heating, 195–196
Polymer melts, 108–109, 113f
Polymer solutions, 442–448
  critical concentration, 445
  rheology, 452
  solvent quality, 444
  thermodynamics, 444
Polymerization methods, 77–80, 96–97
Polymers, 68–94
  classifications and naming, 71t, 72t
Powder blending, 45, 47–48, 85
Powder compaction sequence, 346, 375f, 377–378, 392
  ceramics, 14–15, 345
  metals, 377–378
Powder extrusion, 9t
Powder flow, 8, 36, 48, 343–345, 349–353, 375
Powder injection molding, 9t, 494
Powder metallurgy, 11–15, 35, 39, 46–47, 345, 373, 387f, 395, 408
Powder preparation for pressing, 372–376
  ceramics, 372–373
  metals, 374–376

Powder pressing, 4–6, 5f, 11–12, 408
  die wall friction, 382, 389–390
  isostatic, 9t, 346, 372, 389–391
  lubricants, 353, 375–376, 392
  metals, 11–12
  post-processing, 8
  uniaxial, 9t, 24–25, 346, 372, 376–389, 413
Powder processes, 8, 343
Powder synthesis, 348, 375
  ceramics, 31–32
  metals, 388–389
  polymers, 21
Power law equation (viscosity), 119, 127, 128b
Power law index, 119, 127, 245–246
Pressure casting, 174–175, 175f, 465, 473
Pressure driven flow (Poiseuille flow), 494
Pressure driven flow equations, 137t
Pressure in a bubble, 114b

R

Radio frequency (RF) sputtering, 556, 561
Radius of gyration, 88t, 443–445, 453
Raw material, 1, 7, 15f, 21, 26–27, 63t, 64–65, 66b, 96–97, 183
Reactive injection molding (RIM), 8, 9t, 18f, 223–226, 224f
Reactive sputtering, 546–547, 556, 561–563, 582
Recovery, 259, 267–268, 331, 378
Recrystallization, 251, 267–268, 267f, 334
Recycling, 1–2, 33, 65, 83–84, 217, 502
Redundant work, 284, 286, 289, 296, 303, 334
Residual stress, 267–268
Reverse draw thermoforming, 331
Reynolds number, 131–132, 137, 158, 186, 419–420, 503
Rheology, 65–68, 86–89, 107, 119, 125–126, 418, 448–449, 452, 475, 486
  dispersions, 57–58, 237, 415, 416f, 417–442, 448–455, 464, 489, 505
  glass melts, 15, 108, 123, 239
  metal melts, 121–123, 238
  polymer melts, 16–17, 84–85, 105, 108–109, 125–126, 129, 154, 238, 417, 491
  polymer solutions, 417, 442–448, 486
Rolling, 183, 286, 318–325
  average pressure, 322–323
  contact length, 320–323
  geometry, 320–322, 320f
  roll separating force, 319, 321–323, 334

Rotational molding, 343, 345–346, 396–399, 408
   powder preparation, 397–398
   process steps, 398–399

S

Sand casting, 154f, 156–173
Sedimentation, 37t, 420
Sedimentation velocity, 421b, 439
Selective laser melting, 11–12, 24–25, 401, 405f, 408–409
Selective laser sintering, 11–12, 12f, 343–345, 344f, 400–406, 400f, 495–496, 496f
   laser fusion, 401–403
   post-processing, 406
   powder absorbance, 404t
   powder preparation and flow, 401
Shape casting categories, 154f
Shear rates for processes, 116, 117t
Shear thickening behavior, 451, 452f
Shear thinning behavior, 119, 127, 451
Sheet metal processing, 325
Silicon nitride, 51t, 57, 427
   origin of starting material, 57–58
Single screw extrusion, 205
Sintering, 353–367
   liquid phase, 355, 365–367, 407–408
   mechanisms, 407–408
   solid state, 355, 360, 365–367, 407–408
   stages, 367
   viscous, 367, 407–408
Slag, 26–30, 155
Slide coating, 475–476, 476f, 485
Slip casting, 14f, 416f, 418, 460, 465–467, 465f, 470–473, 490, 504–505
   cast layer thickness, 467–471
   post-processing, 473
   process considerations, 472–473
Softening point, 68f, 69t
Solid processes, 8
Solid state sintering, 355, 365–367, 407–408
Solidification shrinkage, 159–160, 170, 177, 215–216
   injection molding polymers, 177–178
   metals, 161t
   thermoforming polymers, 140t
Solubility parameters, 445, 447t
Spark plasma sintering, 394, 394f
Specific heat, 141, 163–164, 195–196, 285
Spherical pressure vessel, 341–342
Spider die, 211–212

Spray drying, 373, 374f, 375
Springback, 327–328, 378
Sprue, 158, 159f, 216–219
Sputtering, 13f, 551, 555–566
   magnetron sputtering, 559–561
   plasma physics, 557–559
   radio frequency (RF), 561
   rates, 563–566
   reactive, 561–563
Starting materials
   bulk metal, 25–35
   ceramic powder, 50–68
   glass, 60–68
   powder metal, 35–49
   thermoplastic polymer, 74–89
   thermoset polymer, 89–94
Steel, 23, 26–27, 29–30, 33, 177–179, 216–217
   alloying elements, 23, 28–29, 46–47
   impurities, 29–30, 111–112
   origin of starting material, 23, 33, 35
   plant layout, 62f, 100
   powder, 36, 46, 49f
Stereolithography, 17, 415–416, 416f, 495–500
Strain hardening. *See* Work hardening
Strain rate sensitivity, 269–270, 269f, 307–308
Subtractive processes, 6–7, 251
Superplastic forming, 331–334
Surface tension, 107–114, 373, 475, 477–478
Suspensions. *See* Colloidal dispersions

T

Tap density, 39–41, 351–353
Tape casting, 473–491
Theoretical density, 369–370, 372
Thermal annealing, 10t, 45–46, 106, 190–191
Thermal conductivity, 139, 140t, 161–163, 171–172, 216–217, 239, 398–399, 401
Thermal diffusivity, 141, 161–163, 172–173
Thermal expansion coefficient, 140t, 152–153, 540
Thermal properties of materials, 140t
Thermal stress, 15, 178–179, 181, 190–191, 223, 227, 406
Thermoforming, 16–17, 231, 251–252, 287f, 328–331
   polymer properties, 329t
   processing window, 329

Thermoplastic polymers, 16, 21–22, 71–74, 72t, 84–87, 90, 105, 112t, 118t, 140t, 264–265, 328, 397–398
  important temperatures, 89t
Thin film, 7, 12–13, 48, 65, 513–514, 542
  annealing, 578
  epitaxy, 12–13, 532, 539
  microstructures, 532–538
  nucleation and growth, 534–535
  patterning, 579f
  strain, 539–540, 539f
  thermal stress, 15
  wetting, 65, 515
Thixotropy, 452f
Thornton diagram, 532, 533f, 539, 578, 581
3D printing, 6, 344–345, 400, 406–407, 495
Time-temperature-transformation (TTT diagram), 150, 151f
Tresca criterion, 276–278
Triaxial stress state, 264, 273f, 275–276, 279, 281f
True stress-true strain, 255f
Turbulent flow, 130–132, 208–209
Twin screw extrusion, 192, 205–207, 492

U

Ultraviolet (UV) light, 487
Uniaxial tensile test, 252, 254, 256, 279, 326–327
Upsetting, 308
UV cured coatings, 462, 464
UV curing, 8, 462, 463f, 487, 497–498, 504

V

Vacuum, 513, 515–518, 521, 525, 529, 581
  characteristics, 524t
  viscous and molecular regimes, 555–556, 575
Vacuum casting, 174–175, 175f, 471, 473
Vacuum chambers, 12–13, 13f, 394, 529–532, 534, 541, 564, 567
  evacuation sequence, 526f
  pressure gauges, 518, 526
  pumping speeds and conductance, 527
  pumps, 518, 526

van der Waals attraction, 348–350, 422–423, 424f, 425, 441, 504
Vapor processes, 8
Viscoelasticity, 119, 270, 452, 454–455
Viscosity, 68f, 118–119
  effect of shear rate, 126–127, 126f
  temperature dependence, 126–127, 126f
Viscous heating, 116–117, 195–196, 217, 220, 284–285
Vogel-Fulcher-Tammann (VFT) Equation, 124, 127
von Mises criterion, 277–279, 281, 283–284, 291–292, 312–313

W

Waxes, 86t, 232, 237, 375–376
Weber number, 503
Wetting, 65, 111, 111f, 455, 466f, 475, 485, 515, 535–536
Wire drawing, 286–299, 287f, 288f, 291f, 294f, 298f, 301–302, 317, 319, 334
  die angle effects, 296–299, 299f
  draw stress, ideal, 289
  draw stress, with friction, 292, 296–299, 297f, 334
Work hardening, 33–34, 258, 268, 279–281, 289, 301–302, 322, 334, 380–381
Working point, 68f, 69t
Wulff diagram, 538, 538f

Y

Yield criteria, 277f, 312–313
  Tresca, 276–278, 277f, 291–292, 295–296
  von Mises, 277–279, 277f, 291–292, 295–296, 310, 333–334
Yield stress, 33–34, 254–259, 262, 268–269, 275, 277–279, 324, 379–380, 452
Young equation, 110, 535–536
Young's modulus, 254–256, 275, 327–328, 538

Z

Zeta potential, 434–436, 435f, 436f

图书在版编目（CIP）数据

金属、陶瓷和聚合物的加工方法：英文／（美）洛兰·弗朗西斯（Lorraine F. Francis）著. --长沙：中南大学出版社，2017.9
ISBN 978-7-5487-3019-4

Ⅰ.①金… Ⅱ.①洛… Ⅲ.①金属加工－英文 ②陶瓷－生产工艺－英文 ③聚合物－加工－英文 Ⅳ.①TG ②TQ174.6 ③TQ31

中国版本图书馆CIP数据核字（2017）第239001号

金属、陶瓷和聚合物的加工方法
JINSHU、TAOCI HE JUHEWU DE JIAGONG FANGFA

Lorraine F. Francis 著

□责任编辑	胡　炜		
□责任印制	易建国		
□出版发行	中南大学出版社		
	社址：长沙市麓山南路	邮编：410083	
	发行科电话：0731-88876770	传真：0731-88710482	
□印　　装	湖南众鑫印务有限公司		

□开　　本	720×1000　1/16	□印张 39.25	□字数 1017 千字			
□版　　次	2017 年 9 月第 1 版	□2017 年 9 月第 1 次印刷				
□书　　号	ISBN 978-7-5487-3019-4					
□定　　价	186.00 元					

图书出现印装问题，请与经销商调换